AF335678

Frontiers of High-Pressure Research

NATO ASI Series

Advanced Science Institutes Series

A series presenting the results of activities sponsored by the NATO Science Committee, which aims at the dissemination of advanced scientific and technological knowledge, with a view to strengthening links between scientific communities.

The series is published by an international board of publishers in conjunction with the NATO Scientific Affairs Division

A	**Life Sciences**	Plenum Publishing Corporation
B	**Physics**	New York and London
C	**Mathematical and Physical Sciences**	Kluwer Academic Publishers
D	**Behavioral and Social Sciences**	Dordrecht, Boston, and London
E	**Applied Sciences**	
F	**Computer and Systems Sciences**	Springer-Verlag
G	**Ecological Sciences**	Berlin, Heidelberg, New York, London,
H	**Cell Biology**	Paris, Tokyo, Hong Kong, and Barcelona
I	**Global Environmental Change**	

Recent Volumes in this Series

Volume 280—Chaos, Order, and Patterns
edited by Roberto Artuso, Predrag Cvitanović, and Giulio Casati

Volume 281—Low-Dimensional Structures in Semiconductors: From Basic Physics to Applications
edited by A. R. Peaker and H. G. Grimmeiss

Volume 282—Quantum Measurements in Optics
edited by Paolo Tombesi and Daniel F. Walls

Volume 283—Cluster Models for Surface and Bulk Phenomena
edited by Gianfranco Pacchioni, Paul S. Bagus, and Fulvio Parmigiani

Volume 284—Asymptotics beyond All Orders
edited by Harvey Segur, Saleh Tanveer, and Herbert Levine

Volume 285—Highlights in Condensed Matter Physics and Future Prospects
edited by Leo Esaki

Volume 286—Frontiers of High-Pressure Research
edited by Hans D. Hochheimer and Richard D. Etters

Volume 287—Coherence Phenomena in Atoms and Molecules in Laser Fields
edited by André D. Bandrauk and Stephen G. Wallace

Series B: Physics

Frontiers of High-Pressure Research

Edited by

Hans D. Hochheimer and
Richard D. Etters

Colorado State University
Fort Collins, Colorado

Plenum Press
New York and London
Published in cooperation with NATO Scientific Affairs Division

Proceedings of a NATO Advanced Research Workshop on
Frontiers of High-Pressure Research,
held July 15–18, 1991,
in Fort Collins, Colorado

Library of Congress Cataloging-in-Publication Data

Frontiers of high-pressure research / edited by Hans D. Hochheimer and
  Richard D. Etters.
        p.   cm. -- (NATO ASI series. Series B, Physics ; v. 286)
    "Published in cooperation with NATO Scientific Affairs Division."
    "Proceedings of a NATO Advanced Research Workshop on Frontiers of
  High-Pressure Research, held July 15-18, 1991, in Fort Collins,
  Colorado"--T.p. verso.
    Includes bibliographical references and indexes.
    ISBN 0-306-44188-8
    1. High pressure (Science)--Congresses. 2. Materials at high
  pressure--Congresses. 3. Polymers--Congresses. 4. Quantum wells-
  -Congresses. 5. High temperature superconductors--Congresses.
  6. Semiconductors--Congresses. I. Hochheimer, Hans D. II. Etters,
  Richard D. III. North Atlantic Treaty Organization. Scientific
  Affairs Division. IV. NATO Advanced Research Workshop on Frontiers
  of High-Pressure Research (1991 : Fort Collins, Colo.) V. Series.
  QC280.F76   1992
  531'.1--dc20   92-784
                                                          CIP

ISBN 0-306-44188-8

© 1991 Plenum Press, New York
A Division of Plenum Publishing Corporation
233 Spring Street, New York, N.Y. 10013

This conference was convened in memory of

our colleague,

Ian L. Spain

He was a complicated man and endured his share
of difficulties. Those who were privileged to
know him recognize a man who struggled lifelong
in uncompromising pursuit of knowledge and
achievement.

*We wish him smooth sailing and
a gentle wind.*

We wish him well!

PREFACE

     The role of high pressure experiments in the discovery of supercon-
ducting materials with a T_c above liquid nitrogen temperature has demonstrated the importance of such experiments. The same role holds true in
the tailoring of materials for optoelectronic devices. In addition,
much progress has been made recently in the search for metallic hydrogen, and the application of high pressure in polymer research has
brought forth interesting results. These facts together with the success of previous small size meetings (such as the "First International
Conference on the Physics of Solids at High Pressure", held in 1965 in
Tucson, Arizona, U.S.A.; "High Pressure and Low Temperature Physics",
held in 1977 in Cleveland, Ohio, U.S.A.; and "Physics of Solids Under
High Pressure", held in 1981 in bad Honnef, Germany), motivated us to
organize a workshop with emphasis on the newest results and trends in
these fields of high pressure research.

     Furthermore, it was intended to mix experienced and young scien-
tists to realize an idea best expressed in a letter by Prof. Weinstein:
"I think it is an excellent idea. I have often felt that the number of
excellent young researchers in the high pressure field need an opportunity to put forward their work with due recognition." Thanks to the
support of the key speakers, we were able to achieve this goal and had
more than 50% young participants.

     The numerous interactions and lively discussions during the work-
shop showed the success of this concept. Many ideas for new experiments
and improvements of theoretical models were one consequence of these
discussions. As a result, the interesting new results and innovative
ideas which were presented in the oral, poster contributions, and panel
discussions will be of importance for future research, both experimentally and theoretically.

     We would like to express our deep gratitude to the NATO Scientific
Affairs Division for their financial support of the workshop. We are
also indebted to Dr. William J. Bertschy, Virginia Sawyer, and Jennifer
Sterling for their tireless assistance before and during the workshop.

     We also thank Dr. Albert Yates, President of Colorado State Uni-
versity, Prof. John Raich, Dean of the College of Natural Sciences, and
Prof. James Sites, Chairman of the Department of Physics, for their support. The excellent preparation of the raft trip by Prof. David Krue-
ger, was also very much appreciated.

     The editors would also like to thank the staff of Plenum Publish-
ing Corporation, in particular, Patricia M. Vann and Thomas Flood for
their handling of the publication of the Proceedings.

Finally, we would like to thank all speakers and participants, who created an atmosphere conducive to the success of such a workshop by sharing freely both the concepts and the results of their work. We owe special thanks to Profs. Bradley, Ashcroft, Chu, and Weinstein for their excellence as chairmen of the round table discussions.

There is only one sorrow which we have concerning the meeting. We wish that our colleague Prof. Ian L. Spain, who was involved in the planning of the meeting, could have shared the excitement and success of the workshop with us. Unfortunately, Prof. Ian L. Spain passed away at the age of 50 on September 6, 1990 in Fort Collins, Colorado after a two-year struggle against cancer. We therefore dedicate these Proceedings to the memory of Prof. Ian L. Spain, who contributed so much to the field of high pressure research.

Hans D. Hochheimer
Richard D. Etters

Fort Collins, Colorado, U.S.A.

CONTENTS

QUANTUM WELLS AND SEMICONDUCTORS

HIGH TEMPERATURE SUPERCONDUCTORS

POLYMERS UNDER PRESSURE

W. Pechhold

Universität Ulm, Abteilung Angewandte Physik
Albert-Einstein-Allee 11, D-7900 Ulm, Germany

ABSTRACT

Using the bulk Grüneisen Parameter, an equation of state $V(p,T)$ is derived from thermodynamics. It can be successfully applied to crystalline as well as glassy polymers and in a first approximation also to polymer melts. This is shown for high pressure crystallized Polyethylene and - in a contribution by Dollhopf - for Polycarbonate and Polystyrene.

Next, the pressure dependence of the glass relaxation process (in PVAc), as measured by dielectric spectroscopy, is discussed and analysed using the dislocation concept in the meander model and an appropriate Grüneisen parameter $\gamma_{disl.}$. The latter turns out to be about half the value of the bulk parameter, which might be checked by measuring the pressure dependence of the shear modulus $G(p)$.

Finally, the pressure dependence of phase transitions and its transition data will be described for several examples, including the high temperature transition of PE into a CONDIS-phase and the subsequent melting, as well as the nematic/isotropic transition of LC-polymers and - in a contribution by Schwarzenberger - the lamellar/isotropic transition of PDES.

INTRODUCTION

In order to describe phase transitions in polymers on the basis of model-theories, the volume effect and its pressure dependence plays a key role and must be checked carefully. For the glass relaxation the free volume and its pressure dependence is of similar importance. Therefore, to verify any model-theory, the equation of state, $V(p,T)$, of both coexisting phases or the excess volume of appropiate defects have to be well known.

A number of empirical equations were used to determine the volume as a function of pressure[1] and temperature. For metals a universal equation of state[2] has been sucessfully applied. In the first part of this paper a simple $V(p,T)$-expression, based on the quasiharmonic assumption, will be derived and applied to high pressure crystallized polyethylene. Next the pressure dependence of the glass relaxation frequency of PVAc shall be discussed using the dislocation concept in the meander model. The last part will be concerned with applications to some phase transitions.

Frontiers of High-Pressure Research, Edited by H.D. Hochheimer and
R.D. Etters, Plenum Press, New York, 1991

THERMODYNAMIC DERIVATION OF AN ANALYTICAL V(p,T)-EXPRESSION USING THE BULK GRÜNEISEN-PARAMETER γ

For a system in which the inner variables stay constant (e.g. a glass), or can be neglected to a first approximation (e.g. for a crystal), V(p,T) will be determined by the competitive action of thermal pressure and cohesive stress, i.e. by anharmonicity. In this case, a bulk Grüneisen-parameter γ can be defined by

$$(1) \quad d\ln K / d\ln V = -2\gamma,$$

assuming that the bulk modulus K is a direct function of the volume V.

Of course, γ may depend on the state variables p and T, as will be shown in the following. The exact differential of $\ln V$ reads

$$(2) \quad d\ln V = \alpha dT - dp/K$$

with the expansion coefficient α and the compressibility $1/K$, defined as usual. Inserting (2) into (1) yields the exact differential of K

$$(1a) \quad dK = -2\gamma K\alpha dT + 2\gamma dp.$$

Now it can be shown that $K\alpha$ will not depend on p, because

$$\frac{\partial(K\alpha)}{\partial p} = \alpha\frac{\partial K}{\partial p} + K\frac{\partial\alpha}{\partial p} = 2\gamma\alpha - 2\gamma\alpha = 0 \ ,$$

applying Maxwell relation onto (2). A similar Maxwell relation from (1a) yields the differential equation

$$(3) \quad -K\alpha\frac{\partial 2\gamma}{\partial p} = \frac{\partial 2\gamma}{\partial T}, \qquad \text{with } K\alpha = K\alpha(T).$$

Its solution can be written as:

$$(4) \quad 2\gamma = 2\gamma_0\left\{1+a\left[p-p_0-\int_{T_0}^{T}K\alpha dT\right]\right\} \ ,$$

with p_0, T_0, γ_0 as a reference state. The constant parameter a has to be determined by fitting K(p,T) or V(p,T) to experimental results. K(p,T) is derived by integrating (1a) utilizing (4). For this purpose one may introduce as integration variable

$$(5) \quad x = p - p_0 - \int_{T_0}^{T}K\alpha dT \ ,$$

which leads to the simple expressions

$$(1a) \quad dK = 2\gamma dx \qquad \text{and} \qquad (4a) \quad 2\gamma = 2\gamma_0(1+ax).$$

Integration yields

$$(6) \quad K-K_0 = 2\gamma_0\int_0^x(1+ax)dx = \gamma_0(2x+ax^2) \ .$$

In a similar simple manner V(p,T) can be obtained, integrating (2)

$$(7) \quad \ln\frac{V}{V_0} = \int_{p_0,T_0}^{p,T}\frac{K\alpha dT-dp}{K} = -\int_0^x\frac{dx}{K(x)} = -\int_0^x\frac{dx}{K_0+2\gamma_0 x+\gamma_0 ax^2}$$

2

For $\gamma_0 > aK_0$, $V(p,T)$ therefore becomes

$$(7a) \qquad \frac{V(x)}{V_0} = \left[\frac{1+\dfrac{ax}{1+\sqrt{1-aK_0/\gamma_0}}}{1+\dfrac{ax}{1-\sqrt{1-aK_0/\gamma_0}}}\right]^{\dfrac{1}{2\gamma_0\sqrt{1-aK_0/\gamma_0}}}$$

which in the limit $a \to 0$ reads

$$(7b) \qquad \frac{V(x)}{V_0} = \left[1+\frac{2\gamma_0}{K_0}x\right]^{-\frac{1}{2\gamma_0}} \qquad \text{for } a=0$$

Equation (7a) may be compared by least square fitting with $V(p,T)$, e.g. isothermal or isobaric data sets, to determine the parameters K_0, γ_0, a, and $K\alpha=K\alpha(T)$. Since $K\alpha$ turns out to be a weak function of temperature, a first approximation to x will be

$$(5a) \qquad x = p-p_0-(K\alpha)_0(T-T_0)$$

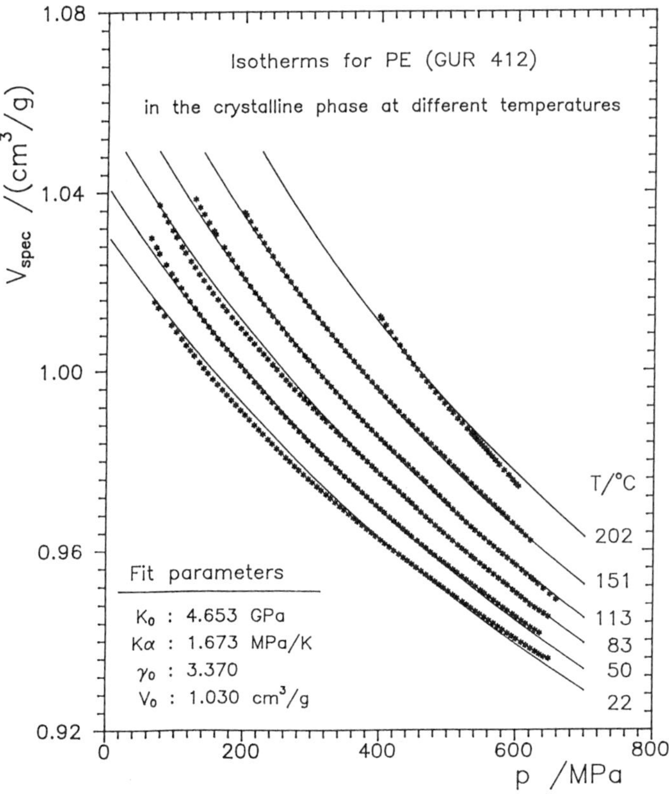

Fig.1 Isotherms measured by high pressure dilatometry for high molecular weight Polyethylene crystallized at high pressure. Data points measured curves drawn according to equation (7b) and (5a) with fit parameters given in the table

In Fig. 1 isothermal measurements on high pressure crystallized high molecular weight polyethylene are shown and fitted to equation (7b) with fit parameters given in Fig. 1. With a similiar fit procedure applied to (7a), the parameter a turns out to be about zero. For the polyethylene crystal $\Delta V/V_0$ data up to 4.6 GPa have been reported by Simha[3], and can best be approximated by equation (7b), i.e. a=0, and the parameters K_0=6.4 GPa, γ_0=3.7.

Since $K\alpha$ may be a weak function on temperature, a second approximation may start with $K\alpha=(K\alpha)_0+(K\alpha)_T(T-T_0)$ and yields

$$(5b) \quad x = p-p_o-(K\alpha)_o(T-T_o) - \frac{1}{2}(K\alpha)_T(T-T_o)^2$$

in which $(K\alpha)_T \equiv \partial K\alpha/\partial T$ at T_0. Since α is fairly constant and K decreases with temperature, $(K\alpha)_T$ will be negative.

THE PRESSURE DEPENDENCE OF THE GLASS RELAXATION FREQUENCY

Many attempts have been made to correlate thermal properties and dynamic experiments in the transition region from glassy to rubbery behaviour in order to get more insight in the structure and dynamics of amorphous polymers. The dislocation concept[4,5,6] is an approach which uses a well defined model of molecular structure (the meander model[4,7], Fig. 2) and represents the free volume by quasi-dislocations. The bundle-like structure is envisaged as an approximately hexagonal packing of macromolecular chains, short range ordered in conformational clusters. Shear fluctuations (micro-Brownian motion) may easily take place by moving quasi-dislocations, provided a dislocation wall structure has been thermally formed throughout. Using Cluster-Entropy Hypothesis (CEH), this will be the case if each segment-line of 3r/d segments in a meander cube (Fig. 3) perpendicular to the dislocation wall has collected at least once the free energy ε_s of wall formation per segment area. Multiplying the intramolecularly activated segmental jump frequency by this probability factor, one arrives at the (dielectric) relaxation frequency of the glass process

$$(8) \quad f_m = \frac{f_0}{\pi} e^{-\frac{Q}{RT}} \left[1 - \left(1 - e^{-\frac{\varepsilon_s}{RT}} \right)^{\frac{3r}{d}} \right]^{3(\frac{3r}{d})^2 \frac{d}{s}}$$

Here f_0 is a vibration frequency, Q a local activation energy, 3r is the side length of a meander cube, d the mean distance between chains, s the length of a segment (one or two monomeric units) and ε_s the free energy per segment area of a dislocation wall, which depends on the elastic properties of the glass and thus on temperature and pressure

$$(9) \quad \varepsilon_s = \frac{0.3\ G\ b^2\ d}{4\ \pi\ (1 - \nu)}$$

here b is the Burgers vector, G the shear modulus and ν Poisson's ratio.

This concept of the glass relaxation frequency must be checked by studying several factors affecting the glass temperature of a polymer[8], e.g. its molecular weight, the degree of crosslinking, swelling and hydrostatic pressure.

Since the pressure dependence is of most interest here one may take advantage of dielectric measurements by Heinrich and Stoll on PVAc[9] in the range 0-500 MPa. Fig. 4 was taken from this work and shows frequency curves of real and imaginary part of the dielectric constant at a given temperature measured at different pressures.

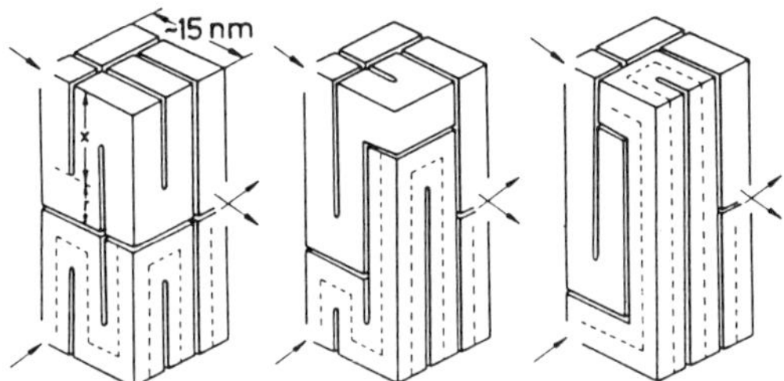

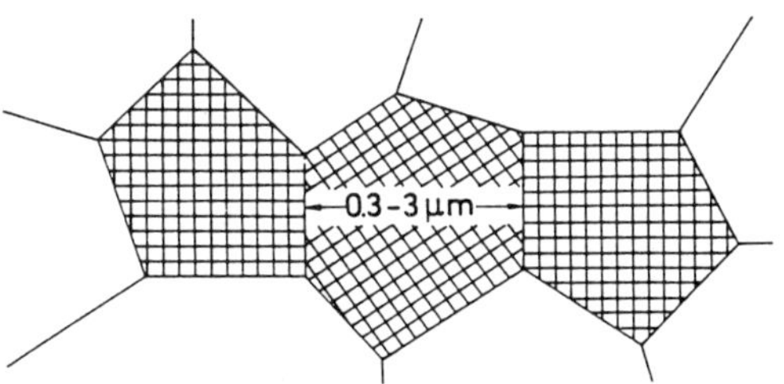

Fig.2 Three levels of order in polymer melts or networks, as described in the meander model[4,7]

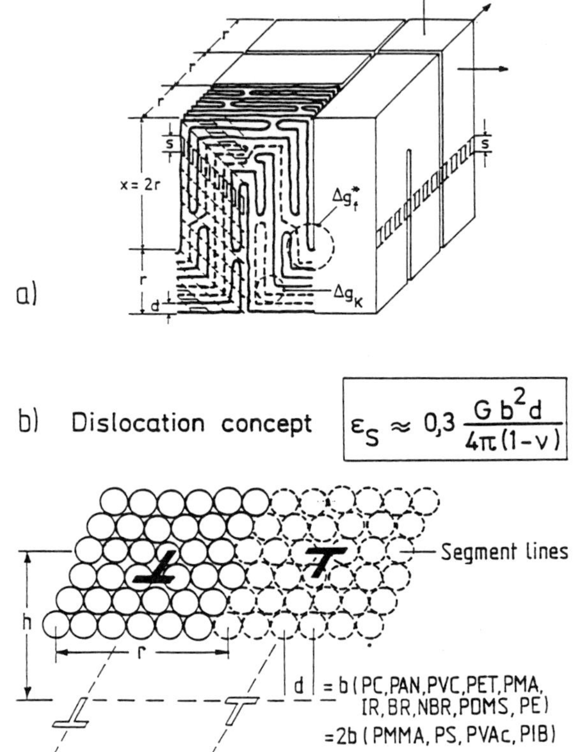

Fig. 3 a) Meander cube to define geometry and free energy
       contributions Δg_f^*, Δg_k of superfolding. Two
       different bundle segment lines and one cube segment
       line are shown. b) Dislocation wall concept of intra-
       meander shear deformation. The cross section of two
       bundles with a step dislocation in each are indicated:
       b=Burgers-vector, d=interchain distance

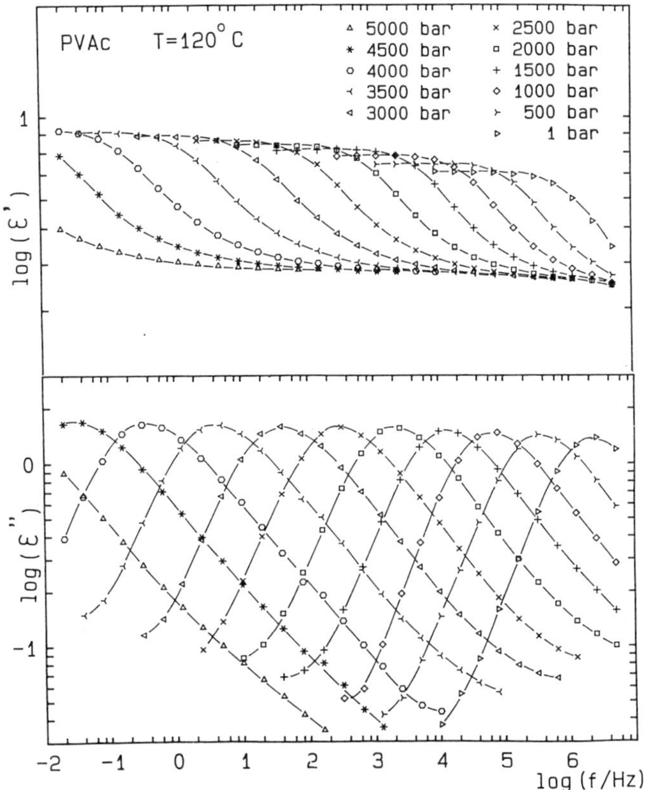

Fig.4 Frequency curves for real and imaginary part of the dielectric constant of PVAc[9], measured at different pressures at 120°C

Plotting the ε"-maximum frequencies f_m from Fig. 4 (and from similar data at other temperatures) double logarithmically versus $10^3/T$, one arrives at the activation diagram of the glass relaxation in PVAc[6]. This diagram, reproduced in Fig. 5 (small symbols), reflects the pressure dependence of the glass process and has been quantitatively discussed by Heinrich and Stoll[6]:

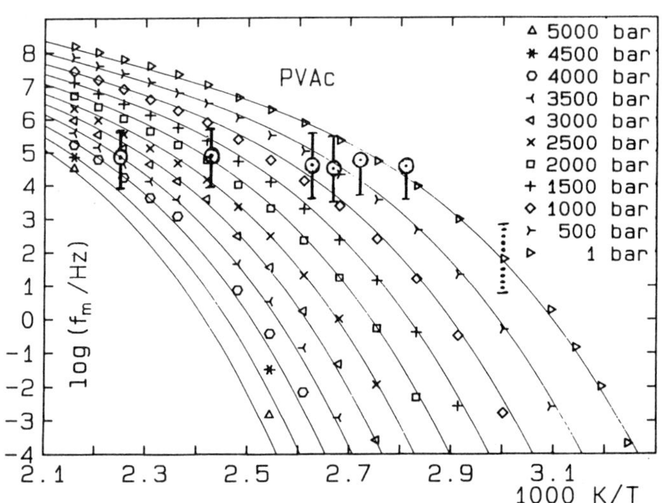

Fig.5 Activation diagram for the glass relaxation of PVAc and its dependence on pressure: small symbols from dielectric measurements[6] at pressures indicated, larger data points from the G"-maxima in Fig. 6 reduced to dielectric f_m (see text). The bars denote the half widths of the relaxation curves

The right spacing between the curves at different pressures can only be accounted for to some extend by WLF- or Adam-Gibbs-Havlicek theory if serious assumptions, violating experimental results, are made ($\Delta\kappa/\Delta\alpha$=210 instead of 470 K/GPa for WLF, or $\Delta\alpha$ being independent on p in the other case). A good fit was achieved[6] instead by applying the dislocation concept, i.e. using formulae (8) and (9).

To this aim one has to find the pressure and temperature dependence of the molecular parameters Q and ε_s (assuming f_0 and 3r/d to be constant). Applying Slaters definition of the bulk Grüneisen parameter γ, i.e. equation (1), to the combination of elastic constants in ε_s (9), one may define a Grüneisenparameter γ_{disl} by

$$(10) \quad d\ln\varepsilon_s/d\ln V = 1 - 2\gamma_{disl} , \quad \text{or} \quad \varepsilon_s/\varepsilon_{s0} = (V/V_0)^{1-2\gamma_{disl}}$$

taking into account $b^2 d$ to be proportional to the macroscopic volume.

To achieve a good fit Heinrich and Stoll[6] were lead to assume a intermolecular contribution to the local activation energy which will behave like ε_s

(11) $\quad Q = Q_0 [0.4 + 0.6 \, (\varepsilon_s / \varepsilon_{s0})]$

The amount of this contribution in (11) has been found recently from swelling data and is somewhat less than assumed by Heinrich and Stoll[6].

Taking advantage of equation (7b), using K_0 and $K\alpha$ data from dilatometric experiments, and the bulk Grüneisenparameter $\gamma_0 \approx 5$ from Brillouin-scattering, the fully drawn activation curves in Fig. 5 resulted from a least square fit of equation (8) taking (10) and (11) into account. The fit parameters became[6]

(12a) $\quad f_0 = 3 \cdot 10^{13} \, Hz$, $Q_0 = 49 \, kJ/mol$, $\varepsilon_{s0} = 3.54 \, kJ/mol$, $3r/d = 17$

which describe the curve for normal pressure and

(12b) $\quad \gamma_{disl} = 2.0$

being the essential parameter for the pressure dependence.

A rigorous check of the dislocation concept must include an explanation of (12b), i.e. the pressure dependence of $G/(1-\nu)$ not being discussed so far. Because the pressure dependence of K is known from equation (6) with $\gamma_0 = 5$, and that of G can be measured by the quartz resonator method[10], one best substitutes the Poisson ratio ν by the well known formula from continuum mechanics

(13) $\quad \nu = (3K - 2G) \, / \, (6K + 2G)$,

and gets from (9), (10) and (13)

(10a) $\quad 2\gamma_{disl} = - \, dlnG \, / \, dlnV + dln(1-\nu) \, / \, dlnV$

$$\approx \left[2\gamma + \left(\frac{K}{G} + \frac{2}{3} + \frac{4G}{9K} \right) \frac{K}{G} \cdot \frac{\partial G}{\partial p} \right] \Big/ \left(\frac{K}{G} + \frac{5}{2} + \frac{4G}{9K} \right)$$

The second expression follows after some calculations with $dlnG/dlnV = -(K/G) \, \partial G / \partial p$.

To check the validity of (12b), $2\gamma_{disl} = 4.0$ for lower pressures, one may put $2\gamma_0 = 10$ from Brillouin scattering and $K/G = 8/3$, i.e. $\nu = 1/3$, in (10a) and obtains $\partial G/\partial p \approx 0.9$ which is rather small. From preliminary measurements of the shear modulus versus pressure reported in Fig. 6 one can estimate that $\partial G/\partial p$ may well be in this range, but more accurate measurements are needed in the glassy state for a rigorous proof. Such investigations are under work with a new network-analyser to measure the high damping of shearquartz resonators more precisely in this range of the shear modulus.

For higher pressures $\partial G/\partial p$ seems to decrease and finally may tend to zero. It might be speculated that the pressure dependence of G may look like

(14) $\quad G = G_0 \, [1 + \beta (\partial G/\partial p)_0 \, tanh \, (p/\beta G_0)]$

Inserting (14) and $K = K_0 + 2\gamma p$ into (10a), $\partial G/\partial p = (\partial G/\partial p)_0 / cosh^2 (p/\beta G_0)$ will strongly tend to zero for $p > \beta G_0$, and $2\gamma_{disl}$ will be decreasing (more

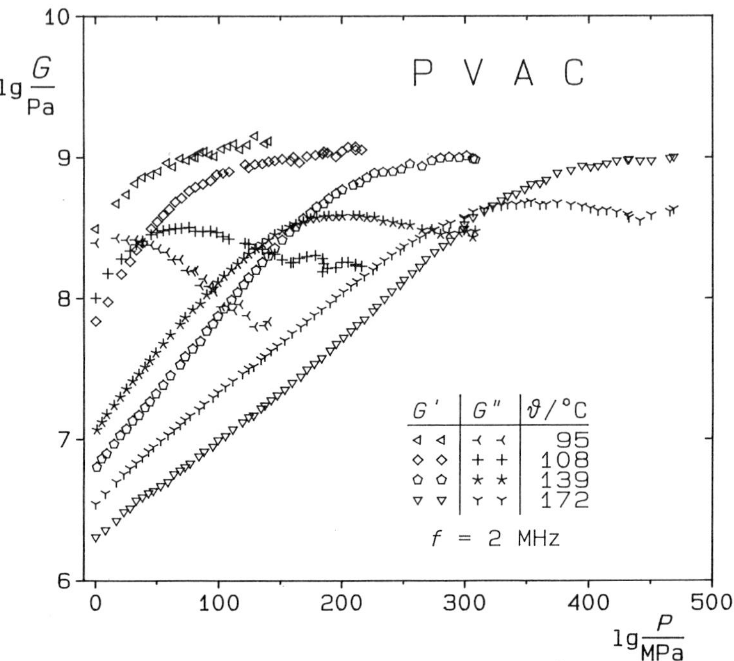

Fig.6 The complex shear modulus of PVAc measured in dependence on pressure
at constant temperatures with the shearquartz resonator method[10]

Fig.7
Glass relaxation fre-
quencies from G"-, ε"-
and J"-maxima at normal
pressure plotted versus
$10^3/T$

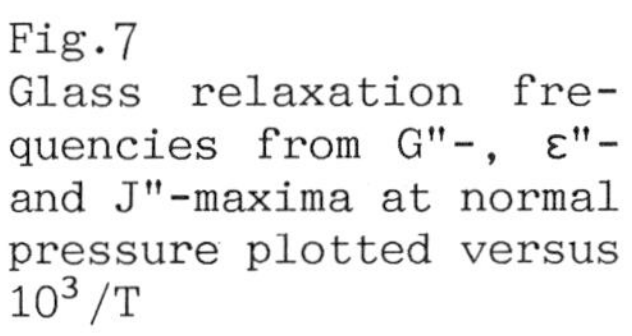

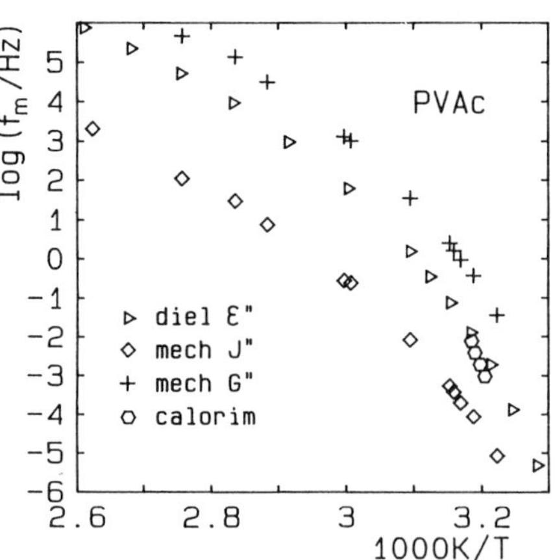

slowly) with pressure to assume values of 1 or less at high pressure. $2\gamma_{disl}$ =1 would cause ε_s from (10) to become independent on pressure as will be the glass temperature in this case.

In spite of the above mentioned scatter in the complex shear modulus when approaching the glassy state, its precision becomes much better in the glass relaxation regime. Therefore, the G"-maximum pressures and half widths at each temperature from Fig. 6 (and two other curves not shown here) were recorded in the activation diagram, Fig. 5, with larger symbols and bars, using the pressure scale given by the dielectric curves. Interestingly its frequency position turned out to be $(6\pm2)10^4$ Hz, instead of 2 MHz, the shear quartz frequency. The simple explanation can be read from Fig. 7, indicating a frequency shift of 1.5 decades between dielectric and shear modulus activation curves (measured at normal pressure). Therefore, the relaxation frequencies from Fig. 6 should appear shifted to $6 \cdot 10^4$ Hz to fit into the dielectric activation diagram. Besides this complete agreement it should be noted, that the half widths of the relaxation maxima on the frequency scale stay at about two decades nearly independent on temperature, eventhough they strongly change on the pressure scale (Fig. 6).

PRESSURE DEPENDENCE OF PHASE TRANSITIONS IN POLYMERS

In order to describe phase transitions in polymers on the basis of model-theories, the transition data including the volume effect must be checked carefully in dependence on pressure. In this lecture there is no time to go into any theoretical detail. Instead one may restrict oneself to present a rough sketch of the principal nature of phase transitions in polymers as described in the meander model.

The melting of atomic crystals is an one step transition in which the long range order is destroyed by a dense fluctuating dislocation wall structure being thermally activated[11]. The pressure dependence of the melting temperatures T_m, i.e. the phase diagram, has been the subject of many investigations, e.g. of the alkali metals[12]. For Na, $\Delta S/k=0.85$ per atom and $\Delta V/V=0.025$ at normal pressure. $\Delta V/V$ decreases with pressure becoming zero at about 1.8 GPa as does dT_m/dp in accord with Clausius-Clapeyron equation. Up to now the pressure dependence has not been quantitatively discussed in the theory of dislocation melting, but it has a good chance to be successful because the dilatational part of a step dislocation may become compressed more strongly than its environment.

The melting of molecular crystals may be a one step transition (e.g. H_2O, H_2Te) or a two step one (e.g. H_2S, H_2Se). In the latter case the entropy change at the second transition ($\Delta S_m/k\approx1.5$) is not far from that of the fcc-metals ($\Delta S_m/k \approx 1.1$) indicating a dislocation type of melting[11] plus some vibrational entropy contribution.

Measurements of the specific volume of high molecular weight polyethylene are reported in Fig. 8a (isobars, heating rates 5K/h) and Fig. 8b (isotherms during loading and unloading with ±0.5 MPa/min). They indicate that melting of polymers turns out to be at least a two step transition too. The intermediate phase appearing above 400 MPa is called CONDIS (conformational disordered) phase after Wunderlich, and may be visualized in a model of conformational clusters (Fig. 9). The first order transition from the all-trans PE-crystal into its CONDIS phase has been described in a cooperative pair-theory taking helical- and trans-sequences into account[13]. Transition temperatures and other transition data ($\Delta S_t/k\approx1.7$ per monomer) including the volume effect were obtained in dependence on pressure. The subsequent melting of the CONDIS phase ($\Delta S_m/k\approx0.7$ per monomer) may be envisaged by incorporating dislocation walls into the conformational domains, with no

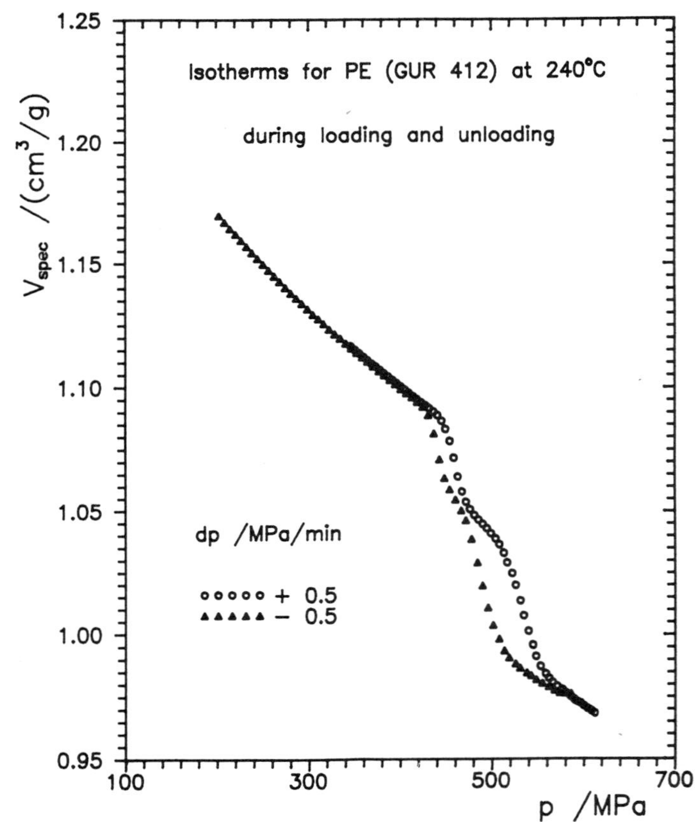

Fig.8a Specific volume versus temperature of high molecular weight polyethylene at different pressures

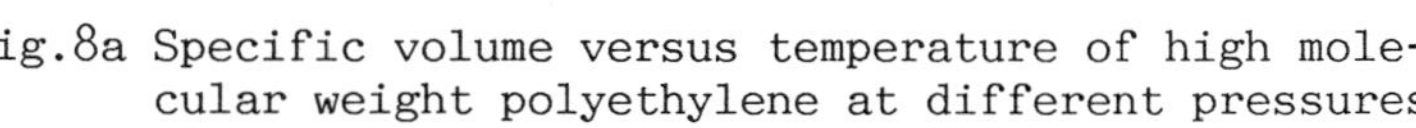

Fig.8b Specific volume versus pressure of high molecular weight polyethylene at 240°C

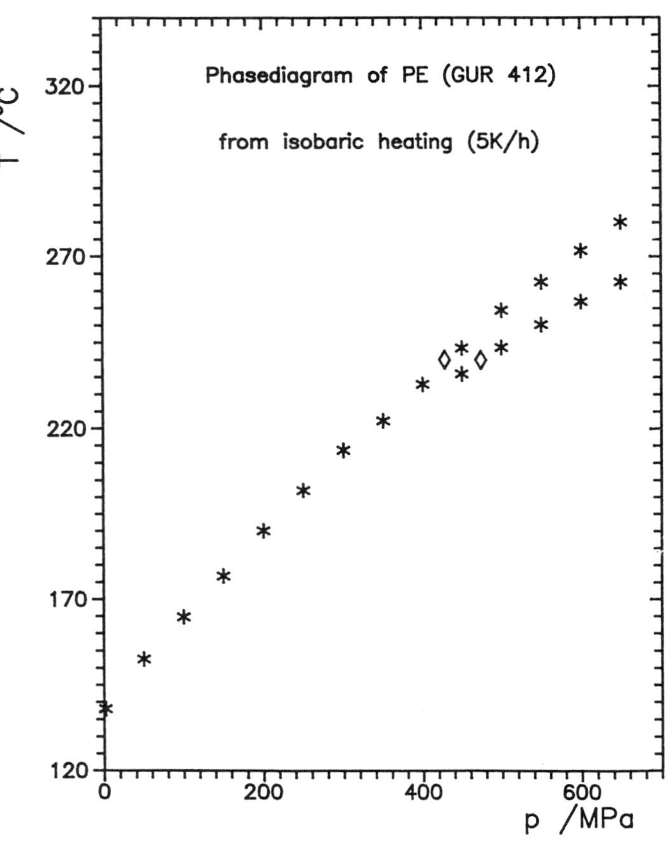

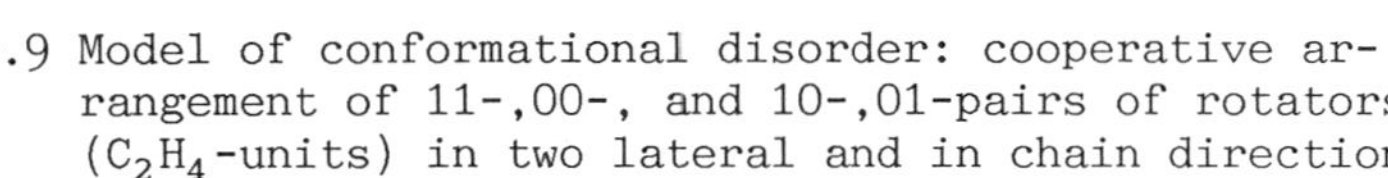

Fig.9 Model of conformational disorder: cooperative arrangement of 11-,00-, and 10-,01-pairs of rotators (C_2H_4-units) in two lateral and in chain direction

Fig.10 Phasediagram of high molecular weight polyethylene showing the intermediate CONDIS phase above 400 MPa

further change in conformation (as indicated by spectroscopy). As with the dislocation melting of atomic crystals, the pressure dependence of this transition with its onset at 400 MPa (cp.phase diagram, Fig. 10) has not been quantitatively discussed so far.

Weak transitions in polymer melts

The phase transitions discussed above are governed by the free energy of conformations and/or of the short range order (including dislocation walls). Therefore a weak contribution to the transition data due to changes in the meander superstructure cannot be detected. These changes are only morphologically observed during the formation of stacks of lamellae, the long period of which amounts to a triple, double or single height of the meander cube in the molten state. Moreover the kinetics of crystallization strongly depends on the superstructure of the melt.

However, two classes of polymers exist which exhibit an intermediate phase in the temperature range between melting and the isotropic melt. Both kinds of intermediate phases are strongly anisotropic indicating a pronounced orientational order of (at least partially) aligned chains, even though its entropy of transition to the melt is very low.

To the first class belong the nematic liquid crystal (LC) polymers having mesogenic groups either in the main chain or as side chains ($\Delta S_{n/i}/k < 0.3$ per monomer of about $M \approx 400 g/mol$). But also polyethylene would show an isotropic to nematic (i/n) transition (at 230K) if it would not crystallize at about 400K, as can be concluded from a $(T - T_{n/i})^{-1}$-behaviour found in orientational correlation measurements (in PE and the n-alkanes). In the meander model the i/n-transition is described by cube rotation across the spacediagonal towards a parallel cube organization, turning the isotropic coarse grains (Fig. 2) of the melt into anisotropic domains (as observed in the polarizing microscope)[14]. The i/n-transition does not show a hysteresis, because only local cube rotation is needed.

The second class of polymers forming thermotropic mesophases include some poly(organosiloxanes) and poly(phosphazenes)[15]. Its transition entropy from the mesophase to isotropic melt is $\Delta S_{m/i} < 0.1$ per monomer. These polymers contain no mesogenic groups and are highly flexible ($T_g < 170K$). In the meander model the i/m-tansition is explained by the same superstructural reorganization which is necessary for lamellar crystallization (e.g. of PE): the lamellae must be nucleated and further grow by simultaneous rotation of 4 cubes aroung its middle cube edge, a process which needs disentanglement and clearly exhibits a hysteresis. The very existence of these mesophases with strongly marked orientational order, formed at the expense of a small amount of entropy only, reveals the presence of orientational order - embedded in the anisotropic meander blocks - already in the isotropic melt.

A rough sketch of the type of superstructural reorganization for both, the i/n- and the i/m-transition can be found in the paper by Pechhold and Schwarzenberger[10] in this volume. There the phase diagram of poly(diethylsiloxane) is discussed. A model theory of the i/n-transition in low molecular and polymer LCs which explains also the pressure dependence of the volume effect has recently been published[14].

ACKNOWLEDGEMENT

Support by the Deutsche Forschungsgemeinschaft and by the Fonds der Chemie are gratefully acknowledged. Thanks are due to Dr. B. Stoll for helpful discussions and to Mrs. Schiffner-Großmann for preparing the manuscript.

REFERENCES

1. P. Bolsaitis, I. L. Spain, in "High pressure technology", Marcel Dekker, New York, I. L. Spain and J. Paauwe Ed. 477-545 (1977)
2. J. H. Rose, J. R. Smith, F. Guinea and J. Ferrante, Universal features of the equation of state of metals, Phys.Rev. B 29:2963 (1984)
3. R. Simha, in "High Pressure Science & Technology", Pergamon Press, B. Vodar and Ph. Marteau Ed. 682-684 (1980)
4. W. Pechhold, E. Sautter, W. v.Soden, B. Stoll, H. P. Grossmann, Mechanical and Dielectric Investigations of Relaxation in Polymers, Macromol.Chem.Suppl. 3:247 (1979)
5. W. Pechhold, B. Stoll, Motion of Segment Dislocations as a Model for Glass Relaxation, Polym.Bull. 7:413 (1982)
6. W. Heinrich, B. Stoll, Description of the Freezing-in Process in Poly(vinylacetate) based on the Meander Model, Progr.Colloid and Polym.Sci. 78:37 (1988)
7. W. Pechhold, M. Böhm, W. v.Soden, Meander Model of Polymer Melts and Networks, Progr.Colloid and Polym.Sci. 75:23 (1987)
8. W. Pechhold, A. A. Mansour, E. Sautter, B. Stoll, Factors affecting glass relaxation in polymers, to be published in Colloid & Polymer Sci.
9. W. Heinrich, B. Stoll, Dielectric investigation of the glass relaxation in poly(vinylacetate) and poly(vinylchloride) under high hydrostatic pressure, Colloid & Polymer Sci. 263:873 (1985)
10. W. Pechhold, P. Schwarzenberger, Phasediagram, Superstructure and Properties of Poly(diethylsiloxane), PDES, this volume
11. W. Pechhold, H. P. Großmann und W. v. Soden, Versetzungstheorie des Schmelzens auf der Grundlage der Cluster-Entropie-Hypothese (CEH), Colloid & Polymer Sci. 260:248 (1982)
12. I. N. Makarenko, A. M. Nikolaenko and S. M. Stishov, in "High-Pressure Science and Technology", Plenum Press, New York, K. D. Timmerhaus and M. S. Barber Ed. 347-356 (1979)
13. H. P. Großmann, W. Pechhold, The high Temperature Transition of Polyethylene into a Condis Phase, described on the Basis of the Cluster-Entropy Hypothesis (CEH), Colloid & Polymer Sci 264:415 (1986)
14. W. Pechhold, H. P. Großmann and E. Sautter, On the Nature of Mesophase Transitions in Polymers, I. The Isotropic-Nematic Transition, Colloid & Polymer Sci. (1991), in press
15. Yu. K. Godovsky and V. S. Papkov, Thermotropic Mesophases in Element-Organic Polymers, Advances in Polymer Sci. 88:129 (1989)

ION TRANSPORT MECHANISMS IN POLYMER ELECTROLYTES

AT NORMAL AND HIGH PRESSURE

B.-E. Mellander, I. Albinsson and J.R. Stevens[*]

Department of Physics
Chalmers University of Technology
S-412 96 Göteborg, Sweden

[*]Guelph-Waterloo Program for Graduate Work in Physics
Guelph Campus, Department of Physics
University of Guelph
Guelph, Ontario N1G 2W1 Canada

INTRODUCTION

Most polymeric materials are excellent electrical insulators with conductivities as low as 10^{-18} S/cm. Nevertheless, both ion and electron conducting polymers have become very active fields of research. A polymer electrolyte is a material which is ionically conducting while the electronic conductivity is negligible. There is a broad current interest in solid polymeric electrolytes for such applications as batteries, electrochromic devices and fuel cells[1-2]. To obtain an ion conducting polymer, a polymer such as poly(ethylene oxide), PEO, or poly(propylene oxide), PPO, is complexed with a salt, e.g., an alkali metal perchlorate or triflate. The low molecular weight glycols (PEO and PPO with hydroxyl terminations) are often used (PEG and PPG). When the salt is complexed with these polymers, the cations co-ordinate with the ether oxygen atoms; a lithium ion may be co-ordinated to approximately four ether oxygens. But these links are in a dynamic equilibrium (breaking and reforming) permitting the cation to move in the complex in response to a very broad distribution of relaxation times for polymer segment motion. The structural relaxations in the polymer chains are thus coupled to the motion of the ions. It is also generally contended that, in most cases, both the cation and the anion contribute to the conductivity. There is, however, considerable disagreement in the literature as to the role of cations, anions, ion clusters, solvent separated ion pairs, contact ion pairs, salt precipitates, etc., in ion transport, especially related to the measurement of transference, transport and diffusion.

High-pressure studies of ion conducting inorganic salts have been a very valuable tool in order to study ion transport mechanisms, crystal defects, etc. For polymer electrolytes, there have only been a few studies under high pressure conditions. In this article, we will make comparisons between inorganic ionic conductors and polymer electrolytes at normal and high pressure and discuss the possibilities for high pressure measurements to contribute to the understanding of the ion transport in polymer electrolytes.

Frontiers of High-Pressure Research, Edited by H.D. Hochheimer and
R.D. Etters, Plenum Press, New York, 1991

IONIC CONDUCTIVITY OF CRYSTALLINE MATERIALS

In crystalline materials, crystal defects such as Frenkel or Schottky defects have to be created in order to make ionic transport possible. When defects are present, the ion transport may occur via vacancy, interstitial or interstitialcy mechanisms, for example. Frenkel and Schottky defects are thermally generated, thus the number of defects is temperature dependent. Another type of defect is caused by the presence of an aliovalent impurity in the lattice. If, for example, an impurity of valency $+2$ replaces an ion with valency $+1$, it may be compensated for by a cation vacancy or an anion interstitial. Solid electrolytes that have the highest cationic conductivities, such as α-AgI, have an open anion lattice where many positions are available for the mobile cation. In these cases, the number of positions available for cations is not temperature dependent but a property of the undisturbed lattice. High ionic conductivity can thus be due to either a very high concentration of thermally created defects, as in AgBr at high temperature[3], or to an inherent defect structure of the material as for α-AgI[4]. As will be shown below, the pressure dependence for these two types of materials will be very different.

For crystalline ionic conductors, the ionic conductivity σ due to the presence of thermally activated defects can be described by

$$\sigma T = (\sigma T)_o \exp[-(\tfrac{1}{2}\Delta H_f + \Delta H_m)/kT] \tag{1}$$

where T is the absolute temperature, ΔH_f and ΔH_m are the formation and migration enthalpies for the fundamental defects which permit ionic motion, and k is the Boltzmann constant. The pre-exponential factor for Frenkel defects is

$$(\sigma T)_o = q^2 a^2 \alpha \nu N^{1/2} M^{1/2} k^{-1} \exp[(\Delta S_f/2k) + (\Delta S_m/2k)] \tag{2}$$

where q is the ion charge, a the jump distance, α a geometrical factor, and ν the jump attempt frequency. N and M are the number of lattice sites and interstitial sites per unit volume, and the ΔS terms are the formation (subscript 'f') and migration (subscript 'm') entropies for the defect. If the number of crystal defects are independent of temperature, the equation for the conductivity will contain only the migration term,

$$\sigma T = (\sigma T)_o \exp[-\Delta H_m/kT] \quad . \tag{3}$$

From studies of the pressure dependence of the ionic conductivity, the formation and migration volumes for the defects can be determined. These volumes are defined through

$$(\partial \Delta G_f/\partial p)_T = \Delta V_f \tag{4}$$

$$(\partial \Delta G_m/\partial p)_T = \Delta V_m \tag{5}$$

where ΔG_f and ΔG_m are the Gibbs free energies for formation and migration of a defect and p is the pressure. It can be shown that the activation volume ΔV can be written as

$$\Delta V = \Delta V_f/2 + \Delta V_m = kT[((\partial/\partial p)\ln R)_T + \kappa\gamma] \qquad (6)$$

where R is the electrical resistance of the sample, κ is the isothermal compressibility, and γ is the Grüneisen parameter. Eq.(6) is valid where thermally generated defects dominate; if the number of defects is constant we get

$$\Delta V = \Delta V_m = kT[((\partial/\partial p)\ln R)_T + \kappa\gamma] \qquad (7)$$

The term $\kappa\gamma$ is usually very small and it may in many cases be omitted; however, it has to be taken into consideration for materials with very low activation volumes. It is expected that the formation volume for a Schottky defect should be of the order of one molecular volume while that of a Frenkel defect should be considerably less. The migration volumes should, in general, be low. The experimental results confirm this, see Table I, where it can be seen that it is relatively easy to distinguish between the different types of defects if ΔV is obtained.

TABLE I. Activation volumes for different types of defects. NaCℓ has Schottky defects where sodium vacancies are the mobile species, AgBr has Frenkel defects with mobile silver interstitials, and α-AgI has an open structure where only the migration term will enter the expression for the pressure dependence of the ionic conductivity.

	ΔV_m cm^3/mol	ΔV_f cm^3/mol	ΔV cm^3/mol	reference
NaCℓ	7	55	34.5	5
AgBr	3.3	16.7	11.6	6
α-AgI	0.8	-	0.8	4

IONIC CONDUCTIVITY OF POLYMER ELECTROLYTES

Polymer electrolytes with high ionic conductivities are mostly non-crystalline and, when these conductivities are plotted in a conventional Arrhenius plot, the ionic conductivity deviates considerably from a straight line, see Figure 1. It has been generally found[7-9] that the ionic conductivity, σ, varies with temperature according to the empirical-phenomenological Vogel-Tammann-Fulcher (VTF) relationship, viz.

$$\sigma = AT^{-1/2}\exp(-E_o/k(T-T_o)) \qquad (8)$$

where E_σ is a pseudo activation energy, A is a constant proportional to the number of carrier ions, and T_0 is a reference temperature usually associated with the ideal glass transition temperature T_0 at which "free" volume disappears, or the temperature at which the configurational entropy becomes zero. In either case, T_0 usually lies 35 to 50 K below the glass transition temperature. The nature of the prefactor in Eq.(8) is discussed from several points of view by Ratner et al.[10] and Papke, Ratner and Shriver[11].

A calculation of the activation volume for amorphous polymer electrolytes should thus be based on Eq.(8) rather than Eq.(1). However, the experiments of Fontanella and co-workers[12] suggest that this leads to unreasonable results if T_0 is related to the glass transition

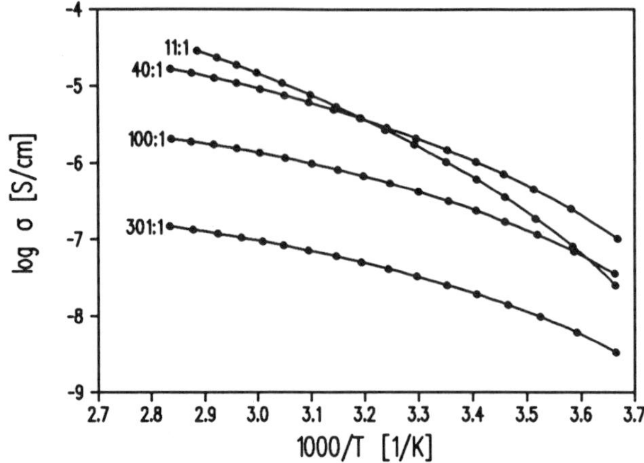

Fig. 1. The ionic conductivity σ versus inverse temperature for complexes of PPG of molecular weight 4000 and $LiCF_3SO_3$. The numbers show the ether oxygen to lithium ion ratio, O:Li.

temperature T_g. It is thus of interest to investigate the VTF equation in more detail at normal and high pressures.

For polymer electrolytes, there are a number of factors that influence the ionic conductivity, such as the concentration, type of cation and anion of the salt that is used when complexing, and properties due to the polymer chain, e.g., chain length, termination and flexibility. An inverse relationship between ionic conductivity and viscosity has long been recognized. Walden[13] proposed that the product of the molar conductivity at infinite dilution, Λ_∞, and the shear viscosity η of the pure solvent is constant, independent of temperature in a given solvent and independent of solvent at a given temperature. At first glance, the viscosity seems to explain most of the variation of the ionic conductivity for different molecular weight of the polymer and for temperature, but not for concentration.

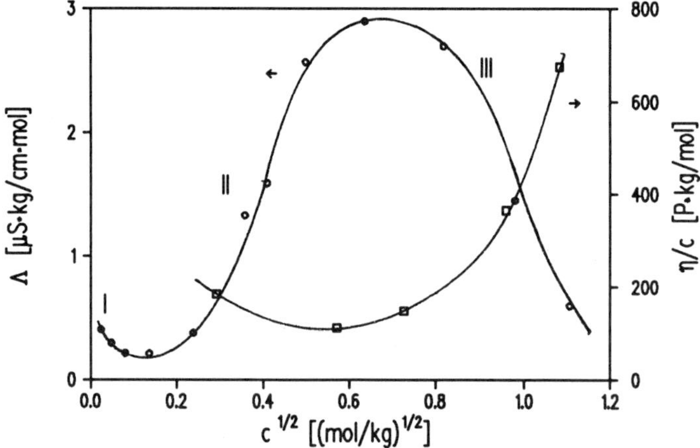

Fig. 2. Molar conductivity and rationalized shear viscosity versus $c^{1/2}$ plot for complexes of PPG of molecular weight 4000 and $LiCF_3SO_3$. The concentration used in this figure will, if multiplied by the density ($\rho = \rho(c,T)$) of the solution at that concentration and at a particular temperature give the molar concentration.

It has been customary to plot the molar conductivity $\Lambda = \sigma/c$ versus c   where c is the molar concentration. This can be seen in Fig. 2 for PPG 4000-LiCF$_3$SO$_3$ complexes at 295 K. Λ_m for a charge carrier of type m is proportional to the product of the mobility μ_m and the fraction of mobile charge carriers α_m since

$$\sigma = \Sigma \alpha_m c_m q_m \mu_m \qquad (9)$$

for charge carriers of concentration c_m and charge q_m. Three regions are identified in Fig. 2, region I at low salt concentrations where Λ decreases with increase in c, region II where Λ increases with increasing c to a maximum and region III where Λ again decreases. The variation of η/c with concentration is also shown in Fig. 2. The fall in Λ at higher concentrations is related to the rapid increase in viscosity and T_g, see Fig. 3.

Polymers such as PPG 4000 have a relatively low permittivity ($\epsilon_r \approx 5$) and it is therefore expected that polymer electrolytes are weak electrolytes although they are much better solvents than one would expect having better solvation properties than acetone[14]. There may be "free ions", ion pairs and larger charged or neutral aggregates in the electrolyte; the number of charge carriers varies with salt concentration, molecular weight and temperature. It has been shown by Raman scattering, for example, that the fraction of "free" charge carriers varies considerably with temperature and concentration[14-17]. In order to explain the shape of the curve in Fig. 2, it is suggested that in region I the fraction of "free" ions decreases with increasing concentration, in region II redissociation causes an increase of α with increasing concentration, and in region III the ionic conductivity decreases due to rapidly increasing viscosity, see Fig. 2, and to an increase of the number of larger aggregates.

If the VTF equation is fitted to the experimentally determined data, a relatively good fit is normally obtained. The VTF parameter T_o for PPG 4000-LiCF$_3$SO$_3$ is shown in Fig. 3 together with results for the glass transition temperature obtained from differential scanning calorimetry measurements. It may be noted that, although the values for high salt concentration seem reasonable, T_o for low salt concentrations is very high, even higher than the glass transition temperature. A similar behaviour has also been found for complexes based on poly(ethylene oxide) modified poly(dimethyl siloxane)[16]. This is in obvious disagreement with the assumption that T_o should be 35 to 50 K below T_g.

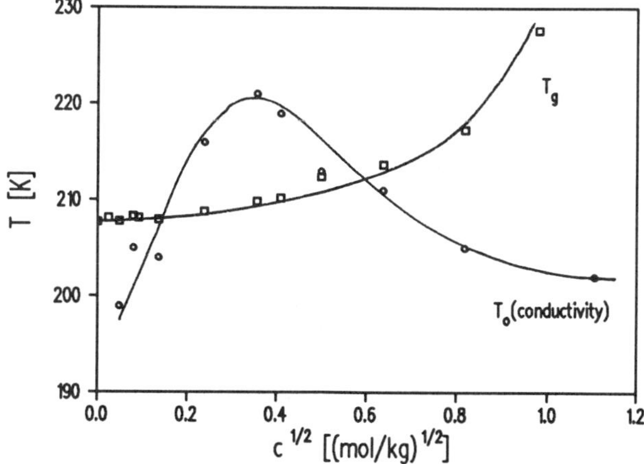

Fig. 3   The glass transition temperature T_g and T_o from the VTF equation for ionic conductivity for PPG 4000-LiCF$_3$SO$_3$

When using the VTF equation, it is assumed that the number of charge carriers is independent of temperature. However, the number of "free" ions in region II decreases with increasing temperature. The parameter A in Eq.(8) is thus temperature dependent. If the VTF equation is modified to take this temperature dependence into consideration, the fitted value for T_o will be lowered to a reasonable value[9]. The VTF equation therefore has to be used with caution, especially when interpreting the fitted parameters.

HIGH PRESSURE MEASUREMENTS ON POLYMER ELECTROLYTES

High pressure studies have been performed on the ionic conductivity of polymer solid electrolytes for pressures up to 0.65 GPa[12,18-24]. These studies have shown that the pressure variation of the ionic conductivity is the same as the pressure variation of the electrical relaxation time for the α relaxation[20-21]. This confirms that the ionic conductivity is controlled by the flexibility of the polymer chains. A typical value for the activation volume for a siloxane-based polymer complexed with $NaCF_3SO_3$ is 29.9 cm^3/mol at 313 K[23]. Using the standard interpretation for crystalline materials, one would conclude that some kind of defects have to be created in order to obtain mobile ions. However, for polymer electrolytes following the work of Angell et al.[25], one might suggest that the reason for the relatively strong pressure dependence of the ionic conductivity may not be associated with the creation of defects but be related to the pressure dependence of T_o as for ionic liquids[25]. This T_o dependence has not, however, been observed for the polymer electrolytes studied by Fontanella et al.[21] who conclude that ionic conductivity in these systems is not "liquidlike". It is interesting to note, nevertheless, that these authors[21] suggest that "an alternative view of the activation volume is that it is approximately the same for all materials at a given temperature interval above T_o". (The material studied was high molecular weight PPO complexed with $LiCF_3SO_3$, $LiC\ell O_4$, LiI and LiSCN for O:Li = 8:1). In a later publication Wintersgill, Fontanella et al.[22] observed that T_o increases by about 5 to 10 K/kbar for some polymer electrolytes while the shift was very small for others.

As can be seen, there is some confusion in the literature which should be clarified. We believe that measurements of the pressure dependence of T_g for various polymer electrolytes should contribute to this clarification. These measurements are in progress. Further, a better understanding of empirical fitting parameters for ionic conductivity will also benefit our understanding of ion transport in these materials by elucidating the role of activation volume.

REFERENCES

1. M.B. Armand, in: *"Polymer Electrolyte Reviews 1"*, J.R. MacCallum and C.A. Vincent, eds., Elsevier, London (1987), p.275.
2. B. Scrosati, in: NATO ASI Series Vol. 217 *"Solid State Micro-batteries"*, J.R. Akridge and M. Balkanski, eds., Plenum Press, New York (1990), p.103.
3. B.-E. Mellander and D. Lazarus, Phys. Rev. **B29**, 2148 (1984).
4. B.-E. Mellander, Phys. Rev. **B26**, 5886 (1982).
5. D.N. Yoon and D. Lazarus, Phys. Rev. **B5**, 4935 (1972).
6. S. Lansiart and M. Beyeler, J. Phys. Chem. Solids **36**, 703 (1975).
7. M.B. Armand, J.M. Chabagno and M.J. Duclot, in: *"Fast Ion Transport in Solids"*, P. Vashishta, J.N. Mundy and G.K. Shenoy, Elsevier-North Holland, New York (1979), p.131; M.B. Armand, Ann. Rev. Mater. Sci. **16**, 245 (1986).
8. G.G. Cameron and M.D. Ingram, in: *"Polymer Electrolyte Reviews 2"*, J.R. MacCallum and C.A. Vincent, eds., Elsevier, London (1989), p.157.
9. I. Albinsson, B.-E. Mellander and J.R. Stevens, to be published.
10. M.A. Ratner, in: *"Polymer Electrolyte Reviews 1"*, J.R. MacCallum and C.A. Vincent, eds., Elsevier, London (1987), p.173; M.A. Ratner and D.F. Shriver, Chem. Rev. **88**, 109 (1988); M.A. Ratner and A. Nitzan, Faraday Discuss. Chem. Soc. **88**, 19 (1989).

11.   B.L. Papke, M.A. Ratner and D.F. Shriver, J. Electrochem. Soc. **129**, 1694 (1982).

12.   J.J. Fontanella, M.C. Wintersgill, J.P. Calame, F.P. Pursel, D.R. Figueroa and C.G. Andeen, Solid State Ionics **9&10**, 1139 (1983).

13.   P. Walden, Salts Acids and Bases: Electrolytes: Stereochemistry, McGraw-Hill, New York (1929), p.283; Z. Physik. Chem. **55**, 249 (1906).

14.   J.R. Stevens and P. Jacobsson, Can. J. Chem., to be published.

15.   S. Schantz, J. Chem. Phys. **94**, 6296 (1991).

16.   I. Albinsson, P. Jacobsson, B.-E. Mellander and J.R. Stevens, to be published.

17.   P. Jacobsson, I. Albinsson, B.-E. Mellander and J.R. Stevens, to be published.

18.   A.V. Chadwick, J.H. Strange and M.R. Worboys, Solid State Ionics, **9&10**, 1155 (1983).

19.   M.C. Wintersgill, J.J. Fontanella, P.J. Welcher and C.G. Andeen, J. Appl. Phys. **58**, 2875 (1985).

20.   J.J. Fontanella, M.C. Wintersgill, J.P. Calame, M.K. Smith and C.G. Andeen, Solid State Ionics **18&19**, 253 (1986).

21.   J.J. Fontanella, M.C. Wintersgill, M.K. Smith, J. Semancik and C.G. Andeen, J. Appl. Phys. **60**, 2665 (1986).

22.   M.C. Wintersgill, J.J. Fontanella, M.K. Smith, S.G. Greenbaum, K.J. Adamic and C.G. Andeen, Polymer **28**, 633 (1987).

23.   S.G. Greenbaum, Y.S. Pak, K.J. Adamic, M.C. Wintersgill, J.J. Fontanella, D.A. Beam, H.L. Mei and Y. Okamoto, Mol. Cryst., Liq. Cryst. **160**, 347 (1988).

24.   S.G. Greenbaum, Y.S. Pak, M.C. Wintersgill, J.J. Fontanella, J.W. Schultz and C.G. Andeen, J. Electrochem. Soc. **135**, 235 (1988).

25.   C.A. Angell, L.J. Pollard and W. Strauss, J. Solution Chem. **1**, 517 (1972).

V(p,T)-MEASUREMENTS ON POLYCARBONATE AND POLYSTYRENE

AND THEIR ANALYTICAL DESCRIPTION

W. Dollhopf, S. Barry, M. J. Strauss

Universität Ulm, Abteilung Angewandte Physik
Albert-Einstein-Allee 11, D-7900 Ulm, Germany

INTRODUCTION

A large number of empirical or semiempirical equations of state (EOS)
were used to determine the volume as a function of pressure[1,2] and sometimes
of temperature, mostly in the form $p=p(V/V_0,T)$. In the field of polymer
physics and polymer manufacturing the form $V=V(p,T)$ is more usefull. Our
aim is to describe the volume of the glassy polymers Polycarbonate and
Polystyrene with an EOS derived by Pechhold[3] using elements of solid state
physics. The parameters of the EOS - the bulk Grüneisen parameter γ, the
bulk modulus K, the product $K\alpha$ (α=thermal expansion coefficient) are determined firstly by fitting the EOS to the experimental data and secondly by
direct evaluation of these data.

EXPERIMENTALS

The polycarbonate (PC) Makrolon® 3208 from Bayer AG, a polyester from
bisphenol A (4,4'dihydroxidiphenol-2,2-propane) and carbonic acid, without
additives, has a molar mass M_n=19000 g/mol and M_w=34000 g/mol. Samples,
22 mm in diameter and appriximately 100 mm long, where melted under vacuum,
kept at 220°C for one hour and than cooled to room temperature at a pressure of 20 MPa. After pressure release, we measured the specific volume at
20°C: V_0=0.8363 cm^3/g.

The polystyrene (PS) PS168N from BASF, also without additives, has a
molar mass M_w=340000 g/mol, $M_w/M_n \approx 2.5$. Samples of the same size were also
melted under vacuum, kept at 180°C for two hours and than cooled to room
temperature at 20 MPa or, in some cases, at 300 MPa.

The pressure dilatometer consists of a piston-cylinder device with two
cylinders made from hardened steel (Fig.1, No.1+2), shrink-fitted to sus-
tain a maximum pressure of 1 GPa. The pistons were sealed by Bridgman-type
unsupported area seals (Fig.1, No.6). The sample (No.7) was wrapped in a
10 μm Al-foil to prevent adhesion to the walls of the steel-cylinder. In
order to be sure that the pressure around the sample was always hydrostatic, the sample was embedded in a small quantitiy of silicon oil (type AK
12500, Wacker Chemie). The volume-change of the sample was determined by
measuring the displacement of the upper piston in relation to the lower one
by an inductive transducer. After calibrating the system with a steel

Frontiers of High-Pressure Research, Edited by H.D. Hochheimer and
R.D. Etters, Plenum Press, New York, 1991

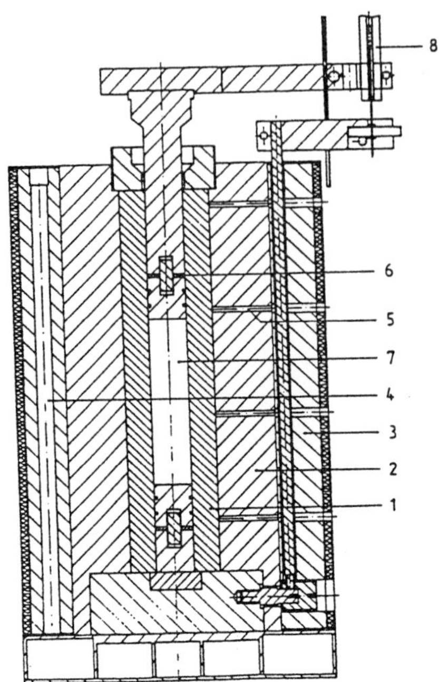

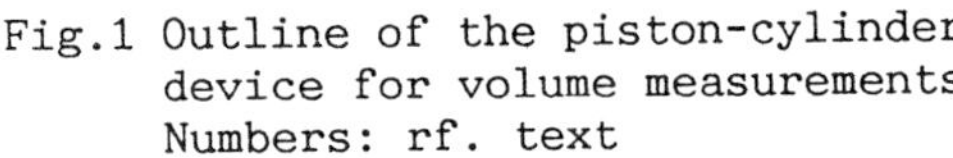

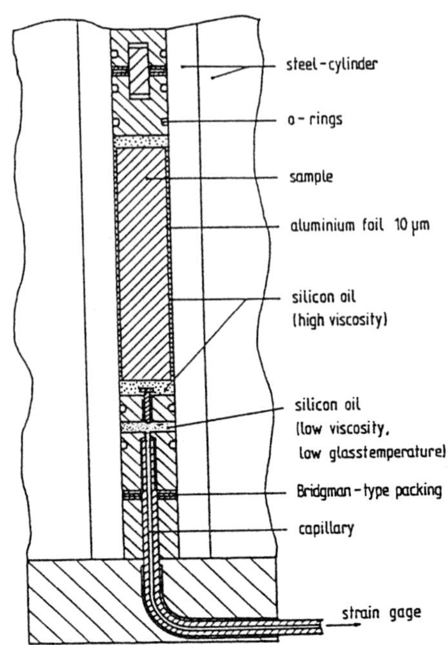

Fig.1 Outline of the piston-cylinder
device for volume measurements
Numbers: rf. text

Fig.2 Sample arrangement with capillary to the pressure measuring
gage

cylinder in order to correct the influence of pressure and temperature on
the pistons and seals, the pressure effect on the oil is determined. After
these corrections the specific volume can be determined with an accuracy of
±0.15%.

For the PC-measurements the pressure was calculated from the force
acting on the pistons. We evaluate the friction between piston and cylinder, due mostly to the packing rings, by measuring the specific volume of
molten PC with increasing and then with decreasing temperature: the force
necessary to obtain the same volume at the same temperature differs by
twice the friction force. The pressure accuracy obtained is ±1%. For the
PS-measurements we replaced the bottom pistons by an arrangement with an
external pressure transducer (Fig.2) for pressures up to 700 MPa. In this
way the pressure accuracy is 0.5%.

The temperature is measured with 4 platinum-resistors (Fig.1, No.5)
with an accuracy of ±0.2°C.

RESULTS ON POLYCARBONATE

Many empirical EOS are used[1,2], mostly in the form $p=p(V/V_0,T)$, but for
polymers very often a modified Tait-equation is successfully fitted to the
pressure-volume-data[4]. Pechhold[3] however, wanted to describe $V=V(p,T)$ in
the glassy state of PC with quantities (γ, K) used in solid state physics.
For this purpose he has derived an EOS thermodynamically.

$$V(p,T) = V_0 \left\{ 1 + \frac{2\gamma_0}{K_0} \left[p - p_0 - K\alpha(T-T_0) \right] \right\}^{-\frac{1}{2\gamma_0}} \qquad (1)$$

The parameter are the physical quantities

V_0 = Volume at p_0=0.1 MPA and T_0=20 °C
K_0 = Bulk modulus at p_0 and T_0
$K\alpha$ = Product of K and the thermal expansion coefficient α
γ_0 = Grüneisenparameter at p_0 and T_0

Fig.3 displays isothermal data: in the range from 20-120 °C the sample is in the glassy state; between 140 and 220 °C the samples are molten at low pressures, the glassy state occurs at higher pressures causing that the so generated glasses have a different degree of inner order. Finally from 240-270 °C the sample is molten in the whole pressure range of our measurements. K can be calculated from our measurements (Fig.3), and the results are shown in Fig.4 indicating that K(p) can be fitted to straight lines outside the glass-transition regions. The intersection of the straight line at 20 °C with the ordinate yields K_0=3.75 GPa, the slope $dK/dp=2\gamma_0$ yields γ_0=4.0. In order to determine the last parameter (K·α) of EOS (1) we habe performed isobaric measurements at 15 different pressures (Fig.5). From this measurements we have determined α and have multiplied this values with the corresponding K. The product is shown in Fig.6 as function of p and T. It can be seen that $K\alpha$ is nearly independent of p and T in the glassy state.

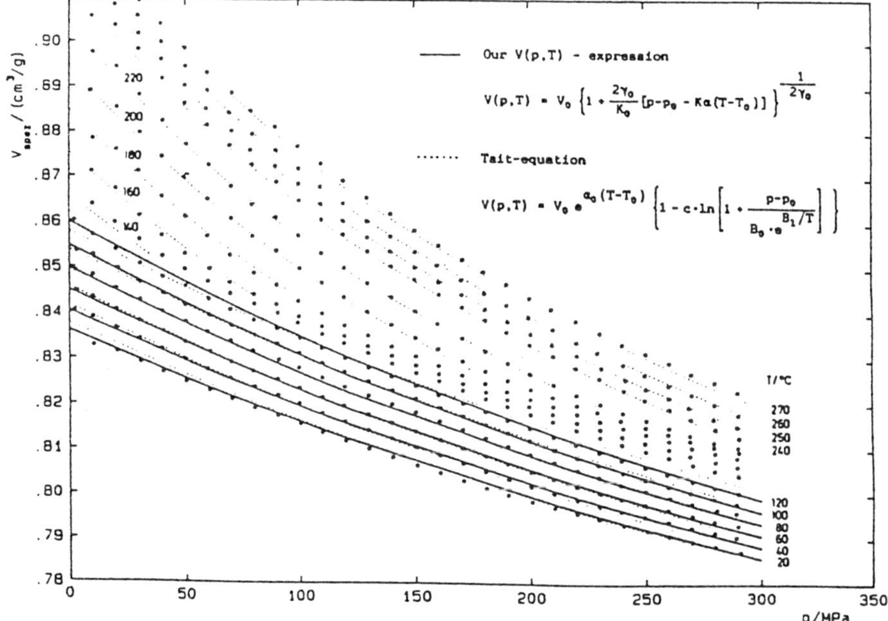

Fig.3 Specific volume of PC as a function of pressure at various temperatures

The data points for the glass with the same degree of order (20-120 °C) shown in Fig.3 were least-square-fitted to EOS (1) and to the Tait-equation. The values of the parameters are compiled in Tabel 1 and Table 2 respectively, together with the standard deviation.

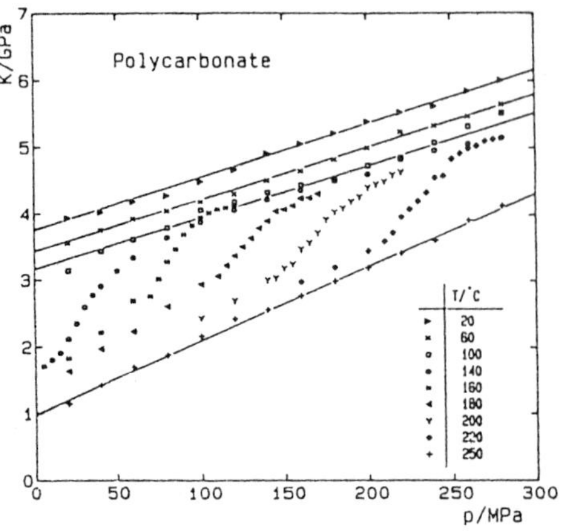

Fig.4 Bulk modulus as a function
of pressure

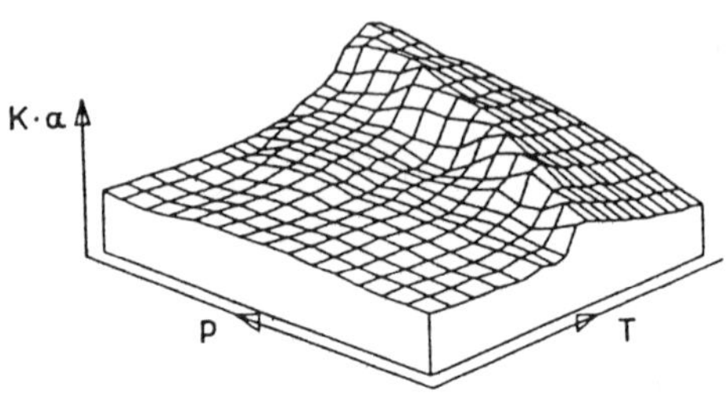

Fig.6 Kα of PC as a function of
pressure and temperature

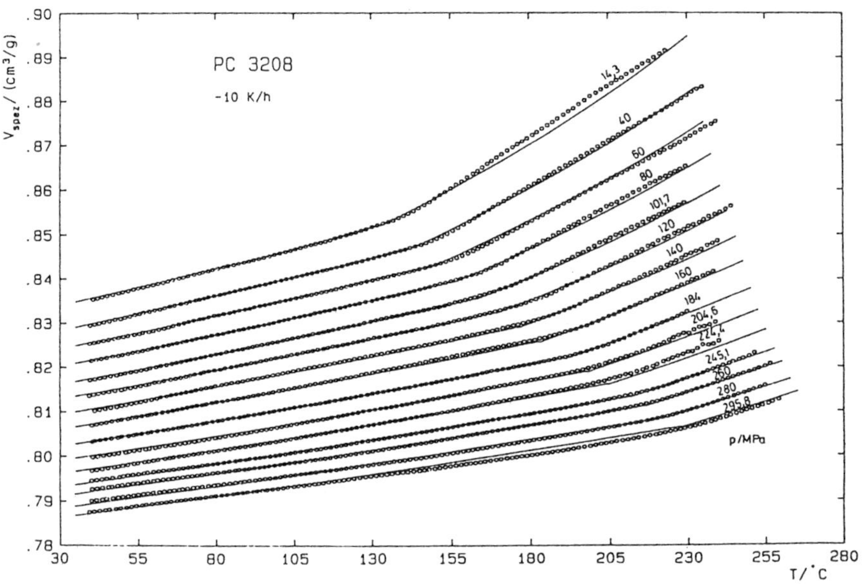

Fig.5 Specific volume of PC measured as a function of temperature at
various pressures with calculated curves

Using values of $\gamma=3.8-4.2$ EOS (1) yields fits of equal quality, but all fits are better than those using the Tait-equation. As the assumptions for the derivation of EOS (1) are not valid in the melt, in this state a better fit is obtained with the Tait-equation. The isobaric melt-data are fitted with EOS (1) [$V_0=0.8232$ cm^3/g; $\gamma_0=4.46$; $K_0=3.50$ GPa; Kα=1.05 MPa/K] as shown in Fig.5. The fit for the isobaric glassy state data of Fig.5 [$V_0=0.8356$ cm^3/g; $\gamma_0=4.29$; $K_0=3.62$ GPa; Kα=0.707 MPa/K] differs a little from the values in Table 1. The reason is the formation of glasses with different degrees of inner order at each pressure. We shall discuss this phenomenon also for polystyrene.

Table 1. Parameter for EOS (1) applied to PC in the glassy state from the experimental data and those obtained from a fit to the EOS (1).

	from experiment	from fitting	
$V_0/cm^3 g^{-1}$	0.8363	0.8362	0.8362
γ_0	4.0	3.88	4.2
K_0/GPa	3.75	3.748	3.703
$K\alpha/MPa.K^{-1}$	0.850	0.950	0.947
$\sqrt{\overline{\Delta_V^2}}/10^{-4}$		6.37	6.49

Table 2. Parameter for the modified Tait-equation applied on PC in melt and glass, obtained from fitting.

	glass	melt
$V_0/cm^3 g^{-1}$	0.8379	0.8118
$\alpha_0/10^{-4} K^{-1}$	2.38	5.39
c	0.1051	0.0922
B_0/GPa	0.3342	0.0687
$B_1/°C$	1.79	142.3
$\sqrt{\overline{\Delta_V^2}}/10^{-4}$	9.73	7.54

RESULTS ON POLYSTYRENE

In order to study the influence of pressure on the volume at the transition from melt to glass (producing so called densified glasses), we measured a series of isobars (Fig.7) at p=20, 120, 300 MPa. It can be seen that the volume in the glassy state depends on the pressure at which this state is reached: V decreases with increasing pressure (less defects, less free volume). In the melt V(p,T) has an unequivocal value, due to thermodynamic equilibrium.

If a PS-glass, produced at another pressure as p is heated under p, its volume tends towards the equilibrium value some 10 K below the glass temperature T_g. We test the EOS (1) with 2 series of samples: PC-glasses cooled under pressures p=20MPa ("PS-20MPa") and p=300MPa ("PS-300MPa") respectively.

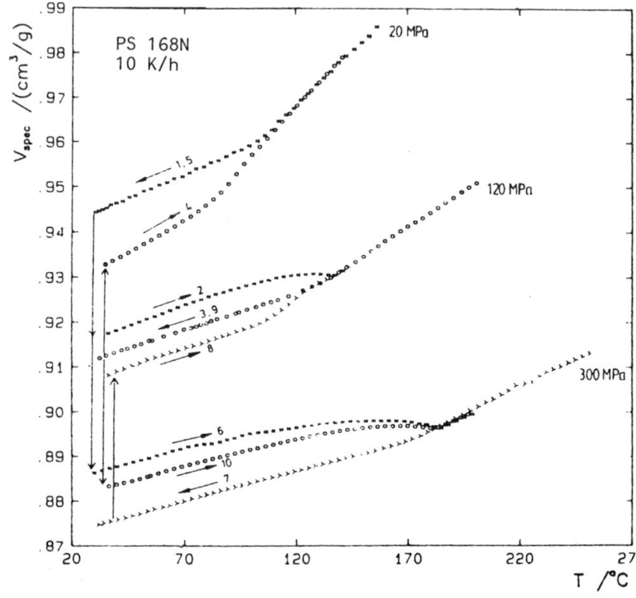

Fig.7
Cooling and heating curves for PS at 3 pressures. The arrows and the numbers indicate the order of succession of the cycles

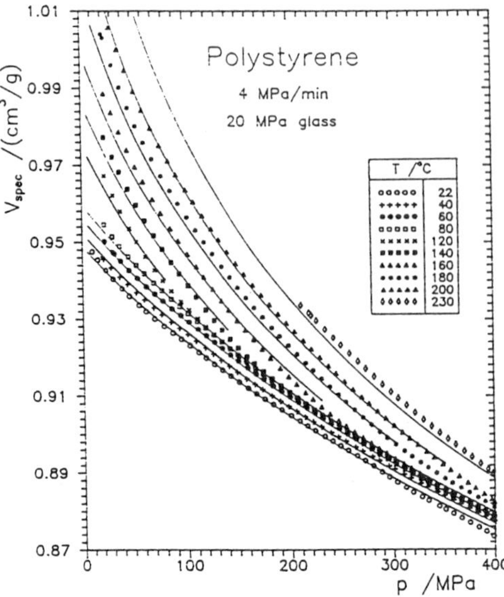

Fig.8 Specific volume of PS-20MPa
as a function of pressure

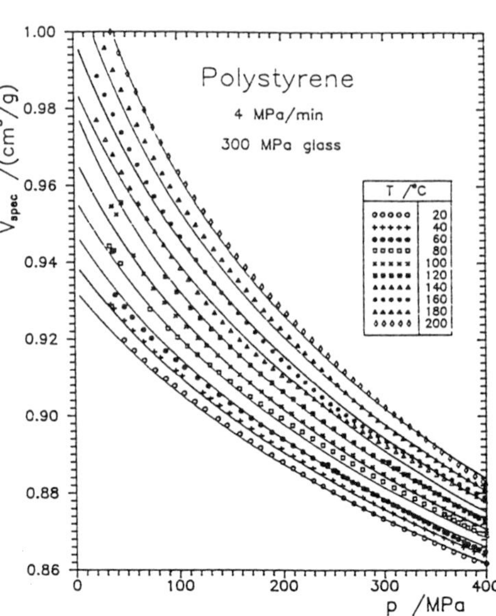

Fig.9 Specific volume of PS-300MPa
as a function of pressure

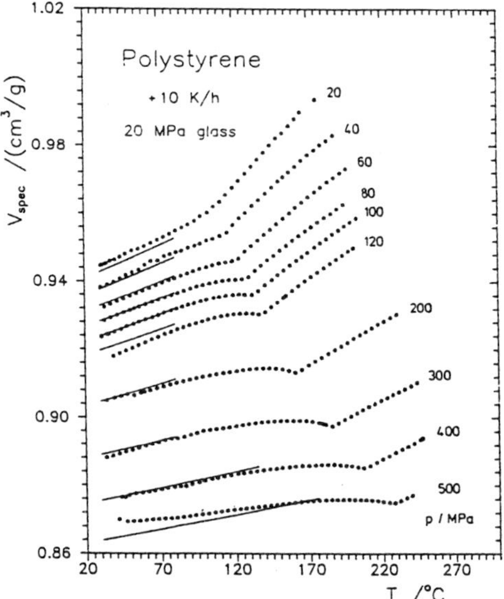

Fig.10 Specific volume of PS-20MPa
as a function of temperature
(heating only)

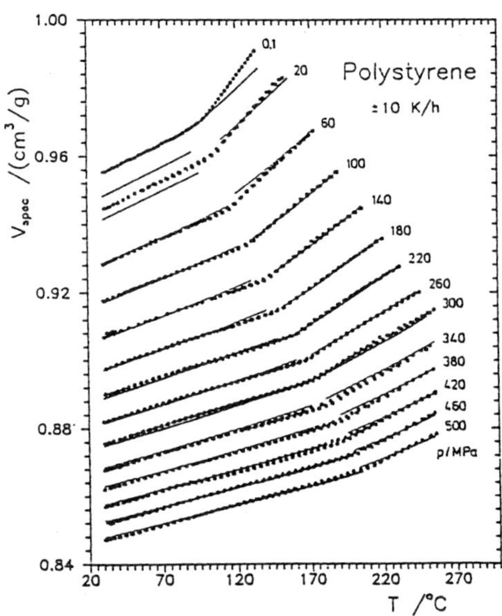

Fig.11 Specific volume of PS
as a function of temperature
(cooling and then heating)

Fig.8 displays isothermal data on PS-20MPa at temperatures up to 80°C; the curves in Fig.8 are calculated with EOS (1) and the parameters listed in Table 3a (the values entitled "from experiment" are obtained in the same way as for PC). At higher temperatures the melt is reached; the transition pressure increase with higher temperature (producing densified glasses with lower volumes). Above 200°C the sample remains in the melt. The curves in the melt are calculated with the parameters in Table 3d. The same parameters from Table 3a and 3d are used for the isobaric curves in Fig.10, plotted together with measured data on PS-20MPa.

Table 3. Parameter for EOS (1) applied to PS from the experimental data and those obtained from a fit to the EOS (1) as well as parameters for equation (2)

	a) PS-20MPa		b) PS-300MPa		c) PS dens. glasses	d) PS-melt	
	from experiment	from fitting	from experiment	from fitting	from fitting	from experiment	from fitting
$V_0/cm^3 g^{-1}$	0.9535	0.9505	0.933	0.9329	-	-	-
$V_{00}/cm^3 g^{-1}$	-	-	-	-	0.9462	-	0.9457
$\Delta V/cm^3 g^{-1}$	-	-	-	-	0.0223	-	0.0278
p^*/MPa	-	-	-	-	200	-	132
γ_0	4.44	4.55	6.2	6.90	3.50	4.8	4.35
K_0/GPa	3.35	3.361	2.9	2.784	4.16	-	3.95
$K\alpha/MPa \cdot K^{-1}$	0.60	0.607	0.99	1.00	0.895	-	1.19
$\sqrt{\overline{\Delta_V^2}}/10^{-4}$	-	6.3	-	8.4	5.6	-	7.7

Fig.9 displays isothermal data on PS-300MPa, only in the glassy state (for the melt, there is no difference to Fig.8); the curves are calculated with the parameters in Table 3b. It strikes that γ_0 and α are larger and that K_0 is smaller as for PS-20MPa.

The last series of measurements (Fig.11) points out isobaric volume values as obtained by cooling from the melt at different pressures. In the melt the degree of internal order is a function of pressure and temperature, in the glassy state a function of pressure during freezing in. This is in contradiction to the assumptions for the derivation of EOS (1) if one applies it to the V_{spec}-data in Fig.11. For this reason the best-fit of the experimental data (Fig.11) with EOS (1) produces a standard deviation of $11.4 \cdot 10^{-4}$ (glassy state) resp. $16 \cdot 10^{-4}$ (melt). As the cycles of Fig.7 suggest, the volume at 20°C and 1 bar is a function of pressure under which the glassy state was reached. In order to describe the measurements in Fig.11 we replaced V_0 in EOS (1) with

$$V_0 = V_{00} - \Delta V \tanh (p/p^*) \tag{2}$$

In this way we obtained the parameters listet in Table 3c and the curves drawn in Fig.11. The deviation at low pressures indicates an additional compressibility probably due to grain boundaries.

The experiments with the different PS samples, prepared at different pressures, show the influence of the preparation-pressures on the volume in the glassy state, but also the practicability of the EOS (1). The parameters depend on the preparation: for PS-20MPa $\gamma_0 = 4.5$, for PS-300MPa $\gamma_0 = 6.9$.

For the melt ($\gamma_0 = 6.0$) and for the different glasses in Fig.11 ($\gamma_0 = 5.3$) the values of γ are in between these two.

The main advantage of the EOS discussed here is that the parameters involved have a physical meaning. Three of them (V_0, K_0, $K\alpha$) may be determined at room temperature and without high pressures. If the bulk modulus K is measured over a range of at least 100 MPa, the Grüneisen parameter γ_0 can be determined from the slope of $K(p)$. If such measurements are not available than we can use a value of $\gamma_0 = 5$. This is justified because measurements with various methods on many amorphous polymers in the literature[5,6,7] yield values between $\gamma = 3$ and $\gamma = 7$. On the other hand the error induced by not exactly determined γ is not large. For instance if we change γ from 5 to 6 in Table 3c the volume at 520 MPa and 20°C changes by 0.007 cm^3/g, which is the largest change in our pressure and temperature range.

In cases where not the highest accuracy is required, for instance in polymer processing, the parameters of the EOS are easily accessable.

REFERENCES

1. I. L. Spain and J. Paauwe, "High pressure Technology", Marcel Dekker, New York (1977) p.485.
2. W. B. Holzapfel, Proceedings of the II. Archimedes Workshop "Molecular Solids under Pressure", ed. R. Pucci, Elsevier Science Publ., Amsterdam (1990).
3. W. Pechhold, Polymers under pressure, this volume
4. P. Zoller, A Study of the Pressure-Volume-Temperature Relationships of Four Related Amorphous Polymers, J.of Polym.Sci.: Polym.Physics Edition 20:1453 (1982).
5. R. W. Warfield, The Grüneisen Constant of Polymers, Makromol. Chemie 157:3285 (1974).
6. B. Hartmann, Ultrasonic Absorption and the Grüneisen Parameter, Acustica 36:24 (1977).
7. U. Leute, H. P. Grossmann, On the pressure dependence and the Grüneisen parameters of the B_{1U}-lattice vibration in high density polyethylene, Polymer 22:1335 (1981).

LOW DIMENSIONAL ORGANIC METALS: STRUCTURAL AND ELECTRONIC PROPERTIES OF Cs[Pd(dmit)$_2$]$_2$

I. Marsden †, M. Allan †, R. H. Friend †, A. E. Underhill §, R. A. Clark §

† Cavendish Laboratory, Madingley Road, Cambridge CB3 OHE, UK
§ Department of Chemistry, University College of North Wales, Bangor
Gwynedd LL57 2UW, UK

INTRODUCTION

Although most organic materials are electrical insulators, with conductivities of the order of 10^{-9} to 10^{-14} Scm^{-1}, the possibility of obtaining organic materials with conductivities comparable to metals was suggested as early as 1911, [1, 2]. Interest was generated in a number of areas of organic materials, including both organic polymers doped with suitable inorganic donor or acceptor ions, and highly conducting organic molecular crystals. It is in the latter of these materials that the highest conductivities, and in some cases superconductivity, has been found. It is these organic molecular materials that will be discussed here.

These materials consist of large organic, or organometallic, planar molecules which tend to form crystals in which the molecules are stacked with the molecular planes parallel, and a smaller distance between molecules along the stack than in any other direction. The electronic properties of these materials are governed by the extent of the intermolecular interactions which in turn depend on the degree of overlap of the molecular orbitals. The orbitals of interest, because they are partially filled in these solids, are the π orbitals. π orbitals are highly directional, and hence due to the nature of the stacking of the molecules the intermolecular interactions are highly anisotropic with generally good contact along the stacks, and much poorer contact between stacks. This in turn leads to highly anisotropic electronic properties.

Work in this field began to flourish in the 1960's with the discovery of the highly conducting organic metal TCNQ (tetracyanoquinodimethane), [3], see figure 1. This was followed quickly by the synthesis of nearly one hundred new organic crystals based on TCNQ. Most of these were highly insulating but some had room temperature conductivities as high as 100 Scm^{-1}. In the 1970's came the first organic molecular salt, with the combination of TCNQ with TTF (tetrathiafulvalene), figure 1, to produce TTF-TCNQ. This proved to be a salt with very strong metallic properties, and a room temperature conductivity of the order of 600-900 Scm^{-1}. The molecules used in the synthesis of this salt, TTF and TCNQ, became the basis for a large family of organic metals, the *charge transfer salts*. These compounds are stabilised by partial transfer of electrons from a donor molecule (e.g. TTF) to an acceptor molecule (e.g.TCNQ), resulting in at least one set of stacks having partially filled electron energy bands. In the case of TTF-TCNQ there is a partial charge transfer of about 0.59 electrons per molecular unit at room temperature, resulting in both stacks possessing partially filled conduction bands. In the case of TTF-TCNQ the interstack interaction is very weak resulting in a quasi-one dimensional system, whilst BEDT-TTF , figure 1, salts are generally quasi-two-dimensional due to the interstack interactions being comparable with the intrastack.

With the exception of a few materials which become superconducting, the conductivity of organic molecular crystals is generally reduced at low temperatures through phase transitions to insulating or semiconducting states. This loss in conducivity is due to the stabilisation of a less conducting ground state through one or more phase transitions. There are a wide variety of possible ground states including the charge-density wave (CDW) state and spin density wave (SDW) state.

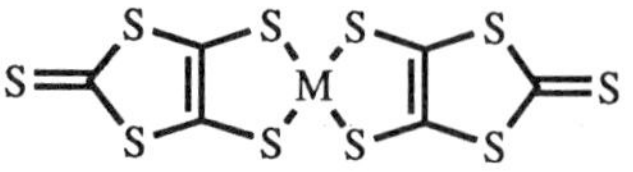

dmit	1,3- dithiol-2-thione-4,5-dithiolate (acceptor);
M=Ni,Pd,Pt,Au	
BEDT-TTF	bis(ethylenedithio)tetrathiafulvalene (donor)
TTF	tetrathiafulvalene (donor)
TCNQ	tetracyanoquinodimethane (acceptor)

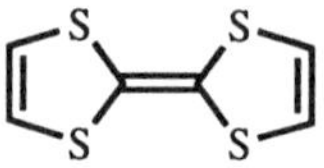

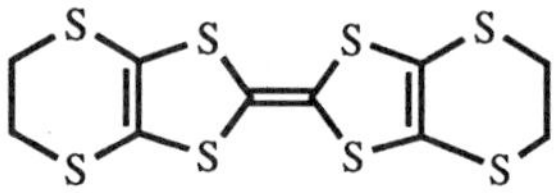

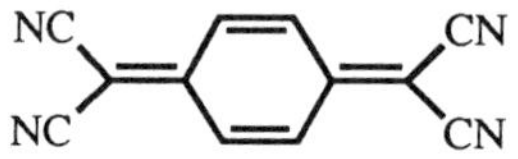

Figure 1.   Examples of organic and organometallic molecules used in the synthesis of charge transfer salts.

LOW TEMPERATURE GROUND STATES

At low temperatures, where almost all the electrons are in states below the Fermi energy E_F, the total energy of a metallic chain can generally be decreased by the creation of a periodic distortion in the electron gas that opens an energy gap at E_F.

This is initiated by the creation of a periodic potential with wavevector q, which will then couple states with equal energy on either side of the energy spectrum. Perturbation theory will remove the degeneracy at the new zone boundaries producing a gap. If the states on either side of a break are occupied by electrons, then the total energy of all the electrons will be unchanged. However, if such a break coincides with the Fermi surface, i.e. $q=2k_F$, then the states displaced downwards are occupied, and the states which are raised are empty. The net result is a reduction of the electronic energy of the system, see figure 2. The wavevector q which couples the two sides of the Fermi surface is also referred to as the nesting vector, and the Fermi surface is then said to be nested. Depending on the shape of the Fermi surface it is possible to have differing degrees of nesting.

There are several ways of introducing a periodic distortion of the electron gas in these materials. Firstly, and the most studied, is where the new periodicity of the electron gas is accompanied by a distortion of the underlying lattice, a Peierls distortion. Peierls first noted that the distorted state is stable at low temperatures due to the fact that the cost in energy of distorting the lattice is offset by the decrease in electronic energy due to the lowering of the states near E_F in the gap formation, [4]. The periodic distortion of the electron gas, in this case, is called a charge-density wave, (CDW), and arises from the electron-phonon interaction.

Another type of gap opening distortion involves the charge density for each spin state of the electron varying with period $2k_F$. This implies that the charge densities for the two spin states will be 180° out of phase and thus the total electron charge density will be constant independant of position i.e. no motion of the molecules occurs. There is however, a net spin polarization that varies with position. This type of distortion is thus known as a spin-density wave, (SDW).

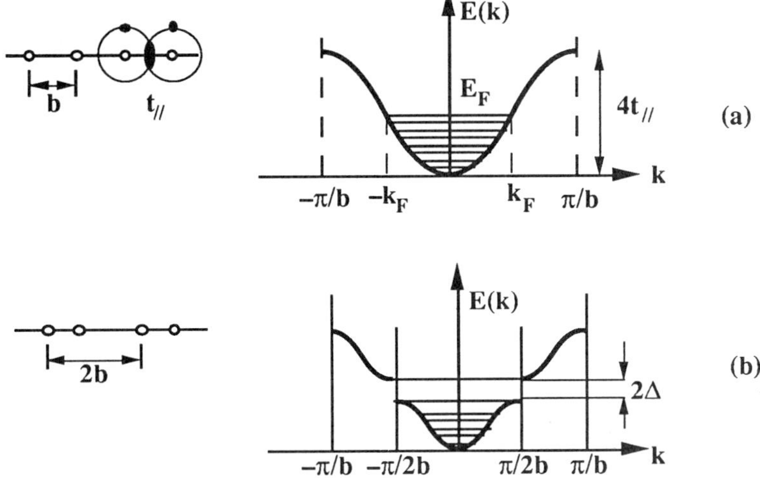

Figure 2. The Peierls Distortion. Schematic of a linear chain of molecules before (a) and after (b) the distortion. Also shown is the one-dimensional electron band associated with the linear chain, half-filled in this example. As can be seen, the dimerisation along the chain causes a gap to open at the Fermi energy causing a reduction in the energy of the system. This is due to the lowering in energy of states near the Fermi level.

Finally, it is possible for a $2k_F$ periodicity of the anion chains to also induce a $2k_F$ distortion. This periodicity is usually due to orientational ordering of the anions, [5].

SUMMARY

It has thus been shown that organic molecular crystals show a wide variety of electronic properties with many possible phase transitions. The use of high pressure has proven itself to be an invaluable tool in this field for finely tuning the electronic properties, in particular the conductivity, of these materials. This is achieved through its effect on the intermolecular spacings and thus interaction strengths. The result of this is generally an increase in conductivity and the suppresion of phase transitions.

An insight into the effect of pressure on these materials will be given below with respect to the series of materials based on the organometallic acceptor molecule $M(dmit)_2$, in particular $Cs[Pd(dmit)_2]_2$.

$M(dmit)_2$ SALTS

The molecular charge-transfer salts formed with $M(dmit)_2$ (dmit=isotrithione-dithiolate), shown in figure 1, are in many ways comparable to those formed with organic donor molecules such as TMTSF and BEDT-TTF, with intermolecular delocalization of the π-electron system and a wide range of metallic behaviour. The metal atoms generally used are M=Ni,Pd,Pt,Au. The one major difference between $M(dmit)_2$ and other organic molecules is that it acts as an electron acceptor rather than donor. The cations (D) and $M(dmit)_2$ molecules tend to form distinct stacks with the stacks making alternating sheets of cations and $M(dmit)_2$. The first superconductor found in this type of material was α-TTF[Ni(dmit)_2]_2, [6, 7, 8, 9]. This material is metallic down to 3K, at ambient pressure, reaching $\sim 10^5$ S/cm at 4.2K, and becomes superconducting under a pressure of 7 kbar with a T_c of 1.62K. With this discovery much of the work on these materials focused on TTF:$M(dmit)_2$ salts which lead to the discovery of superconductivity in α'-TTF[Pd(dmit)_2]_2 with a T_c of 6.5K at 20 kbar, [10, 11, 12].

It is also possible for M(dmit)$_2$ salts to be made with closed shell cations such as (CH$_3$)$_4$N$^+$ to form (CH$_3$)$_4$N[Ni(dmit)$_2$]$_2$, [13, 14]. Due to the fact that closed shell cations cannot contribute to conduction, the electron transferring capability of the Ni(dmit)$_2$ molecule is essential to the functioning of this salt as a conductor. Coupled with the discovery of superconductivity in this material, T$_c$ of 5K at 7 kbar, it can be seen that the metallic and superconducting properties are associated with the delocalization between the acceptor molecules, Ni(dmit)$_2$.

The electronic structure of some M(dmit)$_2$ salts has been investigated by Kobayashi et al [15] and Canadell et al [16, 17]. The dimensionality of the M(dmit)$_2$ sheets is found to be dependant on the intra- and inter-dimer interactions, which can be modified by the donor D and metal M, and the inter-stack interaction. Despite the many intercolumn S...S contacts, in which the S...S distances are shorter than twice the van der Waals distance of 3.70Å, the interchain interactions vary considerably in magnitude compared to the intrastack interactions, depending subtly on the interaction geometry. For example, in the case of α-TTF[M(dmit)$_2$]$_2$, (M=Ni or Pd) and (CH$_3$)$_4$N[Ni(dmit)$_2$]$_2$, the M(dmit)$_2$ sheets were originally thought to have two-dimensional electronic properties. However, based on an extended Hückel band examination [18] it has been shown that the transverse interactions are generally not strong enough to make the system multi-dimensional, and in fact the materials are quasi-one-dimensional consisting of an ensemble of one-dimensional bands with the resultant "multi-Fermi surfaces" . In contrast to these materials δ-TTF[Pd(dmit)$_2$]$_2$ shows strong interstack interactions [16].

Cs[Pd(dmit)$_2$]$_2$

<u>Preparation and Structure</u>

Cs[Pd(dmit)$_2$]$_2$ was obtained as well-formed black plates by electrocrystallisation, [19]. The structure is monoclinic, C2/c, a = 14.490 (4), b = 6.2629 (14), c = 30.601 (7) Å, β = 90.58 (2)°, V = 2777.02 Å^3, Z = 4, and the final R value was 0.062. The crystal structure as viewed along the c-axis is shown in figure 4a. As with other charge transfer salts of this type, the Pd(dmit)$_2$ anions form stacks, along the [110] direction in figure 4a, which form sheets in the ab plane separated from one another by the Cs cations. There are two layers of each species per unit cell which are related by a c-glide plane symmetry so that the Pd(dmit)$_2$ anions of each slab are stacked along [110] and [$\bar{1}$10], in alternate sheets along [001], see figure 4b. These columns have periodicity of four molecules (see figure 4). However because the lattice is C-centred, the real repeat unit of the slab contains two anions, [20].

Along the stacks there is strong dimerisation of the Pd(dmit)$_2$ anions, with the anions within the dimer pair being eclipsed, see figure 3. This is in contrast to the case of [(CH$_3$)$_4$N][Ni(dmit)$_2$]$_2$ for which the anions are slipped sideways. This organisation of the two anions clearly allows interaction between the d orbitals of the metal atoms, and it is well established that the strength of this interaction is in the order Ni < Pd < Pt, [21]. Thus, the change from eclipsed to slipped packing from Cs[Pd(dmit)$_2$]$_2$ to (CH$_3$)$_4$N[Ni(dmit)$_2$]$_2$ follows the expected trend. The very much stronger dimerisation in the Cs[Pd(dmit)$_2$]$_2$ salt has important consequences for the electronic structure.

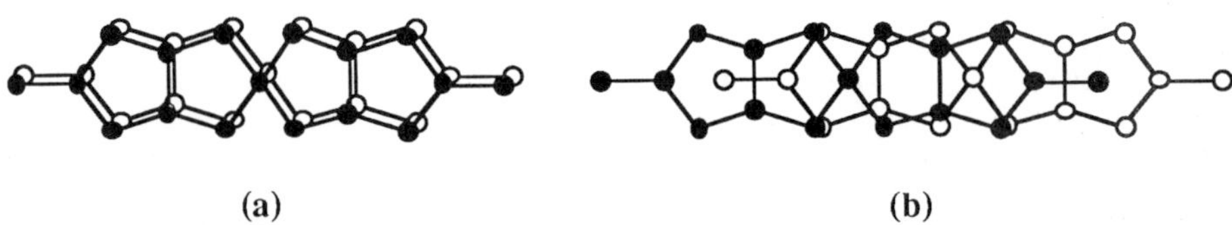

(a)   (b)

Figure 3.   (a) Intra- and (b) inter-dimer overlap in Cs[Pd(dmit)$_2$]$_2$

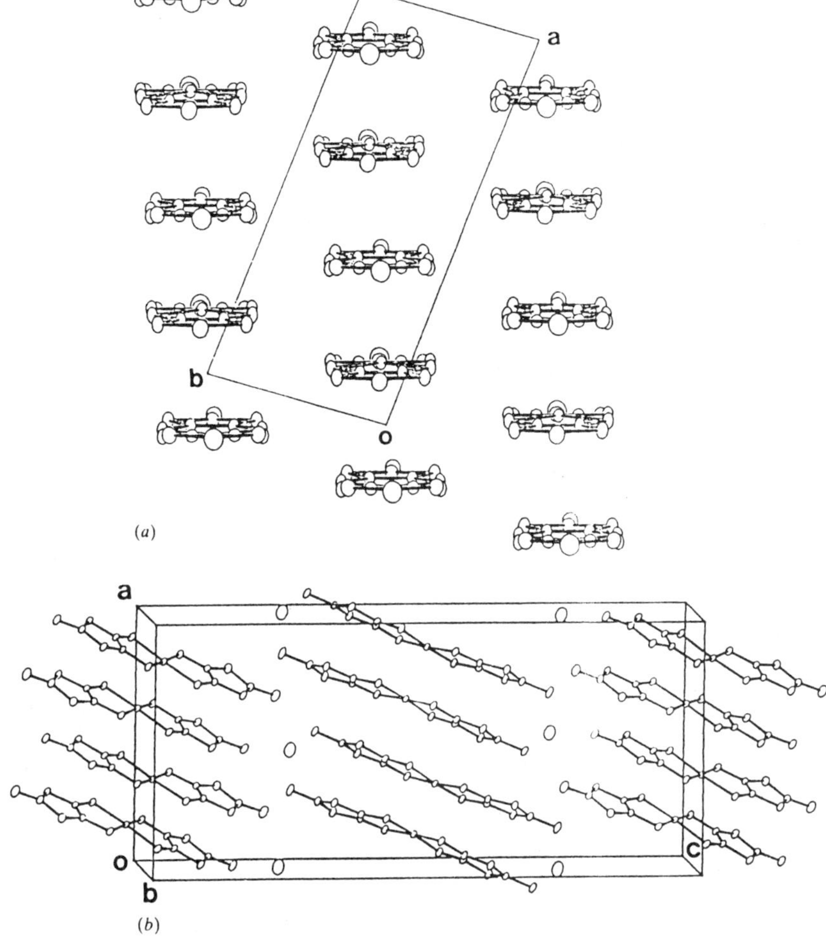

Figure 4.   Crystal structure of Cs[Pd(dmit)$_2$]$_2$, (a) as viewed along the c axis, showing one of the anion slabs, and (b) projection showing alternation of stacking within slabs along the c axis, [20].

<u>Experimental</u>

The synthesised crystals of these organic materials are generally small with typical dimenions of 1.0 x 1.0 x 0.1 mm, and they are very fragile. Four terminal resistance measurements were made using a low frequency AC technique. Electrical contacts were made to the crystals with 10μm gold wire and gold dag. Measurements at pressures up to 11 kbar were made with a beryllium-copper pressure cell in communication with a large intensifier/pressure reservoir, which remains at room temperature, via a stainless steel capillary tube. This system allows accurate control of pressure down to the freezing point of the pressure fluid, (120K at ambient pressure). Measurements at higher pressures, up to 22 kbar are carried out using a maraging steel clamp cell of conventional design.The pressure fluid for both systems consists of a 50:50 mix, by volume, of iso- and n-pentane. The mixture of the two isomers gives a gradual freezing point of the liquid, thus reducing the risk of crystal cracking. The pressure was measured, in both cells, using a manganin wire coil; a 700 Ω coil in the pressure reservoir for the capillary cell, and a 150 Ω coil in the sample space for the clamp cell. The pressure variation on cooling has been studied for both pressure cells using a InSb sample. The capillary cell looses around 1kbar on cooling below the freezing point of the pressure fluid, whilst the clamp cell gradually looses around 2kbar over the whole temperature cycle of 300K to 4K. The greater loss in the case of the clamp cell is due to the fact that this contains a small fixed volume of pressure fluid.

Thermopower is also measured as a function of temperature, 20-350 K. The thermopower is measured by thermally anchoring the crystal to two gold-coated copper blocks which can be seperately heated to 0.5 K above the ambient temperature in the sample space of the cryostat. The thermovoltages developed across the sample and also across a differential thermocouple of gold-iron v chromel are measured with nanovoltmeters.

<u>Electronic Structure- HOMO and LUMO</u>

The energy gap seperating the HOMO (highest occupied molecular orbital) and LUMO (lowest unoccupied molecular orbital) levels of organic molecules is usually large. However, this is not necessarily true for organometallic molecules, such as $M(dmit)_2$, where several ligands are coupled through the metal. The metal d orbital has a very small overlap with the ligand orbitals and consequently the interaction is relatively weak. Consequently the metal-ligand interactions do not lead to a strong HOMO-LUMO splitting ($\Delta\sim0.4$ eV), [16], compared to TTF, TMTSF or BEDT-TTF (~3 eV) or perylene (~1.7 eV).

The dimerization along the $Pd(dmit)_2$ chains causes the HOMO and LUMO orbitals to split, each producing a bonding combination (Ψ^+) of lower energy and an antibonding combination (Ψ^-) of higher energy. In the case of $Cs[Pd(dmit)_2]$, this splitting of the HOMO and LUMO levels is greater than the initial seperation of the HOMO and LUMO levels. The result of this is the energy level of the antibonding combination of the HOMO (Ψ^-_{HOMO}) being above the bonding combination of the LUMO (Ψ^+_{LUMO}). There are five electrons to fill these levels and consequently the singly occupied level is (Ψ^-_{HOMO}), and not (Ψ^+_{LUMO}) as would be expected from the acceptor character of $Pd(dmit)_2$. The dispersion relations for the HOMO and LUMO bands of a $Pd(dmit)_2$ slab, as found in the room temperature structure of $Cs[Pd(dmit)_2]_2$, are shown in figure 5. The half-filled band of these slabs thus originates from the HOMO of the $Pd(dmit)_2$, [20]. This surprising feature was also noticed for δ-TTF[$Pd(dmit)_2]_2$, [16]. In contrast, the partially filled level of the related salt $(CH_3)_4N[Ni(dmit)_2]_2$, which also has dimerized stacks, was found to originate from the LUMO of the acceptor. The origin of this change in nature of the conduction band has been investigated by Canadell et. al. [17].

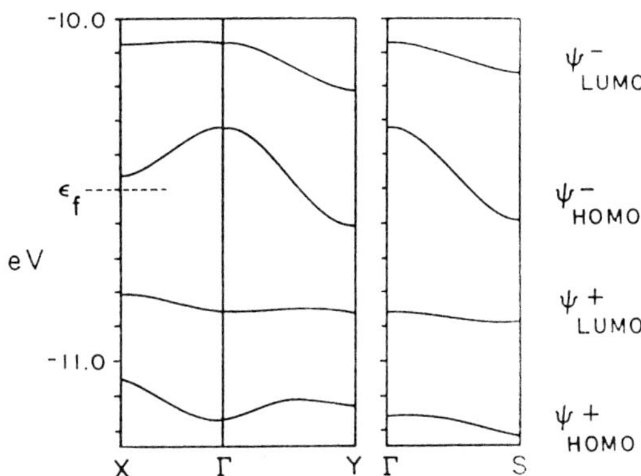

Figure 5.   Dispersion relations for the HOMO and LUMO bands of the $Pd(dmit)_2$ slabs in $Cs[Pd(dmit)_2]_2$. Γ, X, Y and S represent the wavevectors (0,0), (a*, 0), ($a^*/2$, $b^*/2$) and ($-a^*/2$, $b^*/2$), [20]

The partially filled band in figure 5 shows dispersion along Γ→X and Γ→Y and Γ→S. However the dispersion along Γ→X is clearly smaller and this leads to the open Fermi surface of figure 6. As there is no equivalence between the two intraslab directions, the dispersion is different along the Γ→Y and Γ→S reciprocal directions (figure 5) and the Fermi surface for each individual slab has no mirror symmetry. However, the real monoclinic C2/c unit cell contains two $Pd(dmit)_2$ slabs, thus the true Fermi surface is just the superposition of the Fermi surfaces for each slab, figure 6. The best coupling wavevectors for each slab are shown. The observed nesting wavevector has a value intermediate between these two vectors, and is close enough to simultaneously realise the nesting of the Fermi surfaces of the two slabs.

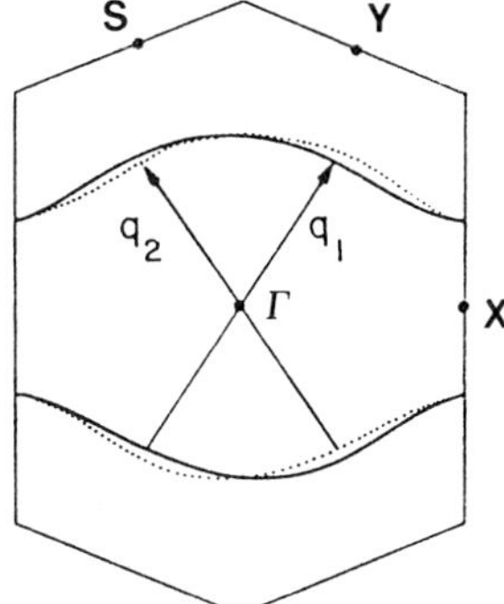

Figure 6.   Superposition of the Fermi surfaces due to the two Pd(dmit)$_2$ slabs related by the c glide plane. q_1 and q_2 are their nesting wavevectors. The Fermi energy contours separate unfilled states around Γ and X from filled states, [20].

<u>Transport Properties</u>

Cs[Pd(dmit)$_2$]$_2$ is metallic at room temperature, with a value for the conductivity about 100 S/cm in the ab plane. The anisotropy within the plane is small,(approx. 1~5), with a higher conductivity in the direction of the a axis. However, due to the small crystal size, the size of the electrical contacts cannot be thought of as relatively small. This introduces considerable uncertainty in the determination of the anisotropy, which was made using the Montgomery analysis, [22].

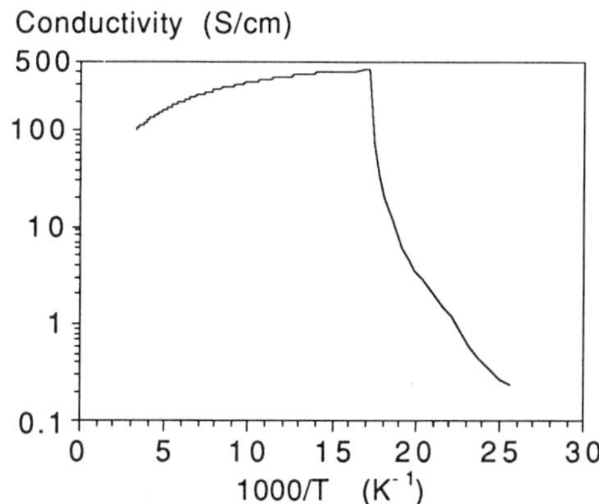

Figure 7.   Conductivity in the ab plane vs. reciprocal temperature.

There is a sharp transition to an insulating state at around 56.5 K, defined as the peak value of $\partial \ln\sigma / \partial(1/T)$. The transition appears to be sharp, but second order, with no evidence for any thermal hysteresis, figure 7.

The temperature dependence of the thermopower as measured along the a-axis is shown in figure 8. In the metallic regime, the thermopower is small and negative, approximately -7 μV/K at 300 K, indicating that the carriers are electron-like. With the Fermi energy in the middle of a single quasi-one dimensional band, as shown in figure 5, the thermopower is expected to be very small, and its sign is not easy to predict. The transition from metal to semiconductor is clearly seen where the thermopower changes sign and rises to a very much larger value, consistent with the opening up of an energy gap at the Fermi energy.

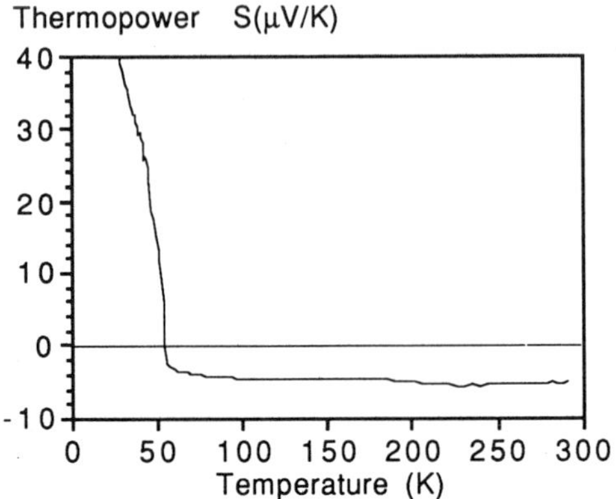

Figure 8.   Thermopower vs. Temperature, measured along the a-axis.

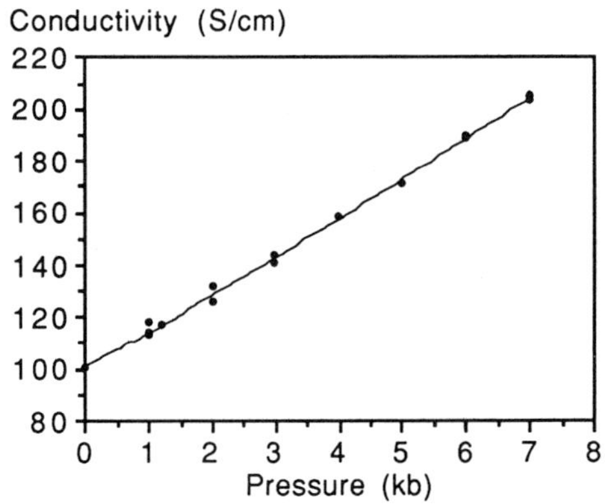

Figure 9.   Room temperature conductivity vs. Pressure.

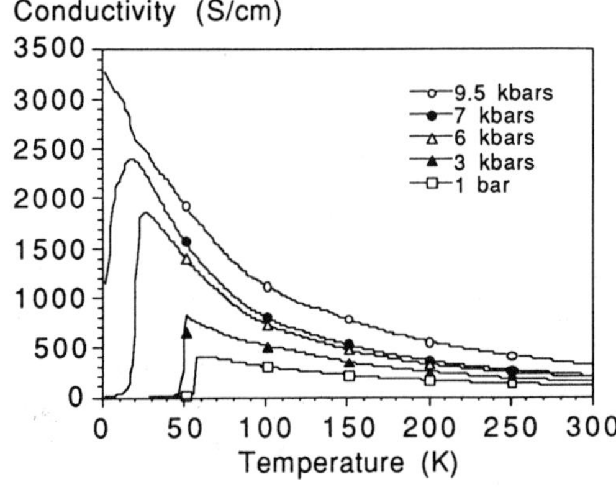

Figure 10. The variation of conductivity with temperature for a range of pressures from
1bar to 9.5 kbars.

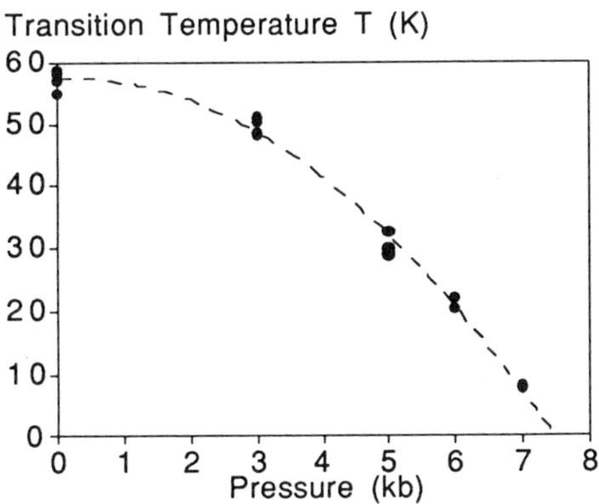

Figure 11. Variation of the metal to insulator transition temperature T_P with pressure. T_P is defined by the temperature at the peak value of $\partial\ln\sigma/\partial(1/T)$, except at 7 kbars where the transition was taken as the downturn in the conductivity.

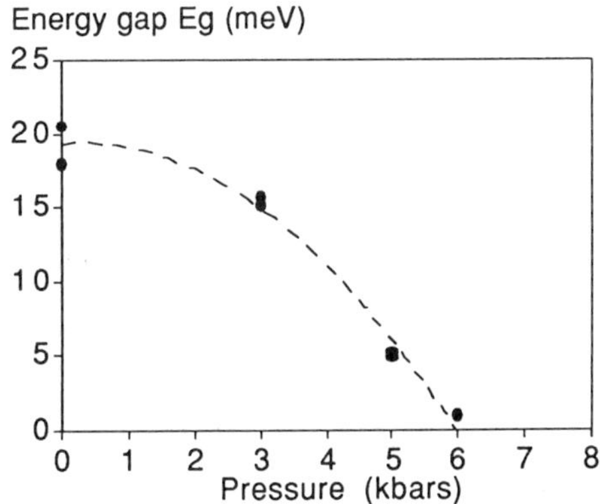

Figure 12. Variation with pressure of the conductivity activation energy E_g, in the insulating phase. E_g is defined by $\sigma=\sigma_0 \exp\{-E_g/kT\}$.

The conductivity increases steadily as pressure is applied, with a pressure of 7 kbars doubling the room temperature conducivity. Figure 9 shows the variation of the room temperature conductivity with pressure. The temperature dependence of the conductivity at several pressures is shown in figure 10, and as can be seen the metal-insulator transition is pushed to lower temperatures as the pressure is increased.

Figures 11 and 12 show the variation of the transition temperature, T_P, and the activation energy for conductivity, E_g, below the transition as a function of pressure. T_P was defined as the temperature at the peak value of $\partial\ln\sigma/\partial(1/T)$ and E_g was taken from fitting the resistivity below the transition to $\sigma = \sigma_0 \exp\{-E_g/kT\}$ in the temperature range a few K below the transition. It is clear that the transition to the insulating state is readily suppressed under pressure, with both T_P and E_g falling towards zero. However, although E_g tends towards zero at 6 kbar, the conductivity at higher pressures can still show a downturn at low temperatures. This is clearly seen in figure 10 for the data at 7 kbar, and it is believed that there is still a transition at this pressure, though no longer to an insulating state, but rather to a semimetallic state. At pressures greater than 12 kbars the resistivity can fall monotonically with falling temperature, as shown in figure 10 at 9 kbar. However, this is dependent on sample condition, and slight downturns in conductivity for other samples even at pressures of up to 20 kbar, have been observed. This downturn is thought to be due to a current distribution that includes a component in the high resistivity c-axis direction perpendicular to the molecular sheets, and is therefore extrinsic in origin. No evidence for superconductivity has been seen at pressures up to 20 kbar and temperatures down to 1.4 K.

The magnetic susceptibility χ was taken to be of the form

$$\chi = \chi_{core} + \chi_{Pauli} + \frac{C}{T} \qquad (1)$$

The value of C is obtained from the low temperature variation of χ, and a value of C = 42. x 10^{-3} emu K/mole was obtained. If this is due to spin 1/2 impurities, it is equivalent to 1.1% per formula unit. This is a high value for an impurity spin concentration, though not exceptional among organic charge transfer salts of this type. The contribution to χ_{core} from the diamagnetic susceptibility of the ions can be estimated from Pascal's constants, and is estimated to be -87.5 x 10^{-6} emu/mole for the dmit group and a total value of -460 x 10^{-6} emu/mole for $Cs[Pd(dmit)_2]_2$. The temperature dependence of the remaining contribution to χ, that due to the delocalised metallic electrons and labelled as χ_{Pauli} in equation 1, is shown in figure 13. In the metallic regime, above the metal to insulator transition, the susceptibility is independent of temperature, and has rather a high paramagnetic value. At the transition, χ falls abruptly, dropping by 370 x 10^{-6} emu/mole between 58 and 55 K, and remains independent of temperature below this. Assuming that χ_{Pauli} drops to zero in the low temperature phase, then a value for χ_{core} of -460 x 10^{-6} emu/mole is required to bring χ_{Pauli} to zero in this temperature regime. This is close to the value calculated from Pascal's constants due to the diamagnetic susceptibility.

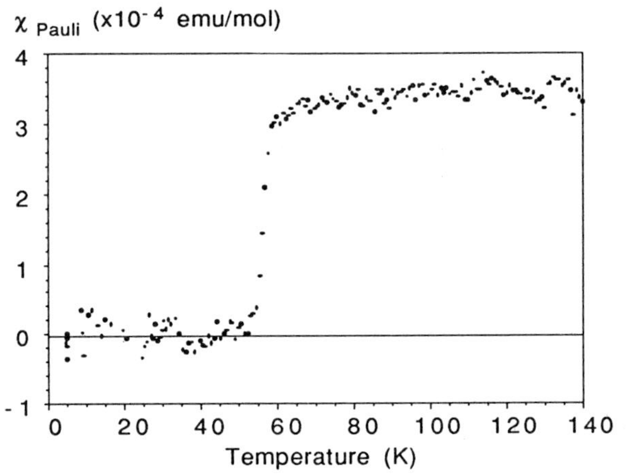

Figure 13. Magnetic Susceptibility.

Nature of the Transition

X-ray measurements indicate that the metal-insulator transition is associated with the setting up of two structural transitions, both appearing near 56.5K, [23]. It is considered that both these transitions couple across sections of the Fermi surface to introduce a gap at the Fermi level, [20, 23]. However, this is not a simple Peierls transition, with the energy gap seen in both the conducivity and susceptibility below 56.5 K being considerably larger than that associated with a distortion driven solely by the Peierls mechanism.

Within the weak-coupling model for the Peierls distortion in a one-dimensional system, the gap at T = 0, $2\Delta(0)$ is related to the mean field transition temperature, T_P, by $2\Delta(0) = 3.5\ k_B T_P$ [24]. Assuming conduction through thermal excitation of electron-hole pairs across the gap, E_g is equal to half the gap, Δ, and with $T_P = 56.5$ K, we expect a value for E_g of 8 meV. The experimentally measured value of E_g, determined from the conductivity data, is very considerably larger than this with a measured value at ambient pressure of about 20 meV (see figure 12).

The high pressure phase diagram is evident in figures 11 and 12, and it is seen that the phase transition (or transitions) are readily pushed to lower temperatures under quite modest pressures. It seems that all trace of the phase transitions as detected in the resistivity is suppressed at 9 kbar, but note that the data shown in figure 10 shows a distinct change in form between 6 and 7 kbar. At the lower pressure the transition is to a semiconducting state, with a well-defined energy gap, and it is believed that , as at ambient pressure, both transitions are present. At 7 kbar, however, there is a distinct phase transition, at ~8 K, but though the resistivity below it rises, it remains metallic. It is possible here that this may be due to a separation of the two phase transitions, with only one transition being present at this pressure. In view of the large modifications of the band energies brought about by these structural transitions, it is surprising that pressure can so quickly suppress them.

REFERENCES

1. McCoy H N and Moore W C *J. Am. Chem. Soc.*, 33 273 (1911)
2. Kraus H J *J. Am. Chem. Soc.*, 34 1732 (1913)
3. Acker D S, Harder R J, Hertler W R, Mahler W, Melby L R, Benson R E and Mochel W E *J. Am. Chem. Soc.*, 82 6408 (1960)
4. Peierls R E, Quantum Theory of Solids, London (1955).
5. Ravy S, Moret R, Pouget J-P and Comès R *Synth. Met.*, 13 63 (1986)
6. Brossard L, Ribault M, Valade L and Cassoux P *Physica B*, 143 378 (1986)
7. Vainrub A, Jérome D, Bruniquel M-F and Cassoux P *Europhys. Lett.*, 12 267 (1990)
8. Bousseau M, Valade L, Bruniquel M-F and Cassoux P *Nouv. J. Chimie*, 8 3 (1984)
9. Schirber J E, Overmyer D L, Williams J M, Wang H H, Valade L and Cassoux P *Phys. Lett. A*, 120 87 (1987)
10. Brossard L, Ribault M, Valade L, Legros J-P and Cassoux P *Synth. Met.*, B27 157 (1988)
11. Brossard L, Ribault M, Valade L and Cassoux P *J. Physique*, 50 1521 (1989)
12. Ravy S, Pouget J P, Valade L and Legros J P *Europhys. Lett.*, 9 391 (1989)
13. Kim H, Kobayashi A, Sasaki Y, Kato R and Kobayashi H *Chem. Lett.*, 1799 (1987)
14. Kobayashi A, Kim H, Sasaki Y, Kato R, Kobayashi H, Moriyama S, Nishio Y, Kajita K and Sasaki W *Chem. Lett.*, 1819 (1987)
15. Kobayashi A, Kim H, Sasaki Y, Kato R and Kobayashi H *Solid State Commun.*, 62 57 (1987)
16. Canadell E, Rachidi I E-I, Ravy S, Pouget J P, Brossard L and Legros J P *J.Physique*, 50 2967 (1989)
17. Canadell E, Ravy S, Pouget J P and Brossard L *Solid State Comm.*, 75 633 (1990)
18. Hoffmann R *J. Chem. Phys.*, 39 1397 (1963)
19. Clarke R A and Underhill A E *Synth. Met.*, B27 515 (1988)
20. Underhill A E, Clark R A, Marsden I R, Allan M, Friend R H, Tajima H, Naito T, Tamura M, Kuroda H, Kobayashi A, et al. *J.Phys.: Condens. Matter*, 3 933 (1991)
21. Clemenson P I, Underhill A E, Hursthouse M B and Short R L *J. Chem. Soc., Dalton Trans.*, 1689 (1988)
22. Montgomery H C *J. Appl. Phys.*, 42 2971 (1971)
23. Ravy S, Canadell E and Pouget J P, Proc. ISSP Internat. Symposium: The Physics and Chemistry of Organic Superconductors (1989) eds. Saito G Kagoshima S, *Springer Proc. Phy.*, 51, 252
24. Friend R H and Jérome D *J. Phys.*, C12 1441 (1979)

NMR STUDIES OF MOTION IN SOLID POLYMER SYSTEMS AT HIGH PRESSURE

A.S. Kulik, K.O. Prins

Van der Waals-Zeeman Laboratory, University of Amsterdam
Valckenierstraat 65-67, 1018 XE Amsterdam, the Netherlands

INTRODUCTION

We investigate the effect of high pressure on the molecular order and the dynamics of solid (crystalline and amorphous) systems of simple synthetic polymers. Pulsed deuteron-NMR techniques[1,2] are used to study the effect of high hydrostatic pressure on the dynamical properties of polystyrene (PS), polycarbonate (PC) and polyethene (PE). Type and time-scale of the motion of polymer chain segments and side groups are studied from the ^{2}H-spin-lattice relaxation rates and solid echo (SE) and spin-alignment (SA) echo spectra, which are dominated by the deuteron quadrupolar coupling. The aim of the application of high pressure is the distinction between the effects of changes in temperature and in pressure. In particular we investigate the dynamics near phase transitions. This paper is a preliminary account of our results.

Deuteron NMR is particularly suited for the investigation of chain order and of type and time-scale of molecular motion in polymers. In deuteron-labelled polymers the local interactions of the ^{2}H spin $I=1$ are dominated by the quadrupolar interaction with the electric field gradient (EFG) tensor at the site of the deuteron. The EFG tensor is axial in aliphatic and nearly axial in aromatic C-H bonds. Therefore, the relation between the NMR spectrum and the C-H bond directions is simple, constituting an excellent tool for monitoring the order and the time evolution of the bond directions.

The ^{2}H-NMR spectrum is obtained in a solid-echo (SE) experiment (using a 90_y-τ_1-90_x pulse sequence) by taking the Fourier transform of the data in the time domain, starting from the echo maximum. Molecular motion on the time-scale of the SE experiment (which is limited by T_2, the spin-spin relaxation time, ≈ 500 μs) results in partial or complete averaging of the EFG tensor, depending on whether the time scale of the motion is comparable or much faster than the time-scale set by the width of the rigid spectrum (≈ 250 kHz). Particular types of reorientational motion result in characteristic types of narrowed lineshapes, as is clearly explained in reference 1. Therefore, SE spectra can be used to distinguish between different types of motion.

The limitation set to the time-scale by T_2, can be overcome in a spin-alignment experiment, using the three-pulse sequence 90_y-τ_1-45_x-τ_2-45_x. The time-scale is increased to the time of decay of spin-alignment, which is of the order of T_1, the spin-lattice relaxation time. As has been shown in reference 2,

Frontiers of High-Pressure Research, Edited by H.D. Hochheimer and
R.D. Etters, Plenum Press, New York, 1991

the alignment echo $S(t,\tau_1,\tau_2)$ represents a correlation function characteristic of the dynamic process, namely $S(t,\tau_1,\tau_2) = \,<\sin\{\Omega(0)\tau_1\}\sin\{\Omega(\tau_2)t\}>$, depending on the quadrupolar frequencies $\Omega(0)$ and $\Omega(\tau_2)$ at the end of the "evolution" period τ_1 and of the "mixing" time τ_2, respectively.

The analysis of the solid echo spectra and spin alignment echo spectra yields information about molecular motion in a very large dynamic range ($\approx$ 1 to 10^7 Hz). This range is increased by including measurements on T_1 which provide information on spectral densities near the Zeeman frequency (10^8-10^{10} Hz).

EXPERIMENTAL

The Fourier-transform NMR-spectrometer used in this investigation is home-built around a 50 mm bore, 6.4 T magnet (Oxford Instruments).

A recent review[3] describes the apparatus used in our NMR experiments at high pressure. In the present investigation, hydrostatic pressure has been applied up to 2500 bar in the high-pressure NMR probe[4] shown in figure 1. This probe has been designed for use in the superconducting magnet at pressures up to 3 kbar. The central part (1) is a cylindrical beryllium-copper (Berylco 25) pressure vessel. The outside diameter is 44 mm, the largest value compatible with the bore diameter (50 mm) of the magnet used. The cylindrical space (2), with a diameter of 15 mm, contains the rf coil, with its axis perpendicular to the axis of the vessel. The main problems to be solved in the design of the probe are the construction of the pressure seal, the rf feedthrough to the coil, and the tuning and impedance matching.

The pressure vessel is closed by the bottom plug (3). The pressure seal between (1) and (3) is achieved by using a BeCu ring (5). The advantage of this type of seal is that it remains tight in a large interval of temperature. The plug (3) contains a stainless-steel high-pressure capillary (6) which is used for the application of pressure. In all our experiments, pressure is generated outside the NMR probe. Pressure is transferred from the pressure-generating equipment via the capillary transfer tube to the polymer sample. The pressurizing medium used for the polymer samples in this study is nitrogen.

The rf feedthrough to the NMR coil is also mounted in plug (3). The conical end of the electrode (7) is pressed against the conically shaped piece of insulating material (8), made of Vespel. The rf connection to the probe is made by the tube (10), supporting the probe from the bottom flange (30); this tube serves as the outer conductor of a coaxial rf transmission line. The inner tube (11) is the central conductor. Tuning and impedance matching of the probe are achieved with two capacitors, which can both be varied from outside the cryostat, as is shown in the right hand side of figure 1.

The high-pressure probe is placed in a cryostat, constructed inside the magnet bore. The construction is also shown in figure 1. Two four-lead platinum resistance thermometers (31) are inserted in the top of the probe and are used for temperature control and measurement. A temperature control unit supplies a current to a heater wire, wound in a groove on the cylindrical surface of the pressure vessel. The temperature can be controlled with a stability within 0.01 K in a temperature interval from 80 to 420 K.

POLYSTYRENE

In PS an analysis is made of the molecular motion as a function of pressure near the glass transition temperature T_g. At normal pressure the glass-transition temperature is about 373 K. High-pressure EOS data on atactic PS with $M_w = 20400$ show that T_g increases with about 40 bar/K[5]. Our study focusses on both the chain motion and the side-(phenyl)-group motion below and above T_g, using chain-deuterated PS-d_3 ($M_w = 229000$) and ring-deuterated PS-d_5 ($M_w = 225000$).

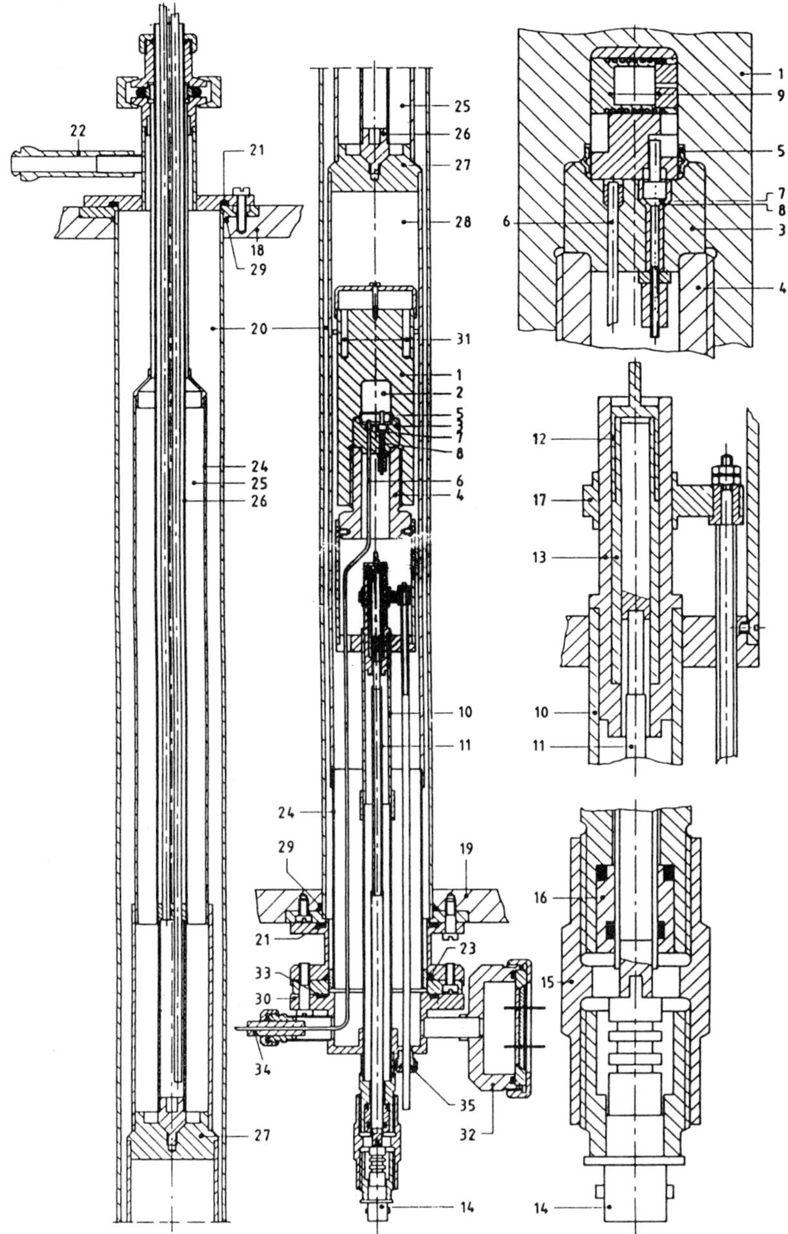

Figure 1. The high-pressure NMR probe

Below T_g the deuteron spin-lattice relaxation in both substances is strongly non-exponential. Figure 2 shows two examples of the relaxation of the magnetization after a saturating pulse sequence in PS-d$_3$, namely at 1 bar and 313 K (i.e. far below T_g) and at 1250 bar and 383 K (i.e. just above T_g). Non-exponential relaxation is observed up to about 15 K above T_g. This behavior is typical for a glass and is evidence that in the glass structure the motion of the C-^{2}H-bonds is restricted, as determined by the local environment[6]; it may also be partly due to spatial heterogeneity. The deuteron-spins in the system do not reach a common spin temperature, because spin-diffusion is very slow.

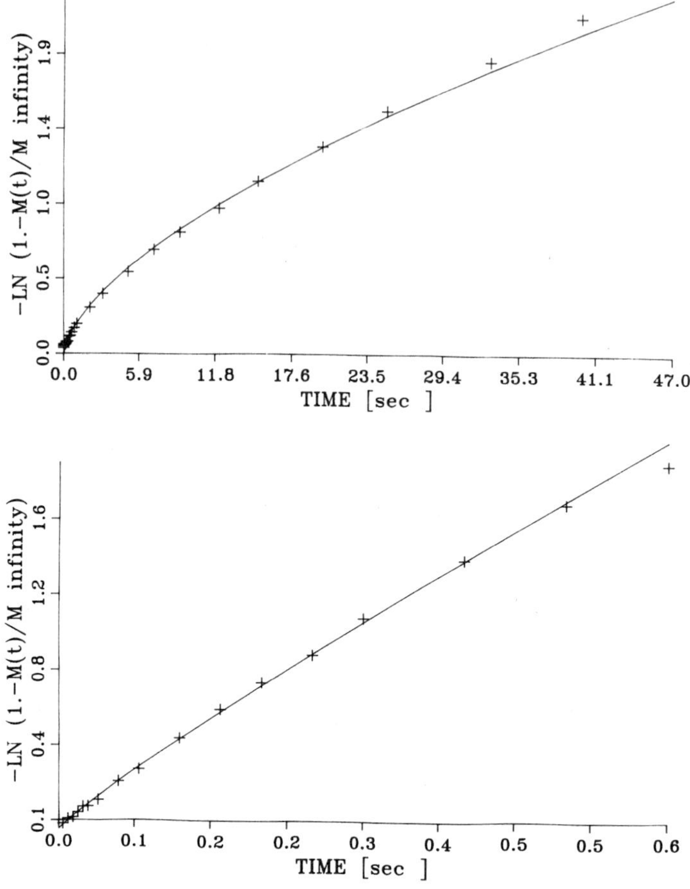

Figure 2. ^{2}H magnetization relaxation in PS-d$_3$; top 1 bar, 313 K; bottom 1250 bar, 383 K.

In order to present a quantitative description of the relaxation curves, we have used a Kohlrausch function

$$M_\infty-M(t)=M_\infty exp[(-t/T_1)^\beta]$$
(1)

to fit the data. This expression is equivalent[7] to using a distribution $f(T_1)$ of relaxation times, according to

$$M_\infty-M(t)=M_\infty \int dT_1\ f(T_1)exp(-t/T_1).$$
(2)

From the distribution one obtains an average relaxation time

$$<T_1> =(T_1/\beta)\Gamma(\beta^{-1}).$$
(3)

Average values of T_1 in PS-d$_3$, obtained in this way, are shown in figure 3. The exponent β increases from 0.6 at low temperatures to 1 at $T \approx T_g+15$ K.

Solid echoes are obtained in PS-d$_3$ and PS-d$_5$ with the usual 90_y-τ_1-90_x pulse sequence. The apparent time constants T_2^* of the SE-amplitude decay (as a function of τ_1) show a sharp drop at T_g (as is shown in

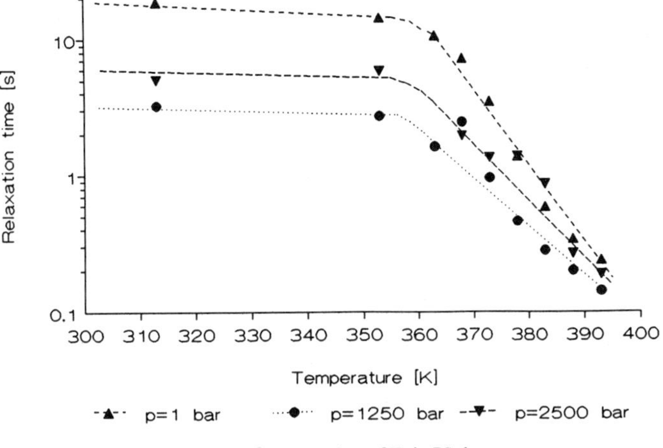

Figure 3. Average values of T_1 in PS-d_3.

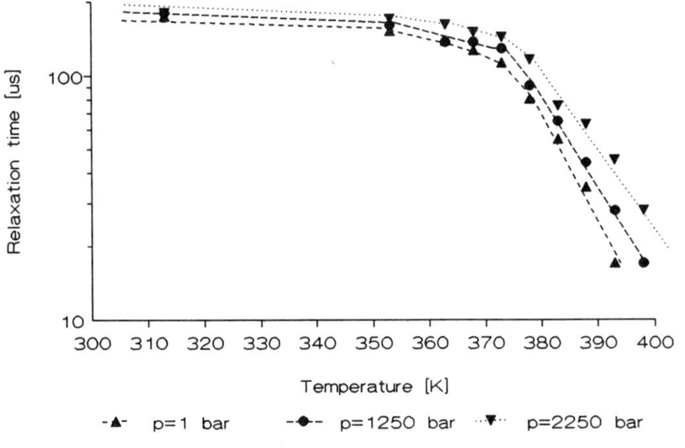

Figure 4. Solid echo decay constants in PS-d_3.

figure 4 for PS-d$_3$). Clearly the SE-decay can be used to monitor the glass transition as a function of pressure.

In PS-d$_3$, below and slightly above T$_g$ almost no motion is observed on the time-scale of the SE experiment. A drastic change in chain mobility on the time-scale of the SA experiment is observed passing through T$_g$. Figure 5 shows SA echo spectra at 373 K. At $\tau_1 = 30$ μs the spectra essentially show the Pake doublet structure, characteristic of the rigid solid. However, at $\tau_1 = 50$ μs, a considerable reduction of intensity in the center of thye spectrum is observed; at 1 bar this effect is stronger than at 2250 bar.

The correlation function S(t,τ_1,τ_2) not only depends on the dynamic parameter τ_2(the mixing time), but also on the geometric parameter τ_1 (the evolution time)[8]. By varying τ_1 one can distinguish between large angle reorientational jumps and small angle, diffusive reorientation. The strong τ_1 dependence in the PS-d$_3$ SA echo spectra is evidence for diffusive reorientations of the C-D bonds on the time-scale of $\tau_2 \approx$ 100 ms. This effect, including the strong τ_1 dependence, is very distinct in the spectra at 388 K (figure 6), now at much shorter mixing times τ_2 of a few ms.

These experimental spectra are compared with model calculations[9]. Figure 7 shows an example: the SA echo spectrum obtained at 1 bar, 388 K, $\tau_1 = 30\mu$s, $\tau_2 = 2$ ms is compared with a simulated spectrum assuming isotropic diffusive motion of the C-D bonds with a single correlation time $\tau_c = 0.2\tau_2$, resulting in reasonable agreement. In reality, the system has a distribution of correlation times, the tail of which is just visible in the SE spectrum. Our results at 1 bar are consistent with results obtained by Kaufmann et al.[10] and by Pschorn et al.[11], who studied the slow motion regime withe the two-dimensional SA exchange technique. They found that the mean correlation time for the diffusive chain motion increases from about 8*10^{-5} s to about 100 s on lowering the temperature from 410 K to T$_g$ (373 K), while the width of the correlation time distribution increases from about 1 decade to about 5 decades. This slow diffusive proces can be identified with the mechanical α-process. Our data obtained at high pressure are being analyzed presently.

It should be noted, that the correlation times derived from T$_1$ (figure 2) are much shorter than the ones derived from the lineshapes. This is evidence for fast motion (superimposed on the diffusive reorientation), involving small angles only. The time-scale is the one of the mechanical ß-process.

The situation is quite different for the motion of the phenyl-groups, as is evident already from the SE spectra of PS-d$_5$, examples of which are shown in figure 8. The spectra show a strong heterogeneity in the mobility. At 353 K (below T$_g$) the spectra are a superposition of a Pake doublet and of the spectrum due to the fraction of phenyl-groups performing 180° flip-motion (which follows[12] from the splitting being reduced by a factor 4), in the fast motion limit. At this temperature, the lineshapes hardly depend on τ_1. At 388 K the fraction of flipping phenyl groups has increased considerably, while as a function of τ_1 a loss of intensity in the center is visible, which, presumably, is due to the diffusive reorientation of the phenyl-flip axes. These effects become more pronounced in the SA echo spectra.

POLYCARBONATE

We have started a study of the effect of high pressure on the motion of the phenyl-rings in polycarbonate by measuring the ^{2}H spin-lattice relaxation rates up to 2500 bar in the temperature interval 233-298 K. The sample is PC-d$_4$, M$_w$ = 35000).

The relaxation is found to be non-exponential. The magnetization recovery curves are similar to the ones obtained in PS. We have used the formalism of equations 1-3 to represent the data. The average relaxation times are shown in figure 9. An increase in pressure of 1 kbar causes a temperature shift of about 30 K in <T$_1$>. This shift is consistent with the result obtained by Walton et al[13]. At 1 bar, the

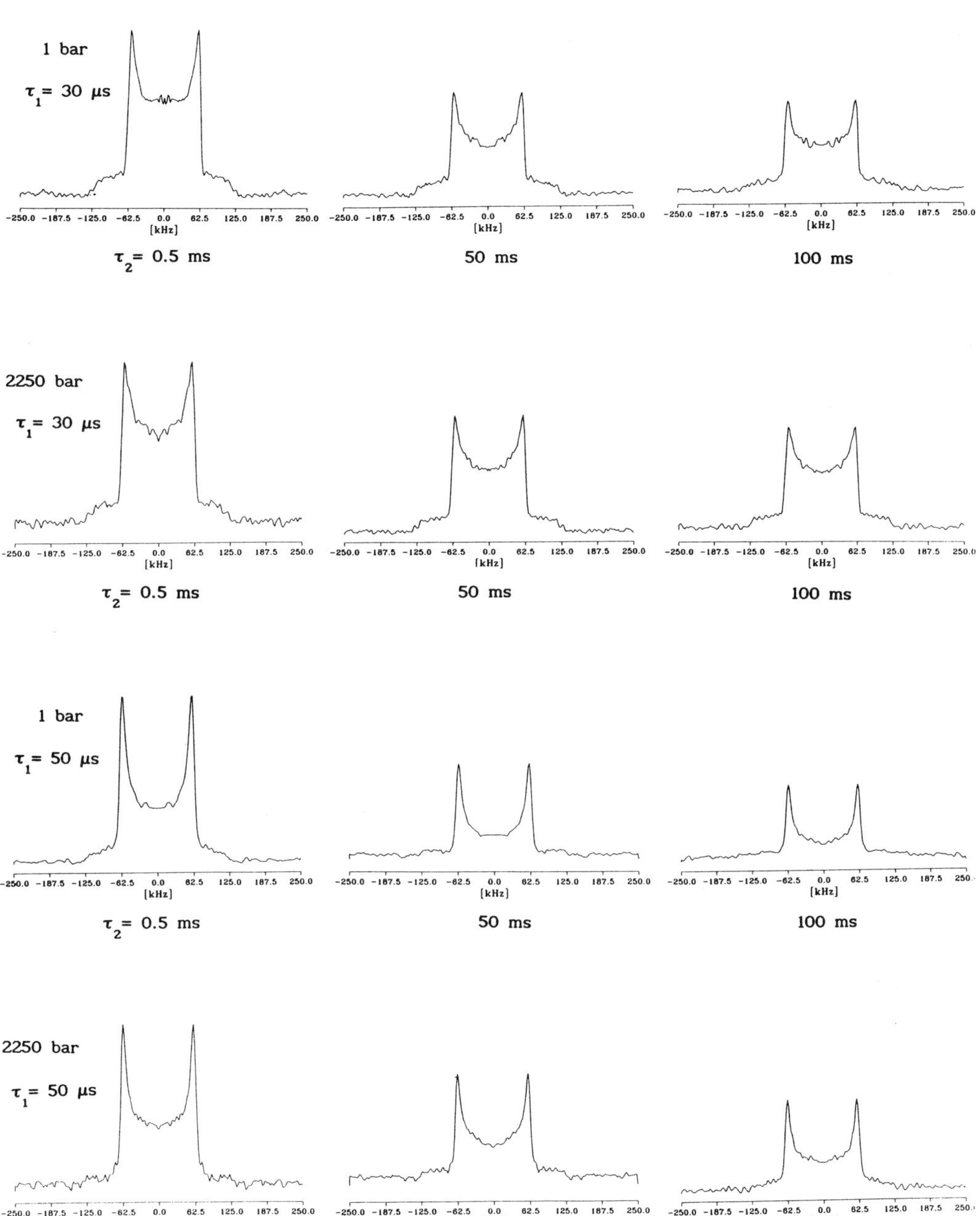

Figure 5. Spin-alignment echo spectra in PS-d$_3$ at 373 K.

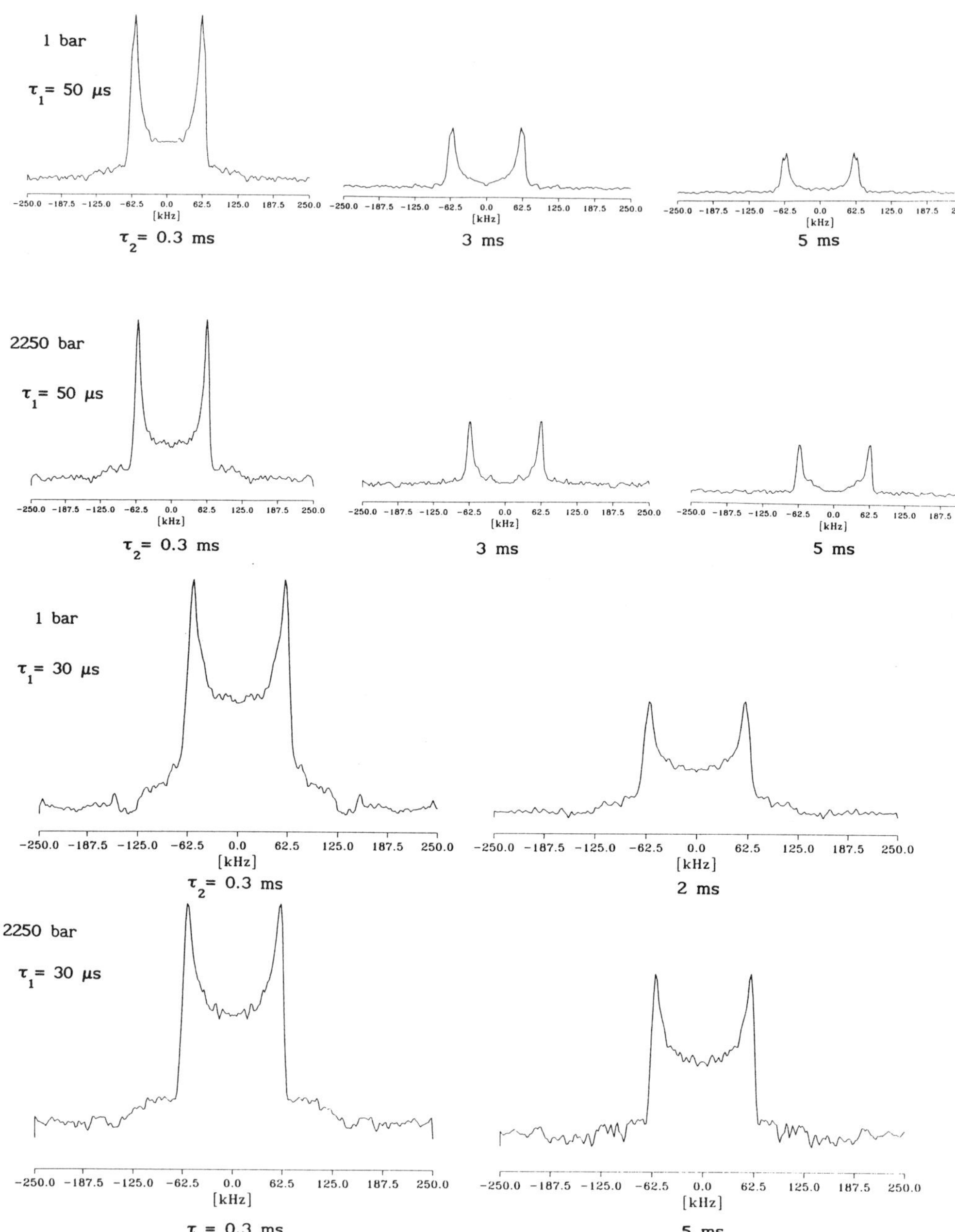

Figure 6. Spin-alignment echo spectra in PS-d$_3$ at 388 K.

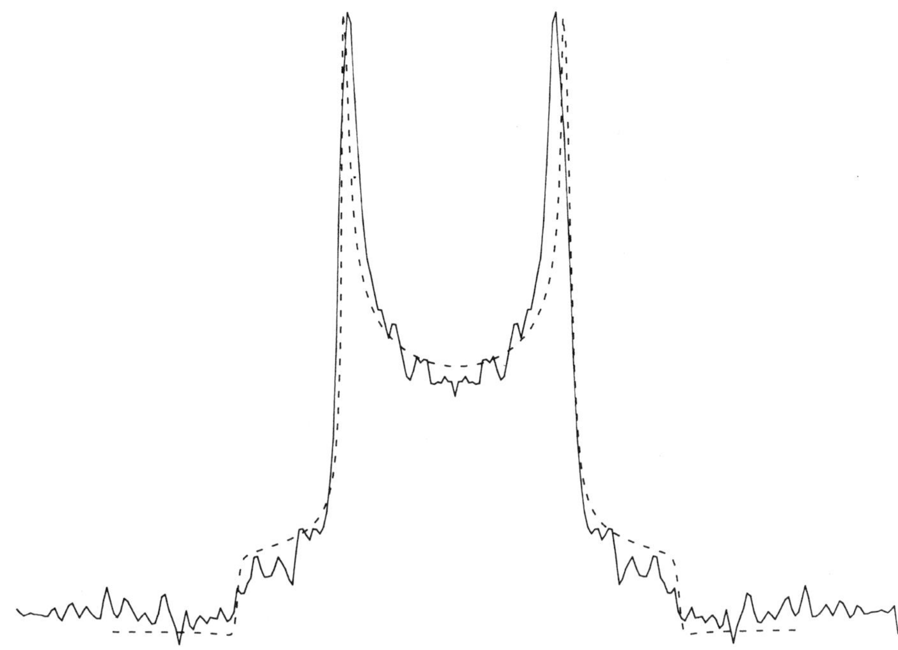

Figure 7. Comparison experimental SA spectrum with simulated spectrum.

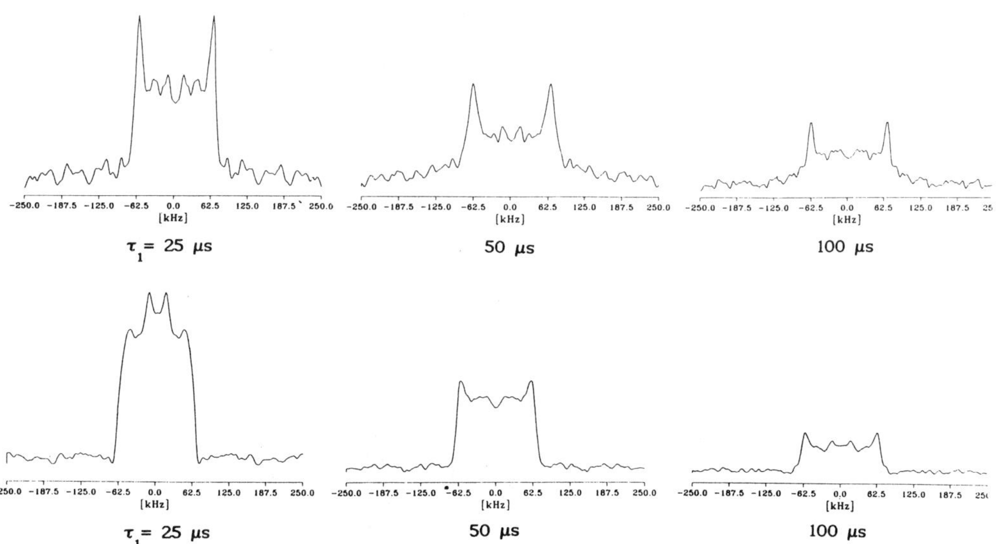

Figure 8. Solid echo spectra in PS-d_5 at 1250 bar, top 353 K; bottom 388 K.

value of ß varies from 0.55 at 233 K to 0.33 at 298 K. This effect can be explained by assuming that the solid contains fast and slow mobile phenyl-groups, and that the slow component fraction decreases from 0.8 at 223 K to 0.1 at 298 K. The magnetization recovery is dominated by the slow component at 233 K, and by the fast component at 298 K.

The analysis of the data is difficult because of the complicating effect of spin-diffusion. Presumably, only the rate constant of the short component in the relaxation curve is directly related to the correlation time of the phenyl flip motion. An approximate calculation of T_1 using the correlation time obtained from ^2H line shape studies[14] yields a value of the order of the short component time.

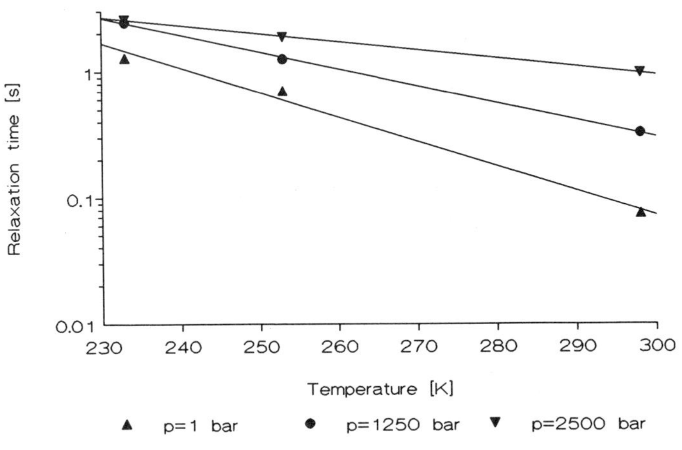

Figure 9. Average values of T_1 in PC-d_4.

POLYETHENE

In PE we have investigated the effect of pressure on the chain motion in the amorphous regions of linear polyethene in the temperature range from 200 to 350 K. The experimental results obtained with the solid echo method have been compared with calculations on a model[15], in which the chain motion is restricted to conformations of the carbon chain that are compatible with a diamond lattice. Entanglements are simulated by randomly distributed fixed points. In this way one can determine to which extent and on which time scale three-, five-, seven-, etc. bond chain motions occur. Roughly speaking, the effect of pressure is similar to lowering the temperature, an increase of the pressure of 1 kbar corresponding to a temperature decrease of about 15 K.

At temperatures above about 320 K the chain motion becomes so fast that the solid echo spectrum narrows to a lorentzian line, from which a reorientational correlation time can be derived, as a function of temperature and pressure. This investigation will be extended to higher pressure and to higher temperature to study the melting region and the solid-solid transition in polyethene.

ACKNOWLEDGMENTS

We gratefully acknowledge the discussions with Dr. F. Fujara and Dr. H.W. Spiess, who also made the deuterated polymer samples available to us. We thank Dr. M. Hansen, who initiated and closely collaborated in the polycarbonate project. We also like to thank Dr. R.L. Vold for providing us with the MXQET program for the solid echo simulations. This investigation is part of the research program of the Stichting voor Fundamenteel Onderzoek der Materie (F.O.M.), supported by the Nederlandse Organisatie voor Wetenschappelijk Onderzoek (N.W.O.).

REFERENCES

1. H. W. Spiess and H. Sillescu, Solid echoes in the slow-motion region, J. Magn. Res. 42:381 (1981).

2. H. W. Spiess, Deuteron spin-alignment: a probe for studying ultraslow motions in solids and solid polymers, J. Chem. Phys. 72:6753 (1980).

3. K. O. Prins, High-pressure NMR investigations of motion and phase transitions in molecular systems, in: High Pressure NMR, Springer-Verlag, Berlin (1991).

4. P. Peereboom, K. O. Prins, and N. J. Trappeniers, High-pressure probe and cryostat for pulsed NMR in a cryomagnet, Rev. Sci. Instr. 59:1182 (1988).

5. G. Rehage and H. J. Oels, Glass-transition phenomena of amorphous polymers under pressure, High Temp. High Press. 9:545 (1977).

6. W. Schnauss, F. Fujara, K. Hartmann and H. Sillescu, Non-exponential ^{2}H spin-lattice relaxation as a signature of the glassy state, Chem. Phys. Lett. 166:381 (1990).

7. C. P. Lindsey and G. D. Patterson, Detailed comparison of the Williams-Watts and Cole-Davidson functions, J. Chem. Phys. 73:3348 (1980).

8. F. Fujara, S. Wefing and H.W. Spiess, Dynamics of molecular reorientations: Analogies between quasielectric neutron scattering and deuteron NMR spin alignment, J. Chem. Phys. 84:4579 (1986).

9. M. S. Greenfield, A. D. Ronemus, R. L. Vold, R. R. Vold, P. D. Ellis and T.E.Raidy, Deuteron quadrupole-echo NMR spectroscopy. III. Practical aspects of lineshape calculations for multiaxis rotational process, J. Magn. Res. 72:89 (1987).

10. S. Kaufmann, S. Wefing, D. Schaefer and H.W. Spiess, Two-dimensional exchange NMR of powder samples. III. Transition to motional averaging and application to the glass transition, J. Chem. Phys. 93:197 (1990).

11. U. Pschorn, E. Rössler, H. Sillescu, S. Kaufmann, D. Schaefer and H. W. Spiess, Local and cooperative motions at the glass transition of polystyrene: Information from one- and two-dimensional NMR as compared with other techniques, Macromolecules 24:398 (1991).

12. H. W. Spiess, Deuteron NMR, a new tool for studying chain mobility and orientation in polymers, Adv. Polymer Science, 66:23 (1985).

13. J. H. Walton, M. J. Lizak, M. S. Conradi, T. Gullion and J. Schaefer, Hydrostatic pressure dependence of molecular motions in polycarbonates, Macromolecules, 23:416 (1990).

14. M. Wehrle, G. P. Hellmann, H. W. Spiess, Phenylene motion in poly-carbonate and polycarbonate/additive mixtures, Coll. & Polym. Sci, 265:815 (1987).

15. K. Rosenke, H. Sillescu, H. W. Spiess, Chain motion in amorphous regions of polyethylene: Interpretation of deuteron NMR line shapes, Polymer 21:756 (1980).

PHASEDIAGRAM, SUPERSTRUCTURE AND PROPERTIES OF

POLY(DIETHYLSILOXANE), PDES

W. Pechhold, P. Schwarzenberger

Universität Ulm, Abteilung Angewandte Physik
Albert-Einstein-Allee 11, D-7900 Ulm, Germany

ABSTRACT

In a first part the properties of PDES and its mesophase behavior are discussed and supplemented by some new volumetric and dynamic mechanical data. The dilatometric measurements indicate a small volume effect $\Delta V/V=0.004$ during isotropization (m/i-transition) at normal pressure, which decreases with pressure and finally vanishes above 80 MPa. To check whether this transition persists at p>80 MPa and to determine the phasediagram of PDES, the improved shearquartz resonator method is described in a second part and successfully applied to measure the complex shear modulus G(p,T) of PDES. Finally the m/i-transition and its pressure dependence is quantitatively discussed.

PROPERTIES OF POLY(DIETHYLSILOXANE)

The phase behavior of PDES was studied with several methods, supplementing the results of Y. K. Godovsky and V. S. Papkov[1]. The main interest was to elucidate the nature of the mesophase (m) between the crystalline melting point (290 K) and the isotropisation temperature (319 K). The sample used was characterized by $M_w=1.7\cdot 10^5$g/mol. From DSC scans (10K/min heating rate) in Fig.1 it can be concluded that different cooling rates do strongly influence the ratio of the two crystal modifications which both undergo a solid state transition around 210 K and melt in the vicinity of 290 K into the mesophase, but have nearly no influence on the m/i-transition (319K).

Even though the heat of the m/i-transition ($\Delta H_{m/i}=3$J/g) and the volume change ($\Delta V/V=0.004$) are about that of a liquid crystal (LC) at its clearing point from the nematic phase, there are several clear differences between PDES and the polymeric LCs:

(i) PDES does'nt show any increase in the fluctuation of orientation correlation above $T_{m/i}$, as measured by Cotton-Mouton effect, whereas LCs obey a $(T-T_{n/i})^{-1}$ law.

(ii) At cooling below $T_{m/i}$ it exhibits a hysteresis (also demonstrated by the dynamic mechanical investigations below), similar to crystallization phenomena, but contrary to the n/i-transition in LCs.

Frontiers of High-Pressure Research, Edited by H.D. Hochheimer and
R.D. Etters, Plenum Press, New York, 1991

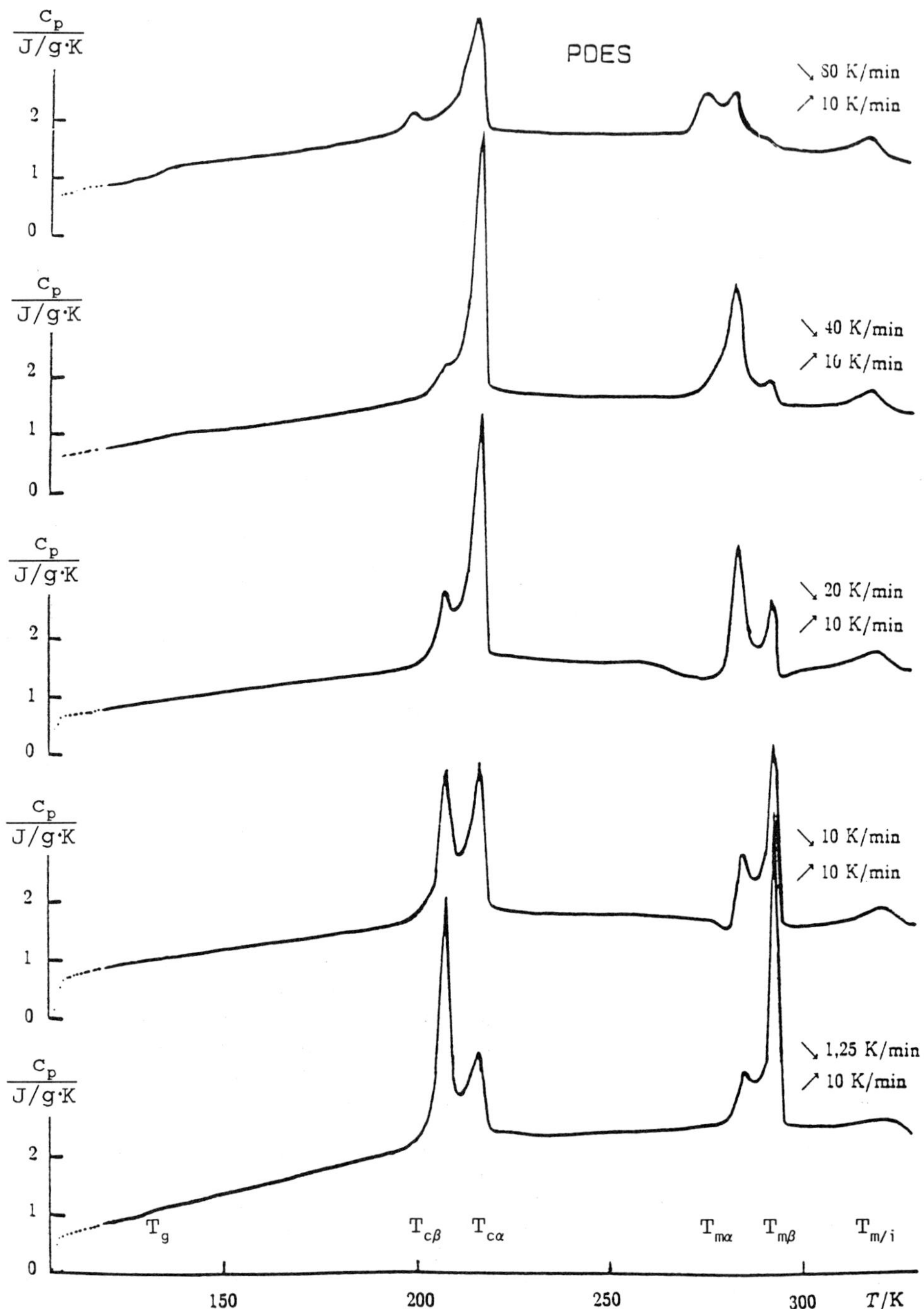

Fig. 1   DSC heating curves at 10 K/min for PDES after cooling with different rates.

(iii) In the polarizing microscope PDES shows light bands
(stacks of lamellae) in the m-phase, reversibly growing and
disappearing during cooling and heating, respectively. Unorient-
ed polymeric LCs exhibit an anisotropic domain (grain) structure.

(iv) This two-phase organization at the PDES mesophase is
also demonstrated by ^{1}H-NMR: there are (at least) two components
in the NMR-line (0.2 and 5-7 ppm width) the intensily ratio of
which is changing reversibly with temperature.

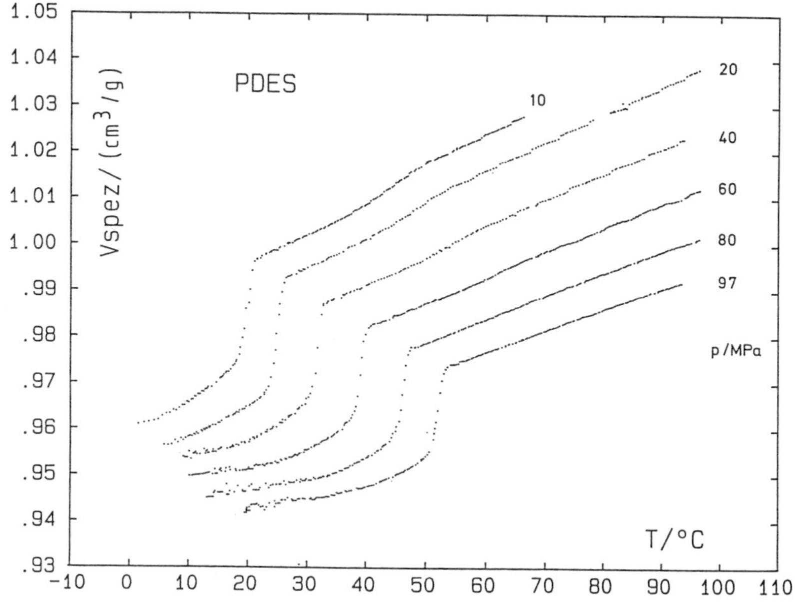

Fig. 2. Dilatometric investigations of PDES. (Dollhopf)

(v) As it is shown in Fig.2 the small volume effect at the
m/i-transition decreases (about linearly) with pressure and
vanishes around 80 MPa, whereas that at the nematic-isotropic
transition of a LC, though being reduced, doesn't vanish (at
least below 500 MPa).

(vi) The shear compliance of the PDES mesophase is up to 30
times lower than the plateau compliance of its melt, whereas in
polymeric LC the shear compliance of the nematic phase is about
equal to that of the melt. In Fig.3 measuremts of the complex
shear compliance of PDES versus temperature are reported which
were taken at 12 kHz with a double torsional resonator apparatus[2]. The open symbols refer to a cycle started after slow cooling to -50°C, showing the solid state transition (-70°C), a
sharp melting (+17°C) as well as the steep slope at the end of

the m/i-transition (+46°C) into the plateau compliance of the
isotropic melt. The closed symbols designate a heating curve
started after slow cooling to only -10°C. After a less sharp
melting into a less perfect mesophase the temperature was lowered again to prove the presence of undercooling which is normal
with crystallization. A similar type of hysteresis is observed
after heating above the m/i-transition (+46°C) followed by cooling below this temperature.

The existence of rather sharp bends in the complex compliance at melting and m/i-transition looks promising for making
use of it to determine the phasediagram at elevated pressures
where dilatometry and light scattering fail to detect the m/i-
transition (because $\Delta V \approx 0$). Since the torsional resonators at 12
kHz are much to large to put them under hydrostatic pressure, we
tried to apply an improved shearquartz resonator method which is
described next.

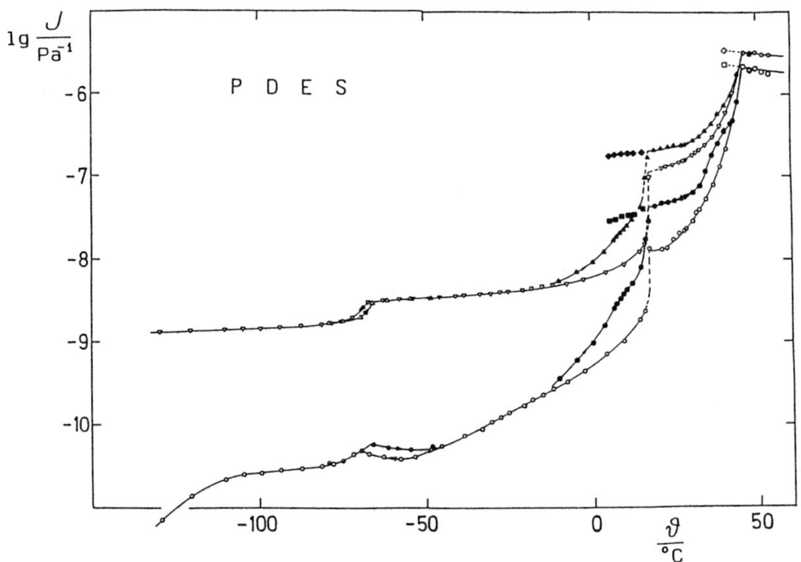

Fig. 3. Dynamic shear compliance of PDES. (12 kHz double torsio
nal resonator apparatus, Pechhold/Bulkin)

QUARTZ RESONATOR METHOD AND ITS APPLICATION IN POLMYER PHYSICS

To measure the complex shear modulus of polymers in the
high frequency range from 50 kHz up to 100 MHz the quartz resonator method is suitable in the whole temperature range of interest, i.e. from the polymer melt down to the glassy or semicry-
stalline regime. The method is based on the damping and frequency shift a resonator undergoes when loading it by the viscoelastic material of interest. The fundamental idea for the quartz

resonator method was put forward by Pechhold, who developed the
kHz double resonator apparatus and the appropriate formula. This
formula connects the change in complex eigenfrequency of the
resonator with the complex shear modulus of the coupled specimen[2]. Sauerbrey investigated the mode of vibration and the amplitude distribution of AT cut quartzes[3], Schilling made first
measurements with a single and a double quartz system[4].

<u>The Quartz-Resonators</u>

Quartz resonators are mostly used in electronics as filter
crystals, but they are even very useful in polymer physics as
mechanical resonators to measure the shear modulus in the high
frequency range. The resonators used in this method are optimized with respect to a high quality factor in air and only transmit shear waves into the coupled sample (quartzes from KVG, Nek-
karbischofsheim, FRG). To master the frequency range from 50 kHz
up to 100 MHz the three following resonator types are useful.

<u>Longitudinal extension resonator, Fig.4(a)</u>. The frequency
range in fundamental resonance is from 50 kHz up to 100 kHz and
harmonic overtones up to 500 kHz are possible. The sample must
cover completely one or both sides of the resonator which carry
the electrodes.

<u>Face shear resonator, Fig.4(b)</u>. These quartzes are used in
the frequency range from 100 kHz up to 1 MHz. The rectangular
quartzes have a rhombic vibrational mode. A sample coupled on
the face of this resonator only gets sheared.

<u>Thickness shear resonator (AT- and Y-cut), Fig.4(c)</u>. These
diskshaped quartzes have a thickness shear vibrational mode. For
this method only optimized shear resonators without any longitudinal component are advisable. These resonators (c) then may be
also immersed fully into the sample which is very important for
measurements under high pressure and in liquids. Different resonator shapes, resonator boundaries and electrode sizes are used
to trap the energy in the middle of the resonator plate[3,5,6].

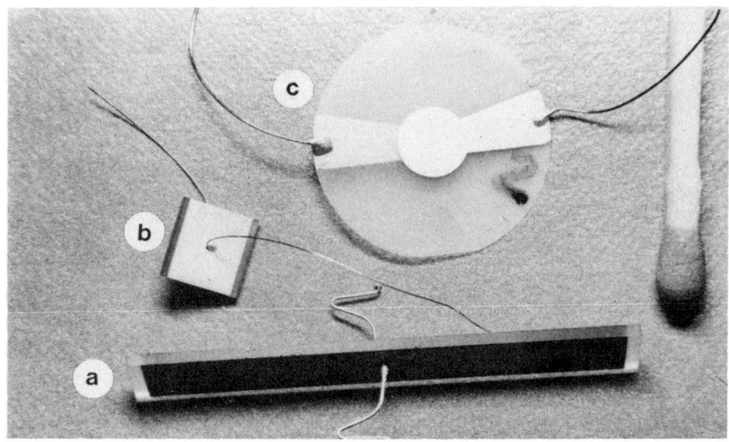

Fig. 4. The three different quartzresonators which are used in
        this method

<u>Short description of the measuring procedure</u>

A networkanalyzer (hp 3570 and hp 3330 b) and a standard serial electrical connection are used to measure the electrical properties of the quartz.

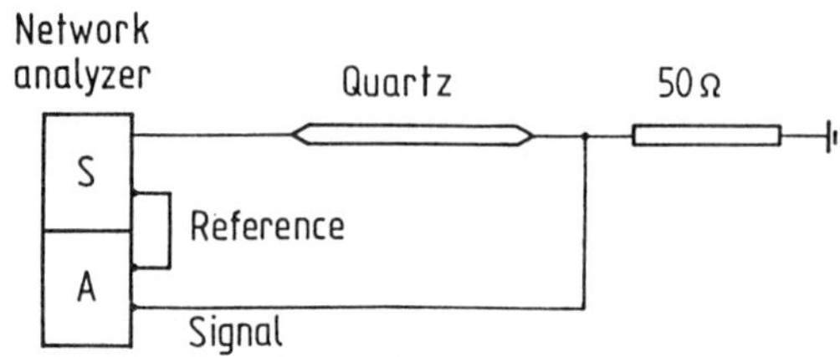

Fig. 5. Electrical connection layout

<u>Process of measurement</u>

After measuring the complex admittance of the resonator the shift in the resonance frequency and the change in half width, caused by the sample, have to be calculated to get the absolute value of the shear modulus of the sample.

Therefore the proceeding is devided into two steps:

i)   Measuring and evaluation of the unloaded resonator (calibration).

ii)   The same proceeding for the loaded resonator, followed by the evaluation of the complex modulus.

To calibrate the measurement the half-width D_0 and the resonance-frequency f_0 of the unloaded resonator must be well known. Therefore, the frequency dependence of the amplitude $U(\nu)$ and of the phase $P(\nu)$ are measured with a networkanalyzer to calculate the electrical admittance $A(\nu)$. Then the electrical admittance is fitted by a six parameter fit program which yields the required parameters f_0 and D_0 (Fig.6 and Fig.7).

The following formula for the theoretical admittance is used in this fit[6]:

$$A^n(\nu) = -a \cdot f_0^2 \cdot \left(\frac{b \cdot (n \cdot f_0 - 2\nu)}{(n \cdot f_0 - \nu)^2 + b^2 \nu^2} + i \frac{\nu - n \cdot f_0}{(n \cdot f_0 - \nu)^2 + b^2 \nu^2} \right)$$

$$+ i\, 2\pi f_0 C_p + s_R + i\, s_I$$

ν : frequency
n : number of the harmonic overtone

fitting parameters:
a   : height of the curve
b   : relative half-width $b_0 = D_0/2 \cdot f_0$
f_0 : resonance frequency
C_p : parallel-capacity
s_R : shift in the real-part
s_I : shift in the imaginary-part

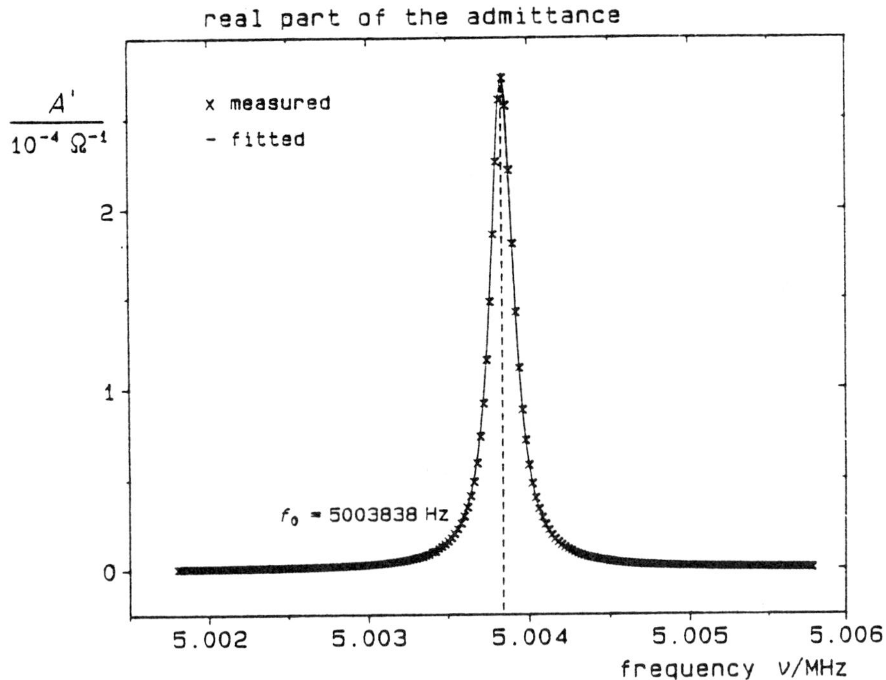

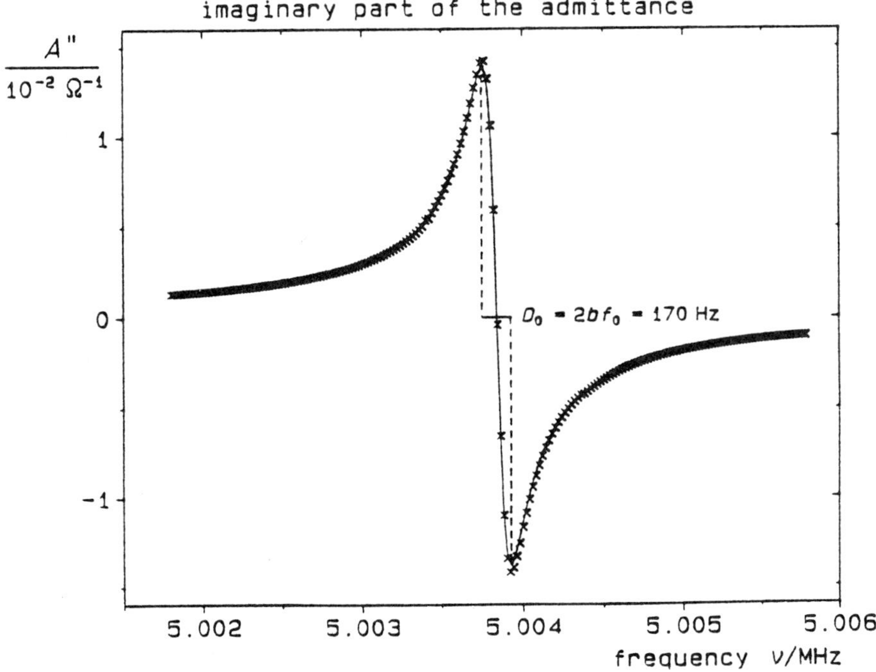

Fig. 6. Measured and fitted admittance of the unloaded quartz.

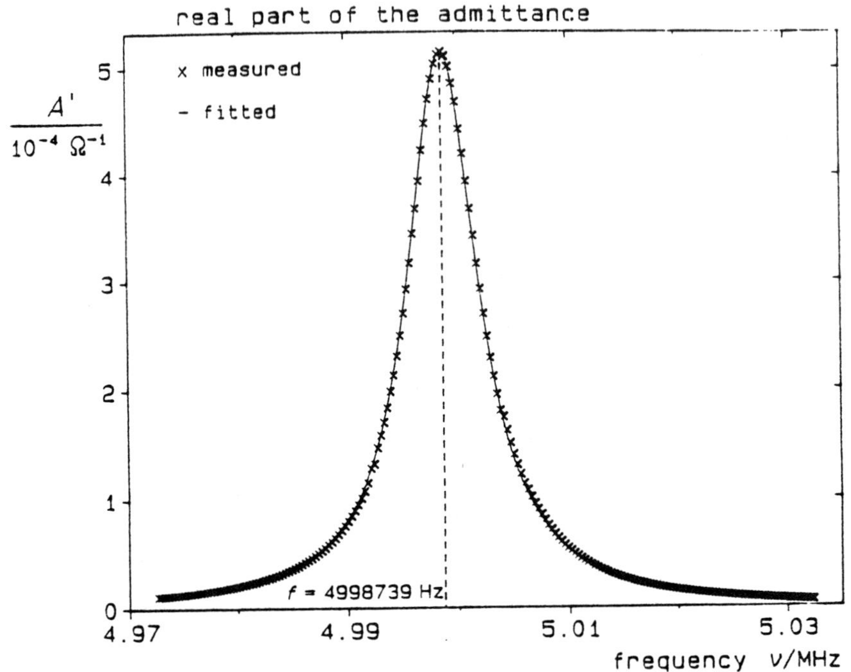

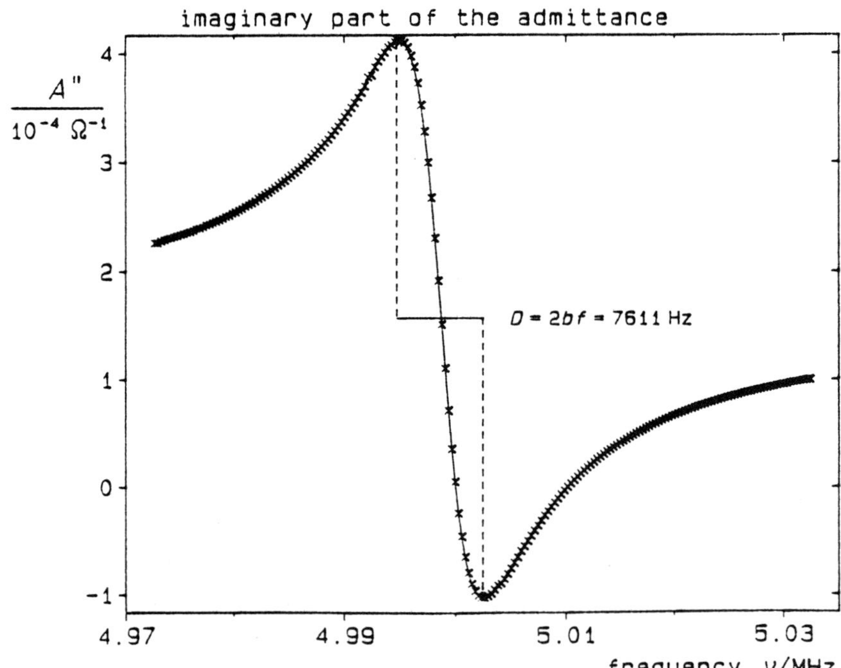

Fig. 7. Measured and fitted admittance of the loaded quartz.

The resonator must be calibrated over the entire temperature range, because the resonance frequency depends on the temperature especially at temperatures higher than 140°C. These data are stored on the hard disk.

The measurements and the fits are automatically controlled by a computer. The same procedure as for the unloaded quartz is now applied for the loaded quartz. The fitting formula used here is the same as used to fit the unloaded quartz. Then the shift in the complex eigenfrequency $\Omega^n = \delta\omega^n/\omega_0^n$ can be calculated[2].

$$\left(\delta\omega^n = \omega^n - \omega_0^n \, , \quad \omega_0^n = 2\pi f_0^n + i\pi D_0^n \, , \quad \omega^n = 2\pi f^n + i\pi D^n \right)$$

Evaluation of the complex shear modulus

The principle of operation of the single quartz system is the emission of a shear wave which is exponentially decreasing within the sample. To get no reflected wave back to the quartz, the height of the sample must be more then 10 wavelengths at least for amorphous polymers. The following formula has been derived to calculate the shear modulus[2,4]:

$$\Omega^n = \frac{i\,\rho_S \cdot \sqrt{\dfrac{G}{\rho_S}}}{\omega_0 \cdot \rho_Q \cdot d}$$

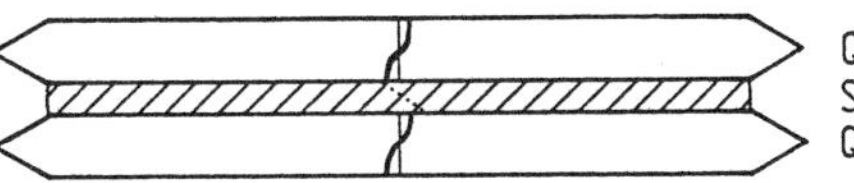

ρ_S : density of the sample
G : complex shear modulus of the sample
ρ_Q : density of the quartz
d : thickness of the quartz

The double quartz system is useful if the shear-modulus of the sample is very high or if there are very thin layers to be measured. The principle of operation is the coupling of two identical resonators by the sample. A parallel and an antiparallel vibrational mode is possible. The height of the sample must be in the range of the wave length.

The following formulae are used to calculate the shear modulus[4]:

$$\text{antiparallel (A)} \qquad \Omega_A = m \cdot \frac{\cot z_A}{z_A}$$

$$\text{parallel (P)} \qquad \Omega_P = -\,m \cdot \frac{\tan z_P}{z_P}$$

$$\text{where} \qquad z_{A/P} = \sqrt{\frac{\rho_S}{G}} \cdot \frac{\omega_{A/P} \cdot h}{2}$$

$$m = m_S/m_Q$$

h : height of the sample
m_Q : mass of quartz
m_S : mass of sample

Antiparallel vibrational mode

<u>Measurements under high pressure</u>

The single thickness shear quartz method also works under high pressure. Now the resonance frequency of the unloaded quartz depends on the temperature as well as on the applied pressure. A low molecular silicon oil (AK 5,Wacker) is used to transmit the pressure and to calibrate the quartz. Although the frequency is in the MHz-range, the silicon oil can be treated as an ideal liquid:

$$G'' = i\omega\eta, \qquad G' = 0$$

Then according to our evaluation formula, the relation

$$\Delta f = -\Delta D/2$$

should hold (and was carefully checked at normal pressure for optimized shear resonators).

Assuming the damping of the unloaded quartz to remain small under elevated pressure ($D_0 \approx 0$), its resonance frequency $f_0(p)$ can be calculated from the calibration measurements in silicon oil

$$f_0(p) \equiv f(p) - \Delta f = f(p) + \Delta D/2$$

yielding a linear relationship at a given temperature

$$f_0(p) = f_0(0) + \frac{\partial f_0}{\partial p}p = f_0(0) + \frac{f_0}{K}\left(\gamma^* - \frac{1}{6}\right)p$$

On the right side, the proportionality factor is expressed in terms of the compression modulus K and a Grüneisen-parameter γ^* for the elastic constant c_{66}, defined by $-2\gamma^* = d\ln c_{66}/d\ln V$.

To carry out G-measurements on polymers under hydrostatic pressure, the quartz is best fully immersed into the sample and then put into the pressure tank. The further procedure is the same as for the loaded quartz outside the tank, but now using the pressure calibration data for $f_0(p,T)$, $D_0(T)$, stored on the hard disk, for evaluating the complex shear modulus $G(p,T)$. Such measurements may be carried out under isobaric condition (p=const.) varying the temperature or isothermally varying the pressure – which can be done much faster except near phase transitions.

PHASE DIAGRAM OF PDES AND PRESSURE DEPENDENCE OF THE VOLUME EFFECT AT THE m/i-TRANSITION

The phase diagram of PDES was determined by measuring the dynamic shear compliance with an AT-quartz (2 MHz) in dependence on temperature and pressure up to 300 MPa. One isotherm and one isobar are shown in Fig.8 as examples. The T,p-values at the sharp bends in J'and J", indicating the melting point and the

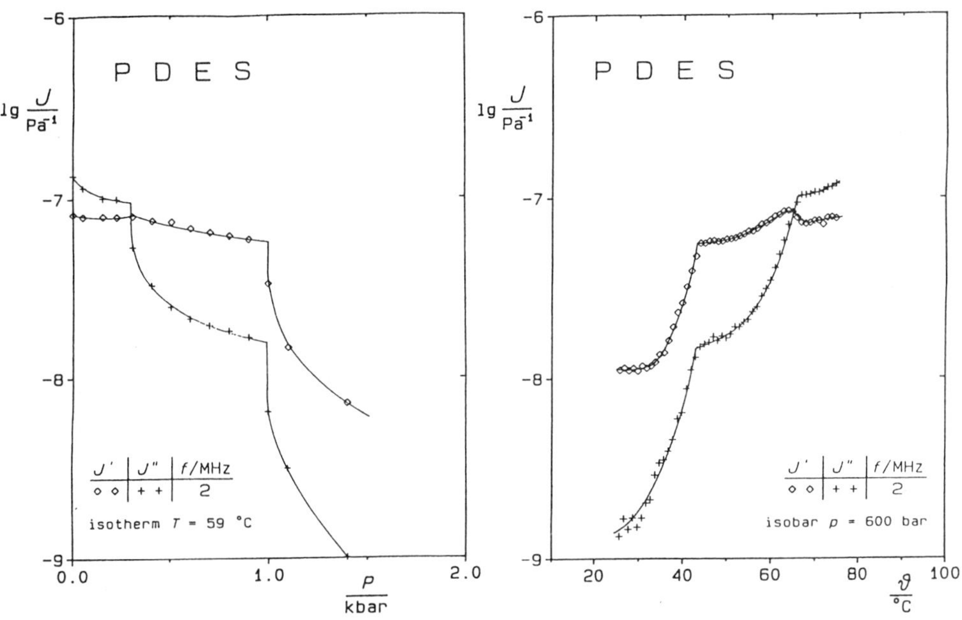

Fig. 8. Isotherm and isobar dynamic mechanical investigations under pressure.

end of the m/i-transition were plotted in the phasediagramm, Fig.10, together with data from dilatometric measurements (small symbols) from Fig.2. This analysis shows that the m/i-transition persists above 80 MPa (even though its ΔV vanishes here) up to 130 MPa when it merges in the melting. In the range 80-130 MPa $dT_{m/i}/dp \approx 0$ as predicted by Clausius-Clapeyron equation.

VOLUME EFFECT AT THE m/i-TRANSITION OF PDES

From $V(p,T)$ one gets approximately the same expansion coefficients for both phases (with p and K in GPa)

$$\alpha_m \approx \alpha_i \approx (6.3-17p) \cdot 10^{-4} K^{-1},$$

but different compression moduli

at 313 K: $K_m = K_{mo} + 2\gamma_m p \approx 1.4 + 12p$

at 343 K: $K_i = K_{io} + 2\gamma_i p \approx 1.2 + 10p$

The isothermally measured specific volume is best approximated (Fig.9)

for $T_m = 313$ K: $v_m(p) \approx 1.0150\left[1 + \frac{2\gamma_m}{K_{mo}}p\right]^{-1/2\gamma_m}$

for $T_i = 343$ K: $v_i(p) \approx 1.0385\left[1 + \frac{2\gamma_i}{K_{io}}p\right]^{-1/2\gamma_i}$

To derive the volume effect $\delta V_s(p)$ per monomer ($M_o=102$ g/mol) when changing from the m-phase into the i-state, one may extrapolate both expressions to a common temperature, e.g. $T_{com}=328K$,

$$v_m(p) \approx 1.0150\left[1 + \frac{2\gamma_m}{K_{mo}}\{p-(K\alpha)_m(T_{com}-T_m)\}\right]^{-1/2\gamma_m}$$

$$v_i(p) \approx 1.0385\left[1 + \frac{2\gamma_i}{K_{io}}\{p-(K\alpha)_i(T_{com}-T_i)\}\right]^{-1/2\gamma_i}$$

By taking the difference one obtains the volume effect $\delta V_s(p) =[v_i(p) - v_m(p)]M_o$, which shows a slightly curved behavior along the p-axis, but can be linearly approximated by $\delta V_s(p) \approx 0.37(1-p/p^*)cm^3$, with $p^* \approx 0.0856$ GPa , in accordance with the step-heights taken directly from the isobars (Fig. 2).

m/i-TRANSITION AND ITS PRESSURE DEPENDENCE

The m/i-transition in PDES is mainly governed by changes in the superstructural entropy during the formation of lamellae from the melt (Fig. 11).

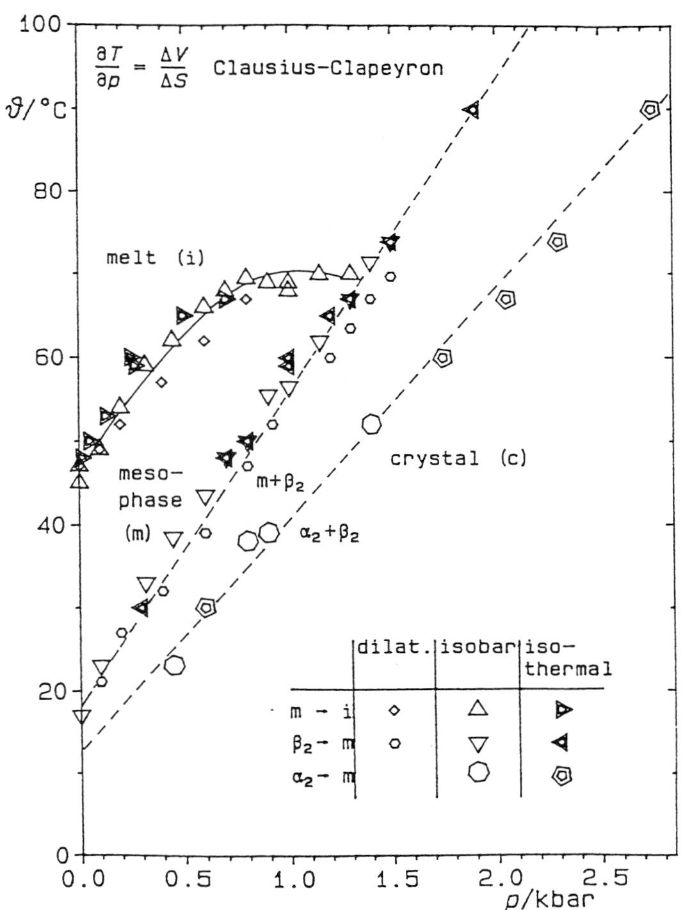

Fig. 9. Isothermal dilatometric measurements
and its thermodynamic description

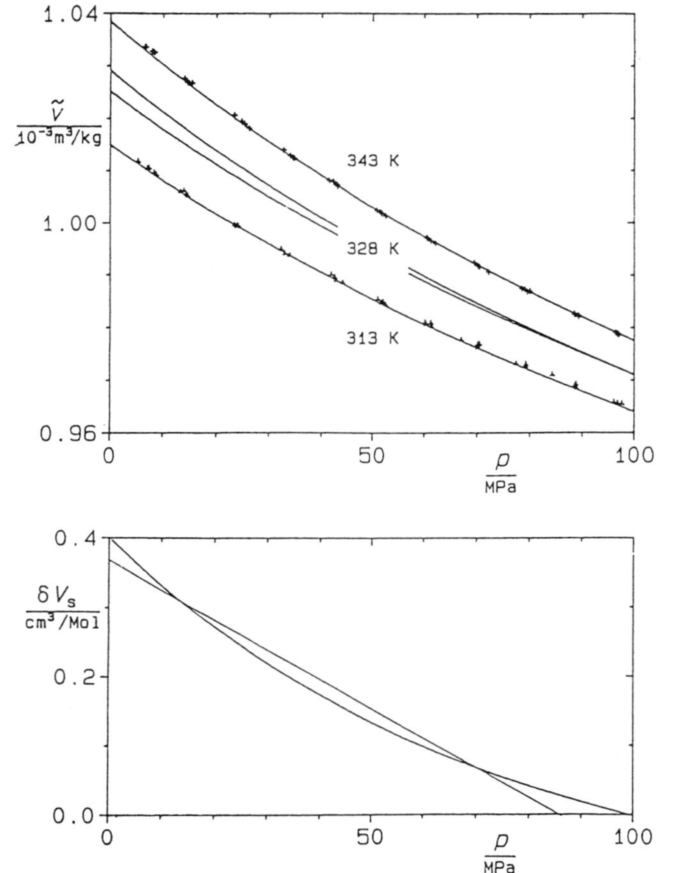

Fig. 10. Phasediagram and pressure dependence
of m/i and melting-transition in PDES

To incorporate the volume effect into the caloric transition data one has to add several terms to the volume independent superstructural parts

$$\delta(\Delta s_s^{sf}) \quad \text{and} \quad \delta(\Delta h_s^{sf})$$

which are described in detail in a subsequent paper[7].

1. At zero pressure, the volume effect $\delta V_s(0)$ leads to

$$\delta\Delta s_s(0) = \delta\Delta s_s^{sf} + K\alpha\delta V_s(0) \qquad \delta\Delta h_s(0) = \delta\Delta h_s^{sf} + B\delta V_s(0)$$

2. At finite pressure

$$\delta\Delta s_s(p) = \delta\Delta s_s(0) + K\alpha[\delta V_s(p) - \delta V_s(0)]$$

$$\delta\Delta h_s(p) = \delta h_s(0) + \int_0^P \delta V_s dp + T_t K\alpha[\delta V_s(p) - \delta V_s(0)]$$

The enthalpy terms are introduced according to dH=Tds+vdp. The thermodynamic relation $\partial\delta\Delta g_s/\partial p = \delta V_s$ is obviously valid.

The transition temperature therefore becomes

$$T_t(p) = \frac{\delta\Delta h_s(p)}{\delta\Delta s_s(p)} \approx \frac{\delta\Delta h_s(0)}{\delta\Delta s_s(0)} \left\{ 1 + \frac{\int_0^p \delta V_s dp/\delta\Delta h_s(0)}{1 - K\alpha[\delta V_s(0) - \delta V_s(p)]/\delta\Delta s_s(0)} \right\}$$

introducing $\delta V_s(p) = \delta V_s(0)(1-p/p^*)$ one obtains

$$T_t(p) \approx T_t(0) + \frac{p(1-p/2p^*)\delta V_s(0)/\delta\Delta s_s(0)}{1 - (p/p^*)K\alpha\delta V_s(0)/\delta\Delta s_s(0)}$$

This pressure dependence of the transition temperature is plotted in Fig. 10.

DISCUSSION OF THE MESOPHASE TRANSITIONS OF POLYMERS IN THE MEANDER MODEL OF POLYMER MELTS

Both mesophases (the PDES lamellar phase and the nematic phase of LC-polymers) and their properties can be described in the frame of the meander model: The nematic structure appears from the isotropic melt, if the cubes get parallel to each other by rotating across their space diagonals (Fig. 11, left side). This tendency to arrange in parallel explains also the orientation fluctuation above $T_{n/i}$. The lamellae of the PDES-mesophase must be nucleated and further grow by simultaneous rotation of 4 cubes around its middle cube edge, a process which needs disentanglement, and also takes place during crystallization from the melt (Fig. 11, right side). The long spacing measured in slowly crystallized PDES, L=49 nm, should also be the average height of a single m-phase lamella, but cannot be observed in the mesophase because of strong bundle-fluctuations coupling the

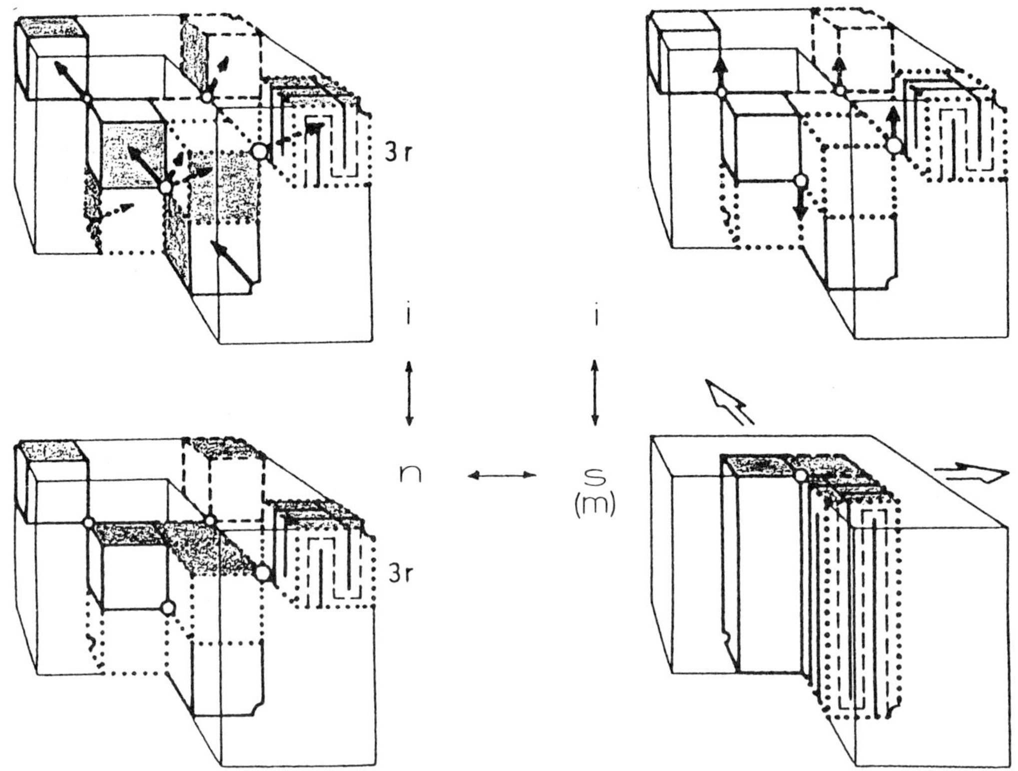

Fig. 11. Consideration of n/i- and m/i-transitions in the meander model of polymer melts.

lamellae. After heating quenched PDES samples L~16 nm was measured, indicating lamellar crystallization by laterally organized
single meander cubes instead of triple meander cubes in the crystallization from the m-phase.

REFERENCES

1.   Y. K. Godosky, V. S. Papkov, Thermotropic Mesophases in
     linear Polysiloxanes, Macrom.Chem., Macrom.Symp. 4:71
     (1986)
2.   W. Pechhold, Messung des komplexen Schubmoduls im Frequenz-
     bereich von 1-100 kHz, Acustica 9:39 (1959)
3.   G. Sauerbrey, Einfluß der Elektrodenmasse auf die Schwin-
     gungsfiguren dünner Schwingquarzplatten, A.E.Ü. 18:617
     (1964)
4.   H. Schilling und W. Pechhold, Zwei Quarzresonatormethoden
     zur Untersuchung des komplexen Schubmoduls von Polyme-
     ren, Acustica 18:244 (1969/70)
5.   W. Shockley, D. R. Curran, D. J. Koneval, Trapped Energy
     Modes in Filter Crystals, L.A.S. 41:981 (1967)
6.   P. Schwarzenberger, Diplomarbeit Ulm (1989)
7.   W. Pechhold et. al., On the nature of mesophase transitions
     in Polymers, to be published in Coll. Polym. Sci..

THE EFFECTS OF PRESSURE ON THE CHARGE DENSITY WAVE
IN LOW DIMENSIONAL SOLIDS: A RAMAN STUDY

Robert J. Donohoe*, Steven P. Love,
Maria A. Y. Garcia and Basil I. Swanson*

Inorganic and Structural Chemistry Group
INC-4; MS-C345
Los Alamos National Laboratory
Los Alamos, NM 87545

ABSTRACT

The Raman spectroscopy of bromide- and chloride-bridged platinum linear chains under isotropic pressure reveals an initial hardening and subsequent softening of the symmetric halide stretching mode as pressure is increased. These results reflect an initial increase in the extent of the metal sublattice charge disproportionation (charge density wave, CDW) followed by CDW weakening and are consistent with previous absorption studies of the band gap. Phase changes that are independent of the CDW strength occur for both materials near 2.5 and 4.5 GPa and, at least for the lower pressure transition, may be due to an indirect effect that leads to interchain ordering.

INTRODUCTION

Low dimensional charge density wave (CDW) materials are currently of great theoretical and practical interest. Inorganic quasi-one-dimensional CDW solids are typified by Wolffram's red salt, which is a halide-bridged mixed valence platinum linear chain structure, and its many variants derived by alteration of the metal, bridging halide, equatorial ligands and counterions. In these materials, the highly correlated CDW phase is formed by charge disproportionation within the metal sublattice and a concomitant dimerization of the halide sublattice.[1] The strength of this Peierls distorted CDW phase is primarily dictated by the relative magnitudes of the competitive electron-electron and electron-phonon interactions.[2] Consequently, the electronic properties of

Frontiers of High-Pressure Research, Edited by H.D. Hochheimer and
R.D. Etters, Plenum Press, New York, 1991

the MX chains can be manipulated by the choice of the metal and halide as well as doping,[3] photolysis[4] and alteration of environment via temperature,[4,5] pressure[6] or applied fields.[7]   As an example, the chloride bridged platinum chain $[Pt^{II}(en)_2][Pt^{IV}(en)_2Cl_2](ClO_4)_4$ (en=ethylenediamine), referred to hereafter as PtCl, is a strongly charge disproportionated insulating solid while the iodide analogue, PtI, is a weakly distorted CDW semiconductor.   Recently, the first report of a non-Peierls-distorted, spin-density-wave (SDW) MX chain, NiBr, was published.[8]   In addition, a report of the behavior of PtI chain solids at high magnetic fields showed evidence for a field induced diamagnetic transition suggestive of a metal-insulator transition in this material.[7]   Thus, the MX solids offer the opportunity to probe ground and local states as the electronic properties are systematically tuned to the phase boundary between CDW, SDW, and possibly conducting states.

One of the primary thrusts of current research on MX chain solids has been theoretical and experimental studies of local gap states (polarons, bipolarons, kinks, and excitons) induced by impurity doping[3] or photoexcitation[4] beyond the IVCT band gap.   Kurita and co-workers first observed photoinduced absorptions, the A and B bands, to grow below the band gap in PtCl single crystals at low temperature.[4a]   On the basis of theoretical expectations and the correlations of these absorption features with photoinduced EPR signals, these metastable mid-gap states were attributed to polarons.   This assignment was verified by the observation of a mid-IR electronic absorption (the C band) that exhibited the same photoinduced behavior and thermal decay.[4b]   Subsequent resonance Raman studies of the near-infrared metastable gap states[9] showed asymmetry in the A and B band absorptions for electron and hole polarons that could be quantitatively modeled using a 3/4-filled, 2-band Peierls Hubbard Hamiltonian where both the halide and metal bands were included.[10]

Recent optical and EPR studies of PtBr showed that changes in the strength of the CDW have a strong effect on the stability and nature of the photoinduced gap states.[11]   Whereas photoexcitation leads to the formation of electron and hole polarons in PtCl, irradiation above the IVCT edge results in the formation of electron bipolarons and hole polarons in PtBr.   Further, EPR studies indicate extensive delocalization of the hole-polaron in PtBr as opposed to that in PtCl and temperature studies reveal that photoinduced valence defects in PtBr are much less stable than those observed in PtCl.

We have initiated a resonance Raman study of MX chains under isotropic pressure.   Part of our interest in high pressure studies of the MX solids is the potential for pressure tuning of the CDW strength which, if true, would provide a method for study of the changes in photoinduced gap states as the CDW is continuously varied. Earlier studies of the high pressure behavior of MX chain solids are limited. The application of pressure has been reported to yield increases as great as nine orders of magnitude in the conductivity of MX chains;[6a] a result which indicates that the insulating gap induced by the Peierls distortion is drastically diminished under pressure. This suggestion is supported by the red shift in the intervalence charge transfer (IVCT) band edge absorption of PtCl under pressure as reported by Kuroda et al.[6b] In this study, a gap state absorption was also observed to grow in continuously as pressure was increased in PtCl; this gap state absorption was observed to disappear slowly (days) following release of pressure.

The high pressure behavior of the MMX solids $K_4\{Pt_2(P_2O_5H_2)_4X\}\cdot3H_2O$ (where X=Cl, Br; hereafter referred to as Pt_2X) has been studied using optical absorption and resonance Raman probes.[12] As with PtCl, the Pt_2X solids show a strong red shift of the IVCT band edge with increasing pressure. The pressure induced shifts of the other electronic transitions were also measured and, while complex, found to be in agreement with expected changes in the electronic structure. Raman results for Pt_2Br at high pressure are complicated by existence of two different structures. The highly crystalline 'trapped valence' form of Pt_2Br can only be followed to approximately 4.0 GPa owing to an irreversible pressure induced transition to the 'delocalized' form. Within this pressure window, trapped valence Pt_2Br and Pt_2Cl both show monotonic increases in the frequencies of the oxidized units (Pt^{III}-Pt^{III}) and the Pt^{III}-X stretching modes and a slight decrease in the frequency of the reduced units (Pt^{II}-Pt^{II}) as pressure is increased. In contrast, the 'delocalized' form of Pt_2Br showed evidence for a reverse Peierls distortion at high pressure.

Raman spectroscopy of MX chains has been exceptionally useful for investigation of both the ground state and defect electronic and vibrational properties. The zone center Raman-active phonon for the ground state CDW is the in phase symmetric X-Pt-X stretching mode, referred to as ν_1. This Raman feature is extraordinarily intense due to the fact that this coordinate is coupled to the Peierls stabilization: upon resonance excitation within the IVCT, more than fifteen quanta of this mode can be observed.[13] For most materials, the application of isotropic pressure results in increased phonon energies due to contraction of the lattice, with typical increases on the order of +1 cm^{-1}/GPa. However, as will be shown, the ν_1 feature in MX chains softens dramatically upon pressurization and reveals the weakening of the CDW phase and the occurrence of two phase changes.

EXPERIMENTAL

The MX chains were loaded as single crystals in diamond anvil cells with a 220 micron diameter drilled gasket and mineral oil used as a pressurizing fluid. Several ruby chips were included to allow calibration of the pressure. The Raman excitation energies were provided by a Ti:sapphire laser pumped by an argon ion plasma laser (Spectra Physics models 3900 and 2045, respectively). The laser power focussed onto the sample was in the 1-3 mW range. The Raman data were acquired with excitations below the IVCT band edge in order to avoid photolysis. Although such energies are often in resonance with defect transitions, we have not observed any photoinduced spectral changes of MX chain samples in either the ground or defect states as a result of irradiation in this regime. In addition, the excitation energy used for examination of the PtCl crystals, 1.71 eV, is within the absorption band reported by Kuroda et al. for a pressure induced defect absorption (room temperature data).[6b] The diamond anvil cells were cooled within a He displex refrigerator and the temperature (measured at the cell) was typically 25 K. The scattered radiation was coupled into either a SPEX model 1403 scanning double monochromator with an RCA C31034 photomultiplier tube (PtCl measurements) or a SPEX 1877D triple stage spectrograph with a charge coupled device detector (Princeton Instruments, 1152 x 298 pixel EEV chip, PtBr measurements). The sample pressure was altered at room temperature, so the cell was warmed and recooled for each measurement.

RESULTS

The Raman data for the v_1 band of PtCl at 25 K versus pressure are shown in Figure 1. The ambient pressure spectrum reveals the fine structure in in the v_1 mode which has recently been unequivocally attributed to chlorine isotope effects.[14] Upon application of pressure, the v_1 mode shifts dramatically to lower energy (-8 cm^{-1}/1.27 GPa), the fine structure components broaden and the spacings collapse by approximately 20%. We note that our results are quite different from those reported by Tanino et al. in their room temperature study of a similar PtCl complex.[6c] The origin of this difference is unclear. An increase in pressure to 1.77 GPa brings about additional softening, complete loss of the fine structure and the initial appearance of a peak (292 cm^{-1}) to lower energy from the main Raman peak. Interestingly, this new feature appears to gain intensity at the expense of the higher energy peak across a broad pressure range until it becomes the predominant feature at 2.86 GPa. This observation suggests the presence of multiple domains and the occurrence of a remarkably sluggish phase change as a function of pressure: typically, pressure-induced phase changes exhibit sharp critical points.

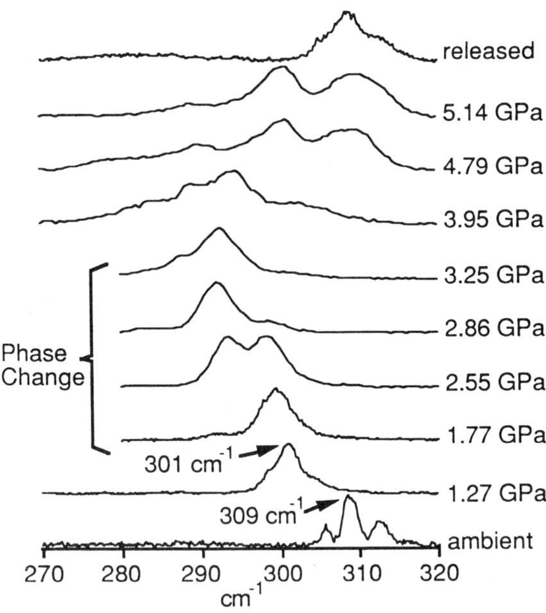

Fig. 1.   Raman of PtCl at 15K with 1.71 eV excitation versus pressure.

Beyond 3 GPa, the Raman-active modes broaden and begin to shift up in energy. A second phase change may occur near 4 GPa, where multiple peaks appear. These peaks appear to diverge between 5.14 and 6.15 GPa. The gasket began to fail at 6.15 GPa, so the pressure was released and the Raman data for the crystal reacquired at ambient conditions. Except for broadening of the fine structure components, the ν_1 mode appears to return to its initial profile and energy. The loss of resolution is most likely due to inhomogeneities caused by a combination of the pressure induced phase changes and the multiple warming and cooling cycles entailed in this experiment.

With 1.71 eV excitation, the first overtone of the ν_1 peak gains intensity relative to the fundamental as the pressure is increased (not shown). This result is in accord with the observation of a red shift in the IVCT band edge between ambient pressure and 2 GPa. In our experiments, the PtCl crystal was examined within the cell under a polarizing microscope before and after pressurization at room temperature. These observations revealed that the transmitted visible light (red at ambient conditions) was diminished until the crystal became opaque with incident light polarized along the chain axis, which is as expected upon a red shift of the IVCT edge. It is important to note that no significant change was observed in the appearance of the crystal under perpendicular polarization: the visible absorption remains polarized along the chain axis at all pressures.

The excitation energy chosen for the PtCl experiments, 1.71 eV, is in resonance with the pressure-induced defect absorption band reported by Kuroda et al.[6b] These authors assigned this absorption to a soliton-to-band transition. In Figure 2, an extended view of the Raman data for PtCl versus pressure reveals a highly structured broad feature near 150 cm^{-1}. This set of peaks appears to change its structure as the pressure is varied, but no clear trends are suggested. These bands most likely arise from resonance enhanced features associated with the pressure-induced defect. A possible explanation for the broad character of this band, consistent with the soliton picture, is that the chain is kinked at the defect, so that environmental inhomogeneity due to random orientation of the kink with respect to counterions and ancillary ligands leads to a range of defect phonon energies. It is interesting to compare the ν_1 peak at 2.55 GPa as displayed in Figures 1 and 2. The extended spectrum was acquired second and the change in relative intensities of the two predominant features indicates that the phase change is ongoing even at static pressure.

As noted in the experimental section, excitation energies were chosen to avoid photolysis of the material as has been observed upon irradiation within or above the IVCT. However, for PtCl, we also conducted a series of experiments with excitation at 2.41 eV. For this experiment, the behavior of the ν_1 band was identical, within experimental error, to that observed with the 1.71 eV probe light. However, with 2.41 eV excitation we did observe a broad, structured peak near 330 cm^{-1} that was not seen with the lower energy irradiation. This peak exhibited a great deal of variability and depended on the location of the laser beam on the crystal, the incident power and the temperature. We have observed a similar peak in mixed bromide/chloride MX chains and attribute it in those systems to edge states created by the interface of distinct CDW domains.

The sub-gap excitation Raman spectra of PtBr as a function of pressure are shown in Figure 3. In this experiment, the initial pressure applied was relatively slight (0.25 GPa) and the ν_1 peak is seen to shift up in energy. The next pressurization (0.93 GPa), however, leads to softening of the ν_1 mode and the appearance of a new peak near 150 cm^{-1}. Further pressure results in an upshift for the weaker peak and continued softening of the predominant feature. There is an abrupt softening of the peak energy between 2.73 and 2.96 GPa and, as is the case for PtCl, multiple features are clearly present in all of the data beyond this phase change.

Plots of the predominant Raman mode energy for the PtCl and PtBr materials versus applied pressure are shown in Figure 4. The initial upshift in the ν_1 mode energy upon slight pressurization as observed for PtBr has not yet

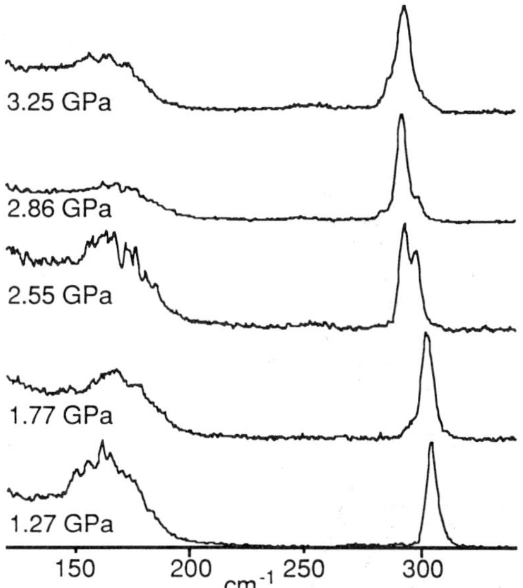

Fig. 2. Resonance Raman of the pressure induced defect modes and ν_1 in PtCl at 15 K versus pressure

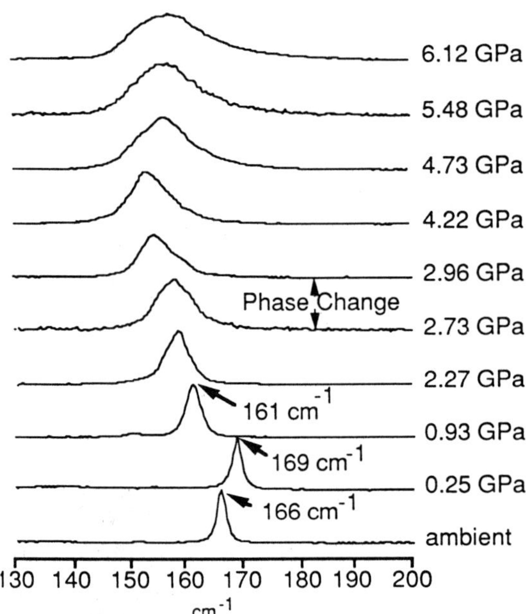

Fig. 3.   Raman of PtBr at 20 K with 1.28 eV excitation versus pressure.

been investigated for PtCl. This upshift suggests that the strength of the CDW is initially increased upon pressurization. Although we have not yet obtained Raman data for PtCl under such low pressures, we note that the absorption data from Kuroda et al. reveal an initial blue shift in the IVCT absorption edge[6b] which is consistent with a preliminary increase in the charge disproportionation.

DISCUSSION

The initial upshift observed in the Raman data for PtBr and the absorption data for PtCl indicate that, at low pressures, the MX materials undergo a slight increase in the strength of the CDW. This may be the result of an anisotropic response in which the crystal is easily compressed along axes perpendicular to the chains, while a slight increase occurs in the metal to metal distance along the chain axis. Such an increase would be expected to occur primarily at the long metal-halogen bond and would enhance the metal sublattice charge disproportionation. Further pressurization can no longer be compensated for solely by reduction of the interchain distance and now the material is compressed along the chain axis, reversing the elongation on the long metal halogen bond and reducing the strength of the CDW phase. An alternative explanation is that the MX chains undergo a phase transformation at low pressures. This possibility can be investigated by pressure dependent structural determinations, which are currently underway in our laboratory.

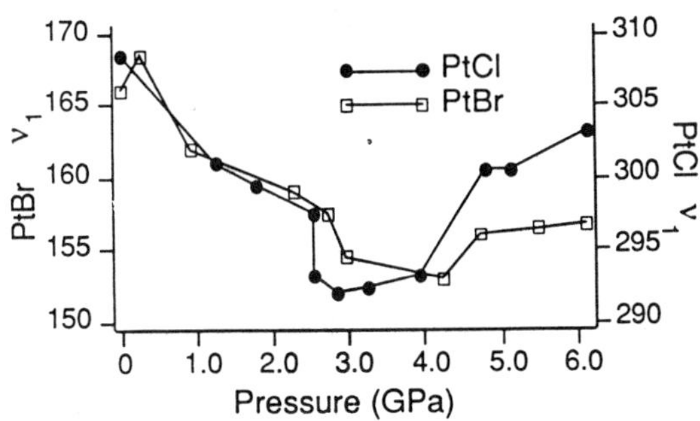

Fig. 4.   Plot of ν_1 energy versus pressure for PtCl and PtBr.

What is quite remarkable about the data as presented in Figure 4 is that both PtCl and PtBr appear to respond to pressure in a very similar manner. Note that both of the vertical axes in the figure span a 20 cm^{-1} range. Abrupt changes in both profiles support the suggestion of phase changes near 2.5 and 4.5 GPa in both materials. In fact, the lower pressure transformation appears to be a result of a change in the predominant form of two coexisting phases: both PtBr and PtCl exhibit peaks that correspond to both phases prior to and after the

2.5 GPa change. In PtBr, the lower energy peak (corresponding to the higher pressure phase) clearly shifts up as pressure is increased toward the phase change, while in PtCl a slight upshift is indicated. Because these two materials behave in such a similar fashion, despite the considerable difference in the strengths of their CDWs, it must be concluded that the primary factor dictating the pressure response is not the strength of the CDW, and that the phase transitions are not driven by direct interactions between CDWs on neighboring chains. Rather, some other interaction is responsible for the phase transitions, and changes in the CDW appear to be an indirect result of this effect. Within this framework, there are several interesting possibilities for the origin of the observed spectral changes associated with the phase changes.

(1). Associated with the phase is an extension of the periodicity of the CDW (e.g., to a period of four Pt sites rather than the original two), and the new Raman features are due to the associated folding of the Brillouin zone. For the 2.5 GPa change, this option can be readily dismissed because the v_1 phonon band is known to increase in energy from the zone center, thus predicting that new features would appear at higher frequency upon zone folding, whereas experimentally the new Raman peaks grow in below the v_1 band. In addition, such an occurrence would be expected to be dependent on the strength of the CDW, and we therefore view this as an unlikely explanation for either of the phase changes.

(2). The phase changes result in an ordering of the chains with respect to each other. Under ambient conditions, x-ray studies show that the monoclinic forms of the crystals are disordered at the bridging halide, that is, that there is no long range chain-to-chain correlation between the phases of the CDWs. Possibly, the first phase change is an disorder/order process in which progressively larger domains of the crystals become ordered with respect to the halide position. This explanation is consistent with the sluggish manner in which this change occurs and the observation of coexistence of the two phases over a large pressure window: such behavior might be expected for a partially frustrated transition in which the ordering of local domains as pressure is applied leads to progressively larger barriers to reorientation for the remaining disordered regions. The ordering might entail the opposition of like valences on adjacent chains; that is, Pt^{IV} surrounded by Pt^{IV} on the neighboring chains. In such an arrangement, the increased interchain electrostatic interactions (compared to ordering the CDW chains out of phase with respect to all neighbors) should lead the metal ions to decrease the extent of the charge disproportionation, thereby weakening the CDW, as is observed. Since we have ruled out direct interactions between CDWs as a driving mechanism, this sort of transition could instead involve interactions between the ligands and/or counterions, which might be expected to exhibit subtle orientational differences between the nodes and antinodes of the CDW. While this picture has many appealing aspects, it is not clear that such a phase change would be reversible, and reversibility is suggested by the PtCl data.

(3). The phase change involves a straightforward reordering of the perchlorate ions. The counterions do appear to mediate the strength of the CDW, as observed by the dependence of the IVCT energy on the type of counterion.[15] However, while the extent of hydrogen bonding between the perchlorate oxygens and the amine hydrogens appears to change during the temperature dependent monoclinic/orthorhombic phase transition near room temperature, no concomitant significant alteration in the strength of the CDW (as monitored by Raman) is observed.

Several other possible explanations for the pressure dependent Raman data can be imagined, including the transition to a new space group, the advent of extended periodicity between the chains, or the influence of pressure-induced defects on the ground state CDW. In order to address this issue we are currently undertaking structural characterization of the MX materials at high pressures. In addition, attention to the behavior of the ligand and counterion modes, as was used by Sakai et al. in a study of the temperature induced phase change,[16] may provide insight in this matter.

SUMMARY, CONCLUSIONS AND FUTURE DIRECTIONS

We have observed the pressure dependence of the ν_1 mode in strong and moderately weak CDW MX chains, PtCl and PtBr, respectively. After an initial increase in energy, possibly due to anisotropic compression, the ν_1 mode softens dramatically, demonstrating that the strength of the CDW weakens as pressure is raised. In both materials, the presence of multiple domains is observed and reversible phase changes appear to occur near 2.5 and 4.5 GPa. These changes do not depend on the strength of the CDW and probably involve interchain reordering. We are attempting to investigate this process in the weak CDW solid PtI. However, in this material, we are currently unable to irradiate the solids below the IVCT band edge. We have acquired data for the PtCl material under pressure with excitation above the band edge and observed new, broad and structured features near 330 cm^{-1} with complicated dependences on the incident intensity. Therefore, such an experiment appears to entail pressure-stabilized photoinduced defects and complicates the interpretation of the response of the ground state CDW. An FT Raman apparatus capable of obtaining data using with 0.94 eV excitation will soon be available, which should allow us to avoid photolysis in the PtI system. For the PtCl and PtBr systems, we are currently pursuing x-ray structural measurements and further Raman studies in order to elucidate the nature of the phase changes and probe photoinduced gap states created at high pressure.

REFERENCES

1. M.-H. Whangbo, Structural and Electronic Properties of Linear Chain Compounds and Their Molecular Analogies, <u>Acc. Chem. Res.</u> 16:95 (1983).

2. D. Baeriswyl and A. R. Bishop, Localized Polaronic States in Mixed-Valence Linear Chain Complexes, <u>J. Phys. C: Solid State Phys.</u> 21:339 (1988).

3. N. Matsushita, N. Kojima, T. Ban and I. Tsujikawa, Photo-Induced Absorption Band in One-Dimensional Halogen-Bridged Mixed-Valence Platinum Complex: $[Pt(en)_2][PtI_2(en)_2](SO_4)_2 \cdot 6H_2O$ and its Au-Doped Complex: $[Au_xPt_{1-x}I(en)_2]SO_4 \cdot 3H_2O$, <u>J. Phys. Soc. Japan</u> 56:3808 (1987).

4. a) S. Kurita, M. Haruki and K. Miyagawa, Photo-Induced Defect States in a Quasi One-Dimensional Mixed Valence Platinum Complex, <u>J. Phys. Soc. Japan</u>, 57:1789 (1988). b) R. J. Donohoe, S. A. Ekberg, C. D. Tait and B. I. Swanson, The Mid-Infrared Signature of Photo-Induced Defects in the Quasi-One-Dimensional Solid $[Pt^{II}(en)_2][Pt^{IV}(en)_2Cl_2](ClO_4)_4$, <u>Solid State Commun.</u>, 71:49 (1989).

5. N. Matsushita, N. Kojima, T. Ban and I. Tsujikawa, Study of Mixed-Valence State in Halogen-Bridged Mixed-Valence Platinum(II,IV) Complexes by Temperature Dependence of Intervalence Charge-Transfer Spectra, <u>Bull. Chem. Soc. Jpn.</u>, 62:3906 (1989).

6. a) L. V. Interrante, K. W. Browall and F. P. Bundy, Study of Intermolecular Interactions in Transition Metal Complexes. IV. A High-Pressure Study of Some Mixed-Valence Platinum and Palladium Complexes, <u>Inorg. Chem.</u>, 13:1158 (1974). b) N. Kuroda, M. Sakai, Y. Nishina, M. Tanaka and S. Kurita, Soliton-to-Band Optical Absorption in a Quasi One-Dimensional Pt^{II}-Pt^{IV} Mixed-Valence Complex under Hydrostatic Pressure, <u>Phys. Rev. Lett.</u>, 58:2122 (1987). c) H. Tanino, N. Koshizuka, K. Kobayashi, M. Yamashita and K. Hoh, Pressure Dependence of Absorption Edge, Luminescence Peak and Raman Frequency in Wolffram's Red Salt, <u>J. Phys. Soc. Japan</u>, 54:483 (1985).

7. M. Haruki and P. Wachter, Magnetization Anomalies in One-Dimensional Halogen-Bridged Platinum Complex: Field-Induced Diamagnetic Transition and Antiferromagnetism, <u>Physica B</u>, in press.

8. K. Toriumi, Y. Wada, T. Mitani, S. Bandow, M. Yamashita and Y. Fujii, Synthesis and Crystal Structure of a Novel One-Dimensional Halogen-Bridged Ni^{III}-X-Ni^{III} Compound, $\{[Ni(R,R\text{-}chxn)_2Br]Br_2\}$, <u>J. Am. Chem. Soc.</u> 111:2341 (1989).

9. R. J. Donohoe, C. D. Tait and B. I. Swanson, Resonance Raman Evidence for Electron and Hole Defect Asymmetry in the Quasi-One-Dimensional Solid $[Pt^{II}(en)_2][Pt^{IV}(en)_2Cl_2](ClO_4)_4$ (en=ethylenediamine), <u>Chem. Mater.</u>, 2:315 (1990).

10. J. T. Gammel, R. J. Donohoe, A. R. Bishop and B. I. Swanson, Electron and Hole Polaron Asymmetry in a Two-Band Peierls-Hubbard Material, <u>Phys. Rev. B</u>, 42:10566 (1990).

11. R. J. Donohoe, L. A. Worl, C. A. Arrington, A. Bulou and B. I. Swanson, On the Nature of the Photoinduced Defects in the Bromide-Bridged Platinum Linear Chain $[Pt^{II}(en)_2][Pt^{IV}(en)_2Br_2](ClO_4)_4$, <u>Phys. Rev. B</u>, submitted for publication.

12. B. I. Swanson, M. A. Stroud, S. D. Conradson and M. H. Zietlow, Observation of a Pressure Induced Reverse Peierls Instability in the Quasi-One-Dimensional Solid $K_4[Pt_2(P_2O_5H_2)_4Br]\cdot3H_2O$, <u>Solid State Commun.</u>, 65:1405 (1988).

13. R. J. H. Clark, Raman and Resonance Raman Spectroscopy of Linear Chain Complexes, in "Advances in Infrared and Raman Spectroscopy, Volume 11", R. J. H. Clark and R. E. Hester, eds., Wiley, New York (1984).

14. S. P. Love, L. A. Worl, R. J. Donohoe, S. C. Huckett and B. I. Swanson, Origin of the Fine Structure in the Vibrational Spectrum of $[Pt^{II}(en)_2][Pt^{IV}(en)_2Br_2](ClO_4)_4$, (en=ethylenediamine): Vibrational Localization in a Quasi-One-Dimensional System, <u>Phys. Rev. Lett.</u>, submitted for publication.

15. N. Matsushita, N. Kojima, T. Ban and I. Tsujikawa, Comparison Between Optical Properties of Sulfates and Hydrogensulfates in Halogen-Bridged Mixed-Valence Platinum(II,IV) Complexes, <u>Bull. Chem. Soc. Jpn.</u>, 62:1785 (1989).

16. M. Sakai, M. Hayakawa, N. Kuroda, Y. Nishina and M. Yamashita, Raman Scattering by Ethylenediamines in Quasi-One-Dimensional Mixed-Valence Complexes $[M(en)_2][Pt(en)_2Cl_2](ClO_4)_4$ (M=Pt, Pd, en=ethylenediamine), <u>J. Phys. Soc. Japan</u>, 60:1619 (1991).

PRESSURE-INDUCED POLYMERIZATION OF CYCLIC MOLECULES: A STUDY OF BENZENE AND THIOPHENE

M. Gauthier, J.C. Chervin and Ph. Pruzan

Physique des Milieux Condensés, CNRS URA 782
Université Pierre et Marie Curie,
Tour 13 4ème Etage, B.P. 77
4 Place Jussieu
F-75252 Paris Cedex 05

Abstract

It has been shown that pressure induces irreversible transformation of cyclic molecules such as benzene and thiophene. Investigations were carried out using various techniques such as Raman scattering, Infrared absorption, Brillouin scattering, refractive index determination and visual observations. We use crystallographic considerations to propose a first step for the reaction path of these transitions.

Introduction

At very high density, the intermolecular interactions become of the same order of magnitude than the intramolecular ones. The molecular character tends to disappear and it is well known that unsaturated hydrocarbons may transform irreversibly at high pressure[1]. Although aromatic bonds are expected to be more stable, benzene has been found to undergo an irreversible chemical process during a pressure cycle up to a minimum of 20 GPa. A similar reaction occurs in thiophene for pressure cycles above 16 GPa[2]. These reactions are due to charge transfers which lead to modifications of the bond scheme of the solids. For benzene and thiophene cases, the final products are white solids. Once they have reacted with water of atmosphere, they are chemically stable and infusible.

Experimental

All experiments were performed in a membrane diamond anvil cell (MDAC) which has been fully described elsewhere[3]. This type of high pressure cell allows the monitoring of the pressure outside of the spectrometer cavity and therefore the recording of the signal during the pressure variations. The diameter of the diamond anvils culets was 500 µm, and then the samples were initially 200 x 50 µm. The pressure was determined from the shift of the ruby fluorescence R_1 line according to the Mao and Bell calibration :

Frontiers of High-Pressure Research, Edited by H.D. Hochheimer and
R.D. Etters, Plenum Press, New York, 1991

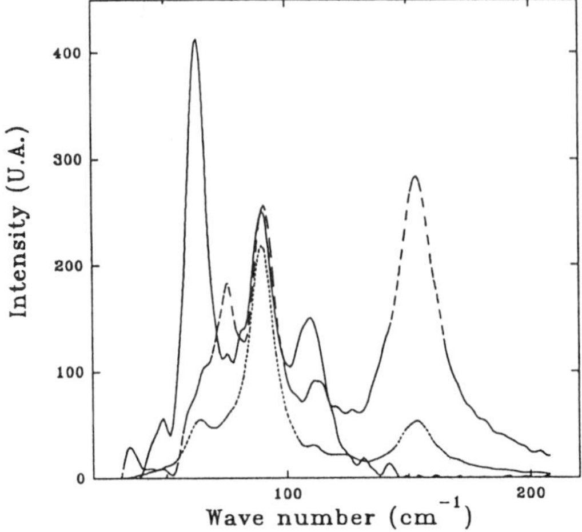

Fig. 1 Polarized Raman spectra of benzene in MDAC. Solid b(cc)b, dashed b(aa)b and dotted b(ac)b

$$p = 380.8 \left[\left(\frac{\lambda(p)}{\lambda_0} \right)^5 - 1 \right] \qquad (1)$$

The Raman spectra were recorded in back-scattering geometry. A microscope objective is used to focus the laser line and collect the signal. A spatial filter at the entrance of a triple monochromator rejects the signal coming from the anvils of the MDAC and leads to a very good signal to noise ratio. Detection is performed by an intensified diode array system.

The low frequency polarized Raman spectra of benzene at ambient conditions in the MDAC are shown in Fig.1. We have used the blue line of an argon ion laser (488. nm). The laser beam was focused down to less than 5 µm with a power of 50 mW to prevent heating of the sample. It is to be noted that the samples are very sensitive to the light irradiation. In particular, the transition pressure of thiophene can be lowered by a few GPa depending of both the radiation used and incident power.

Infrared investigations were performed with a Fast Fourier Transform spectrometer from Bruker (IFS 113V) (Fig.2-3). The first spectra were obtained using Ia diamond stones, and then investigations were carried out with IIa stones. In addition to the absorption in the range 1800-2650 cm^{-1}, present in all types of diamonds, type Ia stones have nitrogen absorption bands in the wave number range 1000-1400 cm^{-1}.

The Brillouin scattering experiment were performed with a five · pass Fabry - Perot

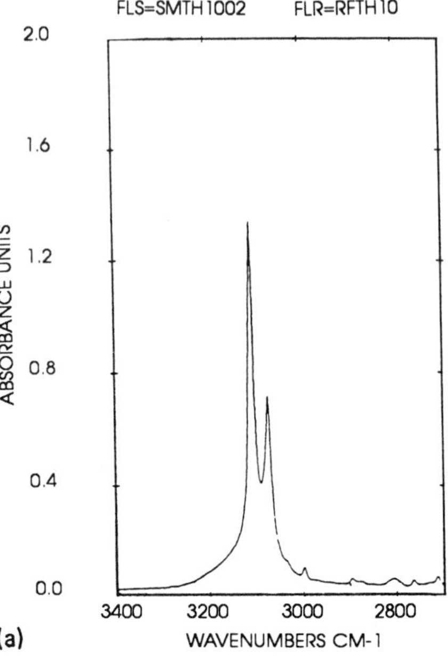

(a)

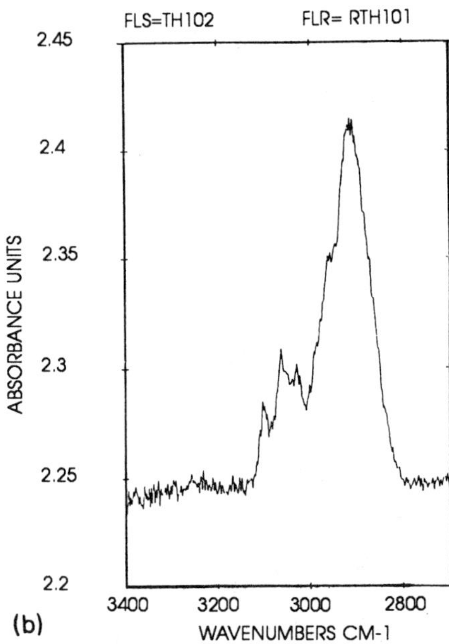

(b)

Fig. 2 Infra-Red spectra of thiophene in the C-H stretching modes region
a) Liquid at ambient conditions
b) Recovered product at ambient conditions

interferometer, where the scanning is produced by piezoelectric transducers. Detection uses a photon-counting system and data accumulation is monitored by a microcomputer. The free spectral range of the spectrometer was chosen either 2.7 or 5.4 cm^{-1} depending on the Brillouin shift and so that the Brillouin lines of the diamond anvils coincide with the Rayleigh lines. The overall finesse achieved was 50 and the contrast was about 10^{10}. Typical spectra are shown in Fig. 4-5.

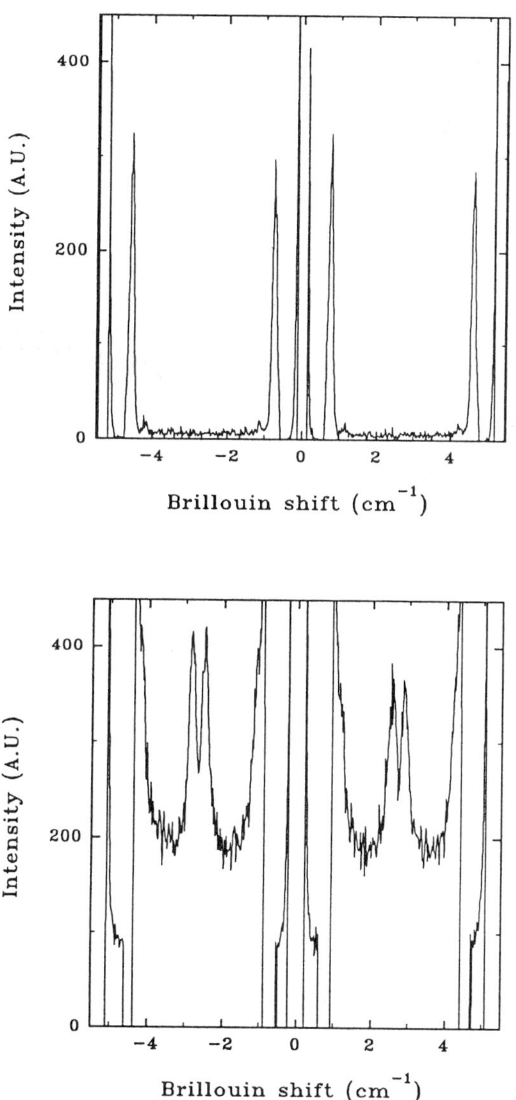

Fig. 4-5 Brillouin shift of benzene in solid phase I (up) and solid phase III (down).
Spectra were obtained at P = 0.38 GPa and 14.1 GPa with 10 and 440 accumulations respectively.

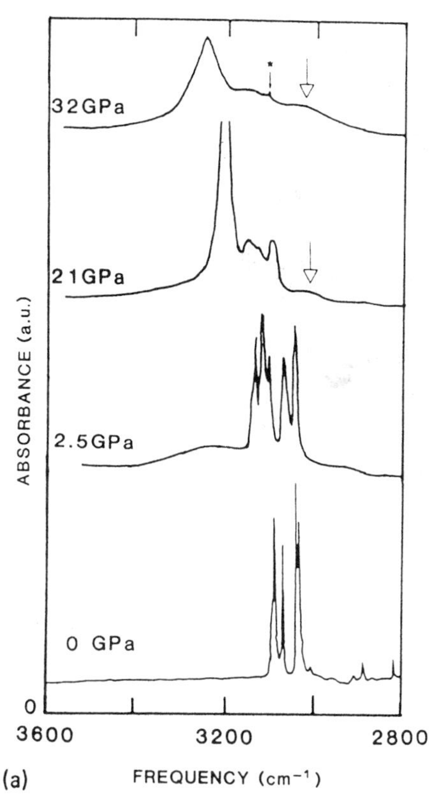

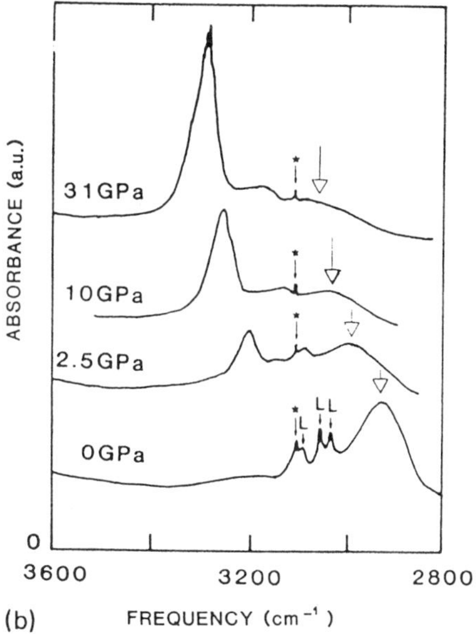

Fig. 3 Infra-Red spectra of benzene
a) Increasing pressure runs
b) Decreasing pressure runs
Stars : diamond peak, L : liquid benzene, arrows : "polymer peak". P=0 GPa is obtained outside the MDAC.

The Brillouin shift is given by:

$$\Delta\sigma = \frac{2nv}{\lambda c}\sin\left(\frac{\theta}{2}\right) \qquad (2)$$

where n is the refractive index of the sample parallel to the polarization direction of the laser beam, λ the wavelength of the exciting line, c the speed of light in vacuum, v the sample sound velocity and θ the angle between incident and scattered wave vectors. In the back-scattering geometry the longitudinal waves are more intense. As we shall see later, due to the existence of a preferential orientation in the MDAC, equation 2 may be used to compute the elastic constant through the relation:

$$C_{ij}(p) = \rho(p)\, v_L^2(p) \qquad (3)$$

where ρ is the density at pressure p.

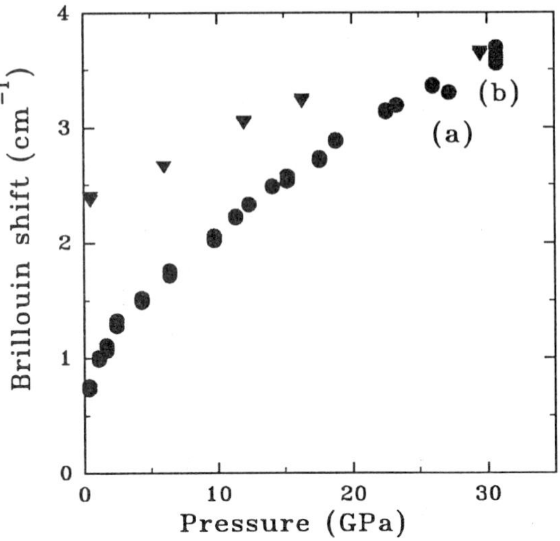

Fig. 6 Brillouin shift of benzene under pressure
Circles : increasing run; triangles : decreasing run
(a) : After 12 hours; (b) 2 hours later

Results

Raman scattering

When single crystals are grown in the MDAC along the melting curve by controlling both temperature and pressure, preferential orientation of the single crystal was found thanks to polarized Raman scattering. Fig.1 shows the spectra obtained in the MDAC. From these spectra, group theoretical analysis and comparison with low temperature data[4-6], it is shown that in our experiments the b axis of orthorhombic benzene phase I is parallel to the symmetry axis of the MDAC.

From Raman scattering it was not possible to obtain information about the high pressure chemical transformation. It is to be noted that 1) the transition is incomplete and 2) the recovered product does not exhibit significant Raman peak.

The only information we get during pressure runs comes from the untransformed part of the sample. At the opening of the MDAC, the untransformed benzene evaporates and no Raman signal could be recorded.

Infrared absorption

The Infrared investigations have shown that both thiophene (Fig.2) and benzene (Fig.3) undergo an irreversible process occurring at high pressure. In the case of thiophene the minimum pressure to obtain irreversibility is around 16 GPa, whereas in benzene this pressure is around 20 GPa. From upward and downward runs, mode frequencies versus pressure are shown to be reversible. However, in the C-H stretching region (3000 cm^{-1}), a new faint band appears respectively above 21 GPa and 16 GPa in benzene and thiophene. This band can be assigned to C-H vibrations where C is tetracoordinated. Actually in the recovered samples this band is dominant. Apart this band, the spectrum of the recovered product from benzene, outside the MDAC, shows only a few broad bands which are characteristic of highly cross-linked polymers. Thus the transformation involves a ring opening transformation. On the other hand the spectrum obtained with samples recovered from thiophene exhibits a significant number of peaks. This suggests in this case that the molecular cycle is still present but obviously altered.

Brillouin scattering

As Raman scattering experiments have shown, benzene crystallizes in the MDAC with a preferential orientation. This can be easily checked in Brillouin scattering experiment by rotating the polarization of incident light. Equation 2 shows that the Brillouin shift is proportional to the refractive index, so the knowledge of the refractive index anisotropy[7] (Table I) allows to establish the preferential orientation by a second independent method. These experiments lead to the conclusion that we have obtained the C_{22} elastic constant.

A quantitative interpretation of our results (fig.6) needs the equation of state (EOS) and the refractive index versus pressure.

Increasing pressure runs

The EOS of solid benzene[6,8] (fig.7) may be used during upward runs. We have found $B_0 = 5$ GPa for the bulk modulus and B'=6.8 for the first derivative using the first order Murnaghan EOS :

$$V(p) = V(0)\left(1 + \frac{B'}{B_0}p\right)^{-1/B'} \qquad (4)$$

Both values are consistent with the molecular character of the compound.

The refractive index has been fitted to the data using the Vedam model[9-11] :

$$n(\rho_0) - n(\rho) = A\varepsilon$$
with
$$\varepsilon = \frac{1}{2}\left[1 - \left(\frac{\rho}{\rho_0}\right)^{2/3}\right] \qquad (5)$$

where ε is the Eulerian strain. Our calibration points are the transition point Liquid-Solid I and the refractive index matching with ruby obtained at $p = 6.3$ GPa (Table I and Fig.8).

Decreasing pressure runs

During decreasing pressure runs, the refractive index matching with ruby at low pressure (p = 1.8 GPa) (Fig.8) has been used. To obtain the EOS of the recovered sample we have made the two following assumptions : 1) The volume change occurring during the high pressure transformation is negligible. 2) Relation 5 is still valid; actually this can be checked for hydrocarbon compounds. Assumption 1 obviously leads to an underestimate of the room pressure density of the new product. Nevertheless, we can estimate an EOS for this product and we have found $B_0 \approx 80$ GPa and B'$\approx$ 4. At room pressure, the density of the new product is around 1.39 g/cm^{-3} (Fig. 8).

The values obtained for the bulk modulus and its first derivative are no longer consistent with usual values obtained with molecular compounds. The chemical transformation involves a lost of the molecular character and leads to a rather isotropic

Table I Refractive index of benzene. $p = 0.08$ GPa values are obtained at melting at $p = 0$ GPa and -3°C [7]

p	Refractive index		
	n//a	n//b	n//c
increasing p			
0.08 GPa	1.544	1.646	1.55
6.3 GPa	1.755		
decreasing p			
1.8 GPa	1.76		

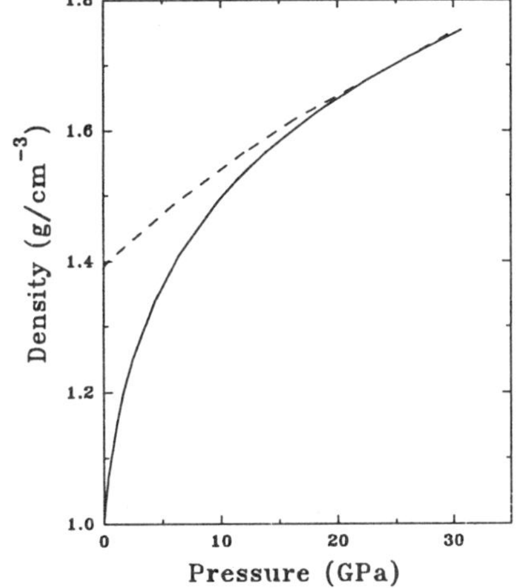

Fig. 7 Murnaghan equation of state of benzene.
Solid : increasing pressure run[6,8]
Dashed : decreasing pressure run (this work)

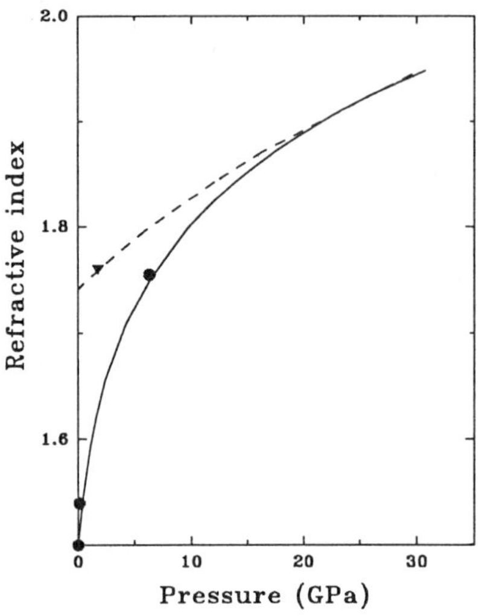

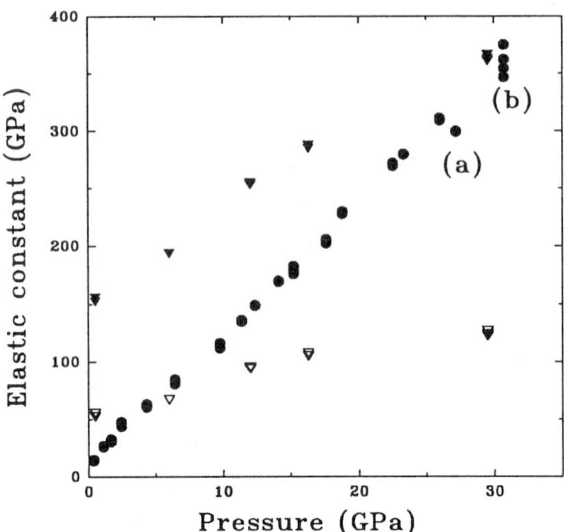

Fig. 8 *Refractive index of benzene. lines are calculations using the Vedam model. Pressure data points are obtained by comparison with ruby.*

Fig. 9 *Elastic constants of benzene and recovered sample Circles : increasing pressure C_{22}; Triangles : decreasing pressure C_{11} full, C_{44} empty (a), (b) : see Fig. 6*

material. We have used this property to calculate the elastic constants of the compound. If our product can be considered as an amorphous sample, there are only two independent elastic constant: C_{11} directly related to the longitudinal waves which depends on the compressional modulus, and C_{44} which is related to the shear modulus of the sample. Using equation 4 and our results we obtained C_{44} by $C_{44} = 3/4\ (C_{11} - B)$. Final results are shown in Fig.9. The values obtained for the recovered sample at ambient pressure are extremely different from those of solid benzene (Table II). Both C_{11} and C_{44} are at least of the same order of magnitude as strongly covalent compounds. C_{11} is 30 times larger than in the molecular solid.

Table II. Characteristic values of solid benzene and the recovered sample at p=0 GPa

C_6H_6	$\Delta\sigma$ cm^{-1}	n	v km.s^{-1}	ρ g.cm^{-3}	B_0 GPa	B'
Solid I	0.75	1.55	3.7	1.006	5.0	6.8
Recovered Product	2.35	1.75	10.0	1.39	80.0	4.

Discussion

As we have seen in the previous section, the molecular behavior of our compounds vanishes. Visual observation indicates that the optical anisotropy of the sample disappears under pressure at least for thiophene. The Infrared investigation of the recovered product gives evidences for the destruction or the alteration of the molecular cycles. The more significant information is given by the Brillouin spectroscopy : Recovered product behaves as a tridimensional strongly bonded network. At this point the question is "What is the reaction path?".

Table III *Crystallographic data of benzene solid I (4.2 K)[12] and benzene solid III (room temperature)[13]. Assumptions used for the computation at 15. GPa are given in the text. Column 4 : Results for the shortest intermolecular distances ($\perp$, // respectively between perpendicular and parallel molecules).*

C_6H_6		atomic coordinates				First neighbors		
p/GPa	benzene I					H..H	H..C	C-C
	Pbca	C(1)	-.0612	1412	-.0052			
0.	a= 7.355	C(2)	-.1402	0447	.1272			
	b= 9.371	C(3)	-.0777	- 0969	.1326			
	c= 6.699	H(1)	-.1085	2505	-.0119			
	Z= 4	H(2)	-.2491	0768	.2260			
		H(3)	.1085	- 2505	.0119			
	benzene III							
	P2$_1$/c	C(1)	-.2556	- 0397	-.1213	2.19	2.745	3.475
2.5	a= 5.417	C(2)	-.0604	- 2124	-.1119	2.66	2.77	3.51
	b= 5.376	C(3)	.1951	- 1726	.0094			
	c= 7.532	H(1)	-.4321	- 0672	-.2051	type		
	β= 110.00	H(2)	-.1022	- 3590	-.1892			
	Z= 2	H(3)	.3299	- 2918	.0159	$\perp$	$\perp$	//
	benzene III							
	P2$_1$/c	C(1)	-.2753	-.0437	-.1283	1.87	2.37	3.04
15.	a= 5.030	C(2)	-.0650	-.2340	-.1184	2.19	2.44	3.07
	b= 4.880	C(3)	.2101	-.1901	.0099		2.53	3.15
	c= 7.120	H(1)	-.4653	-.0740	-.2170	type		
	β= 110.00	H(2)	-.1101	-.3955	-.2001			
	Z= 2	H(3)	.3553	-.3215	.0168	$\perp$	$\perp$	//

Table III summarizes the known crystallographic data available for benzene[12-14]. Solid I is orthorhombic Pbca, whereas solid III which appears at 4 GPa is monoclinic P2$_1$/c. We have used the atomic coordinates redetermined by Fourme et al[14] from the previous work of Piermarini et al[13]. Assuming incompressible molecule and constant molecular orientation in the lattice, the atomic coordinates have been computed at 15 GPa using equation 4. Computation of the shortest intermolecular H..H, H..C and C..C distances suggest that couplings between molecules may occur by two different ways :

a) The shortest H..H (1.87 and 2.19 Å) and H..C (2.37 and 2.44 Å) distances are found for roughly perpendicular molecules. These types of interaction would involve H_2 formation as in the usual chemical polymerization. H_2 formation was not found during our experiments.

b) The shortest C..C (3.04, 3.07 and 3.15 Å) distances are found for parallel molecules. The spacing between molecular planes becomes shorter, which very likely causes the π orbital overlap and bond formation[15-16]. This is supported by the infrared data which indicate the formation of tetracoordinated carbon.

In conclusion case b is more likely.

The scheme (Fig.10) represents the shortest C..C intermolecular distances and possibly the bonds formed at high pressure. The shortest C..C distance (dashed lines) is along the b direction of the crystal and is close to 3.05 Å. If molecular couplings occur during this first step a fibrous material is expected. The second shortest distance (3.15 Å, dotted lines) is found in the a direction and close to 3.05 Å. Therefore, fiber couplings is likely and may induced a lamellar material. These couplings concern the same family of molecule i.e. between molecules which are roughly parallel and lying at the corner of the crystallographic cell. The second family is defined by the molecules lying in the 011 faces. This second family polymerizes in the same way that the first one but formed an independent network. The shortest C..C distance between the two networks is close to 3.25 Å. This suggests a plausible bridging between the networks.

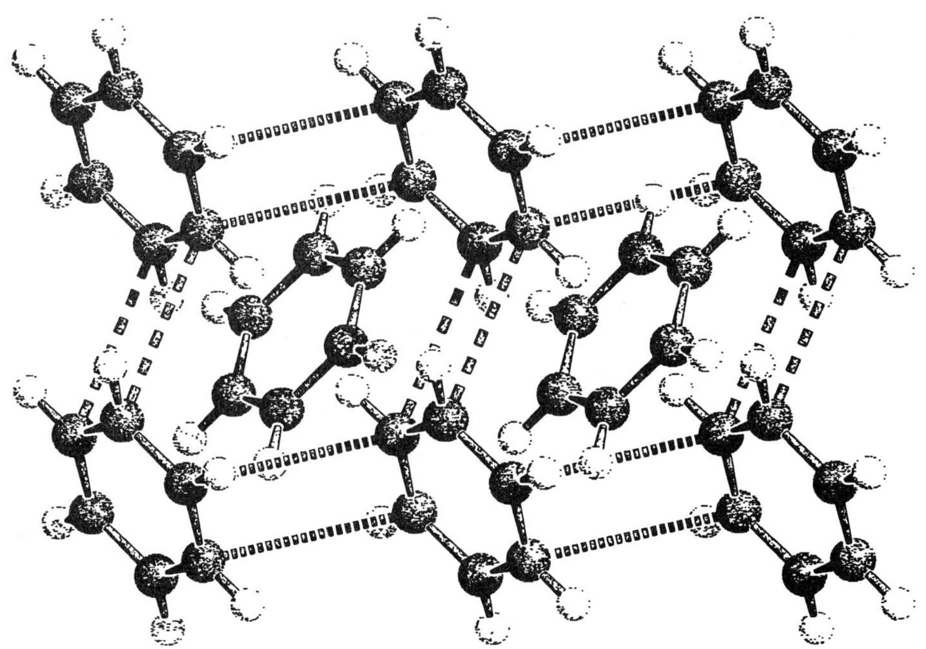

FIG. 10 Plausible bond formations at high pressure between benzene molecules.
The view is oriented in the 101 direction. For clarity only one benzene bonds family is shown.
Dashed lines : Shortest C..C distances in the 010 direction
Dotted lines : Shortest C..C distances in the 100 direction

References

1) Ph. Pruzan, J. C. Chervin, M. M. Thiéry, J. P. Itié, J. M. Besson, J. P. Forgerit and M. Revault, J. Chem. Phys. **92**, 6910 (1990) and reference therein.
2) Ph. Pruzan *et al.*, to be published.
3) R. Le Toullec, J. P. Pinceaux and P. Loubeyre, High Pressure Res. **1**, 77 (1988)
4) H. Bonadeo, M. P. Marzocchi, E. Castellucci and S. Califano, J. Chem. Phys **57**, 4299 (1972).
5) G. R. Elliott and G. E. Leroi, J. Chem. Phys. **58**, 1253 (1973).
6) M. M. Thiéry and J. M. Léger, J. Chem. Phys. **89**, 4255 (1988) and reference therein.

7)   R. M. Hochstrasser, G. R. Meredith and H. P. Trommsdorff, J. Chem. Phys. **73**, 1009 (1980).

8)   J. M. Léger, Sol. State Commun. **66**, 245 (1988).

9)   K. Vedam and P. Limsuwan, J. Chem. Phys. **69**, 4762 (1978).

10)   K. Vedam and P. Limsuwan, J. Chem. Phys. **69**, 4772 (1978).

11)   M. Gauthier, Ph. Pruzan, J. C. Chervin and A. Polian, Sol. State Commun. **68**, 149 (1988).

12)   Isis annual report, Rutherford Appleton Laboratory (UK), p43 (1989).

13)   G. J. Piermarini, A. D. Mighell, C. E. Weir and S. Block, Science **165**, 1250 (1969).

14)   R. Fourme, D. André and M. Renaud, Acta Cryst. **B27**, 1275 (1971). (parameter c must be read 7.532 Å and not 7.352 Å).

15)   H. G. Drickamer, Science **156**, 1183 (1967).

16)   H. G. Drickamer and C. W. Frank, Electronic transitions and the high pressure chemistry and physics of solids. Chapman and Hall, London (1973).

HIGHLIGHTS OF THE ROUND TABLE DISCUSSION ON POLYMERS AT HIGH PRESSURE

R. Mark Bradley

Department of Physics
Colorado State University
Fort Collins, Colorado 80523

The panelists were Professor Dr. W. Pechhold of the University of
Ulm, Dr. K. O. Prins of the University of Amsterdam, Dr. B.-E. Mellander of
the Chalmers University of Technology, and Mr. Ian Marsden of the University of Cambridge. The chair of the discussion was Professor R. M. Bradley
of Colorado State University. The session began with a brief statement
from each of the panelists regarding what they consider to be the frontiers
of polymer research at high pressure.

Professor Dr. Pechhold pointed out that pressure dependence provides
a unique opportunity to test model theories for a number of processes occurring in polymers. He identified two processes of special importance:
the glass transition and phase transitions. There is actually a strong
connection between these two topics. Dislocations play a dominant role in
the simplest theories of glass transitions. They are also the mediating
defect in the melting transition, both in low and in high molecular weight
materials. Pressure will be an important tool in testing theories of these
phenomena. Moreover, the insights gained in the study of polymers may have
wider application.

The aim of Dr. Prins' research effort is to understand the pressure
dependence of phase transitions in polymers. This is because pressure provides a ready means of separating the effects of temperature and density,
and also clarifies the nature of the interactions responsible for a particular phase transition. In the near future, his group will continue to
study the dynamics of the glass transition in polystyrene and polycarbonate
from ambient to high pressure. They also plan a high pressure study of the
relaxational motion in the CONDIS phase of polyethylene up to 10 kbar using
^{13}C NMR.

Dr. Mellander remarked that in addition to providing tests of theories of phase transitions in polymers, experiments at high pressure can be
used to test transport theories. He also expressed concern that interactions between the polymer sample and the pressure-transmitting medium may
have an important effect in some cases, and therefore the results may not
always be truly representative of the bulk properties of the polymer. He
suggested that this effect could in fact be turned to advantage: one could
learn about the properties of polymer interfaces under pressure. Of special interest would be the interaction between polymers and polar liquids
or gases.

Frontiers of High-Pressure Research, Edited by H.D. Hochheimer and
R.D. Etters, Plenum Press, New York, 1991

Mr. Marsden specializes in transport measurements of low dimensional organic charge transfer salts. Phase transitions and electrical transport in these materials are strongly influenced by the application of hydrostatic pressures. Pressure gives a direct means of changing the intermolecular distance in these quasi-one-dimensional materials, and so directly alters the intermolecular interactions and electrical transport.

The session was opened to comments from the floor at this point. Prof. Dr. K. H. Michel asked the panel whether the glass transition in polymers occurs at lower or higher temperatures when pressure is applied. Prof. Dr. Pechhold stated that in most polymers the glass transition temperature T_g increases with pressure. He believes that this has something to do with the value of the Grüneisen constant and the behavior of the compressional dilation zone of a dislocation. Interestingly, in PMMA it has been found that T_g starts to level off at pressures of 8 kbar. If the curve actually becomes flat at still higher pressures, the compressional dilation zone of a dislocation would have no excess volume. An interesting avenue for future research would be to see if this phenomenon occurs in other polymers.

Prof. R. D. Etters stated that he has been trying to learn polymer theory during the past year, but that the treatments he had seen so far seem somewhat unsatisfactory since they do not begin from a fundamental, atomistic viewpoint. He asked the panel for recommendations on polymer theory treatises. Prof. R. M. Bradley strongly recommended P. G. de Gennes' *Scaling Concepts in Polymer Physics* because it adopts the viewpoint of a theoretical physicist--in fact, that of a statistical mechanician. The development begins with the simplest case of a single polymer chain in solution, and then progresses to the more complex problems of concentrated polymer solutions and polymer melts. Even for single chains, the theory yields scaling laws which can be compared with light scattering and hydrodynamic experiments on dilute polymer solutions. Prof. N. Ashcroft remarked that in theories of this type, the degrees of freedom for short length scales are summed over to yield a theory valid only for polymers of sufficiently high molecular weights. As stressed by Prof. Bradley, an important advantage of such theories is that they can be applied to many different types of polymers--only the parameters in such phenomenological theories depend on the particular polymer in question. These parameter values need not be calculated, but instead can be taken from experiment. Thus, the "universality" of semi-macroscopic theories make them much more appealing than atomistic theories which are valid only for specific polymers.

Prof. H. D. Hochheimer expressed his concern that polymer physics appears to be driven entirely by applications and so is not fundamental. Prof. Bradley replied that in his opinion, polymer theory is related to a number of issues of fundamental interest in statistical mechanics. As an example, he cited the problem of the theta point in very dilute solutions of polymer chains. The simplest model of a single polymer chain in solution is self-avoiding walk. (Polymer chains cannot self-intersect because two monomers cannot occupy the same point in space at the same time.) This model is not always adequate, however, since it only takes into account the hard core part of the van der Waals potential which acts between unbonded pairs of monomers. The weak attractive part of the van der Waals potential gives rise to a collapse transition (or "theta point") as the temperature is lowered. In the general theory of critical phenomena, critical exponents are used to characterize the divergence of physical quantities as a phase transition is approached. The simplest type of critical point has only two independent critical exponents, but more complex types of critical points do exist. The next in complexity is a tricritical point, which is described by three independent critical exponents. The theta point of a

polymer chain is the simplest example of a tricritical point. Polymer
theory is therefore of fundamental interest, as well as being important in
applications.

Prof. Bradley asked the panel what the prospects are for improving
the properties of polymers through the application of high pressures.
Prof. Dr. Pechhold noted that the application of high pressures to polyethylene results in a substantial improvement in its strength. This phenomenon is apparently restricted to polyethylene, however. The special
feature of polyethylene that is responsible is the existence of the high
pressure, high temperature CONDIS phase. The strongest polymer fibers are
presently obtained through gel spinning, remarked Dr. Prins. This is because the polymer chains are all aligned with the fiber axis. Prof. I.
Silvera noted that one would expect that in carbon rich polymers, the
application of high pressures should result in increased crosslinking, and
therefore there might be the prospect of forming super-hard materials.
Prof. Dr. Pechhold pointed out that in fact there is evidence to the contrary: natural rubbers are normally weakly crosslinked, and the introduction of additional crosslinks actually reduces the breaking stress.

Prof. Etters remarked that it has been shown that the addition of
small amounts of polymer to water leads to substantial drag reduction. The
amount of polymer which must be added to see an effect is on the order of
parts per billion. This effect is of great interest to the military because it could be used to reduce drag on submarines, and could also be useful lubrication. The experiments are still rather confusing and inconsis-
tencies do exist, but it is now clear that the effect is real. There is
presently no consensus on how to explain polymer drag reduction, although
numerous suggestions have been advanced. Interestingly, the role of other
impurities in the water seems to be substantial. No one has looked to see
if polymer coatings reduce drag on an airfoil. Prof. Ashcroft suggested
that one end of the polymer chains could be bound to the surface of the
moving object. If the unbound ends of polymer chains were hydrophobic, the
results might be significant drag reduction. Prof. Etters stated that this
is definitely a possibility. It has also been suggested that significant
amounts of the polymer are present in the boundary layer, and that its presence inhibits the formation of vortices. The polymer chains might act as
"energy sinks", and delay the onset of turbulence. The problem with this
scenario is that experiments show that the drag is reduced even in the tur-
bulent regime. Prof Dr. Pechhold speculated that the drag reduction might
be the result of polymer extension caused by shear flow. It is known that
polymer chains in a shear flow experience an instability at a critical
shear rate. Above the critical shear rate, the chains are almost fully
stretched. In extrusion experiments it has been shown that the polymer
chains will even break at sufficiently high shear rates. Prof. Bradley
suggested that high pressure studies of polymer drag reduction might be
helpful in elucidating the phenomenon, and that it might be possible to
build small toroidal cells which would make high pressure hydrodynamics
accessible to experiment.

SOLID HYDROGEN AT ULTRA HIGH PRESSURE

Isaac F. Silvera

Lyman Laboratory of Physics
Harvard University
Cambridge, MA 02138

INTRODUCTION

In recent years the diamond anvil cell (DAC) technique has been developed to statically compress matter, with pressures reaching into the hundreds of gigapascals (100 GPa = 1 megabar). One of the most interesting substances to study under these conditions is solid hydrogen, which is predicted to have a number of phase transitions, including the insulator-metal (IM) transition. At very high pressures of order 300-400 GPa, hydrogen is predicted to become a metallic atomic solid (Wigner and Huntington, 1935; Ceperley and Alder, 1987). However, it is expected that hydrogen first becomes a metal at somewhat lower pressures, within the molecular solid phase, by an electronic band overlap mechanism (Ramaker et al., 1975; Friedli and Ashcroft, 1977). Phase transitions in hydrogen have been studied by a number of methods, including Raman scattering and optical reflection and absorption. An unexpected new phase has been shown to exist for pressures above 149 GPa by study of the pressure-temperature phase line (Lorenzana, Silvera, and Goettel, 1989). This phase is called the hydrogen-A (H-A) phase and it is suspected, on a number of grounds that the H-A phase is metallic. However, at this time there is no direct evidence to support this interpretation. On the theoretical side calculations have predicted band gap closure, or metallization, at a pressure of 150-180 GPa for a structure with the hcp lattice with molecules oriented along the c-axis (Garcia et al., 1990). Again, uncertainties arise, as recent calculations find a different structure to have a lower energy and a larger gap and metallization pressure. In this article I shall discuss the recent rapid developments in this challenging area of research.

REVIEW OF MOLECULAR HYDROGEN

At low pressures and temperatures the ortho-para concentration of hydrogen plays a

Frontiers of High-Pressure Research, Edited by H.D. Hochheimer and
R.D. Etters, Plenum Press, New York, 1991

determining role in the structures and phases of solid hydrogen; this has been reviewed by Silvera (1980). The para-molecules are in the ground rotational state with non-degenerate rotational quantum number J=0, M=0 so that the orientational distributions of single molecules are described by the spherical harmonics $Y_{JM}=Y_{00}$, which are spherically symmetric. The ortho-molecules are in the J=1 states which have orientational distributions resembling a figure eight. In equilibrium with its vapor pressure hydrogen solidifies at about 14 K into an insulator with a large valence-conduction band gap of about 16 eV; the crystal structure is hcp. Pure para-hydrogen remains in the hcp phase down to T=0 K. This is called a symmetric hcp phase as the molecules themselves are in almost pure J=0 states which are essentially spherically symmetric. The pure ortho-hydrogen molecules are orientationally disordered in the hcp phase at higher temperatures; the solid has an interesting orientational order-disorder phase transition at lower temperatures. Ortho-para mixtures have an orientational order transition for ortho concentrations higher than 0.56 at zero pressure. This critical concentration for ordering has been found to decrease with increasing pressure (Silvera and Jochemsen, 1979).

Para-H_2 remains in the symmetric hcp phase at very high pressures. It is noted that the symmetry of hcp allows for an orientational ordering of molecules with a symmetry axis along the crystalline c-axis (Lagendijk and Silvera, 1981). However, this orientational order will be very small and would be zero if the ground state single molecule distributions were described by pure Y_{00} states. Due to the weak anisotropic intermolecular interactions there is a small admixture of higher-lying rotational states into the ground state, which otherwise would be a product state of Y_{00} wave functions. This admixture allows for a small amount of order which has not been experimentally detected, but is not symmetry forbidden. At very high pressures the anisotropic interaction potential becomes sufficiently large, relative to the splitting of the rotational states, so that the ground state distorts and is characterized by a strong admixture of Y_{2M} and higher rotational states. The lattice spontaneously orders at T=0 K into a structure which minimizes the anisotropic interaction energy. This transition is called the BSP (broken symmetry phase) transition and has been observed in deuterium at 28 GPa (Silvera and Wijngaarden, 1981) and hydrogen at 110 GPa (Lorenzana et al., 1990). However, the crystal structures have not been determined.

Pure ortho-hydrogen at zero pressure has a transition to an orientationally ordered phase at 2.8 K. This minimizes the electric quadrupole-quadrupole (EQQ) anisotropic interaction. The new structure is Pa3 which has molecules located on the sites of an fcc lattice and oriented along one of the body-diagonals to form a four-sublattice structure. The phase line was followed up to 5 kbar (0.5 GPa) by Silvera and Jochemsen (1979), who found the critical temperature to increase with density ρ as $\rho^{5/3}$, consistent with the density dependence of the EQQ interaction. This phase line has not yet been followed to the much higher pressures which are now available.

NEW PHASES AT MEGABAR PRESSURES

In the past few years a number of phase transitions have been reported in hydrogen and deuterium at ultra high pressures. In this section we discuss these results.

The Hydrogen-A Phase

In 1988 Hemley and Mao observed a discontinuous shift in the frequency of the Raman active vibron (internal vibrational mode). This was at a pressure of 150 GPa and a temperature of 77K. They explained the discontinuity as being caused by a phase transition; they proposed this as an extension of the phase line of EQQ orientational order, known from lower pressures and temperatures. Their studies covered a pressure extending up to 250 GPa and temperatures between 77 and 130 K. However, only a single pressure-temperature point was determined for the transition.

Lorenzana, Silvera, and Goettel (1989) studied the high pressure discontinuity in the vibron mode frequency in the same pressure-temperature regime and arrived at a very different result. They observed a phase line rising abruptly with an onset at P=149 GPa at T≈0 K, shown in figure 1. They found the phase line to be independent of ortho-para concentration to within their experimental uncertainty. Since such a line could not be an extension of the low pressure phase line (which depends strongly on ortho-para concentration), the transition must have been to a new, unexpected phase, which they named the hydrogen-A (H-A) phase.

Subsequently, Hemley and Mao (1989) concurred with Lorenzana et al's conclusion that the H-A phase was not related to the low pressure phase of quadrupolar orientational

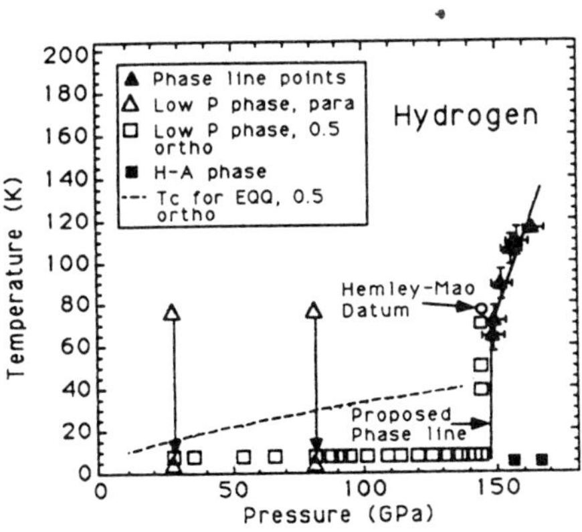

Fig. 1. The phase diagram of hydrogen (Lorenzana et al, 1989) showing the phase line for the H-A phase, rising sharply at a pressure of about 150 GPa. The solid symbols are points on the phase line or within the H-A phase; the open symbols are points in the low pressure phase. The dashed line is a theoretical curve representing the extension from low pressure of the phase line for EQQ ordering.

order. They observed a vibron discontinuity in deuterium, similar to that in hydrogen, but at a higher pressure. They argued on the basis of this higher critical pressure that the transition in hydrogen could not be due to orientational order. However, Silvera et al. (1991) have shown that their reasoning used some specious arguments. The importance of the observation in deuterium is that a phase transition similar to the H-A, which one could call the D-A transition, occurs in deuterium. However, to establish this unequivocally, the phase line should be determined in deuterium by means similar to the determination in hydrogen.

At the T=0 K onset of the phase transition the discontinuous shift in vibron frequency was approximately $100 \, \text{cm}^{-1}$, much larger than would be expected for an orientational order transition. It was suspected that the H-A phase might be the long sought IM transition in hydrogen. The phase-line was similar to what had been proposed earlier Silvera, 1989). More important, the decrease in the vibron frequency can be understood as a transfer of electrons from molecular bond states to crystalline band states. Since the interproton electronic charge density in hydrogen is responsible for the binding forces, if this charge density were reduced, it would result in a weaker bond and lower vibron frequency.

<u>Orientational Order at High Pressure</u>

There have been two reports of orientational order in hydrogen at high pressure. Lorenzana et al. (1990a) studied the pressure dependence of the J=0 $\rightarrow$ 2 roton spectrum in para-hydrogen at high pressures and low temperatures. At 110 GPa and 5 K they observed the collapse of the intensity of the roton bands. This occurred along with a small discontinuous shift in the vibron frequency. They interpreted these observations as a transition to the BSP in hydrogen, similar to the BSP observed earlier, at lower pressures, in deuterium. Unfortunately, due to technical experimental problems they only observed one point on this phase line, shown in figure 2. Moreover, the spectroscopic observations differ from those in deuterium in that the roton intensity collapsed, whereas in deuterium there was

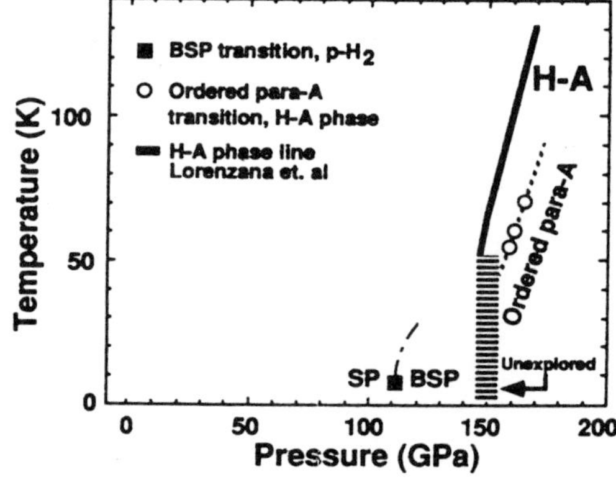

Fig. 2. The phase diagram of high pressure para-hydrogen showing a single point on the BSP phase line and a new phase of orientationally ordered hydrogen in the H-A phase region (after Lorenzana et al, 1990).

an abrupt change in line shape to a broad doublet. This transition merits further probing in the future.

In this same study Lorenzana et al. also found that the rotational intensity of para-hydrogen returned in the H-A phase. A study of the temperature dependence of the intensity of the rotational modes showed another abrupt change as temperature was increased. They were able to determine the P-T line shown in figure 2, which they reported as a phase line for para-hydrogen within the H-A phase. The nature of this phase is unknown.

Lorenzana et al. also studied the rotational Raman transitions in the H-A phase for equilibrium hydrogen at 77 K (50-50 ortho-para). They found an abrupt collapse (or severe broadening) of the rotational Raman intensity when they raised the temperature so that the sample departed from the H-A phase. Since this is the behavior expected when the molecules go from a state of order to disorder (in the disordered state the rotational lines undergo severe broadening), they interpreted this to mean that the H-A state can also be characterized by orientatiuonal order, whereas the low pressure or high temperature phase is disordered.

Hemley et al. (1990a) also studied the rotational Raman transitions as a function of pressure at 77 K. They reported no change in the intensity of the rotational transitions when entering the H-A phase. We point out here that the study by these researchers and by Lorenzana et al. were at different P-T points so that the measurements are not in conflict, but indeed show that this phase line is quite complicated and should be further studied in greater detail. In this research, Hemley et al. also studied the Raman active lattice phonon on the same P-T trajectory. When crossing the H-A line they found no discontinuity in the phonon frequency, in contrast to the discontinuity in the vibron frequency. Since the phonon is sensitive to the translational motions of the center of mass of the molecules this suggests that at the H-A phase transition the positions of the molecular centers do not undergo a distortion or a structural change so that the lattice of molecules remains hcp. We note that this does not mean that the H-A phase necessarily has hcp symmetry. This would depend on the state of order of the molecules. If the molecules are oriented along the c-axis then the hcp symmetry is conserved. A more complex ordering would lower the lattice symmetry.

The Critical Point in the H-A Phase Line

With increasing pressure the discontinuity in the vibron frequency at the transition was observed to go to zero. This was interpreted as a critical point termination on the H-A phase line (Hemley and Mao, 1990b; Lorenzana, Silvera, and Goettel, 1990b). Below the critical point the transition from one phase to the other is first order so that the vibron frequency has a discontinuous change; above the critical P-T point, the transition is continuous.

Silvera (1991) has shown how this might be understood in terms of an IM transition with two order parameters, one for the orientational ordering of the molecules and the other for the IM transition. The Hamiltonian consists of parts for the orientational degrees of freedom, the electronic degrees and a coupling between the orientational order parameter and the charge density (since the valence-conduction band separation or overlap is reduced by orientational order, the charge density will depend on the orientational order). The orientational states can be modeled by a three-states (for molecular orientation along x,y,z) Potts model. This results in a first order phase transition for orientational order; this in turn closes the gap so that that the charge density changes discontinuously. Lorenzana et al. showed how the spectroscopic observations might be interpreted in terms of an order parameter for the charge density of metallic hydrogen. However, as we shall see, due to recent experimental findings some uncertainty has developed for the support for the notion that the H-A phase is metallic, so that some of the results discussed above may be premature.

EXPERIMENTAL EVIDENCE FOR METALLIZATION OF HYDROGEN

In the past year and a half a number of optical experiments have been carried out to determine the nature of the H-A transition. Initially these experiments indicated that the H-A phase was the metallic molecular phase of hydrogen. However, in view of more recent experimental results some of the initial claims of direct evidence of metallization appear to have been lacking of critical consideration. Prior to the identification of the H-A as a unique high pressure phase, Mao and Hemley (1989) reported evidence for the metallization of hydrogen. They had pressurized hydrogen to 250 GPa at 77 K and suggested that it gradually became metallic above 200 GPa. The hydrogen was in the interstices of a compact of ruby powder and they reported darkening of the hydrogen at these extreme high pressures. This claim of metallization was criticized by Silvera (1990a). Optical data was inconclusive as the reflection measurements may have been consistent with a metallic behavior, while transmission was more characteristic of an insulator. Darkening itself at these high pressures is interesting, but not evidence that a material is metallic, and could easily be due to a narrowing of the band gap so that the materials behave as semiconductors. It has been suggested by Ruoff and Vanderborgh (1991) that ruby (chromium doped Al_2O_3) may react with hydrogen at very high pressures to liberate free aluminum. It is thus also possible that the observed darkening might be due to this effect.

<u>Dielectric Dispersion of Hydrogen</u>

Eggert, Goettel, and Silvera (1990) studied the pressure dependence of the dispersion of the index-of-refraction in hydrogen to 74 GPa. They related the dispersion to a "single oscillator model" (SOM) representing the valence-conduction band gap optical transition, which decreases with increasing pressure. They originally extrapolated the frequency of the single oscillator with pressure to find the value at which the oscillator frequency goes to zero. To the extent that the SOM is valid, this should identify the average direct gap. However, since the metallization in hydrogen is expected to occur at a lower pressure due to indirect band gap closure, a calculational step is required to determine the density or pressure for metallization. Hemley (1990c) suggested that it would be more appropriate to extrapolate the gap closure with density. When Eggert et al. did this they were able to determine by extrapolation and use of an electronic band structure calculation that the indirect band gap should close in the pressure region of 140 to 215 GPa, lending some indirect support to the proposal that the H-A was the metallic phase of hydrogen.

Hemley et al. (1991) as well as Garcia et al. (1991a) subsequently measured the dispersion of hydrogen to higher pressures. As expected it was found that the single oscillator frequency did not go to zero in the 150 GPa region. Hemley et al. have actually used dispersion data to determine single oscillator frequencies to pressures as high as 170 GPa. This is an inappropriate use of the model as it is not meaningful to use it in the region where the probing light frequency and the oscillator frequency are similar. Furthermore, whereas at low densities the joint density of states (JDS) for valence-conduction band transitions is expected to be quite narrow so that a SOM is meaningful, at high pressures the JDS becomes very broad so that the SOM loses usefulness (Garcia et al., 1991a).

Other criticisms can be mounted against the use of the SOM. First, we are apparently trying to determine the metallization pressure of hydrogen, with particular focus on the H-A phase. However, this is a first order phase transition and it is perhaps inappropriate to extrapolate low pressure results to high pressure for such a transition. For example, Ashcroft (1990) has shown that orientational order can reduce the band gap in an insulator. Garcia et al. (1990) have carried out a detailed calculation to show that this can be a very large effect (indirect bandgap closure at 180 GPa for an ordered lattice and 400 GPa for a lattice with spherical charge distributions). Since there is some evidence that the H-A phase is orientationally ordered and the lower pressure phase is not, then extrapolating data from the low pressure disordered phase should indicate the band gap closure for the disordered solid, which can differ substantially from that for the ordered solid.

<u>Optical Absorption and Reflection in Hydrogen at High Pressures</u>

The most rigorous means of establishing the IM transition is to measure the dc electrical conductivity and show that it remains finite in the limit that T$\rightarrow$0 K. This is difficult to do in a DAC at megabar pressures due to the small sample size (approximately 10-20 microns in diameter and 1-4 microns thick). Experimenters have opted to study the transition by optical means. In the metallic state the optical reflectance and absorption can be reasonably well described by a Drude free electron model, as has been demonstrated for the IM transition in xenon (Goettel et al., 1989). The Drude model is a three parameter model (plasma frequency ω_p, electron relaxation time τ, and $\Delta\varepsilon'$, the difference in the real part of the dielectric constant at the diamond sample interface). In this model the plasma frequency is proportional to the square route of the charge carrier density. The reflectance and absorption rise below the plasma frequency and fall to their high frequency values above. The reflectance approaches one as $\omega_\rho\rightarrow$0 and the absorption coefficient becomes very large. The details of the changes in the absorption and reflectance in the region of the plasma frequency depends on τ and $\Delta\varepsilon'$.

Mao, Hemley, and Hanfland (1990) studied the reflectivity of hydrogen in the visible and IR down to an energy of 0.5 eV and for pressures to 177 GPa at room temperature. At the highest pressures they observed a rising reflectivity at their low frequency spectral limit. This was interpreted as free carrier reflection. They fit the optical data to the Drude model to find ω_p and τ. The real part of the dielectric constant is not known at high pressures so they modelled this with $\Delta\varepsilon' = \varepsilon'_{hydrogen} - \varepsilon'_{diamond}$ being positive at their highest pressures. This was reasonable since the index of refraction of diamond, which has a value of 2.42 in the visible at zero pressure, is expected to change very little with pressure, and to actually decrease in the low pressure range. The index of hydrogen has been studied to 37 GPa (van Straaten et al., 1982; Shimizu et al., 1981), and is found to rise rapidly with pressure from its zero-pressure value of 1.13 and index matches diamond at 130 GPa (Hemley et al., 1991), so that for higher pressures, its index is greater than that of diamond. With these assumptions they could fit their data to a simple Drude model and plot their plasma frequency as a function of volume to determine that ω_p becomes zero at a pressure of 149 GPa. They presented this as direct evidence of the IM transition in hydrogen.

In their studies they loaded the samples with ruby powder, both for pressure determination and to suppress interference fringes which they felt hampered their measurements (interference fringes arise from the Fabry-Perot cavity formed by the two diamond culets; the ruby powder, interior to the cavity, shifts the phases of the multiply reflected light to destroy the coherence.) In one of their samples, at 161 GPa, there was only a small amount of ruby present and fringes were apparent throughout the frequency range of measurement.

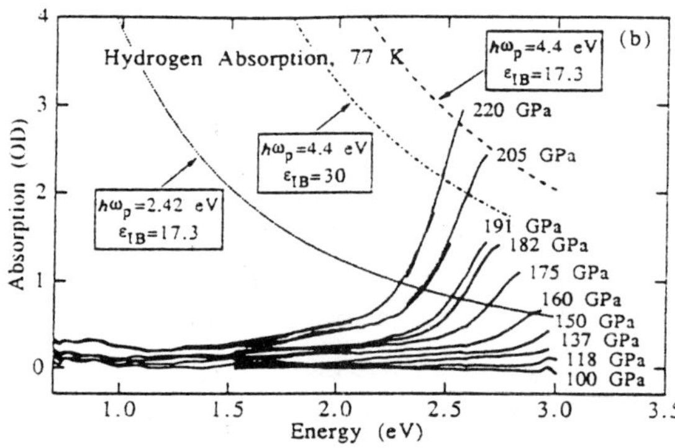

Fig. 3. The absorption coefficient as a function of photon energy in hydrogen for several pressures, going as high as 220 GPa (after Eggert et al., 1991). The rising absorption on the high energy side is due to the diamond anvils. The most striking aspect is the lack of an absorption edge at low energy, inferred from the reflectivity measurements of Mao et al. (1990). The drawn curves with rising edges at low energies are the calculated absorptions that should have been observed, using the Drude model with the parameters given by Mao et al, or extrapolations of their values to the higher pressures of these measurements.

Concurrent with the publication of Mao et al.'s (1990) results, Eggert et al. (1991) were studying a sample of hydrogen under pressures as high as 230 GPa. Their spectral range of study went down to a frequency of 0.7 eV, slightly higher than the lowest energy used by Mao et al. They measured both absorption and reflection, but <u>did not observe a rising Drude edge,</u> at temperatures which ranged from 77 to 295 K, as seen in figure 3. It should be remarked that these measurements were compatible with those of Mao et al. in the lower pressure region of overlap. However, Mao et al. presented specific fits to the plasma frequencies which could be extrapolated. Extrapolation of the results of Mao et al. to the high pressures of Eggert et al. implied that extremely strong absorption and reflection should be present, which was not observed. Such a result could arise from the samples of the two groups being different. For example it might be possible that hydrogen has a metastable insulating state. Since Eggert et al. prepared their sample at low temperatures and Mao et al. at room temperature, perhaps Eggert et al.'s sample was stuck in a metastable state. Another possibility could be that Eggert et al.'s sample had a much higher degree of crystalline disorder so that the electron relaxation time was substantially shorter, which could seriously shift the optical specrum, or even give rise to Anderson locallization. From the interpretational point of view, the model used by Mao et al. is highly oversimplified. It is believed that hydrogen in the metallic phase has a structure with the molecules on the sites of an hcp lattice. This will have a highly anisotropic fermi surface and different electron masses for motions along or perpendicular to the c-axis. Thus, a Drude model with tensor components for the electron and hole masses, plasma frequencies, etc. may well be necessary to understand the optical properties of this system.

To distinguish between these possibilities, Eggert et al. reanalyzed the data of Mao et al. for their 161 GPa sample, which demonstrated fringes. They showed that, with the absorption that could be calculated from the plasma frequency of Mao et al., the fringes

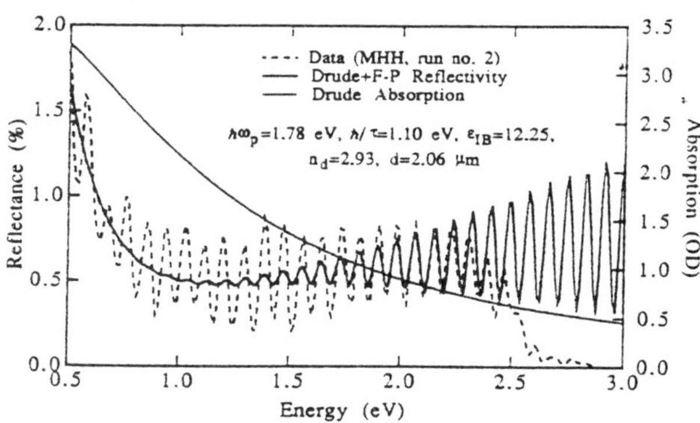

Fig. 4. The reflection of hydrogen at 161 GPa (after Eggert et al., 1991). The dashed curve is the data of Mao et al (1990). The solid, wiggling curve is a solution of the Fabry-Perot equation for an absorbing sample, using the absorption calculated from the parameters of Mao et al., shown by the light, solid curve. This shows that the reflectivity data is internally inconsistent with the absorption that is predicted from a Drude model fitting of the rising edge.

should be strongly attenuated due to free carrier absorption of the multiply reflected waves in the Fabry-Perot, as shown in Figure 4. Thus, the data of Mao et al. apparently was inconsistent with their interpretation in terms of a Drude model. As a result there was no optical evidence for the metallization of hydrogen. There remains the problem of understanding the reflection observed by Mao et al.

Ruoff and Vanderborgh (1991) have presented an interesting idea. They suggest that the ruby, which was packed with the hydrogen sample, may have undergone the chemical reaction, $Al_2O_3 + H_2 \rightarrow Al + AlOH$, so that a layer of free aluminum forms on the surface of the ruby. They estimate that this pressure dependent reaction might take place starting at 130 GPa, and that due to the dilution of Al in a dielectric medium the Drude parameters for Al would be shifted in a manner which would be compatible with the results of Mao et al. The fringes observed in the sample would thus arise from the open spaces in the sample and the reflectivity from the regions rich in ruby impurity. It would be very interesting experimentally to test this hypothesis of aluminization of the ruby.

In their first article Mao et al. stated that they had absorption data, supportive of their interpretation of metallization. Hanfland et al., (1991) have now published an article, presenting their absorption data. They find a very weak absorption coefficient which starts to rise at the lower end of their spectral range of study. This absorption is much weaker than would be expected from their original fit and Drude parameters, and is consistent with the data of Eggert et al., who did not observe a rising edge (but did not study the sample in this spectral region). However, Hanfland et al. again claim that their data is supportive of their interpretation of metallization of hydrogen, in terms of a Drude model. By making the assumption $\varepsilon'_{diamond} > \varepsilon'_{hydrogen}$, so that $\Delta\varepsilon'$ is negative, contrary to what seems to be the case, they find that the apparent optical absorption is much weaker. They now find new values for ω_p and τ, but do not give them, so that it is difficult to make a critical analyses here, and compare to the higher pressure results of Eggert et al. They also state that they can fit the observed optical absorption, but do not show the fit to the measured values.

Since this new claim is based on the assumption that $\Delta\varepsilon'$ is negative, it is useful to examine the dielectric properties of diamond and hydrogen in more detail. In an earlier paper, Eggert et al. (1990) tried to model the dielectric properties of diamond with a SOM and found that the index increased with pressure at high pressure. Syassen (1990) pointed out that the index of diamond actually decreased with increasing pressure at low pressure. Eggert et al. (1991) then used the more conservative low pressure value of the index of diamond. Surh et al. (1991) have calculated the stress dependence of the index of diamond using a stress tensor which simulates the stresses of a diamond culet in a DAC and find that the index remains below the zero pressure value of 2.41.

The index n of hydrogen has been measured to 37 GPa (van Straaten et al., 1982) and it would be useful to extrapolate these results to higher pressure. We present a means of doing this here. From the index of refraction one can determine the polarizability α as a function of density, ρ, using the Lorenz-Lorentz relation

$$(n^2-1)/(n^2+1) = (4\pi/3)(N_A/V)\alpha(\rho)$$

where N_A is Avagadro's number and V is the volume. To extrapolate to high pressure it would be useful to have a megabar pressure point. Since it has been observed that hydrogen and diamond index match at 130 GPa we shall make the assumption that hydrogen as an index of 2.41 at this pressure, to supplement the data set. These values of the polarizability are then least squares fit to a power series expansion of the form

$$\alpha = \alpha_0(1 - \alpha_1\rho - \alpha_2\rho^2),$$

giving values α_0=8.51107x10^{-25} cm^3, α_1=0.31806, and α_2=0.48769. This form of the polarizability can be used to generate the index as a function of density, using the Lorenz-Lorentz relation, again. The results are shown in figure 5. In all of this work the equation of state of van Stratten and Silvera (1988) has been use for the pressure-density conversions. In the absence of very high pressure data, this appears to be the best estimate of how the index of refraction behaves. One of the most important inputs is the observation of index matching at 130 GPa. Considering the low pressure behavior, it seems rather unphysical that the two indexes would match at this pressure and then "repell" each other, as is implied by the non-crossing assumption of Hanfland et al. It might be possible that at higher stresses, behavior of diamond takes an unusual change, but at this time there is no supportive evidence of such a change in the 150 GPa pressure region. As a result of these considerations we do not believe that there is optical evidence for the metallization of hydrogen.

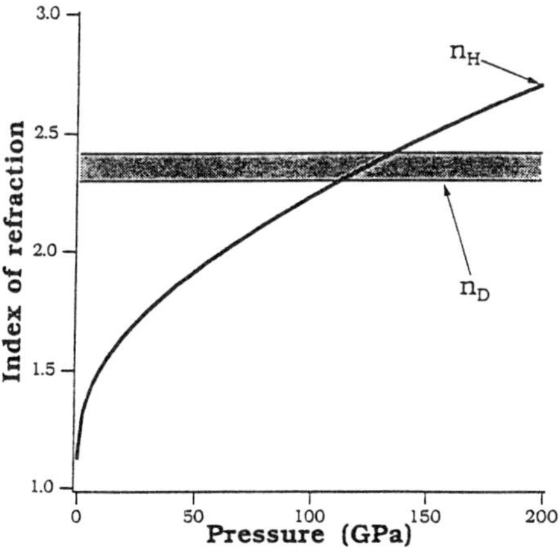

Fig. 5. The index of refraction of diamond and hydrogen using a semi-empirical approach to extrapolate hydrogen to high pressure. The index of diamond is expected to fall within the cross hatched region, according to calculations.

The theoretical picture also has some interesting twists. Earlier calculations (Ramaker et al., 1975; Friedli and Ashcroft, 1977) of the pressure for the indirect band gap closure used the Pa3 structure. Critical pressures in the 150-250 GPa region were given; however, techniques such as the local density approximation (LDA) were used. The LDA is known to be quite accurate for calculating ground state properties, such as stable structures, but it predicts gaps about a factor of two smaller than experiment. In the use of the LDA to determine stable structures, the main limitation is that the results give the energy of the structure which is input; this is not necessarily the lowest energy structure which exists in nature. It is thus necessary to calculate for a number of structures and find the one with the lowest energy. This does not guarantee that the correct structure has been found as it may not have been considered.

Barbee et al. (1989) used the LDA and found that a structure, hcp with molecules oriented along the c-axis, had a much lower crystal energy (more favorable structure) than Pa3, and had a critical pressure of about 60 GPa, using the LDA. Garcia et al. (1990) did a semi-empirical calculation for the hcp-c (molecules oriented along the c-axis) structure, to correct for the LDA deficiencies and found a critical pressure of 180 (20) GPa. Garcia et al. also did a detailed calculation for the hcp lattice using spherical charge distributions for the hydrogen molecules to simulate orientational disorder. This increased the gap closure pressure from 180 to 400 GPa, so that it would seem likely that metallization and orientational order-disorder are highly correlated.

Chacum and Louie (1991), have carried out a calculation of the band structure of hcp hydrogen, to improve on the results of Garcia et al., using the G.W. technique, known to give accurate gap results. They find a metallization pressure of 150 GPa, coinciding with the pressure of the H-A transition. More recently Kaxiras et al. (1991) have reported an orientational structure on the hcp lattice which differs from c-oriented and has a lower crystal energy but a larger gap. This has been confirmed by Garcia et al. (1991b). Thus, at this time the theoretical expectations for the hydrogen metallization pressure have come a long way, but there is no clear picture as to what the structure of metallic molecular hydrogen may be. Important problems which have not been carefully incorporated into the electronic structure calculations are the orientational distributions of the molecules, which seems to have profound effects and the zero-point motion of the molecules and the atoms.

CONCLUSIONS

Although there has been great progress in the last few years in the study of the IM transition in hydrogen, at this time both the experimental and theoretical situations are unclear. There are a number of possibilities. It may be that the H-A phase of hydrogen is metallic, but the experiments to date have been inadequate. It would be useful to extend the spectral range of optical study to lower energies by an order of magnitude. Another possibility is that hydrogen may not enter a molecular metallic state, and simply go directly from the insulating molecular state to an atomic metallic state at a higher pressure than has yet been attained experimentally. If so, what is the nature of the H-A phase?

One possibility is that hydrogen is undergoing some form of frustration. At low density the anisotropic pair interaction is dominated by the EQQ term which has the lowest energy for the T configuration (the mutual orientations of the molecular axes form a T). Calculations of pair interactions at short range corresponding to high density (Ree and Bender, 1979) show that in this regime the X configuration becomes favorable. Thus, one might imagine that at low density the Pa3 structure, which minimizes the EQQ interaction is favored, but at high density another structure for orientational order has lower energy. In the transition region the lattice becomes frustrated and no order occurs at all until one of the interactions becomes dominant. Thus, perhaps H-A is a purely orientationally ordered phase. It is interesting to note that using the potential of Ree and Bender at densities corresponding to 150 or so GPa, Silvera (1989) found a structure with a lower energy than Pa3 and hcp-c; this structure is the one recently proposed by Kaxiras et al. as having a lower energy than hcp-c.

Another possible explanation comes from a recent study by Freiman et al (1991). They considered the phase diagram for orientational ordering of a quantum crystal of diatomic molecules with an orientation hindering potential, as a function of the ratio of the strength of this potential to the molecular rotational constant B. They found that orientational order exists for small and large values of this ratio, but at intermediate values the critical temperature goes to zero and there is no orientational order. In hydrogen as the density is increased this ratio starts out very small and increases in value. Thus, one might anticipate a reentrant behavior where at $T=0$ K ortho-hydrogen is ordered at low density and with increasing density it disorders and then orders again. Perhaps there is a relationship to the H-A transition. A counter arguement is that the D-A phase occurs at a slightly higher pressure than the H-A phase. Yet, the rotational constant B of deuterium is half that of hydrogen, whereas the anisotropic hindering potentials are essentially the same at a given density. Thus, one would expect the behavior predicted by Freiman et al. to occur at very different pressures for H_2 and D_2. Clearly it is important to have more detailed experimental studies in this high density region.

ACKNOWLEDGMENTS

I would like to thank the U.S. Air Force Phillips Laboratory Grant F04611-89-K-0003, for support of this research.

REFERENCES

Ashcroft, N. W., 1990, *Phys. Rev. B*, 41:10963.
Barbee, T. W., III, Garcia, A., Cohen, M. L., and Martins, J. L., 1989, *Phys. Rev. Lett.* 62:1150.
Ceperley, D. M. and Alder, B. J., 1987, *Phys. Rev. B* 36:2092.
Chacum, H., Louie, S. G., *Phys. Rev. Lett.* 66:64.

Eggert, J. H., Goettel, K. A., and Silvera, I. F., 1990, High pressure dielectric catastrophe and the possibility that the hydrogen-A phase is metallic, *Europhys. Lett.* 11:775; Addendum, *Europhys. Lett.* 12:381.

Eggert, J. H., Moshary, F., Evans, W. J., Lorenzana, H. E., Goettel, K. A., and Silvera, I. F., 1991, Absorption and reflectivity in hydrogen up to 230 GPa: implications for metallization, *Phys. Rev. Lett.* 66:193.

Freiman, Y. A., Sumarkov, V. V., Brodyanskii, A. P., and Jezowski, A., 1991, preprint.

Friedli, C. and Ashcroft, N. W., 1977, *Phys. Rev. B* 16:662.

Garcia, A., Barbee, T. W., III, Cohen, M. L., and Silvera, I. F., 1990, Bandgap closure and metallization of molecular solid hydrogen, *Europhysics Lett.* 13:355.

Garcia, A., Cohen, M. L., Moshary, F., Eggert, J. H., Evans, W. J., Goettel, K. A., and Silvera, I. F., 1991a, to be published.

Garcia, A., Chacum, H., Louie, S. L., and Cohen, M. L., 1991b, private communication.

Goettel, K. A., Eggert, J. H., Silvera, I. F., and Moss, W. C., 1989, Optical evidence for the metallization of xenon at 132(5) GPa, *Phys. Rev. Lett.* 62:665.

Hanfland, M., Hemley, R. J., and Mao, H. K., 1991, *Phys. Rev. B* 43:8767.

Hemley, R. J. and Mao, H. K., 1988, *Phys. Rev. Lett.* 61:857.

Hemley, R. J. and Mao, H. K., 1989, *Phys. Rev. Lett.* 63:1393.

Hemley, R. J., Mao, H. K., and Shu, J. F., 1990a, *Phys. Rev. Lett.* 65:2670.

Hemley, R. J. and Mao, H. K., 1990b, *Science,* 249:391.

Hemley, R. J., 1990c, private communication.

Kaxiras, E., Broughton, J., and Hemley, R. J., 1991, preprint.

Lagendijk, A., and Silvera, I. F., 1981, Roton softening in the solid hydrogens, *Phys. Rev. Lett.* 84A:28.

Lorenzana, H. E., Silvera, I. F., and Goettel, K.A., 1989, Evidence for a structural phase transition in solid hydrogen at megabar pressures, *Phys. Rev. Lett.* 63:2080.

Lorenzana, H. E., Silvera, I. F., and Goettel, K.A., 1990a, Orientational phase transitions in hydrogen at megabar pressures, *Phys. Rev. Lett.* 64:1939.

Lorenzana, H. E., Silvera, I. F., and Goettel, K. A., 1990b, The order parameter and a critical point on the megabar pressure hydrogen-A phase line, *Phys. Rev. Lett.* 65:1901.

Mao, H. K. and Hemley, R. G., 1989, *Science,* 244:1462.

Mao, H. K., Hemley, R. J., and Hanfland, M., 1990, *Phys. Rev. Lett.* 65:484.

Ramaker, D. E., Kumar L., and Harris, F. E., 1975, *Phys. Rev. Lett.* 34:812.

Ree, F., and Bender, C. F., 1979, *J. Chem. Phys.* 71:5362.

Ruoff, A. L. and Vanderborgh, C. A., 1991, *Phys. Rev. Lett.* 66:739.

Silvera, I. F., and Jochemsen, R., 1979, Orientational ordering in solid hydrogen: dependence of critical temperature and concentration on density, *Phys. Rev. Lett.* 43:377.

Silvera, I. F., 1980, The solid molecular hydrogens in the condensed phase: Fundamentals and Static Properties, *Rev. Mod. Phys.* 52:393.

Silvera, I. F. and Wijngaarden, R. J., 1981, New low temperature phase of molecular deuterium at ultra high pressure," *Phys. Rev. Lett.* 47:39.

Silvera, I. F., 1989, The phase diagram and excitations in solid hydrogen: prospects for metallization, *in*: "Simple Molecular Systems at Very High Density," Polian and P. Loubeyre, eds., Plenum Press, New York.

Silvera, I. F., 1990a, *Science,* 247:863.

Silvera, I. F., Eggert, J. H., Goettel, K. A., Lorenzana, H.E., 1990b, Towards metallic hydrogen at high pressure; Pucci, R., Picaito, G., 1991, Molecular Systems under Pressure, eds., North Holland Publ. Co., Amsterdam, 181.

Silvera, I. F., 1991, The insulator-metal transition in molecular hydrogen and the three-state potts model, *Physica B* 169:551.

Shimizu, H., Brody, E. M., Mao, H. K., and Bell, P. M., 1981, *Phys. Rev. Lett.* 47:128.

Surh, M., Louie, S., Cohen, M. L., 1991, to be published.

Syassen, K., 1990, private communication.

Van Straaten, J. , Wijngaarden, R.J., and Silvera, I. F., 1982, The low temperature equation of state of molecular hydrogen and deuterium to 0.37 megabar: Implications for metallic hydrogen, *Phys. Rev. Lett.* 48:97.

Van Straaten, J., and Silvera, I.F. , 1988, Equation of state of solid molecular H_2 and D_2 at 5 Kelvin, *Phys. Rev. B* 37:1989.

Wigner, E. P. and Huntington, H. P., 1935, *J. Chem. Phys.* 3:764.

DENSE HYDROGEN AND ITS STATES OF ORDER

N. W. Ashcroft

Laboratory of Atomic and Solid State Physics
Cornell University, Clark Hall
Ithaca, NY, U.S.A.

INTRODUCTION

The system with which we are concerned, a dense neutral canonical assembly of protons (masses m_p, spin 1/2, coordinates $\vec{r}_{1p},\ldots,\vec{r}_{np}$) and electrons (masses m_e, spin 1/2, coordinates $\vec{r}_{1e},\ldots,\vec{r}_{Ne}$), has a remarkably simple and symmetric Hamiltonian. If $\hat{\rho}_\alpha^{(1)}(\vec{r}) = \sum_{i=1}^{N} \delta(\vec{r} - \vec{r}_{i\alpha})$ (α = e, or p) is the one-particle density for either Fermion system, and

$$\hat{\rho}_{\alpha\alpha'}^{(2)}(\vec{r},\vec{r}') = \hat{\rho}_\alpha^{(1)}(\vec{r})\hat{\rho}_{\alpha'}^{(1)}(\vec{r}) - \delta_{\alpha\alpha'}\delta(\vec{r} - \vec{r}')\hat{\rho}_\alpha^{(1)}(\vec{r}) \tag{1}$$

the corresponding two-particle density, then with $v_c(r) = e^2/r$ the fundamental Coulomb interaction (and with $Z_p = 1$, $Z_e = -1$), this Hamiltonian is

$$\hat{H} + \sum_\alpha \left\{ \hat{T}_\alpha + \frac{1}{2}\sum_{\alpha'} \int_\Omega d r \int_\Omega d r' \, v_c(\vec{r} - \vec{r}')Z_\alpha Z_\alpha' \hat{\rho}_{\alpha\alpha'}^{(2)}(\vec{r},\vec{r}') \right\} \tag{2}$$

where Ω is the volume of the assembly of particles and $\hat{T}_\alpha = \sum_i - \hbar^2 v_{i\alpha}^2/2m_\alpha$ their kinetic energies. For deuterium the mass m_p is replaced by $m_d\,(\sim 2m_p)$ and the spin by unity (the system is then a mixture of Fermions and Bosons). It might already be enquired whether the evident simplicity of $\hat{H}$ can be explored in a quite general fashion; we will return to this question below.

Given (2) the link to the thermodynamic functions is through the partition function

$$Z = Tr \exp(-\beta\hat{H}) \, , \qquad (\beta = 1/k_B T) \tag{3}$$

where the trace is over the states of (2) arising from the solution of a many-body Schrödinger equation with boundary conditions reflecting the volume Ω. But before discussing these functions it is necessary to examine the physical _nature_ of the states that enter into the trace. It is here that "states of order," can be endowed with especial meaning insofar as the hydrogen problem is concerned.

ORDER WITHIN BORN-OPPENHEIMER SEPARATION

To begin with it is usual to proceed on this issue by exploiting the very significant difference in time scales implied by the mass difference

Frontiers of High-Pressure Research, Edited by H.D. Hochheimer and
R.D. Etters, Plenum Press, New York, 1991

of the two fermion systems ($m_p/m_e \sim 2 \times 10^3$). Thus, it is common, but in fact no longer essential[1] (see the section entitled "Band Overlap and Random Potentials"), to invoke the adiabatic, or Born-Oppenheimer approximation, where the electronic part of the problem is solved in principle for essentially static values of the proton coordinates $\{\vec{r}_{jp}\}$. The resulting form of the combined wave-function then leads to

$$Z = Z(\Omega, T) = Tr_p\left\{Tr_{e(p)}e^{-\beta\hat{H}}\right\} \tag{4}$$

where the electronic trace $Tr_{e(p)}$ is over states determined with fixed proton positions, these later being freed to complete the overall trace (in the proton-dynamical problem). A reinterpretation of (4) then leads to an effective Hamiltonian for proton motion

$$H_{eff}\left(\{\vec{r}_{jp}\}, \beta, \Omega\right) = -k_\beta T \log Tr_{e(p)}e^{-\beta\hat{H}} \quad . \tag{5}$$

It is to be noted that for the wide band systems we are dealing with, the *volume* (or *state*) dependence of H_{eff} is fundamental; it is intrinsic to an underlying <u>electron</u> problem which is sensitive to an overall boundary condition established in the volume Ω. If this electron system, by choice of thermodynamic conditions, can be taken close to its ground state, then (5) simplifies to

$$H_{eff}\left(\{\vec{r}_{jp}\}; \Omega\right) = \hat{T}_p + v\left(\vec{r}_{1p}, \ldots, \vec{r}_{Np}; \Omega\right) \tag{6}$$

where v is a many-site potential energy whose form is far from trivial, though approximations to it can be introduced that are intuitive, at least at low and moderate densities.

For an arbitrary, but fixed arrangement of protons, the electron trace required in the construction of (5) or (6) is, by itself, a formidable problem in the full many-body sense. As has been emphasized earlier[2], the electron structure is sensitive to each and every arrangement in the configuration space of the protons; these arrangements are both complex and (given the simplicity of (2)) even surprising, as will be seen below. The problem is made somewhat more tractable by appealing to further approximation, depending on purpose. For example, if the intent is to discuss the entirely fundamental change in electron structure that can accompany global variations in density, the model of Mott,[3] originally introduced in 1949, can be invoked. Here the protons are taken as fixed, and moreover, are assumed to occupy the sites of a perfect *Bravais* lattice ($\vec{r}_{ip} \equiv \vec{R}_{ip}$); the lattice constant of this crystal is then expanded or contracted to achieve any required change in density. Mott's Hamiltonian is just a special case of (2), namely

$$\hat{H} = \hat{T}_e + \frac{1}{2}\sum_{\alpha\alpha'} \int_\Omega d\vec{r} \int_\Omega d\vec{r}' \, v_c\left(\vec{r} - \vec{r}'\right) Z_\alpha Z_{\alpha'} \hat{\rho}^{(2)}_{\alpha\alpha'}\left(\vec{r}, \vec{r}'\right) \tag{7}$$

but with the simplification that

$$\hat{\rho}^{(1)}_p(\vec{r}) \equiv \rho^{(1)}(\vec{r}) \equiv \sum_{j=i}^{N} \delta\left(\vec{r}_{jp} - \vec{R}_p\right) \quad , \tag{8}$$

and with the $\vec{r}_{jp}$ entering as c-numbers (the lattice sites).

At very low densities (i.e., at large lattice constants a, or at large values of the standard electron spacing parameter r_s defined by $(4\pi/3)r_s^3 a_o^3 = \Omega/N$) it is physically clear that the electron problem cannot depart significantly from a periodic arrangement of hydrogen atoms.

116

Nevertheless, on account of fluctuating multipole interactions[4] arising in the electron density $<\rho_c^{(1)}(\vec{r})>$, the subsequent electron trace leads to an overall energy which is negative and approximately, per proton, $-(1 + \gamma\alpha^2/r_s^6)$ where $\gamma = 27A_6/128\pi^2$ with A_6 the sum of inverse sixth powers over a cubic lattice (here a simple cubic) with unit lattice constant, and α the polarizability (in cubic Bohrs) of the hydrogen atom. Though (7) and (2) are independent of spin, the states do not necessarily share this property; thus at low densities a simple spin-Hamiltonian argument[5] shows that the atomic arrangement is antiferromagnetic in electronic spin. The system is clearly an insulator. But at high density (small values of r_s) matrix elements of the attractive electron lattice interaction (i.e., $(N/\Omega)4\pi e^2/K^2$, where K is a reciprocal lattice vector of the chosen lattice) can be made an ever declining fraction of a paramagnetic band-width $(9\pi/4)^{2/3}(1/r_s^2)$ $(e^2/2a_0)$ simply by choosing r_s to be sufficiently small $(r_s \sim 1)$. In this limit a Bloch-Wilson description must eventually hold; and since in simple structures there is a single-electron per unit cell, the system will be found in a highly conducting state. This picture assumes that electron-electron interactions, so crucial in the low density phase, are now playing a relatively minor role.

Between the two limits there is necessarily a metal-insulator transition, a quite fundamental change, originally estimated by Mott to occur for the assumed Bravis lattice case at

$$r_s \approx 2.48 \tag{9a}$$

or

$$(N/\Omega)^{1/3}a_o = 0.25 \ . \tag{9b}$$

The reader is referred to the recent revision[6] of Mott's monograph for a review of the theory of the metal-insulator transition based on (7). For the present it suffices to remark that by actually allowing the protons to relax under the action of (6), and hence to move as necessary *off* the sites of a Bravais lattice, the physics is changed radically and the problem of real physical hydrogen is quite different. There will eventually be a metal-insulator transition, but it is likely to have a completely different character. And electron-electron interactions can also play a quite different role, though still a decreasing one at very high densities where, eventually, a monatomic arrangement is predicted to occur.

A second example of significant approximation in the task of actually carrying out the electron trace in (4), is to invoke from the beginning a model of independent electrons, and this is again made tractable by the assumption of a crystalline arrangement, though the protons are not necessarily arrayed on the sites of a Bravais lattice. Indeed, let there be an m-atom basis in a structure with lattice vectors $\{\vec{R}\}$; then the Bloch-problem is defined by the one-electron Hamiltonian

$$h = \frac{-\hbar^2}{2m}\nabla^2 + \sum_{\vec{R}_i} \sum_{\ell=1}^{m} v(\vec{r} - \vec{R}_i - \vec{b}_\ell) \tag{10}$$

where $v(\vec{r})$ is an optimum one-electron potential constructed by density-functional techniques, or their variants. For a variety of chosen structures $\{\vec{R}\}$ and cell contents $\{\vec{b}\}$, hydrogen has been shown[7] to persist in a filled-band (insulating) state up to quite high densities. The associated structures correspond to arrangements in which the protons are in paired-states, as might be readily inferred from the nature of the vapor phase. However, it is also known[7,8] that while this essential pair structure is retained, there can be a transition to a band-overlap state, the phases on either side of this transition being by assumption in fully paramagnetic states. Band overlap is projected by such calculations to occur at densities corresponding to values of $r_s \sim 1.5$ (and pressures in the vicinity of 150 GPa) but the precise density is expected to be a reasonably sensitive function of structural choice. In this density range there are indeed signs of optical activity suggesting that low energy electronic excitations are taking place. The evidence for band overlap is otherwise fairly controversial at this time, and is reviewed by Silvera.[9]

The persistence of pairing to extremely high densities, also confirmed by Horsfield, Moulopoulos, and Ashcroft,[10] is perhaps not surprising given the known strength of the covalent bond in H_2 (or D_2). But the nature of the paired state at high density is not completely intuitive. To illustrate this more clearly, consider the case of hydrogen at approximately five-fold solid compression ($r_s = 1.83$), a density that corresponds to a known experimental condition. Now, if the discrete proton charge were to be replaced by a *uniformly charged* but rigid continuum ($\hat{\rho}_p^{(1)}(r) = \rho_b = N/\Omega$), then (2) becomes

$$\hat{H} = \hat{T}_e + \frac{1}{2} \int_\Omega d\vec{r} \int_\Omega d\vec{r}\,' \ v_c(\vec{r} - \vec{r}\,')\left\{\hat{\rho}_{ee}^{(2)}(\vec{r},\vec{r}\,') - 2\hat{\rho}_e^{(1)}(\vec{r})\rho_b + \rho_b^2\right\} \tag{11}$$

and this is immediately recognized as the standard interacting-electron gas problem. At low enough values of density (11) is known to lead to crystalline states (the broken symmetry phase normally described as the Wigner crystal). However, at this chosen value of r_s ($r_s \approx 1.83$), which happens also to coincide with the average electron density found in Be (one of the highest electron density simple metals found under normal conditions) the electron system is in a continuous phase, is deeply *metallic*, and is certainly paramagnetic.

The change from (11) to (2), at fixed average density (say $r_s = 1.83$), is of course hardly a benign or trivial one; the continuum charge is aggregated into units of the fundamental charge, and each is endowed with a mass m_p and is also, of course, fully dynamic. Preceding such changes, the system is highly metallic; succeeding such changes, experiment declares that the system is insulating (even though the electron bands are already approximately 15 eV wide) and the discrete charges exhibit order of a strongly pairing character. As noted, because the H_2 bond in the isolated molecule requires more than 2 eV per proton to sever it, this circumstance may perhaps seem not so surprising. The pairs guarantee an even number of electrons per primitive cell, and then, by the Bloch-Wilson argument, a possible insulator in a non-band overlap state providing the system is truly crystalline.

Far more intriguing, however, is the experimental observation[11] that the pairs may persist in rotational states--as found in gas and liquid phases--and persist to remarkably high densities. The implications of this observation for the electronic problem are both immediate, and also rather severe. The rotational energy is typified by the rotational constant $\hbar^2/2m_p d^2 = (m_e/m_d)(a_o/d)^2 Ry$ or approximately 0.007 eV. That rotational states are observed at all in turn implies that such possible barriers to rotational motion as will inevitably be present (in a crystalline environment) must also be of this size or smaller. These barriers represent the scale of energy variation that accompanies the selection or placement according to angular disposition, of a pair within a unit cell. A faithful representation of such energies, such as might be hoped for, in principle by a total energy calculation, will therefore be required to be accurate, absolutely, to the same level. This is a very significant challenge to current computational methods.

For free and independent rotors it is reasonable to classify the states according to the standard representations of the rotation group. In a crystalline environment, however, crystal field splitting of such states must be expected; a splitting (or mixing) of the rotational states then ensues. Thus as emphasized by Silvera,[11] a system of pairs in putative $J = 0$ states will have mixed into them components of $J = 2$ (and above) which grow with increasing density. It is important to emphasize, however, that for the case of hydrogen, the very existence of predominantly $J = 0$ states is tied to an important element of self-consistency, since a given pair, surrounded by a cage of pairs in similar states will experience a crystal field splitting that, owing its origin to largely spherical objects, possesses far less strength and variation than might be expected from such objects in oriented (say $J = 2$) states. In such an environment, the pair will tunnel through a small but self-consistently determined rotational barrier which nevertheless is expected to rise with density and must *eventually* force order, in an orientational sense, on the pairs. There is no sound argument, so far, that this orientational order would be rigorously periodic.

From the standpoint of proton order, this is a simple enough picture. However, its consequences for the electronic problem are again quite profound, as can be seen already at the single-particle level by returning to (10 and rewriting it in a form appropriate to a rotational state. The meaning to be attached to an assembly of pairs in $J = 0$ states, from the electronic point of view, is that there is a uniform <u>random</u> probability of a given pair axis taking on any orientation: there is no average inter-site correlation. It follows that the equivalent of (10) is[2]

$$h = \frac{-\hbar^2}{2m}\nabla^2 + \sum_{\vec{R}} \sum_{l=1}^{m} v\left(\vec{r} - R - \vec{b}_{\vec{R},l}\right) \tag{12}$$

where the m basis vectors are now grouped into m/2 pairs according to the choice of cell. Since the axes of these pairs are uniformly distributed, it follows that although the $\vec{R}$ occupy the sites of a lattice, the problem is obviously no longer periodic for electrons because of the random content of each unit cell. However, a reference periodic problem can easily be constructed, as follows: first rewrite (12) as

$$h = \frac{-\hbar^2}{2m}\nabla^2 + \frac{1}{\Omega} \sum_{\vec{R},\vec{k}} e^{-i\vec{k}\cdot\vec{r}} \sum_{l=1}^{m} e^{-i\vec{k}\cdot\vec{b}_{R,l}} v(\vec{k}) e^{-i\vec{k}\cdot\vec{r}} \tag{13}$$

where $v(\vec{k})$ is the Fourier transform of the self-consistent electron-proton interaction. Next, introduce h_o as a single-particle Hamiltonian constructed from (13) by averaging the pair orientations over uniform distributions: this leads to

$$h = h_o + \nabla V$$

where

$$\nabla V = \frac{1}{\Omega} \sum_{\vec{R},\vec{k}} e^{-i\vec{k}\cdot\vec{R}} \sum_{l=1}^{m} \left(e^{-i\vec{k}\cdot\vec{b}_{r,l}} - \left\langle e^{-i\vec{k}\cdot\vec{b}_{R,l}}\right\rangle\right) v(k) e^{-i\vec{k}\cdot\vec{r}} \tag{14a}$$

and

$$h_o = -\frac{\hbar^2}{2m}\nabla^2 + \frac{1}{\Omega}\sum_{\vec{R},\vec{k}} e^{-i\vec{k}\cdot\vec{R}} \sum_{\ell=1}^{m} \left\langle e^{-i\vec{k}\cdot\vec{b}_{R,\ell}}\right\rangle v(k)e^{-i\vec{k}\cdot\vec{r}}$$

$$= \frac{-\hbar^2}{2m}\nabla^2 + \frac{N}{\Omega}\sum_{\vec{K}} S(\vec{K})v(\vec{K})e^{-i\vec{K}\cdot\vec{r}}$$

(14b)

which clearly defines a periodic problem. In (14) $S(\vec{K})$ is the structure factor <u>per-proton</u> of the rotationally averaged structure (for example, for m = 2, we find S(K) = sin Kd/Kd, where 2d is the static average spacing between two protons). It is important to note that ΔV is not necessarily a small perturbation, a point of some consequence, as we will see, in the context of possible band overlap transitions.

LATTICE STRUCTURE

From the standpoint of achieving states of protonic order that are favorable in the sense of electronic energies, it has already been noted[8] for the three-dimensional case that bandgaps can be maximized (and electrons placed at higher state density but at <u>lower</u> energies) by adopting a virtual rotational crystal as the preferred structure. Within a strictly crystalline context, equation (3) obviously requires that a crystalline *approximant* be found with a <u>large</u> unit cell whose periodically repeated basis vectors are found directed with reasonable uniformity.[2] Given some guidance from experiment on the likely choice of structure, i.e., a specification of the set $\{\vec{R}\}$ for dense hydrogen, this principle can actually be extended. For, assume for the moment that the structure is hexagonal, as interpretations of the scattering probes have proposed may be the case.[13] This structure naturally impels us to think of layered arrangements, and indeed such structures are favorable from the electronic standpoint, as will now be argued:

We begin with an example containing four protons in a primitive hexagonal cell (primitive vectors $\vec{c}\cdot\vec{a}_1,\vec{a}_2$, with $\vec{c}\cdot\vec{a}_i = 0$, and $\vec{a}_1\cdot\vec{a}_2 = \frac{1}{2}a^2$). Let the basis vectors be $\pm\vec{d}_1$, and $\vec{b}\pm\vec{d}_2$. Then the structure factor per atom is

$$S(\vec{K}) = \frac{1}{2}\left(\cos \vec{K}\cdot\vec{d}_1 + e^{-i\vec{K}\cdot\vec{b}}\cos \vec{K}\cdot\vec{d}_2\right)$$

(15)

from which it follows that for $\vec{K} = \frac{2\pi}{c}(0,0,1)$

$$S(\vec{K}) = \frac{1}{2}\left(\cos \vec{K}\cdot\vec{d}_1 + e^{i2\pi b_z/c}\cos \vec{K}\cdot\vec{d}_2\right)$$

(16)

$$= \frac{1}{2}\left(\cos \vec{K}\cdot\vec{d}_1 - \cos \vec{K}\cdot\vec{d}_2\right)$$

providing only that $b_z = c/2$, i.e., that the vector $\vec{b}$ terminates midway between planes separated by c. If, further, the <u>pairs</u> themselves are in planes perpendicular to $\vec{c}$, then $S(\vec{K} = \frac{2\pi}{2}(0,0,1)) = 0$ exactly as is claimed in experiment. Note that the vanishing of the intensity of such a line in the scattering probes then merely requires $b_z = c/2$: it does not by itself fix the structure as hexagonal <u>close-packed</u>. Indeed, consider the structure of Fig. 1. Two layers are shown, where proton pairs are inclined, in a periodic way, within planes and the entire structure repeats parallel to the page. This structure might naturally be viewed as hexagonal, with a 16-point basis. But notice that the choice of a rather different cell leads to another class entirely. For example, also shown is the base of unit cell which is <u>not</u> hexagonal, but is *orthorhombic*, and which contains an 8-point basis. This alternative viewpoint is physically appealing simply because it is known already that the orthorhombic structure is consistent with the existence of a <u>paired metallic state</u>: the classic example is gallium[14] where the ratio "c/a" $\approx \sqrt{3}$, very nearly.

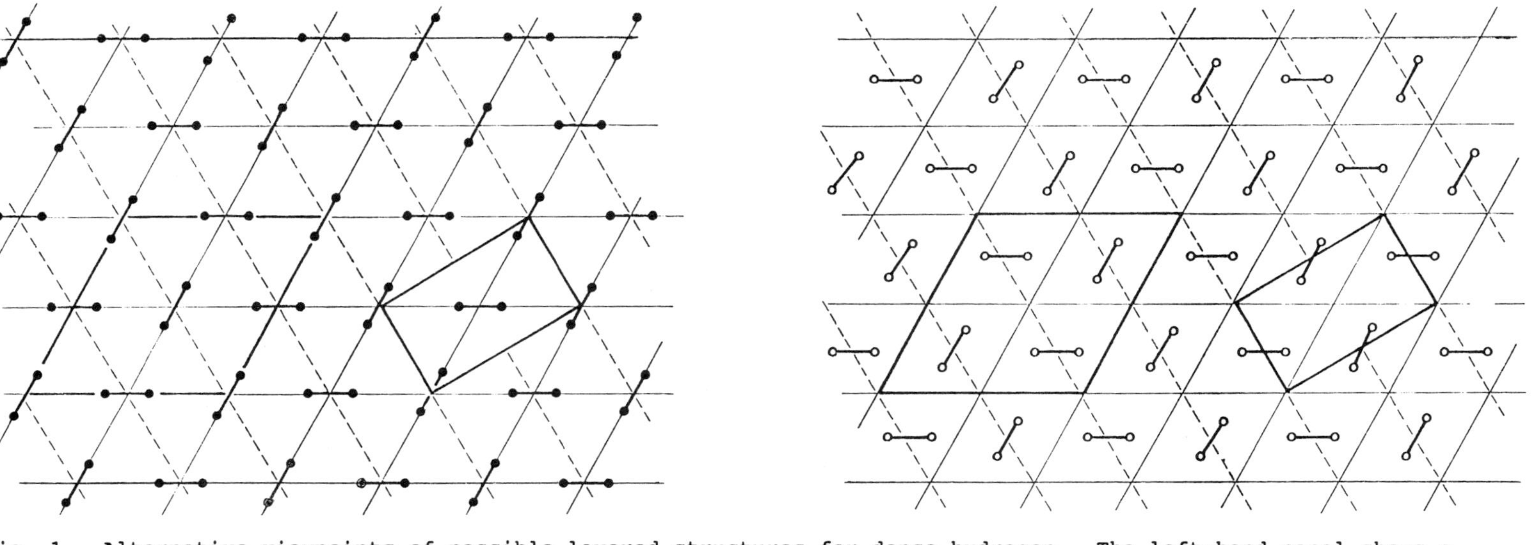

Fig. 1. Alternative viewpoints of possible layered structures for dense hydrogen. The left-hand panel shows a hexagonal unit cell for arrangements that can also be represented as an orthorhombic structure (very similar to the gallium structure). The right-hand panel shows the proton arrangements in the adjacent layers (above and below), again with cell contents representing hexagonal or orthorhombic viewpoints.

Because the disposition of its ions is so very similar to Fig. 3, where
pairing is highly evident, gallium has long been considered a _molecular_
metal. Pairing is also found in the group V semimetals (and, of course,
in group VII); if all of these odd-valent systems are described in terms
of orthorhombic unit cells, then the progression to group I of the
molecular metal concept is seen as quite plausible. Indeed, given that
the electron density in gallium is already quite high ($r_s \sim 2.1$) we might
well expect that in the absence of dynamical effects of hydrogen could
choose to persist in such a state to quite high pressures. Given that the
"hexagonal" c/a ratio is anomalous in hydrogen, it may well be that in the
absence of further refinement its structure is orthorhombic rather than
hexagonal, or that a _transition_ to orthorhombic or a related structure accompanies the tilting of the molecules into the layer planes. Obviously
more structural data is needed to resolve this issue. It should be noted
that in Ref. 12, only two diffraction lines are identified at high pressure, and in Ref. 13, three lines. This level of information is not sufficient to rigorously establish the structure.

Nevertheless, if the structure is _layered_, the following argument can
be made: In order to _maximize_ geometric structure factors in either a
hexagonal or orthorhombic context, it will be necessary to choose planar
arrangements that most closely approximate $\langle \cos \vec{k} \cdot \vec{d} \rangle$, where $\langle \rangle$ now
indicates a uniform planar average. This average is simply $J_o(Kd)$, where
J_o is the ordinary Bessel function; it is compared in Fig. 2 with $j_o(Kd)$
the spherical Bessel function arising from a uniform three-dimensional
average. Given that Kd scales as (d/r_s), and given that J_o can significantly exceed j_o at large values of their mutual arguments, it is apparent
that at high densities a structure should be sought which again, in a
crystalline context of layers, most closely approximates a uniform planar
average. What this implies, once again, is that both within and between
planes, pairs should be sequentially oriented (rather than aligned parallel). This principle can lead to large and very complex unit cells: the
corresponding structures are then crystalline approximants[2] to a planar
rotational state.

Finally, the example discussed here has taken a two-plane repeat distance in the z-direction. The data, so far, does not really fix this
either. As a counter example, consider a three-plane repeat distance with
cell ($\vec{a}_1$, $\vec{a}_2$, and $\vec{c}$). If three pairs are found in such a cell with basis
vectors $\pm \vec{d}_1, \vec{b}_2 \pm \vec{d}_2$, and $\vec{b}_3 \pm \vec{d}_3$, then the geometric structure factor per
atom is

$$S_g(\vec{K}) = \frac{1}{3}\left(\cos \vec{K} \cdot \vec{d}_1 + e^{i\vec{K} \cdot \vec{b}_2} \cos \vec{K} \cdot \vec{d}_2 + e^{i\vec{K} \cdot \vec{b}_3} \cos \vec{K} \cdot \vec{d}_3 \right) \ . \tag{17}$$

Again, choosing $\vec{k} = \frac{2\pi}{c} (0,0,1) = \frac{2\pi}{c} (0,0,\{ \frac{c}{c} = \frac{2\pi}{c} (0,0,2/3\}))$, we find that
S_g vanishes for $\vec{b}_{2z} = x$, or $\vec{b}_{2z} = x$ providing $\vec{k} \cdot \vec{d}_i = 0$. It follows that
neither the content of the hexagonal unit, nor its fundamental dimensions
are actually fixed by the experiment with the limited number of reflections observed so far. With this in mind it seems appropriate to attempt
energy minimization procedures for the electronic part of the problem free
from any prejudice on the _expected_ content of the unit cell, such as an
expectation that the system has crystallized in a hexagonal _close-packed_
arrangement. One approach is to adopt a basic lattice, place, for example
m protons at _arbitrary_ positions within its primitive unit cell, and to
carry out a minimization with respect to continuous adjustment of _all_ m
basis vectors. The results of such a procedure for the hexagonal class
will be reported in detail elsewhere,[10] but to summarize here it may be
noted that for m = 4, the protons do indeed pair, and the Abrikosov structure (hcp) is preferred at very low densities. Apart from this however,
the hcp type of structure appears not to be the lowest energy state. As
observed by others (e.g., Kaxiras et al., Ref. 7), at higher densities the

pair axes prefer to tilt over towards the basal plane. But in addition
there can be relative translation of pairs between neighboring planes.
More importantly it is also found[10] that m = 4 is actually a quite unrea-
sonably low constraint on cell content: we can do much better by taking a
far larger cell, of upwards of 18 protons in the basis. In these
calculations convergence is exceedingly slow precisely for the reason
stated above, namely that on electronic scales, the energy differences
associated with relative pair re-orientation is extremely small.

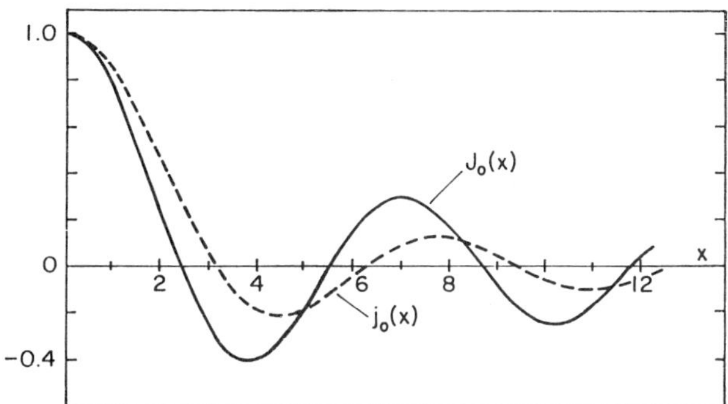

Fig. 2.   Pair structure factors per atom according to fully
          rotational (J_o) and planar rotational (j_o) averaging
          procedures. Reciprocal lattice intercepts on these
          curves occur at values of d/r_s favoring the planar
          average over the fuly rotational average at high
          densities, but not necessarily at low. Accordingly
          planar layering structures should be sought among
          crystalline *approximants* to the planar rotational
          average state.

BAND OVERLAP AND RANDOM POTENTIALS

The total energy studies for a static lattice just described suggest
that the likely insulating states of hydrogen at high compression are
either rotational, or alternatively are states in which the pair axes
adopt random uncorrelated average directions and about which large vibra-
tional/librational motions subsequently take place. In either case the
instantaneous single particle Hamiltonian has the form (see (13) and (14))

$$\hat{h} = \hat{h}_o + \Delta V \ .$$

Now, as has already been demonstrated[7] h_o leads to a band-structure and,
at sufficiently high compression, to a band overlap state. In such a
state there will be one or more electron pockets of width w_e and corresponding hole pockets of width w_h, these widths determining by the relevant
effective masses.[2] The state would clearly be a conducting phase in the
Bloch-Wilson sense: the question at hand, however, is simply whether the
state __is__ conducting when the disordering effect of ΔV are subsequently
reintroduced. The point is that ΔV is not a trivial perturbation, and its
effect may be to produce extended, and possibly significantly polarizable
states, that are *not* however diffusive. The argument behind this tenta-
tive conclusion is a variant of Anderson's argument,[15] which shows that

even in the absence of electron-electron correlation effects, electron diffusion can be completely inhibited by disordering potentials whose scale rises beyond a certain minimum fraction of an appropriate band width found in the absence of such disorder. For nondirectional site-disorder this ratio appears to be[16] approximately 1.5. For directional-disorder, the equivalent figure is not at present known, but a Gaussian (rms) estimate leads to figures in the same vicinity. Accordingly for small w_c, or w_b, which arise just after band overlap in the reference periodic system, the condition should be very easily met. In dense hydrogen, the issue of the onset band overlap conduction can then be seen as an Anderson delocalization transition. Immediately prior to such a transition the states will be quite extended and are likely to be optically active because the polarizability of such states is expected to scale with the cube of the corresponding localization length; on the other hand, since the states are not diffusive they might well lack the Drude form of absorption which will be expected on the basis of the itinerant states (of h_o) from which they derive. As pressure increases, we anticipate a rise in both w_c, w_b, a corresponding drop in the _relative_ disordering interaction, and an eventual delocalization. Whether this can occur _before_ the inevitable dynamic instability[8] of the lattice to a monatomic state is an interesting question; if it does not, then we may never actually witness a genuine signature of the band-overlap metallic state![17]

THERMODYNAMIC FUNCTIONS

We have been guided to date in our analysis of the nature of the states of the fundamental Hamiltonian (2), by experiment. Returning to (4) the anticipated breaking of symmetry by proton-pairing (absent, by assumption, in the model introduced by Mott) can be used to discuss the general characteristics of, for example, the Gibbs energy

$$G(p,T) = k_B\left\{\log Z - V\left(\frac{\partial}{\partial V} \log Z\right)\right\} = Ng \quad . \tag{18}$$

This can be carried out with specific reference to a Bravais lattice (no proton-pairing) which, merely by way of example, will be taken as a simple cubic system, and which will be the end result of a process to be introduced of prescribed atomic motion. Begin with one of the low pressure phases of ortho hydrogen, i.e., the $\alpha - N_2$ or Pa3 structure. This is itself a simple cubic lattice with an eight-point atomic basis grouped in four pairs, the axis of each pair being aligned along one of the four body diagonals of the cube. Take the protons to be separated by $\underline{2d(r_s)}$; take the lattice constant to be a. For the _special choice_ $2d = \sqrt{3}a/4$ it is easy to see that the ensuing structure is then just the simple cubic Bravais lattice introduced above. At this value of 2d, any given atom has the choice of four directions in retracing its steps to the paired phase. There is obvious degeneracy (as there is in the one-dimensional equivalent of this operation--atoms can be paired in two ways, as can be seen in Fig. 3).

Now, for fixed density ρ, imagine the form taken by the Gibbs energy per proton (say $g(p,T,\lambda)$) when atoms undergo a pairing condensation, this condensation being represented by a state of pairing order measured, say, by the obvious order parameter <2d>. It is more convenient to frame the discussion in terms of the mean proton distance λ, measured from the symmetry position (see Fig. 3) and related for a given structure to <2d> by simple geometry. Since it is energetically favorable to pair over a wide density range, the situation must be approximately as shown in Fig. 4. It is only approximate because as noted above, Mott's model (the SC Bravais lattice) has itself a metal-insulator phase transition and this must also be involved in the progression with λ. In the _metallic_ phase

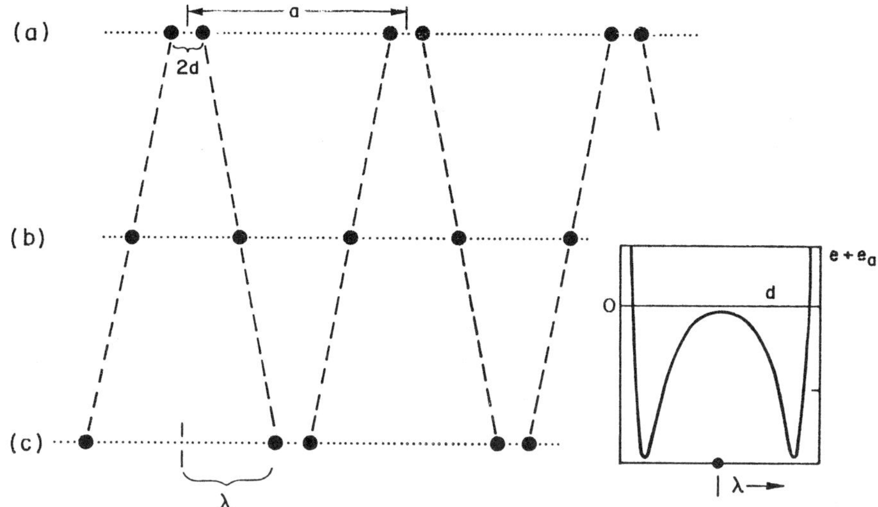

Fig. 3. Degenerate pairing alternatives (a) and (c) in a one-dimensional representation of dense hydrogen. (b) shows the symmetric monatomic arrangement and defines the symmetry point from which, in a crystalline case, the pairing order-parameter is most naturally defined. The inset shows the expected behavior of the internal energy e per atom (measured relative to the atomic limit e_a) as a function of this order parameter for relatively low densities. Here the energy scale is typically 0.1 Ry.

the monatomic arrangement may well have a <u>local</u> minimum in its free energy as a function of λ. For the paths just described, it is clear for fixed a that either for small or large values of <2d>, protons can be found in close proximity and the energy must rise (see Fig. 3). The point of symmetry in this example is the simple cubic arrangement. The minima on either side represent the equilibrium positions according to the directions chosen for motion after departing the simple cubic locations. At very low densities the depths of these minima are simply the binding energies of hydrogen molecules. Let the system first be found in one of these minima. Suppose that at fixed T the system is compressed by the application of an external agency. This agency will supply positive work $-\int_{V_0}^{V} p(V)dV$ to the internal energy, thereby progressively raising the curves, as shown in Fig. 4 which actually reflect the corresponding pp contribution to g. The results is a classic Landau description of the <u>possible</u> eventual onset (at $p = p_c$) of a first-order transition to a state in which the original nonpaired structure actually is now the state of lowest energy, which is reached by a corresponding jump in λ (or <2d>). This is the Wigner-Huntington state normally associated with the known alkali-metal phases of the group I elements.

The curves indicated in Fig. 4 for $g(p,\lambda,T)$ are given for low temperatures; the argument is readily generalizable to higher temperatures. They correspond to a sequence of increasing pressures, and each can be expanded in the form

$$g(p,\lambda) = \sum_{n=0} a_{2n}(p)\lambda^{2n} \ .$$

(19)

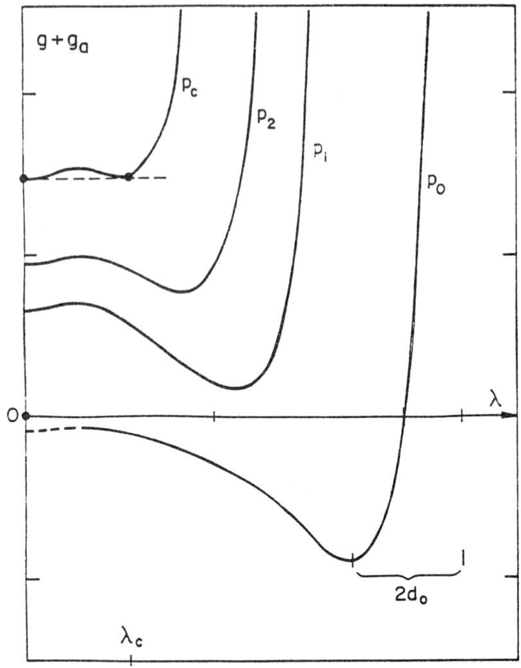

Fig. 4. General forms of the low temperature Gibbs energy per atom g (again measured relative to the atomic limit of the Hamiltonian H (Eq. (2)) as a function of the order parameter. Again the scale is typically 0.1 Ry, and the pressures in atomic units might correspond typically to $p_0 = 0$, $p_1 = 0.01$, $p_2 = 0.02$, and $p_c = 0.03$. It can be seen that at p_c a discontinuous jump in order (i.e., interproton spacing) takes place to a three-dimensional monatomic phase.

The rapidly rising character of g at small <2d> (larger λ) suggests that a minimal description of the eventual transition to a monatomic (λ = 0) phase can be sought from

$$g(p,\lambda) = a_o(p) + a_2(p)\lambda^2 + a_4(p)\lambda^4 + a_6(p)\lambda^6 + \ldots \tag{20}$$

with a_6 now necessarily positive. Note that (19) fully incorporates dynamics, including rotation.

If a transition to a monatomic (unpaired, λ = 0) phase, does exist at a certain pressure, say p_c, Fig. 4 demonstrates that certain conditions will be placed on the a_i. As noted, at such a transition, the order parameter will make a discontinuous jump from a value λ_c to zero, where λ_c satisfies both

$$a_2(p_c) + 2a_4(p_c)\lambda_c^2 + a_6(p_c)\lambda_c^4 = 0$$

and

$$a_2(p_c) + a_4(p_c)\lambda_c^2 + a_6(p_c)\lambda_c^4 = 0 \quad .$$

The critical interproton spacing will therefore be found from

$$\lambda_c = (-a_4(p_c))/2a_6(p_c) = 2a_2(p_c)/(-a_4(p_c)) \tag{21}$$

and the geometry of the system structure. It is clear from the form of the curves in Fig. 4 that a_4 is required to be negative. Positivity of a_2 simply reflects the fact that <u>if</u> a monatomic phase does form, it must be stable against all physically permissible distortions, including those that would lead to pairing within a dense phase environment. This latter point is interesting in that the one-dimensional view of this problem would always lead to pairing by the standard Peierls argument. In three dimension, where the band overlap inhibition of one-dimension is less effective, then as emphasized with the structure factor arguments above, orderings can usually be expected for which a move towards pairing will *always* be accompanied by local increase in free energy. The major point, however, is that the Landau type of argument, through (21), suggests that the critical values of pairing separation will actually be contained (in the sense of expansion) in the displacive characteristics of the eventual high symmetry phase.

The progressive weakening of the minima in Fig. 4 corresponds to the physical transfer of electronic charge from the intra-molecular region (the "bond") to the inter-molecular region. To the extent that $\upsilon(\vec{r}_{p1}\ldots)$ can be represented in terms of sums of both inter- and intra-molecular bonds, both <u>are</u> involved in the detailed form of G and its minima: however, it is apparent form the physical significance attached to <2d> that at least one small oscillations problem is also implied by the quadratic expansion of potentials associated with these minima for <u>identical</u> displacive processes (i.e., independent of relative orientations) in each and every cell. This is simply a q = 0 excitation, and is measurable by Raman effect. In fact, it has been measured for both H and D and the progression of the corresponding shifts can be used to confirm the weakening of an effective potential used to approximate the intra-molecular part of the interaction. Since as noted, the system can exist in rotational states (the rotational motions replacing the corresponding small oscillations problem, that is, the appropriate optic modes, in the lattice dynamical equivalent), there is substantial decoupling of the intra- and inter-molecular displacements, as has been observed earlier.[8]

MANY-PARTICLE VIEWPOINT: BEYOND BORN-OPPENHEIMER SEPARATION

Finally, is there a way of approaching the problem of electron and proton ordering more generally? After all it was pointed out in the "Introduction" that the Hamiltonian is exceedingly simple. A partial answer to this can be given by examining the system in a metallic or semi-metallic range. Formally, our fundamental Hamiltonian (see Eq. (2)) is a function of coordinates and gradients for both electrons and protons. However, at the level of Coulomb interactions the Hamiltonian is also bi-linear in density operators. This fact (an immediate consequence of working at the fundamental level) can be exploited in a rewriting of the partition function[3] as a path integral, and by introducing the corresponding action, where the pair density structure then leads immediately to a Gaussian form. A Hubbard-Stratonovich transformation can now be carried out to reduce the partition function further; so far this is completely without approximation. The latter only arises when, with guidance from experiment, we are eventually obliged to choose auxiliary fields. It is

apparent that for much of the density range of interest in dense hydrogen,
the protons are in states we have described as "paired." Accordingly it
is useful to consider fields which reflect such pairing, but necessarily
pairing of _all_ possible kinds (pp, ee, and e - p). But as shown in
Ref. 2, this choice and a subsequent stationary phase approximation leads
to four temperature dependent gap equations of the BCS type: one of
these, the gap equation for protons, can be shown to be the many-particle
generalization of the familiar Heitler-London state. The electron-proton
(and the companion proton-electron) equations are the generalization of
the excitonic-insulator problem, the limiting form being hydrogen atoms;
the electron-electron gap equation is of a recognizably BCS form.

Two quite interesting results emerge from this approach: the first
is that under reasonably general conditions, the pairing equations which
have been derived from fundamental interactions can be cast in terms of
effective pairing interactions and in any given case, (p - p) for example,
these are reduced (from the direct interaction) as a consequence of the
presence of the second component. This result emerges without any re-
course to an adiabatic approximation, and is, therefore dependent on _both_
electron and proton masses. The second result is that the equations are
all interdependent, and as might be physically expected no solutions arise
above a certain critical temperature. What this immediately implies is
that if pairing of, say, a superconductive nature is established in the
electron system, the Heitler-London type of pairing that is present in the
proton system is itself dependent in detail on this emergent electronic
order. Thus, if the electron system is driven normal by increase of temperature, the fact should be immediately reflected in the proton problem,
the most likely route being through the proton-pair dynamics; in other
words there should be an associated critical point in a dynamical probe of
proton pairing. Such an effect has been observed in dense hydrogen at a
pressure (or density) range where there _are_ band-structure indicators of
band-overlap or at the very least, enhanced optical activity. Yet, there
is no _direct_ evidence that hydrogen is semimetallic in this range: the
possibility therefore remains that electron pairing is present in _extended_
but nonetheless Anderson localized states, a notion that is being pursued
further.

It is a pleasure to thank Dr. K. Moulopoulos for many stimulating and
helpful discussions. This work has been supported by the National Science
Foundation under Grant DMR-9017281.

REFERENCES

1. K. Moulopoulos and N. W. Ashcroft, Phys. Rev. Lett. 66, 2915 (1991).
2. N. W. Ashcroft, "Molecular systems under high pressure," R. Pucci and
    G. Piccitto, eds., Elsevier Science Publishers, BV-North Holland,
    1991) 201.
3. N. F. Mott, Proc. Phys. Soc. A 62, 416 (1949).
4. J. C. Slater and J. G. Kirkwood, Phys. Rev. 37, 682 (1937); N. W.
    Ashcroft, Phil. Trans. Soc. Lond. A 334, 407 (1991).
5. N. W. Ashcroft and N. D. Mermin, "Solid State Physics," (Holt
    Saunders, 1976).
6. N. F. Mott, "Metal-insulator transitions," 2nd ed., Taylor and
    Francis, Ltd. (1990).
7. See, for example; D. E. Ramaker, L. Kumar, and F. E. Harris, Phys.
    Rev. Lett. 34, 812 (1975); C. Friedli and N. W. Ashcroft, Phys.
    Rev. B 16, 662 (1977); B. I. Min, H. J. F. Jansen, and A. J.
    Freeman, Phys. Rev. B 30, 5076 (1984); H. Chacham and S. G. Louie,
    Phys. Rev. Lett. 66, 64 (1991); E. Kaxiras, J. Broughton, and
    R. J. Hemley, Phys. Rev. Lett. 67, xxx (1991).

8.   N. W. Ashcroft, Phys. Rev. B.
9.   I. Silvera, these proceedings.
10. A. Horsfield, K. Moulopoulos, and N. W. Ashcroft (to be published).
11. For a general discussion of rotational states in hydrogen see
         I. Silvera, Rev. Mod. Phys. $\underline{52}$, 393 (1980).
12. H. K. Mao, A. P. Jephcoat, R. J. Hemley, L. W. Finger, C. S. Zha,
         R. M. Hazen, and D. E. Cox, Science $\underline{239}$, 1131 (1988).
13. V. P. Glazkov, S. P. Besedin, I. N. Goncharenko, A. V. Irodova, I. N.
         Makarenko, V. A. Somenkov, S. M. Stishov, and S. Sh. Shil'stein,
         Pisma. Zh. Eksp. Teor. Fiz. $\underline{47}$, 661 (1988).
14. A. J. Bradley, Z. Kristallog. $\underline{11}$, 302 (1935).
15. P. W. Anderson, Phys. Rev. $\underline{109}$, 1492 (1958).
16. M. Schreiber, J. Non-Cryst. Solids $\underline{97/98}$, 221 (1987); P. Edwards and
         D. Thouless, J. Phys. $\underline{C5}$, 807 (1972).
17. I. Silvera, private communication.

HIGH PRESSURE NMR: HYDROGEN AT LOW TEMPERATURES

Aziz Müfit Uluǧ, Mark S. Conradi, R.E. Norberg

Washington University, Department of Physics
St.Louis, MO 63130

1. INTRODUCTION

We report NMR of solid hydrogen in a diamond anvil cell at temperatures 1.85 to 100 Kelvin. We examine the pressure dependence of the orientational ordering phase transition temperature.

2. HIGH PRESSURE EQUIPMENT

We use a home-built piston-driven diamond anvil cell (DAC) to obtain high pressures [Figure 1]. It is made out of non-magnetic materials (titanium, BeCu, stainless steel) and is designed to accomodate the RF coil and resonance circuit for NMR experiments. There is an optical access to the sample region which is mainly used to measure pressure through optic cables and is also used in the process of alignment of the diamonds.

The whole diamond anvil cell with its mechanical-advantage levers was designed to fit in the 2.37 inches inner diameter of the low temperature dewar.

The diamonds used are 1/4 carat, flawless, type 1a diamonds with culet tip diameter of 1 mm. They are mounted on BeCu seats with brass holders.

Pressures are measured using the ruby fluorescence spectra. The source is a 15 milliwatt argon-ion laser.

We use BeCu and rhenium as gasket materials. With a 0.010 inch thick gasket, one can get about 40 kbar (BeCu) and 70 kbar (rhenium) easily. With preindentation (0.005 inch) one can get about 130 kbar with a rhenium gasket.

Our samples are loaded into the diamond anvil cell at 77 Kelvin, from a commercial 6000 psi (400 bar) hydrogen bottle. Thus hydrogen has the density of normal liquid H_2 upon loading.

The experiment is done in a low temperature flow-style dewar where temperatures as low as 1.85 Kelvin can be achieved by pumping on liquid helium.

This work is supported in part by NSF DMR 90-24502 and 90-02857

Frontiers of High-Pressure Research, Edited by H.D. Hochheimer and
R.D. Etters, Plenum Press, New York, 1991

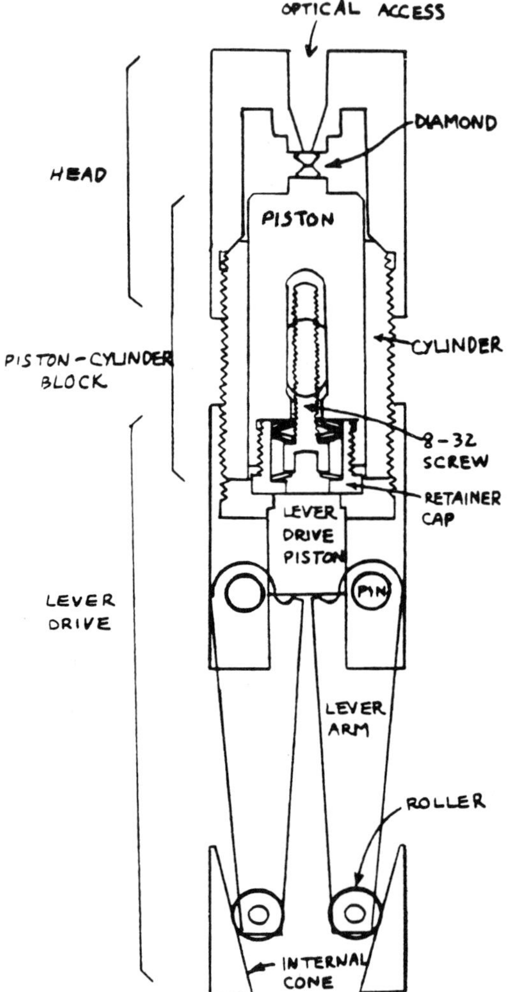

Figure 1. Schematic drawing of DAC (from Lee [5]).

3. NUCLEAR MAGNETIC RESONANCE AND RF COIL

Classically one can consider nuclear spins as vectors precessing about the external magnetic field [1,2] [Figure 2]. The classical rotation frequency, the Larmor frequency of the nuclear spins, is given as,

$$\omega_0 = \gamma H_0.$$

An alternating magnetic field of frequency ω_0 which is perpendicular to the static field can be used to tip the classical spin vector into the xy-plane, where the H_0 field is parallel to the z-axis [Figure 2].

Now, one can talk about two relaxation processes [1,3]. The longitudinal relaxation process describes the return of z-magnetization to its Boltzmann equilibrium value and and is characterized by a time constant T_1. The transverse relaxation process is the process where the tipped spin vector in the xy-plane gets smaller in magnitude due to local magnetic fields effecting individual spins .

In order to do NMR experiments in a DAC, we need an appropriate coil to provide an alternating magnetic field H_1 at the sample. Because the gaskets used are metal, a difficulty arises. An alternating magnetic field cannot pass through a hole in a metal sheet because the induced currents try to cancel the magnetic field [4].

In our experiment we use a "hair-pin" resonator [Figure 3]. It is made out of 0.010 inch copper and is a 3/4 turn coil that produces a magnetic field parallel to the gasket surface [5]. Some of this magnetic field bends into the sample region in the gasket hole without threading the hole (no net flux through the hole); this is sufficient for NMR experiments. The loop has two access holes for the diamonds. The resonator has two trimmer capacitors: one for fine adjustment of the resonance frequency and the other for impedance matching of the resonance circuitry to the transmission line.

4. SAMPLE (SOLID HYDROGEN)

Being a quantum solid, hydrogen molecules are in free rotational states in the solid phase, due to the weak interaction among neighboring molecules [6,7]. The anisotropic force arises from the electric quadrupole quadrupole (EQQ) interaction. Because a hydrogen molecule has almost spherical charge distribution, the EQQ is weak (about 0.8 K for neighbor pairs). Due to the requirement of symmetry of the wave function, molecular hydrogen has two species: ortho and para. Total nuclear spin (I) in hydrogen is

$$1/2 + 1/2 = 1 \quad \text{triplet state} \quad (I = 1), \quad \text{ortho-}H_2$$
$$1/2 - 1/2 = 0 \quad \text{singlet state} \quad (I = 0), \quad \text{para-}H_2.$$

Parahydrogen: $I = 0$, $J = 0, 2, 4, \ldots$ and orthohydrogen: $I = 1$, $J = 1, 3, 5, \ldots$ where J is the quantum number of the molecular angular momentum.

At room temperature and above the equilibrium ratio of ortho to para is 3:1. Parahydrogen does not give any NMR signal (no net spin), but orthohydrogen does.

At low temperatures the molecules are in $J = 0, 1$ states. The $J = 1$ state is a metastable state and converts to the $J = 0$ state if one waits long enough. In gas, hydrogen conversion takes months; in condensed phases it may take days [6]. In two dimensional systems reported conversion rates are much faster.

Solid hydrogen at high temperatures has a hexagonal closed packed (hcp) structure with orientationally disordered hydrogens. In the NMR spectrum this structure appears as a Gaussian line shape [1,7] with about 80 KHz width, due to averaging of the dipolar interaction between the protons of the rapidly rotating molecules. Molecules are free to rotate, so the proton-proton axis does not have any preferred orientation. When the temperature drops there is a phase transition from disordered hcp to the orientationally

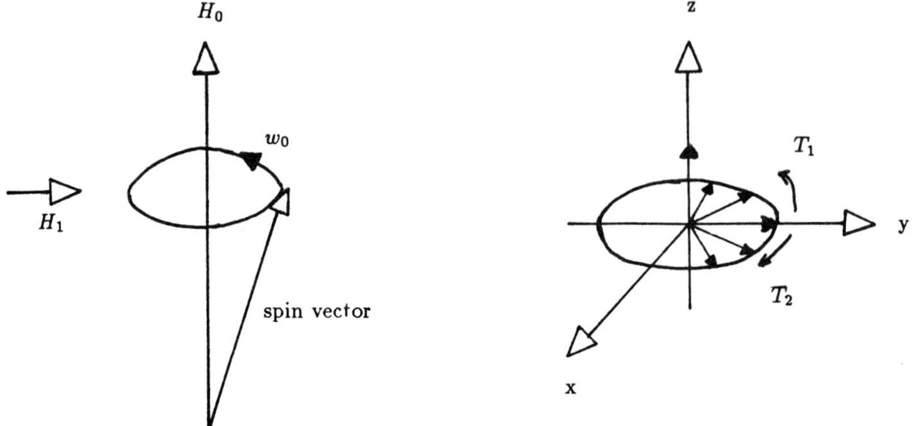

Figure 2. Classical spin vector.

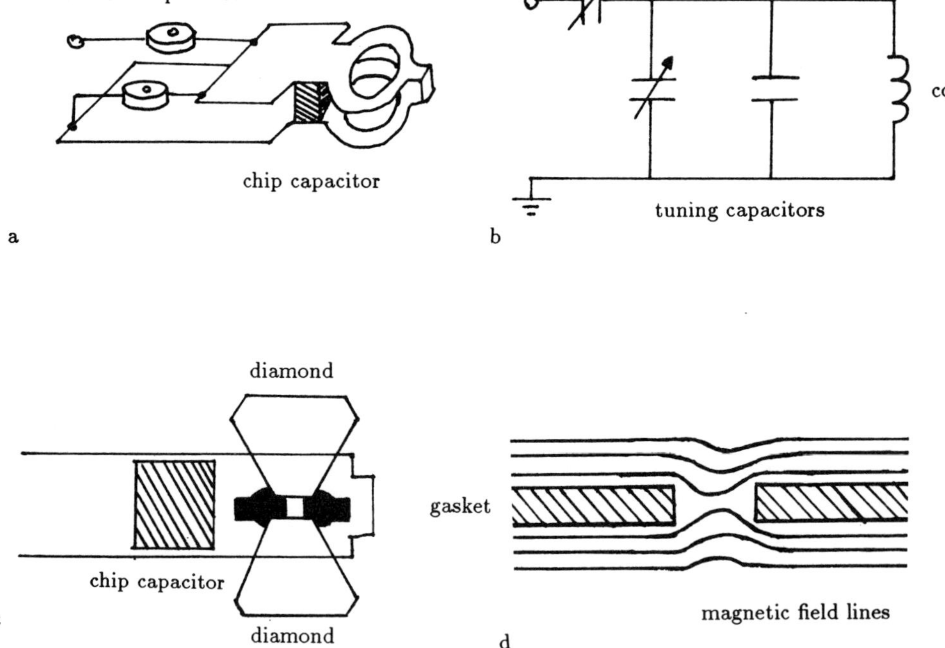

Figure 3. a) hair-pin resonator, b) circuit diagram, c) side view, d) field lines produced by the coil.

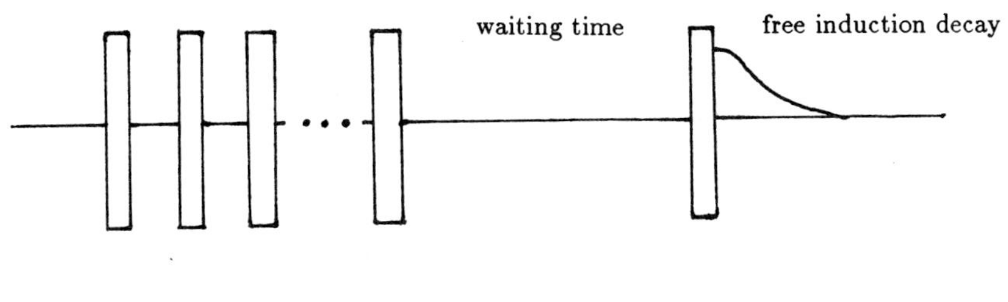

Figure 4. Pulse sequence.

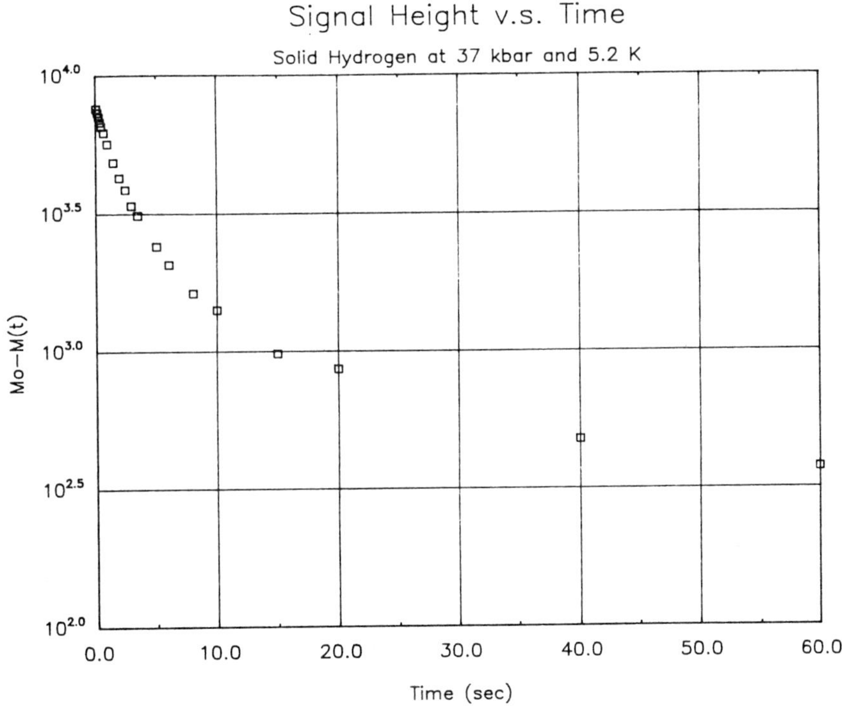

Figure 5. Signal height versus time.

ordered face centered cubic phase (fcc) [6,8]. This is a Pa3 structure with 4 sublattices with proton-proton axes oriented parallel to the body diagonals of the cube. Molecules are not free to rotate in this phase (but they librate) resulting in a Pake doublet (170 kHz width) NMR spectum. A small amount of hcp phase coexists with the ordered phase.

The longitudinal relaxation time T_1 in the disordered phase is short because of the rapid rotational movement. By comparison, in the ordered phase T_1 is found to be longer. Orientationally ordered molecules librate and all the librations are at higher frequencies (the libron spectrum does not extend to zero frequency).

The major anisotropic interaction that drives the ordering transition is the electric quadrupole quadrupole interaction [10]. The EQQ interaction strength scales as R^{-5} where R is the distance between quadrupoles:

$$\Gamma = \frac{6(eQ)^2}{25R^5}$$

Because density is inversely proportional to the third power of the distance;

$$R \propto \rho^{-1/3},$$

the EQQ interaction strength is expected to scale as $\rho^{5/3}$. With the same argument the transition temperature curve is expected to scale as $\rho^{5/3}$. Our purpose in this experiment is to test this prediction.

5. EXPERIMENTAL RESULTS AND DISCUSSION

In our experiment, we measure longitudinal relaxation times (T_1) using the saturate-wait-inspect pulse sequence [Figure 4].

Through the T_1 process, magnetization parallel to the static magnetic field H_0 should relax according to [3,11]

$$M_z(t) = M_0 \left(1 - e^{-t/T_1}\right). \tag{1}$$

Since

$$\log(M_0 - M_z(t)) = -\frac{t}{T_1},$$

when $M_0 - M_z(t)$ is plotted on a semi-log paper, a straight line is obtained. The slope of this line yields the T_1 value.

In disordered solid hydrogen the T_1 data measured are generally found to obey the equation (1) [12,13]. But ordered solid hydrogen is found to have more than one longitudinal relaxation constant T_1 [13]. One can write the magnetization in the z direction as:

$$M_z(t) = \sum_i M_i \left(1 - e^{-t/T_{1i}}\right).$$

Therefore plotting $M_0 - M_z(t)$ versus time (where $M_0 = \sum_i M_i$) on a semilog paper [Figure 5] will not give a straight line.

This phenomenon was seen by Meyer [13] in ordered-phase solid hydrogen and explained as spins instead of coming to an equilibrium and then relaxing to the lattice, independently relax to the lattice. The time to reach a common spin temperature increases as temperature decreases below the transition temperature.

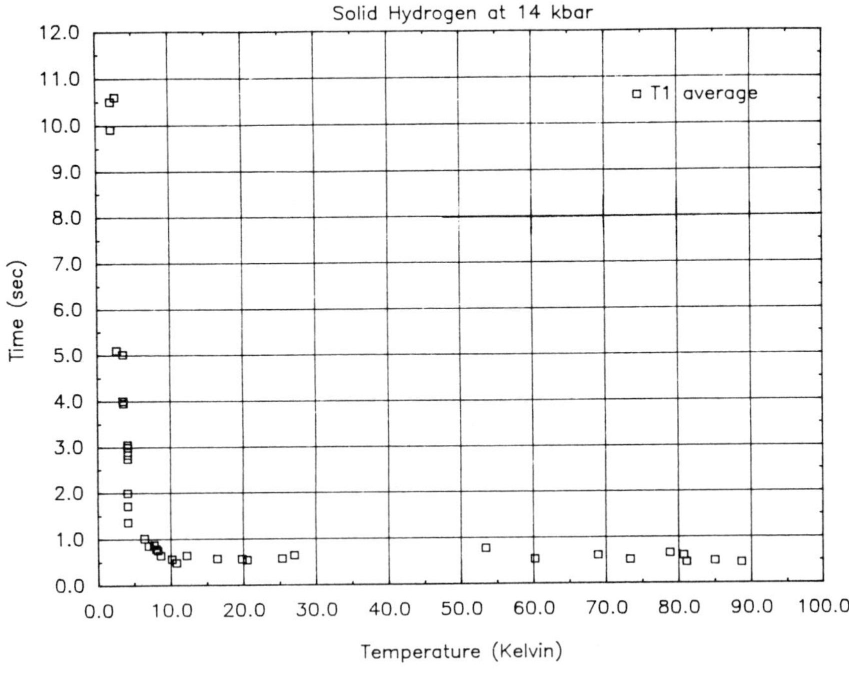

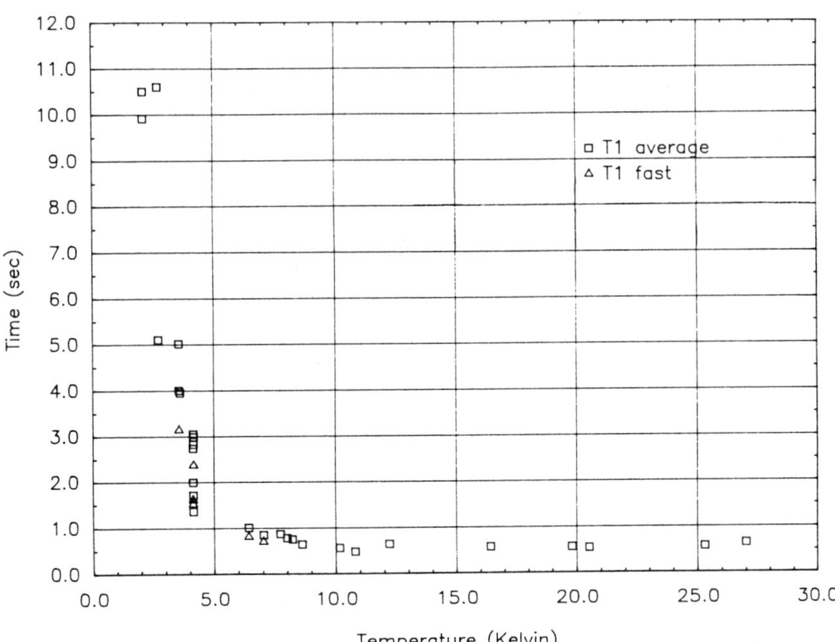

Figure 6. T_1 measurements at 14 kbar.

T1 v.s. Temperature

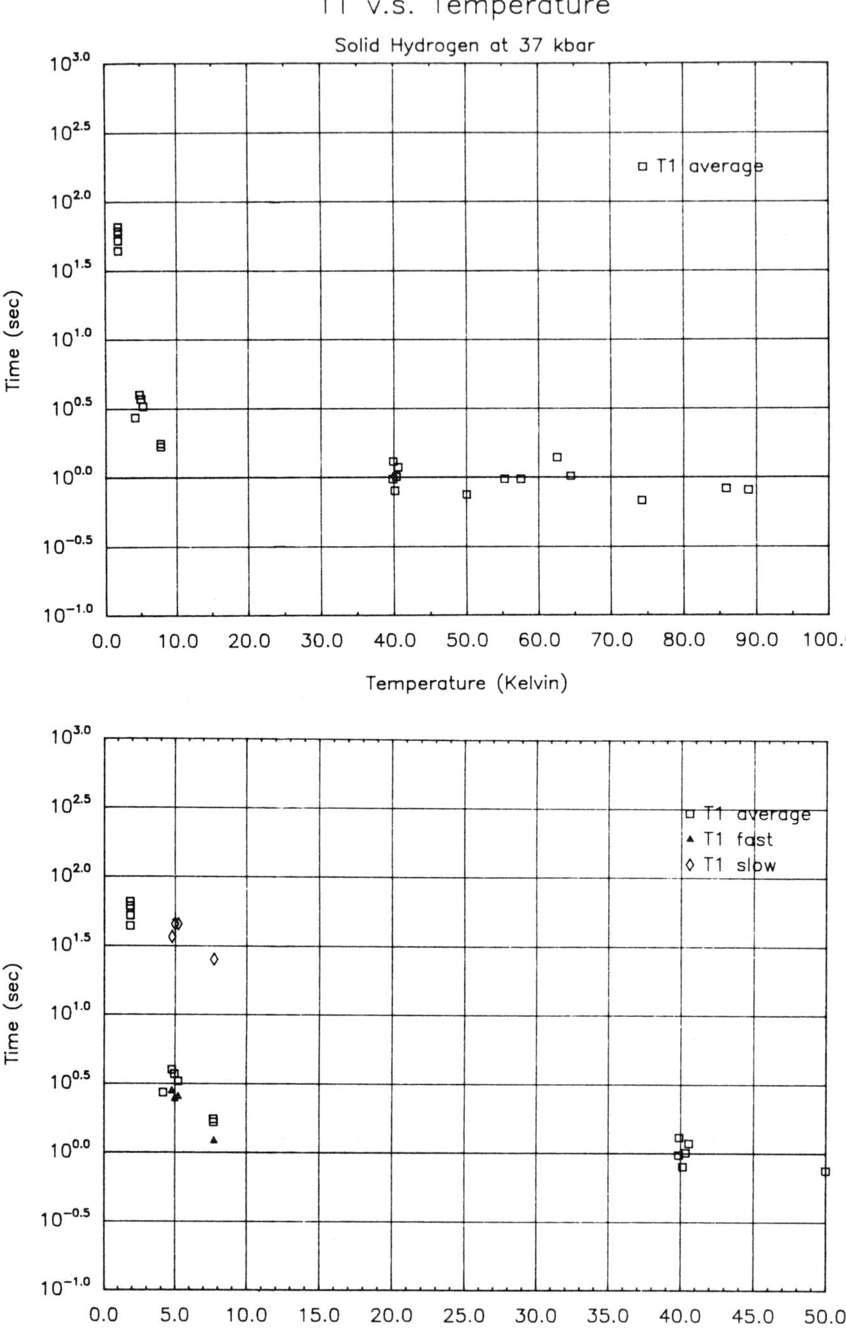

Figure 7. T_1 measurements at 37 kbar.

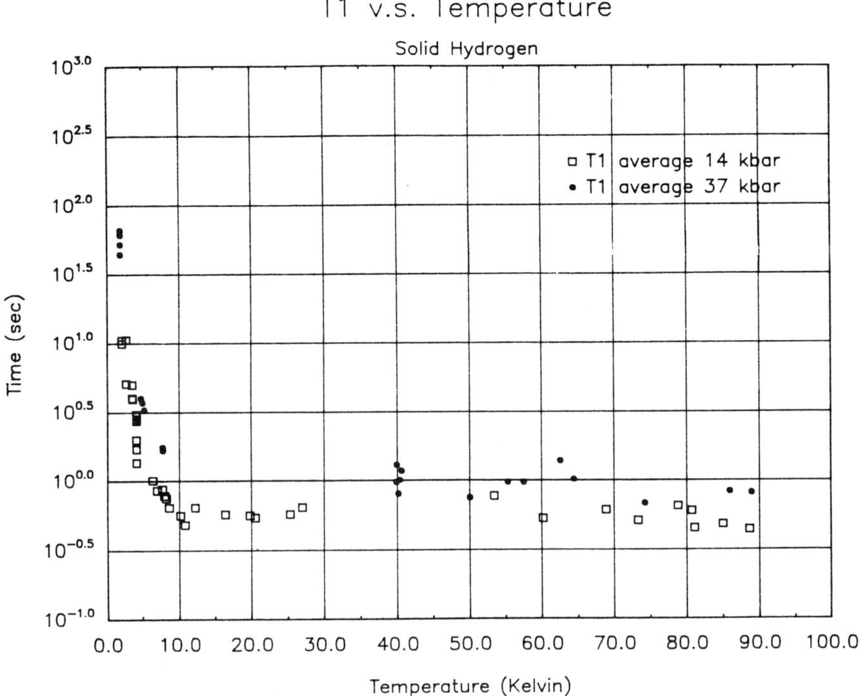

Figure 8. $\overline{T}_1$ measurements at 14 kbar and 37 kbar.

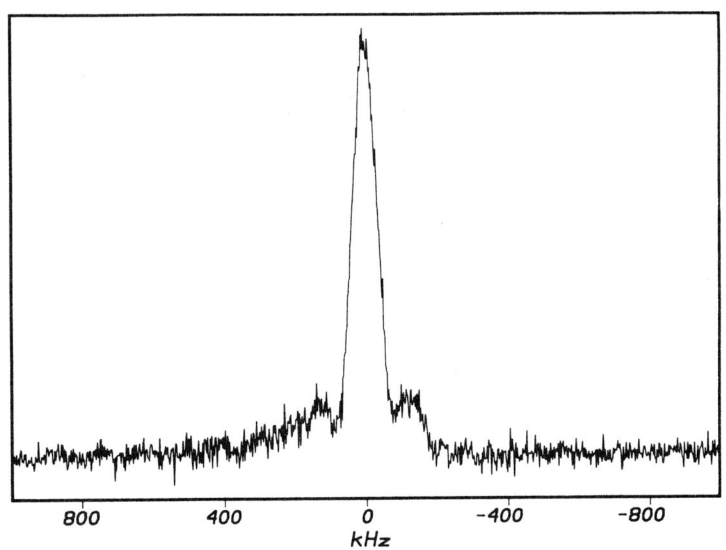

Figure 9. NMR line shape at 37 kbar and 1.85 Kelvin.

It is difficult to evaluate this kind of T_1 data. One can look only at the initial slopes of the T_1 curves and use this as a T_1 average value [11] or one can try to fit more than one exponential to the T_1 curve.

In the analysis of our data we used the initial slope of the T_1 curve to estimate an average T_1 ($\overline{T}_1$), and also when possible we tried to fit two exponentials to the T_1 curve to get two time constants, namely $T_1 fast$ and $T_1 slow$ where $T_1 fast < \overline{T}_1 < T_1 slow$.

We have run our experiment four times at two different pressures: 14 kbar where $\rho = 2.1\rho_0$ and 37 kbar where $\rho = 2.5\rho_0$. Assuming the transition temperature curve scales as $\rho^{5/3}$, at 14 kbar the transition temperature should be about 5.5 Kelvin and at 37 kbar the transition temperature should be about 7.4 Kelvin for normal hydrogen with 75% ortho concentration (at $\rho = \rho_0$ the transition temperature is 1.7 Kelvin [6]).

Figure 6 shows longitudinal relaxation times for 14 kbar and Figure 7 shows longitudinal relaxation times for 37 kbar. All the data are plotted together on Figure 8. It can be seen that there is an increase in T_1 values with an increase in density. We can also see that above the expected transition temperature, T_1 is temperature independent and below the expected transition temperature there is an increase in T_1 values with T_1 increasing while temperature is decreasing. This $1/T$ behaviour is seen by Sullivan and Pound [14] in the ordered phase and is also seen by others [15].

The least squares fit to the T_1 data in the temperature dependent region is used to determine a transition temperature. The transition temperature is taken as the temperature at which the mean value of T_1 in the temperature independent region equals the extrapolation of the least squares fit of the T_1 data in the temperature dependent region. Or simply, it is the temperature at which the $1/T$ least squares fit curve of the temperature dependent region intercepts the $T_1 = constant$ curve of the temperature independent region. At 14 kbar the transition temperature is found to be 5.4 Kelvin and at 37 kbar where there are not many data points in the temperature dependent region, the transition temperature is found to be 6 Kelvin.

The line shapes of our NMR data, which are obtained by Fourier transforming the time domain data (FID's), do not show the Pake doublet that we expected. This might be because our FID's do not contain the very early times, due to receiver delay (about $2\mu sec$). A line shape from our data [Figure 9] shows a large center peak and a doublet asymmetrically located around the center peak.

To get better line shapes we plan to decrease the receiver dead-time and use solid-echo [3] techniques instead of free induction decays.

REFERENCES

1. A. Abragam, "Principles of Nuclear Magnetism," Oxford Science Publications, Oxford University Press, (1983).

2. G.E. Pake, Magnetic resonance, Scientific American, 199, 58 (1958).

3. C.P. Slichter, "Principles of Magnetic Resonance," Springer-Verlag, (1990).

4. S.H. Lee, K. Luszczynski, R.E. Norberg and M.S. Conradi, NMR in a diamond anvil cell, Rev. Sci. Inst., 58, (1987).

5. S.H. Lee, A Study of molecular Diffusion and Reorientation in Solid Hydrogen At Pressures 18 kbar–68 kbar, Washington University, Physics Department, Ph.D. Thesis, (1989).

6. Isaac F. Silvera, The solid molecular hydrogens in condensed phase: Fundamentals and static properties, Rev. of Mod. Phys., 52, (1980).

7. F. Reif and E.M. Purcell, Nuclear magnetic resonance in solid hydrogen, Phys. Rev., 91, 631, (1953).

8. Isaac F. Silvera and R. Jochemsen, Orientational ordering in solid hydrogen: Dependence of critical temperature and concentration on density, Phys. Rev. Lett., 43, 377, (1979).

9. L.I. Amstutz, H. Meyer, S.M. Myers and D.C. Rorer, Study of nuclear magnetic resonance line shapes in solid H_2, Phys. Rev., 181, 589, (1969).

10. L.I. Amstutz, H. Meyer, S.M. Myers and R.L. Mills, Longitudinal nuclear relaxation measurements in hcp H_2, Phys. Chem. Solids, 30, 2693, (1969).

11. J.R. Gaines and R.C. Souers, The spin-lattice relaxation time T_1 in mixtures of hydrogen isotopes, Adv. in Mag. Res., 12, 91, (1988).

12. F. Weinhaus and H. Meyer, Nuclear longitudinal relaxation in hexagonal close packed H_2, Phys. Rev. B, 7, 2974, (1973).

13. P.L. Pedroni, R. Schweizer, H. Meyer, NMR relaxation times and line shapes in solid H_2 at elevated densities, Phys. Rev. B, 14, 896, (1976).

14. N.S. Sullivan and R.V. Pound, Nuclear spin relaxation of solid hydrogen at low temperatures, Phys. Rev. A, 6, 1102, (1972).

15. H. Ishimoto, K. Nagamine and Y. Kimura, NMR Studies of solid hydrogen at low temperatures, J. Phys. Soc. Japan, 35, 300, (1973).

SULPHUR AT HIGH PRESSURE AND LOW TEMPERATURES

B. Eckert, H.J. Jodl, H.O. Albert, P. Foggi*

Fachbereich Physik, Universität Kaiserslautern, FRG
* European Laboratory for non Linear Spectroscopy (LENS),
Firenze, Italy

1. INTRODUCTION

Molecular crystals[1-3] are in general characterized by strong intramolecular forces and much weaker intermolecular forces. The molecular structure remains unaltered and the internal vibrations split off under the influence of the lattice field. One important property of such crystals is their large compressibility, a result of the weak external forces. Therefore pressure dependent investigations on vibrational modes are a good tool to test theoretical models, e.g. forms of potentials, approaches in lattice dynamics. As can be seen later temperature variation gives additional information about anharmonicity.

Sulphur at ambient conditions is the molecular crystal $\alpha-S_8$. The molecules are S_8 zig–zag rings and classified by space group D_{4d}. Group theory yield 11 fundamental modes. The well known crystal structure is orthorhombic with space group D_{2h} and 4 molecules per primitive unit cell.[4-6]. Correlation of molecular point group with crystal factor group predicts 36 raman, 26 infrared and 10 inactive modes.

$\alpha-S_8$ is not a typically molecular crystal with regards to its vibrational energies. There is no pronounced gap between external and internal mode frequencies as it is for e.g. $\alpha-N_2$ ($\sim$ 20 cm^{-1} libration, $\sim$ 2300 cm^{-1} vibration). All excitations are grouped together below 500 cm^{-1}.

Another speciality of sulphur is the large variety of allotrope forms stable under certain conditions and produceable by various procedures.[7-9] Especially at higher temperatures all sorts of molecular species in the liquid and vapor phase can occur.[7-10]

Regarding only vibrational excitations we will now list the state of literature. Theoretical data of the free molecule came from Scott et al.[11], Domingo et al.[12], Cardini et al.[13]. Lattice dynamical calculations were reported by Pawley et al.[14], Rinaldi et al.[15], Luty et al.[16], Gramaccioli et al.[17], Dows[18]. In the case of spectroscopic data we focus only on publications of about the last twenty years. Raman and infrared spectroscopy on $\alpha-S_8$ reported Anderson and Loh[19], Anderson and Wong[23], Ozin[21], Arthur and Mackenzie[22], Gautier and Debeau[25], Anderson and Boczar[24], Harvey and Butler[20], Becucci et al.[27].

Pressure dependencies of modes in Raman spectra were investigated by

Zallen[28], Slade et al.[29], Wang et al.[30] and Häfner et al.[31]. Theoretical pressure dependencies are given from Pawley and Mika[32], and Kurittu[33].

Phase diagrams of sulphur, of its stable and unstable forms, are reviewed by Meyer[8] and by Liu and Bassett[34]. These diagrams are either limited in parameters (T > 300 K) or strongly inconsistent. Figure 1 shows a combined version. Additionally   Raman data at high pressures and low temperatures are not known in the literature and there are no high pressure infrared data at all.

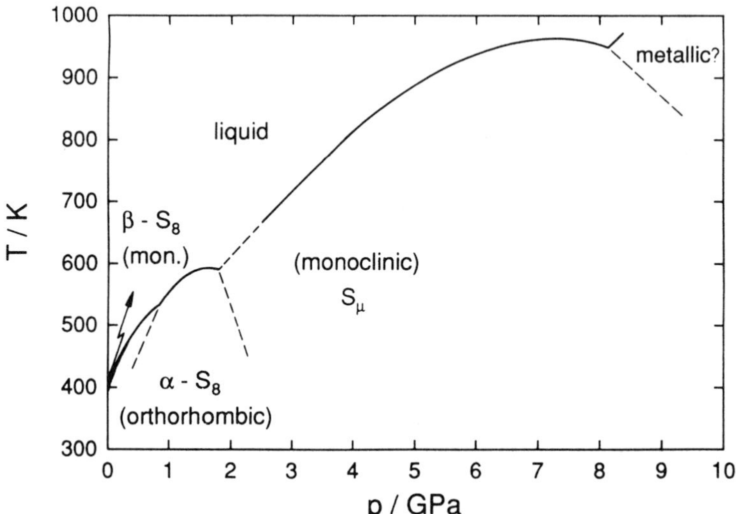

Figure 1.   p–T diagram from recent literature[34].

Therefore our aims here were to discover the low temperature − high pressure section in the p–T–diagram by Raman studies. We looked for new solid state phases of sulphur and their thermodynamic and spectroscopic properties.

In addition we finished low temperature FTIR studies[26] completing the Raman data.

2. EXPERIMENTAL

Most experimental data were obtained with isotopic pure (99.95 %) $^{32}S_8$ single crystals, further purified by repeated crystallisation from saturated $CHCl_3$–solution at LENS, Florence, Italy. Additionally we used natural α–S₈ single crystals, oriented, faced and optically polished by Dr. G. Lamprecht, Neuhausen, FRG.

High pressure measurements were performed by a diamond anvil cell according to Huber et al.[35] The local pressure was determined from the pressure shift of the ruby fluorescence line R_1.[36] To avoid heating of ruby and sample by absorption laser power was reduced to 1 %, i.e. < 0.5 mW compared to the Raman measurements, when the laser was focused on rubys.

Two different methods of loading the cell were applied. The first method consisted in preparing the gasket with a small α–S₈ single crystal (typical Ø 50 μm) surrounded by several rubys (typical Ø 10 μm). Pressure transmitting medium was He gas. The second method consisted in filling the gasket by crystallites of dimensions compareable to gasket dimensions, after the rubys were placed upon the diamond flat. Pressure transmitting medium was sulphur itself. Sample cooling at ambient pressure and in the DAC were achieved by a closed cycle cryostat. Temperature could be stabilized electronically within $\pm$ 2 K. Determination of local temperature ensured on one hand by measuring the current of a calibrated Si–diode and the voltage of a gold–iron–chromel thermoelement mounted at different positions of the DAC; on the other hand by registration of the Stokes–Antistokes ratio of α–S_8 phonon bands and the intensity ratio of the ruby fluorescence lines R_1 and R_2[37].

Spectra were obtained by a standard Raman spectroscopic equipment. The 514.5 nm line of an argon ion laser (Spectra Physics 165) was mostly used for exciting the sample. Spectra were registered through a double grating monochromator (SPEX 1301) with a further monochromator (TTM 1442) for discrimination the laser wing. Data were acquisitated with a Peltier–cooled RCA photomultiplier tube and a standard photon–counting system (ORTEC) connected to a MCA and a PC.

For the low temperature high–pressure experiments the following procedure was used: At room temperature a certain elevated pressure was adjusted. Then the cell was slowly cooled down and warmed up again; in each cycle the spectra were taken at about five to seven temperatures. We were able to describe hysteresis in the cell and in the sample by this procedure. In cases where we had spectra of another sulphur than α–S_8 we released the pressure such that we started each individual low temperature cycle with clearly fixed conditions i.e. a spectrum of α–S_8.

3. RESULTS AND DISCUSSION

First we will describe and discuss the details in pressure and temperature dependencies of α–sulphur. Second we will describe spectra of a high–pressure low–temperature amorphous sulphur, found for the first time. Third we will present pressure shifts of mode frequencies of a photosensitive sulphur allotrope and its transformations into other forms while decreasing pressure. Finally we will extend the p–T diagram towards 15 K and < 7 GPa.

3.1 Pressure and Temperature Dependency of α–Sulphur

a) High Pressure Properties at T = 300 K.

The evolution of the Raman spectrum under increasing pressure can be seen in fig. 2. Mode energies as a function of pressure are presented in fig. 3. In both representations one can notice the following features: the internal modes ν_1 and ν_7, ν_{11}, ν_6, ν_8 split under increasing pressure. The lattice modes b_{2g} (26 cm^{-1}) and a_g (52 cm^{-1}) show some strange splitting. We can extrapolate these splitting to p = 0 GPa but they are not observed in Raman spectra at 300 K. Becucci et al.[27] published high resolution Raman spectra at 4 K and reported a doublet at 32 and 33 cm^{-1}. They assigned the low frequency to the libration with symmetry b_{2g}, and the other with b_{3g}. These doublet was not seen in our low temperature Raman spectra because of spectral resolution. The mode around 70 cm^{-1} extrapolated to p = 0 cannot be recognized clearly in the spectra of Becucci et al[27]. Line crossing can be observed in the case of the a_g and b_{3g} modes of ν_9, of external b_{2g} (63 cm^{-1}) and b_{1g} of ν_9. If the assignment of Becucci et al.[27] is correct, the doublet near 30 cm^{-1} shows a crossing also.

Relative intensity changes are a further significant feature at different pressures. The gross intensity of the lattice mode band decreases with increasing pressure. The strongest mode ν_2–a_g becomes less intense at high pressures, whereas the ν_1 now becomes strongest one. There are also intensity changes between the high and low energy components of ν_8.

In general the frequency shifts of the external modes as expected are on the whole larger than the internal ones. The shifts are well reproduceable and show no hysteresis effects between loading and unloading. We observed no significant discrepancies between spectra of samples loaded by the different techniques mentioned above. We also can state from evaluation of ruby fluorescence shifts observed from three rubys at different positions in the sample, that the pressure distribution over the sample is homogeneous within the experimental error of a few kbar in both cases.

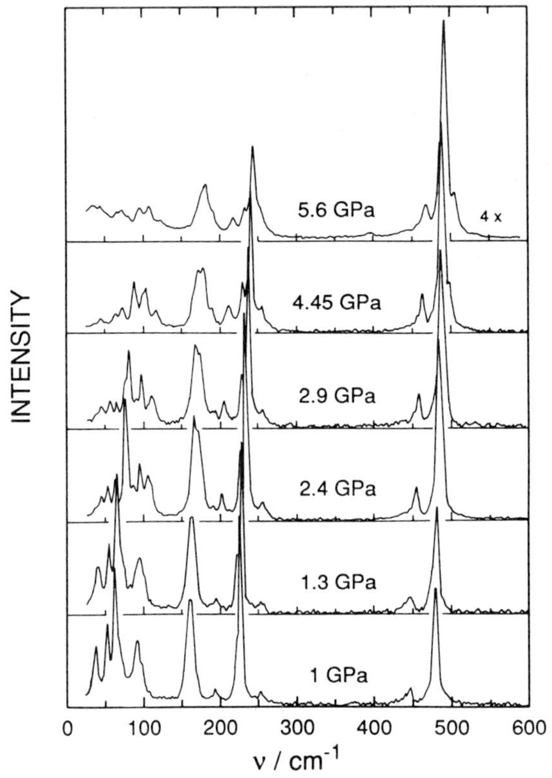

Figure 2. Some selected Raman spectra at various pressures and at room temperature.

It is usual to describe the pressure dependence of mode frequencies ν_i in terms of the mode Grüneisen parameter

$$\gamma_i = -\frac{\partial \ln \nu_i}{\partial \ln V} = \frac{1}{\nu_i \beta_T} \left. \frac{d\nu_i}{dp} \right|_T , \qquad (1)$$

where V is the volume and β_T is the isothermal compressibility. This parameter is a useful tool to compare different measurements and theoretical calculations. However, the experimental value for γ_i depends not only on the observed shifts of mode energies, but also on the compressibility. Wang et al.[30] and Häfner et al.[31] reported pressure dependent lattice constants determined by X–ray spectroscopy. From their data one can estimate the compressibility (at 300 K) to be 0.0953/GPa (Wang et al.) and 0.139/GPa (Häfner et al.). Both values differ distinctly. Saunders et al.[38] published ultrasonic measurements. From their data one can derive $\beta_T = 0.1287$/GPa.

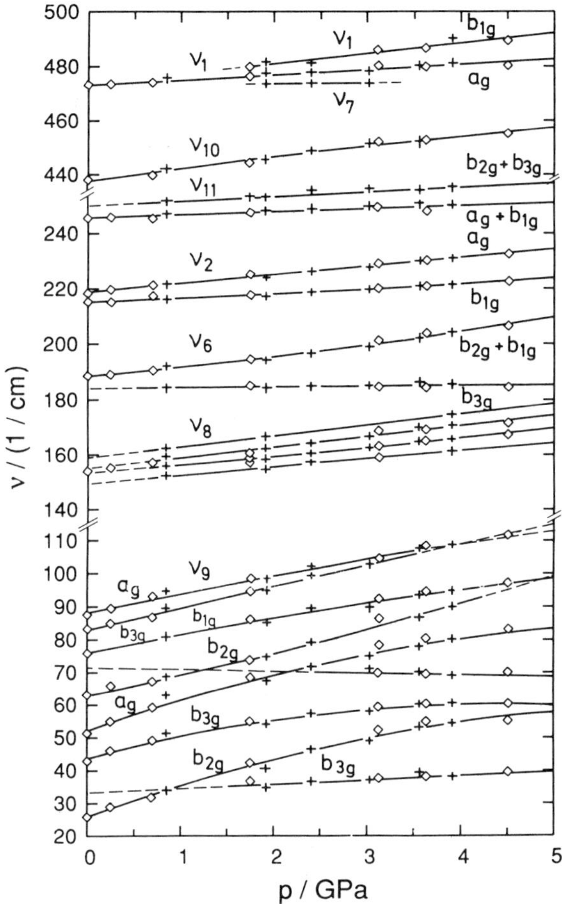

Figure 3. Pressure dependence of α–S_8 mode energies at room temperature. Assignment due to Harvey and Butler[20] and Gautier and Debeau[25].

Values of other literature differ too, but this will not be considered here. It is obvious, that a comparison of our mode Grüneisen parameters with that of different authors depends strongly on underlying values for isothermal compressibility.

These difficulties in interpretation of the different values γ_i form each author is demonstrated in table 1. The left side lists our experimental results: the mode frequency at zero pressure and the parameters of quadratic regression $(\nu(p) = \nu_0 + a \cdot p + b \cdot p^2$, p in kbar). The right side compares the γ_i obtained from different values for the isothermal compressibility. I means the value by Häfner et al., II by Wang et al. Zallen[28] used an older β_T, which corresponds approximately to that of Wang et al. In spite of the problem with β_T differences may come from different pressure ranges: Zallen $0 - 0.9$ GPa, Wang et al. $0 - 5$ GPa, Häfner et al. $0 - 12$ GPa, this work $0 - 6$ GPa. We used the two different values from Wang et al. and Häfner et al. as limits of two extrema.

Table 1. Comparison of our results on pressure shifts with literature

Mode	Sym. (all g)	ν_0 extrapol.	a	b	γ from experiment this work I	II	Häfner[31] (I)	Wang[30] (II)	Zallen[28]	γ theory Kurittu[33]
ν_1	b_1	473.8	0.373	—	0.06	0.08	0.14	0.05	0.067	—
	a	473.1	0.218	0.0006	0.03	0.05				
ν_7		473.3	0.025	—	0.004	0.006	0.19	—	—	—
ν_5		—	—	—	—	—	—	0.06	—	—
ν_{10}		437.9	0.5	−0.0021	0.08	0.12	0.33 / 0.28	0.07	0.1	—
ν_{11}	b_2+b_3	250.2	0.15	—	0.04	0.06	0.17	0.06	0.08	—
	$a+b_1$	246.0	0.09	—	0.03	0.04	0.15	0.05		
ν_2	a	219.1	0.32	—	0.11	0.15	0.41	0.2	0.23	0.30
	b_1	215.6	0.103	0.0014	0.03	0.05	0.35	0.1	0.25	0.20
ν_6	b_2+b_1	188.5	0.286	0.0029	0.11	0.16	0.70	0.02	—	—
	b_3	183.9	0.045	−0.0004	0.02	0.03		0.02	—	—
ν_8	b_2	158.6	0.397	—	0.18	0.26	—	0.59	—	0.60
	a	154.6	0.403	−0.0003	0.19	0.27	0.72	0.40	—	0.34
	b_3	152.2	0.333	—	0.16	0.23	0.54	0.35	0.4	0.51
	b_1	148.9	0.34	−0.0008	0.16	0.24	0.75	0.33	—	0.30
ν_9	a	88.0	0.59	−0.0018	0.48	0.70	1.57	0.8	1.2	1.5
	b_3	82.7	0.69	−0.001	0.60	0.88	1.49	0.8	1.05	1.5
	b_1	75.9	0.59	−0.0022	0.56	0.77	1.4	0.8	2.8	—
external	b_2	63.3	0.56	0.0032	0.64	0.93	1.95	0.8	1.7	1.8
	a	52.4	0.99	−0.0072	1.36	2.0	2.33 / 1.44	2.2	3.1 / 2.5	2.4
	b_3	44.0	0.75	−0.0087	1.23	1.8	1.49	1.9	2.0	2.4
	b_2	25.9	1.05	−0.0081	2.9	4.3	3.23	4.2	4.5	4.4
	b_3	33.7	0.128	—	0.27	0.4	1.88	1.9	—	—

When we compare our spectra with the ones of recent literature there are some similiarities and some differences in addition to the uncertainty coming from different β_T. Most spectral features described above are in accordance to Zallen[28], Slade et al.[29], Wang et al.[30] and Häfner et al.[31]

Theoretical calculated phonon frequency shifts came from Pawley and Mika 1974[32] and Kurittu 1980[33]. Comparison between them and experimental results including ours are quite unsatisfactory. For the external and internal modes we found in general smaller pressure shifts than calculated by Kurittu. Except for the ν_{10} components the experimental shifts are larger. When comparing the mode Grüneisen parameter including those theoretical from Kurittu one can recognize only a gross similar tendency. It is an older finding that the external modes are much more pressure sensitive than the internal modes. This can be seen immediately in fig. 3 or by comparison of the mode Grüneisen parameters. But a sharp distinction between external and internal modes is not possible because of the speciality of α–sulphur modes.

Concluding this section we point out three main aspects: 1. In spite of the problems involved with an exact value for β_T a set of mode Grüneisen parameters is

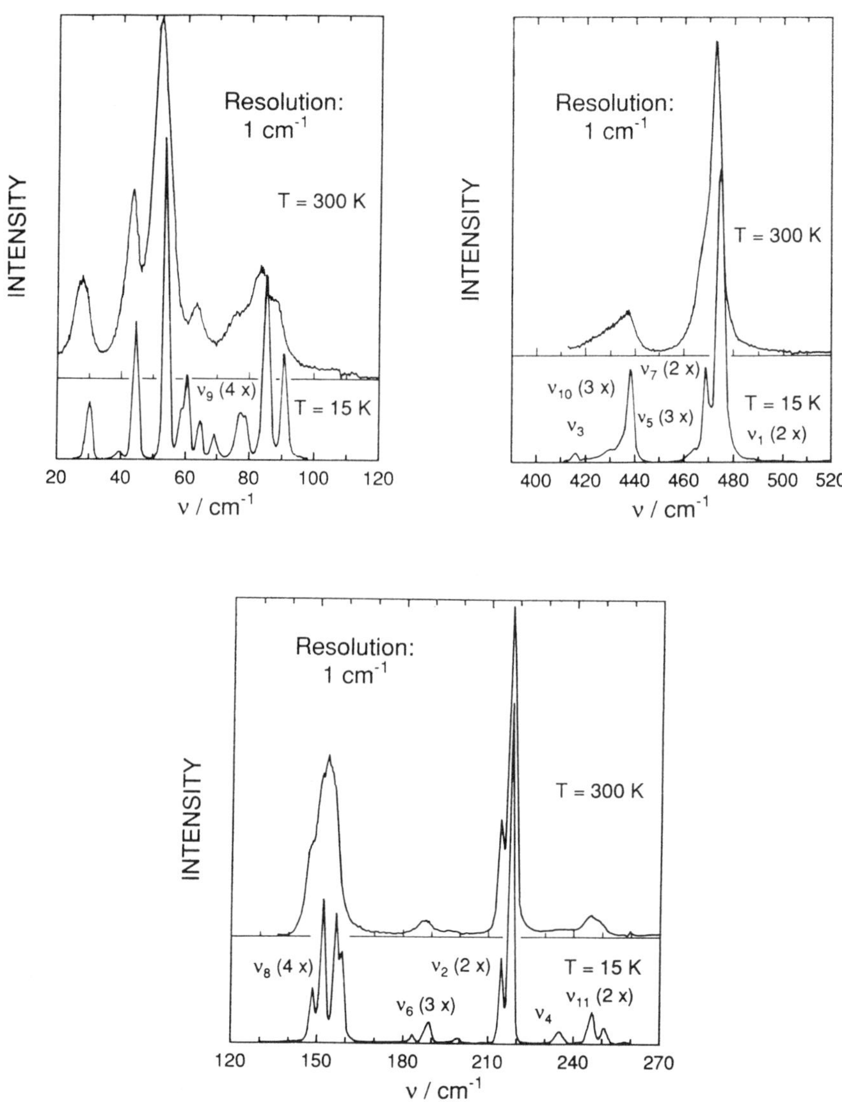

Figure 4. Raman spectra of α–S$_8$ at 300 and 15 K. The assignment is indicated and the degeneracy is numbered in parenthesis. Same assignment as fig. 3.

now available which gives a reliable tendency. 2. As expected γ_i are larger for smaller ν_i. 3. Theoretical values are a factor two larger than experimental ones.

b) Low Temperature Properties at p ≈ 0 GPa.

As mentioned already only incomplete and inconsistent informations are available about the temperature dependence of mode frequencies. The exact knowledge of this total temperature shift $(d\nu/dT)_p$ is in addition important, to discuss anharmonicity in terms of pure thermal volume expansion and of phonon—phonon interaction.

Therefore we investigated the mode frequency shifts of α–S_8 in the temperature range 15–300 K. The local temperature of the crystal was checked by

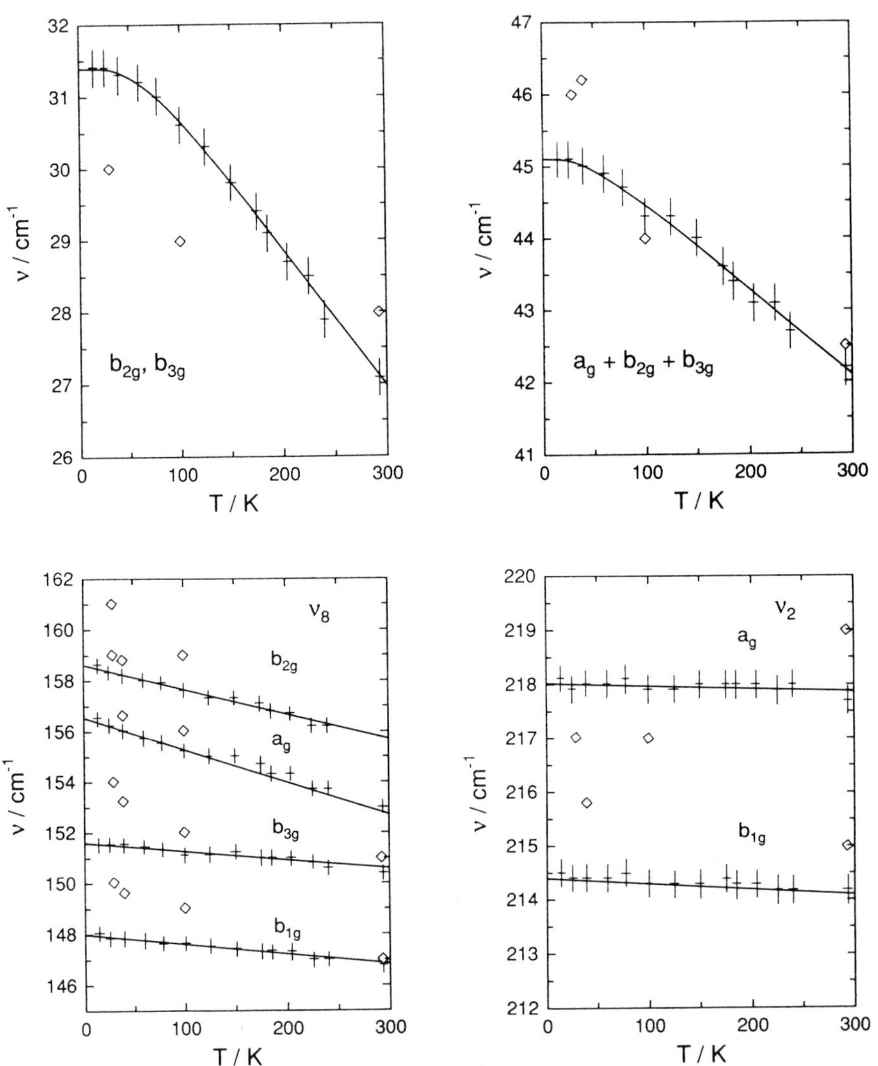

Figure 5.   Temperature dependence of eight selected modes of α–S_8 at ambient pressure. In addition data points from literature: 30 K[25], 40 K[20], 100 and 300 K[19].

Stokes–Antistokes intensity ratios. The temperature accuracy of the Si–diode was better than $\pm$ 5 K. The spectral resolution was 1 cm^{-1} and wavenumber accuracy was $< \pm 0.5$ cm^{-1}. For better assignment of the multicomponent fundamentals we recorded also polarized Raman spectra at low temperatures and at different scattering geometries. Figure 4 shows Raman spectra and figure 5 shows a selection of the temperature dependence of mode energies. Data points from literature are added for comparison. The experimental raw data follow in general a linear regression in the high temperature limit; for most external modes and below < 50 K the fitting curves are bent as expected. In such cases the solid lines in figure 5 are not simply polynomial fit functions but are generated assuming pure phonon–phonon interaction as a first guess. In this model [39,40] a phonon ν_i interacts with an average phonon ν' giving the temperature dependent frequency shift

$$\Delta\nu_i(T) = -B_i \cdot n(\nu',T) \tag{2}$$

where B_i is a temperature independent coupling constant, directly related to cubic and quartic terms of the suited interaction potential; $n(\nu',T)$ is the thermal phonon occupation number

$$n(\nu') = \{\exp(h\nu'/kT)-1\}^{-1}. \tag{3}$$

In case of straight lines this red shift is of the order of 0.001 to 0.01 K/cm^{-1}. In case of bent curves (eq. 2) the coupling constants B_i lye between 0.5 and 4 cm^{-1}; the hath phonons ν' are mostly in the vicinity of modes which are spectroscopically confirmed (a more detailed discussion will be prescribed[41]).

c) Pressure and Temperature Dependence.

Some exemplary results of high pressure studies at various temperatures below 300 K are displayed in figure 6. It shows some external mode frequency shifts and some of components of internal modes. The modes have pressure shifts of about 0.1 –10 cm^{-1}/GPa and temperature shifts of 0.001 $-$ 0.03 cm^{-1}/K. It was mentioned above that the external modes and the low frequency internal modes are much more sensitive against pressure than the remaining high frequency internal modes. In figure 6 one can see the influence of temperature at a given pressure is small and sometimes within the experimental error ($\sim \pm 0.3$ cm^{-1}), sometimes showing the correct trend (e.g. ν_1). The total temperature shift between 12 and 300 K at zero pressure can be recognized by filled circles and squares for better comparison.

It was not possible in the experiment to obtain data over the full pressure and temperature ranges because of the occurance of another sulphur allotrope. However, figure 6 shows an evident discrepancy between pressure and temperature dependence within the experimental error, the latter indicated only for the 300 K values. One can therefore deduce that the mode Grüneisen parameter is qualitatively independent of temperature in the range from 100 to 300 K. We can summarize: There are now a lot of data due to the mode Grüneisen parameter available. The problem for further evaluation is the uncertainty in isothermal compressibility β_T.

If one knows α_p and β_T one is able to separate phonon–phonon interaction and thermal volume expansion. Exactly herein lies the problem.

The different recent values for β_T are discussed above. For α_T we found two values in the literature. $1.03 \cdot 10^{-4}$/K reported Coppens et al.[42] These value was used by Slade et al.[29] to discuss the explicit and implicit terms. Recently Wallis et al.[43] investigated $\alpha_p(T)$ in the temperature range 98–297 K by X–ray powder diffractometry. Their fitted parameters suggest an anomalous behaviour for the lattice constants a and b and for the volume. At 300 K their value for α_p is equal to

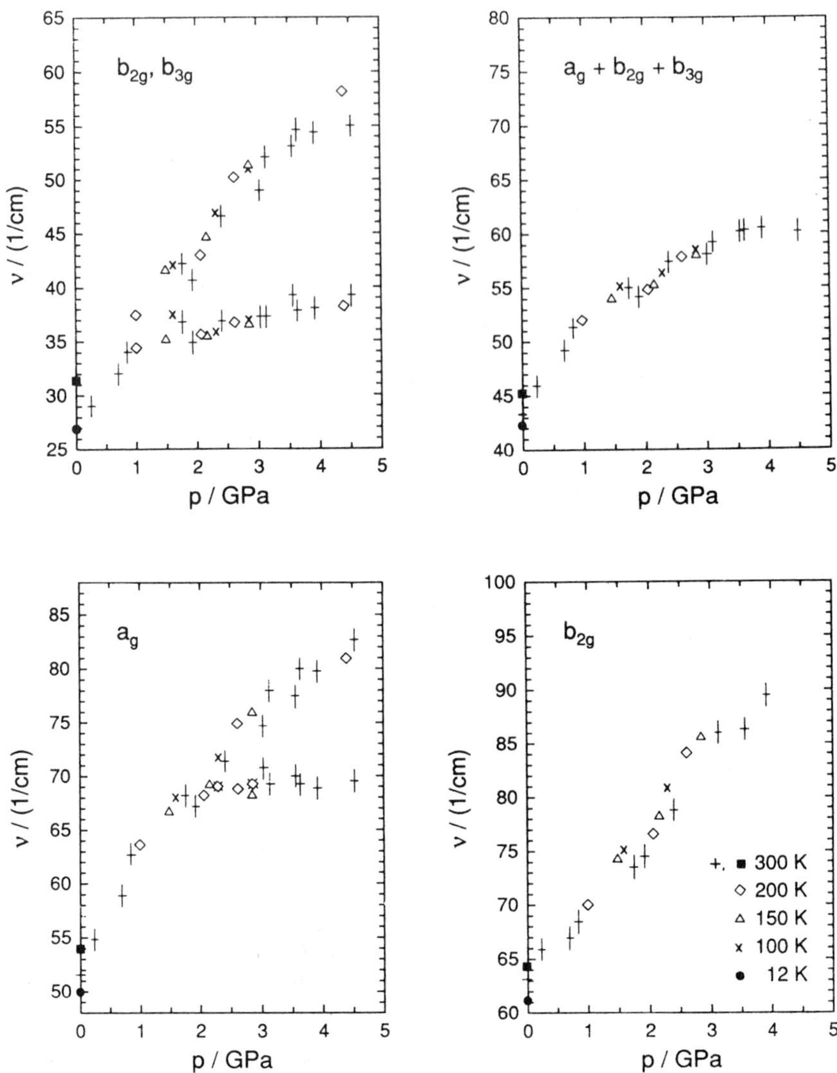

Figure 6a. Pressure shifts of some selected α–S$_8$ external modes at different temperatures (see legend). The data points marked by filled circles and squares came from our separate low temperature — zero pressure runs as a check of possible inconsistency.

2.56 · 10^{-4}/K. There is a large difference to the Coppens value. In view of all different values for thermal expansion α_p and compressibility β_T the reciprocal quotient $(\alpha_p/\beta_T)^{-1}$ enclosed the range 372 − 1350 K/GPa. For comparison, our experimental mode temperature–pressure compensation ratio[29] yields 400 − 600 K/GPa for modes below 100 cm^{-1}, and extremely different ones for the remaining modes, e.g. 27.5 K/GPa for the ν_7 and 7600 K/GPa for the ν_2–α_g mode. Hence there is a large uncertainty to seperate the two parts of the temperature shift. We must conclude that it is very necessary to know more about $\alpha(p,T)$ and $\beta(p,T)$ and to get more consistency. Sulphur is neither difficult to handle nor novel and exotic. Finishing we can state that the gross trend in low temperature and high pressure behaviour of α–S$_8$ is established and exhibits typical features of a molecuar crystal.

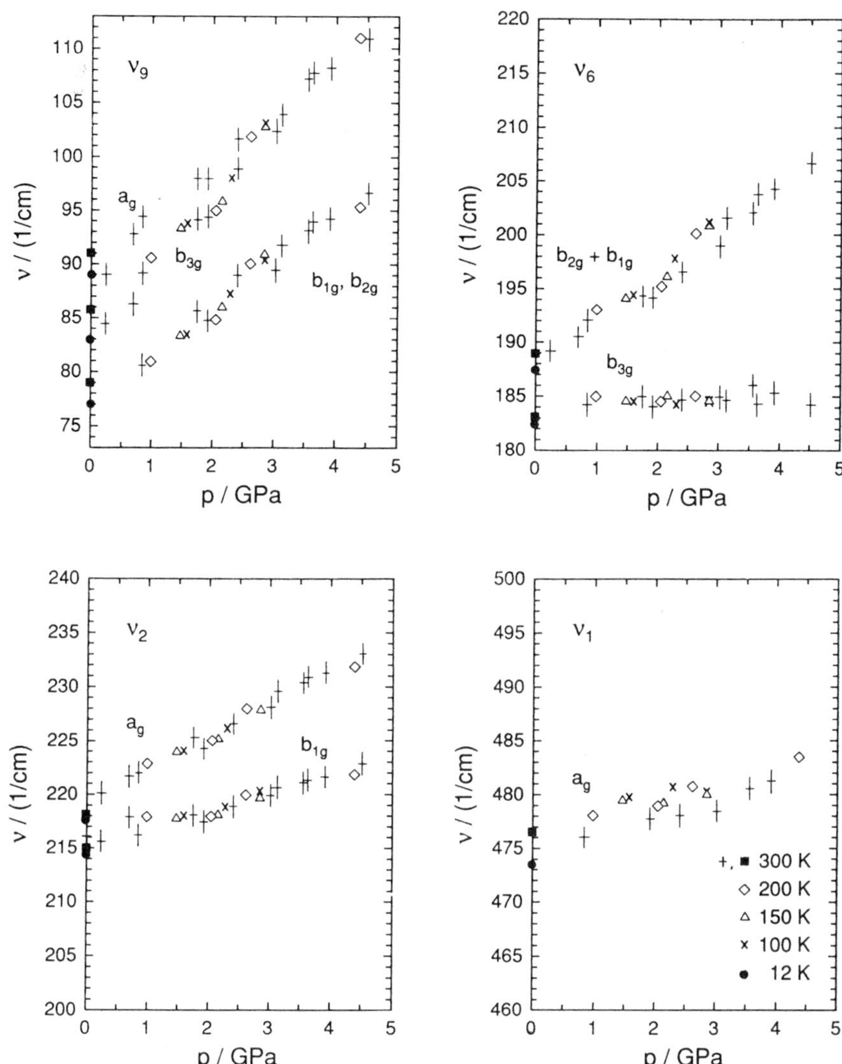

Figure 6b. Same as figure 6a for some internal modes.

3.2. High Pressure — Low Temperature Amorphous Sulphur

We made all high pressure — low temperature variations on different samples in the one way explained at the end of the experimental section.Cooling down the sample (below 200 K) at certain pressures (above 5 GPa) we observed characteristic changes in the Raman spectra (like figure 7). At higher temperatures and lower pressure we observed complicated spectra which can be explained qualitatively by a mere sum of spectra of α–S_8 and of spectra such as figure 7. Either varying temperature or pressure hysteresis effects were dominant when we tried to fix these characteristic changes in spectra. At the end (p → 0 GPa, T ≈ 300 K) spectra associated with α–S_8 always reappeared completely. As can be seen in figure 7, the intensities decreased two orders of magnitude and two prominent features can be identified. In detail: In the mixed phase (α–S_8 and this unknown S) the intensities between ν_2 and

ν_1 change weakly in favour of a broad band close below the ν_1 mode. In addition a mode around 420 cm^{-1} appears, coincides in energy with the inactive ν_3 of S_8. These two bands together show a width of about 100 cm^{-1}. At pressures below 3–4 GPa we observed a broad band below 300 cm^{-1} with a maximum of about 50–100 cm^{-1}. At higher pressures this band is more flat, losses intensity and possess an edge at 300 cm^{-1}.

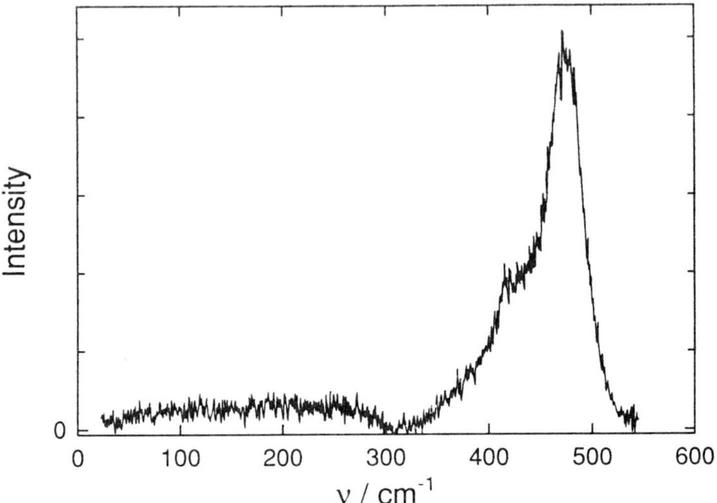

Figure 7   Raman spectrum of pressurized and cooled α–S_8. Apparent peaks lie at 418 and 475 cm^{-1}, latter possing a FWHM of $\sim$ 40 cm^{-1}.

We call this form of sulphur amorphous (a–S) in consequence of the following considerations. The line at 475 cm^{-1} is typically for the S–S stretching mode in the most sulphur allotropes[8]. The broadening of this mode can be explained by loss of one characteristic S–S distance. The external and the internal mode intensities specific for α–S_8 vanish extremely except of the ν_1. This indicates a strong distortion of both the molecular and the lattice structure. So what is dominant in the Raman spectra reflects S–S bondings with averaged interatomic distances.

It is well known for molecular crystals that the covalent character of binding becomes more important when they are affected by pressure.[2] In the case of crystalline sulphur s–p hybrids form the bonding between the S atoms resulting in a strong orientation. So we suppose a threadwork like a chain. For amorphous solids it was shown by Shuker and Gammon[44] that the selection rule $\vec{k} = 0$ in the Raman scattering breaks down. If our assumptions are correct, the broad band below 300 cm^{-1} in our spectra reflects the density of states, because all phonons contribute now.

A comparison with other well known amorphous solids, e.g. Si, Ge, SiO, As_2S_3, As_2Se_3 yields some similar features. Also amorphous Se and Te, which possess the same bonding character as sulphur, yield Raman spectra like ours.

If we compare the spectrum of a–S (figure 7) with the one of $S\mu$ (figure 8, section 3.3) − with respect to the broad band and the asymmetric doublet at 418 and 475 cm^{-1}, we recognize a certain coincidence; as a hypothesis we consider a–S may be a broken S_8 ring, forming a helical like structure. In addition there are very small differences in S–S bond distances and bonding angles between S_8 and $S\mu$.[8]

Dultz et al.[45] calculated the density of states of a planar zig–zag chain as a model for the S_μ helix and compared the result to their measured Raman spectrum (see their figure 3). Their calculated spectrum exhibits an optical phonon doublet like S_μ and a broad band of acoustic phonons with an edge at 220 cm^{-1}, whereas the cut–off band in S_μ spectra lies close below 300 cm^{-1}. Broading a spectrum of S_μ the result is very similar to our a–S Raman spectra.

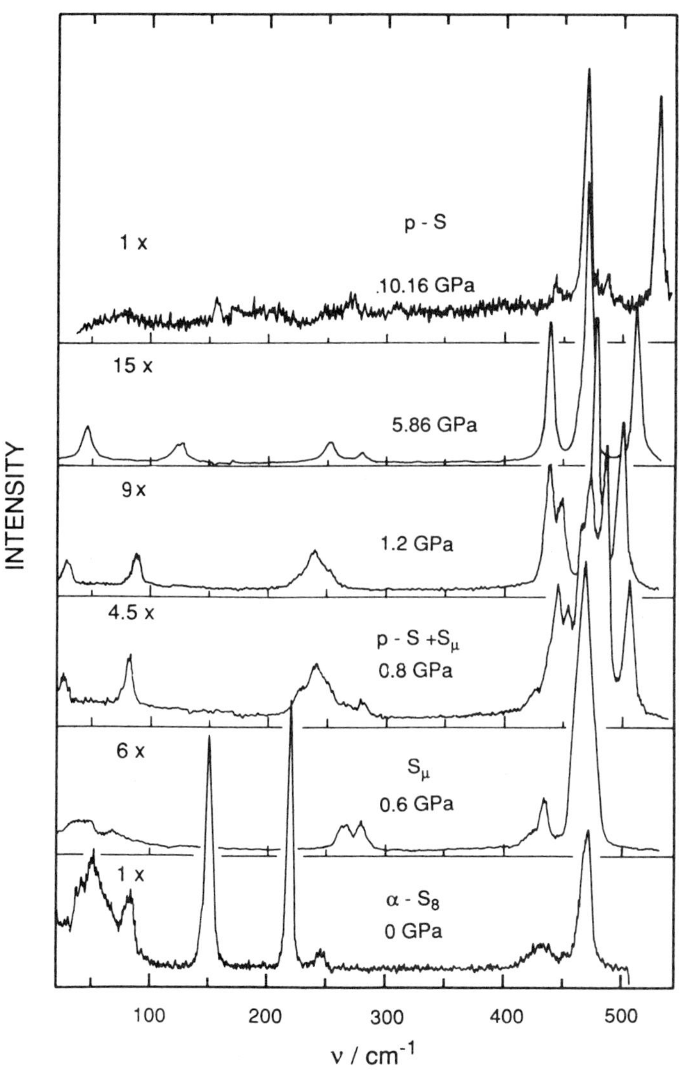

Figure 8.   Raman spectra of p–S while decreasing pressure at ambient temperature. Spectra of different phases are marked.

3.3. Pressure Dependence of p–Sulphur

Both, α– and a–sulphur can transform to a modification first reported by Wolf et al.[46] They investigated this photoreaction to form this modification by varying pressure and laser wave–length. They excluded chemical reactions of sulphur with the gasket or rubys and also heating by laser irridation. The authors reported on Stokes–Antistokes measurements which failed and which maybe a consequence of resonance or absorption effects. They supposed this photosensitive sulphur (p–S) to be S_μ, the helical chain structured sulphur allotrope.

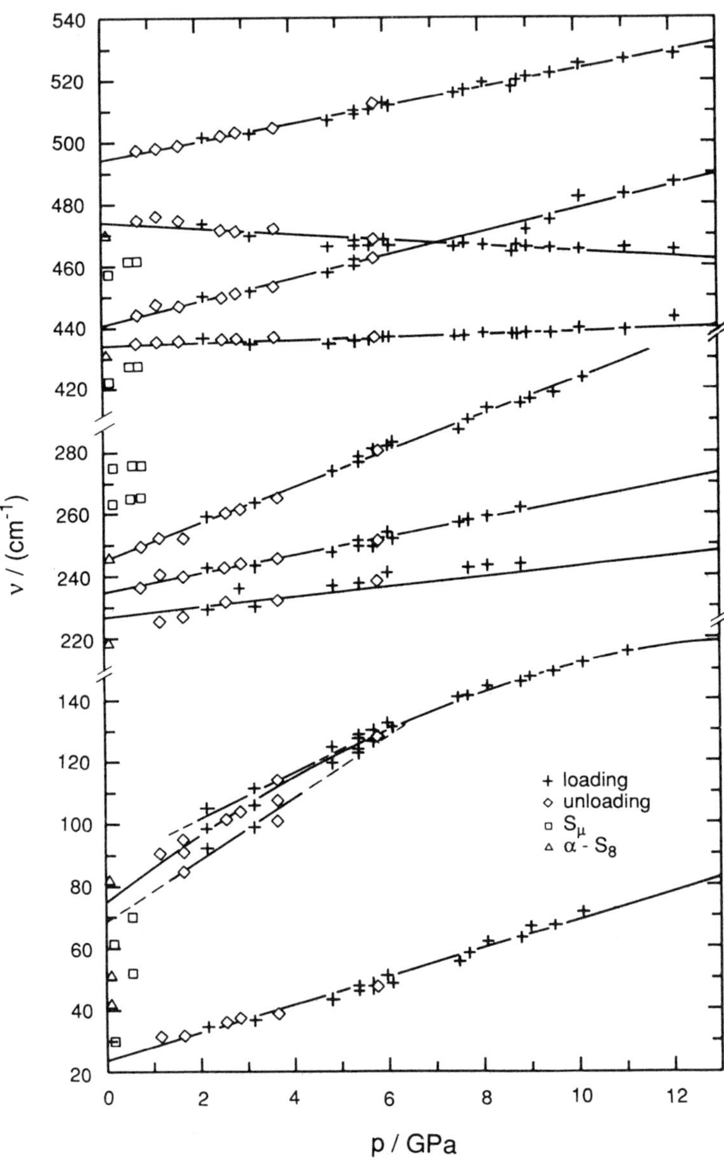

Figure 9.   Pressure dependence of p–S mode energies, additionally occuring S_μ and α–S_8 mode energies during further unloading.

We ourselve produced p–S from a α–S$_8$ single crystal at pressures of about 5.5 GPa at ambient temperature in a DAC by 514.5 mm laser irridation. The spectral features (see figure 8), e.g. frequencies and intensities of the modes, are in agreement with the ones reported by Wolf et al. We could obtain the pressure dependency of modes from 1 up to 10 GPa. p–S was under this condition in a stable form for many weeks. We conclude from our observations, that p–S was not produced each time by laser irridation. Also we observed a formation of p–S by irridation of a–S at pressures near 5 GPa and temperatures above 200 K. At the low temperature limit p–S is not stable at all. After irridation and when laser light is absent p–S reverts to the amorphous phase within less than one hour.

Figure 8 displays Raman spectra at various pressures at T $\approx$ 300 K. Between 10 and 1.2 GPa one can see spectra of pure p–S. Between 0.6 and 0.2 GPa spectra occured known in literature [45,9,47] as S$_\mu$. The spectrum at 0.8 GPa owns features of p–S and S$_\mu$ together. The spectrum at 0 GPa is that of α–S$_8$. We summarize: After producing p–S from α–S$_8$ one can observe its Raman spectra in the range $1 - 10$ GPa. When unloading the DAC p–S reverts to S$_\mu$ at about 0.8 GPa and total unloading transforms S$_\mu$ into α–S$_8$.

Figure 9 shows the pressure dependence of p–S mode energies for the first time. Except the line around 80 cm^{-1} all modes exhibit a linear pressure dependency (see table 2). The 475 cm^{-1} line possess a negative shift. Within an experimental error of about $\pm$ 0.5 cm^{-1} we observed no differences between loading and unloading the sample in the DAC.

Table 2. Mode frequencies of p–S extrapolated to p = 0 GPa and linear and quadratic regression coefficients.

$\nu[\mathrm{cm}^{-1}]$,p=0GPa	a$[\mathrm{cm}^{-1}/\mathrm{GPa}]$	b$[\mathrm{cm}^{-1}/\mathrm{GPa}^2]$
24	4.5	—
68	10	—
74	12	–0,37
82	7.8	—
223	2.7	—
235	3.0	—
244	5.9	—
435	0.48	—
441	3.8	—
475	–0.96	—
492	3.2	—

According to Wolf et al.[46] we assume p–S having a definite molecular structure because of sharp distinct lines in the Raman spectrum (fig. 8). The same consequence is valid for the low lying energetic lines, suggesting a certain crystal structure. In contrary to Wolf et al. we cannot equalize p–S with S$_\mu$ in view of specific spectral features and their evolution under increasing pressure. In addition the number of lines indicates a more compelx structure for p–S than the S$_\mu$ helix. On the other hand p–S is not a sulphur form found by Häfner et al.[31] with regard to the same argument.

Up to now there is not enough information known about this photosensitivity. Meyer [48,8,49] emphasized several times an absorption of α–S$_8$ in the green light, which is responsible for photoreactions and for a breaking of the S–S bond. In addition the absorption edge is measured as a function of temperature[48] and of pressure[50]. This fact cannot directly explain this photosensitivity. For example there is a gap of about 0.25 eV between the 514.5 nm laser light energy and the absorption

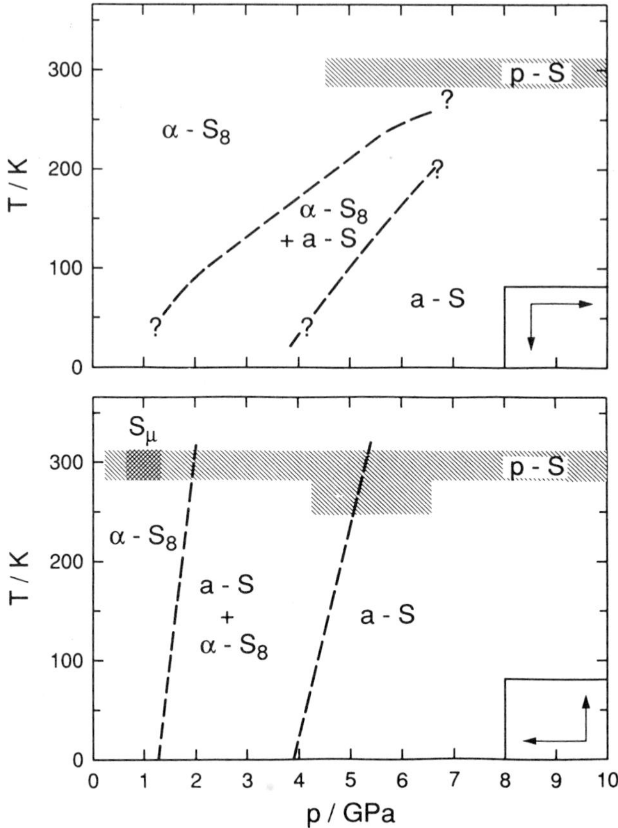

Figure 10. Preliminary p—T diagramm summarizing our results. Upper part: Occurrance and stability ranges of sulphur forms while increasing pressure <u>or</u> cooling down the sample (see arrows). Lower part: Duration and occurrance of sulphur forms as a symbol for hysteresis effects while warming up <u>or</u> decreasing pressure (see arrows).

edge at 5 GPa. Moreover, an absorption for the green can occure at pressures above 8 GPa.[50]

Compiling all the informations about p—S we suggest a structure of this allotrope consisting in S_x fragments, where x is smaller than 8 and larger than 2. There exist a large amount of possible S_x forms already. For example Raghavachari et al.[51] published ab initio quantum chemical calculations on the stability of sulphur clusters. But comparison of our spectra and exptrapolated zero pressure mode energies with their calculated data shows no satisfactory results. So the mechanism of creation and the structure of p—S are still open questions.

4. CONCLUSION AND PRELIMINARY PHASE DIAGRAM

Now one can identify various regions in the low temperature — high pressure phase diagram between 0—7 GPa and 0—300 K (figure 10):
1.   the orthorhombic sulphur region,
2.   a broad band of mixed amorphous and orthorhombic sulphur,
3.   the amorphous range,
4.   some photosensitive areas, which allow the creation of allotropes by laser power (Häfner et al.[31]) and/or laser energy (Wolf et al.[46]).

α–S_8 exhibits large pressure but small temperature effects. This was shown in the first low temperature – high pressure experiments on α–sulphur. Within small experimental uncertainties the mode Grüneisen parameter should be temperature independent. Because of the lack of certain values $\alpha_p(p,T)$ and $\beta_T(p,T)$ one is unable to draw any final conclusion about the phonon–phonon fraction and the volume driven fraction in mode temperature shifts.

We found a new phase, whose spectroscopic features indicated an amorphous structure.

For p–S we reported mode pressure dependencies for the first time. We found also various transition channels between α–S_8, a–S, S_μ on one hand and p–S on the other hand.

REFERENCE LIST

1. S. Califano, V. Schettino, and N. Neto, Lattice Dynamics of Molecular Crystals, Springer, Berlin (1981).
2. B.A. Weinstein, and R. Zallen, Pressure–Raman Effects, in: Light Scattering in Solids, M. Cardona, and G. Guntherodt, eds., Springer, Berlin (1984).
3. A.J. Kitaigorodski, Molekülkristalle, Akademie, Berlin (1979).
4. B.E. Warren, and J.T. Burwell, J. Chem. Phys. 3 (1935) 6.
5. S.C. Abrahams, Acta Cryst. 8 (1955) 661.
6. S.J. Rettig, and J. Trotter, Acta Cryst. C43 (1987) 2260.
7. B. Meyer, Sulfur, in: The Structure of the Elements, J. Donohoe, ed., Wiley, New York (1974).
8. B. Meyer, Chem. Rev. 76 (1976) 367.
9. H.J. Mäusle, and R. Steudel, Z. anorg. allg. Chem. 156 (1981) 125 and 177.
10. R. Steudel, Homocyclic Sulfur Molecules, in: Topics in Current Chemistry, Vol. 102, F.L. Boschke, ed., Springer, Berlin (1982).
11. D.W. Scott, J.P. McCullough, and F.H. Kruse, J. Mol. Spectry. 13 (1964) 313.
12. C. Domingo, and S. Montero, J. Chem. Phys. 74 (1981) 862.
13. G. Cardini et al., to be published (1991).
14. G.S. Pawley, R.P. Rinaldi, and C.G. Windsor, The Lattice Dynamics of Orthorhombic Sulphur, in: Proc. of the Int. Conf. on Phonons, M.A. Nusimovici, ed., Flammarion, Rennes (1971).
15. R.P. Rinaldi, and G.S. Pawley, J. Phys. C 8 (1975) 599.
16. T. Luty, and G.S. Pawley, Phys. Stat. Sol. 69 (1975) 551.
17. C.M. Gramaccioli, and G. Filippini, Chem. Phys. Lett. 108 (1984) 585.
18. D. Dows, priv. comm. (1989), and (1991).
19. A. Anderson, and Y.T. Loh, Can. J. Chem. 47 (1969) 879.
20. P.D. Harvey, and I.S. Butler, J. Raman Spectr. 17 (1986) 329.
21. G.A. Ozin, J. Chem. Soc. A (1969) 116.
22. J.W. Arthur, and G.A. Mackenzie, J. Raman Spectr. 2 (1974) 199.
23. A. Anderson, and L.Y. Wong, Can. J. Chem. 47 (1969) 2713.
24. A. Anderson, and P.G. Boczar, Chem. Phys. Lett. 43 (1976) 506.
25. G. Gautier, and M. Debeau, Spectrochim. Acta 30 A (1974) 1193.
26. R. Bini, B. Eckert, H.J. Jodl, and P. Foggi, to be published.
27. M. Becucci, E. Castellucci, P. Foggi, S. Califano, and D. Dows, to be published.
28. R. Zallen, Phys. Rev. B 9 (1974) 4485.
29. M.L. Slade, R. Zallen, and B.A. Weinstein, Bull. Am. Phys. Soc. 27 (1982) 163.
30. L. Wang, Y. Zhao. R. Lu, Y. Meng, Y. Fan, H. Luo, Q. Cui, and G. Zou, High Pressure Raman and X–Ray Studies of Sulfur and its new Phase Transition, in: High Pressure Research in Mineral Physics, M.H. Manghnani, and Y. Syono, eds., Terra Scientific Publishing, Tokyo (1987).
31. W. Häfner, J. Kritzenberger, H. Olijnyk, and A. Wokaun, High Pressure Research 6 (1990) 57.

32. G.S. Pawley, and K. Mika, Phys. Stat. Sol. 66 (1974) 679.
33. J.V.E. Kurittu, Physica Scipta 21 (1980) 200.
34. L.G. Liu, and W.A. Bassett, Elements, Oxides, and Silicates, Oxford University Press, Yew York (1986).
35. G. Huber, K. Syassen, and W.B. Holzapfel, Phys. Rev. B 15 (1977) 5123.
36. H.K. Mao, P.M. Bell, J.W. Shanner, and D.J. Steinberg, J. Appl. Phys. 49 (1978) 3276.
37. B.A. Weinstein, Rev. Sci. Instrum. 57 (1986) 910.
38. G.A. Saunders, Y.K. Yogurtçu, J.E. Macdonald, and G.S. Pawley, Proc. R. Soc. Lond. A407 (1986) 325.
39. K.S. Viswanathan, Can. J. Phys. 41 (1963) 423.
40. F.D. Medina, and W.B. Daniels, J. Chem. Phys. 64 (1976) 150.
41. B.Eckert, H.J. Jodl, H.O. Albert, and P. Foggi, to be published.
42. P. Coppens, Y.W. Yang, R.H. Blessing, W.F. Cooper, and F.K. Larsen, J. Am. Chem. Soc. 99 (1977) 760.
43. J. Wallis, J. Sigalas, and S. Hart, J. Appl. Cryst. 19 (1986) 273.
44. R. Shuker, and R.W. Gammon, Phys. Rev. Lett. 25 (1970) 222.
45. W. Dultz, H.D. Hochheimer, and W. Müller–Lierheim, One–phonon density–of–states from the Raman spectrum of disordered linear chains: fibrous sulphur, in: Proc. 5th Int. Conf. Amorphous and Liquid Semicond., J. Stuke, and W. Brenig, eds. Taylor and Francis, London 1974.
46. P. Wolf, B. Baer, M. Nicol, and M. Cynn, preprint (1990).
47. A.T. Ward, J. Phys. Chem. 72 (1968) 4133.
48. B. Meyer, M. Gouterman, B. Jensen, T.V. Oommen, K. Spitzer, and T. Stroyer–Hansen, Sulfur Research Trends (1972) 53.
49. B. Meyer, Sulfur, Energy and Environment, Elsevier, Amsterdam 1977.
50. K. Syassen, priv. communication, (1990).
51. K. Raghavachari, C. McMichael Rohlfing, and J.S. Binkley, J. Chem. Phys. 93 (1990) 5862.

PRESSURE DEPENDENCE OF RAMAN LINEWIDTHS OF MOLECULAR CRYSTALS

M. Jordan, H. Däufer, H.-J. Jodl

Fachbereich Physik,
Universität Kaiserslautern, FRG

Introduction

Investigation of relaxation processes by means of high resolution Raman spectroscopy (HRRS) or coherent antistokes Raman spectroscopy (CARS) has been developed in recent years for testing and modelling potentials of molecular crystals[1,2] like N_2[3], CO_2[4], naphtalene[5], anthracene[6], benzene[7] or CS_2[8]. Until now the linewidth (or equivalently the lifetime) of the optical internal and external phonons were measured at different temperatures and at constant (ambient) pressure. Even though relaxation theory assumes constant volume for all temperatures, experimental and calculated data are in good agreement as has been demonstrated on CO_2[4] or naphtalene[5].

The relaxation of a phonon is governed by the third and/or higher order derivatives of the crystal potential with respect to the normal coordinates at equilibrium position. Consequently potential parameters can be determined only for this configuration.

Application of pressure to the sample by means of a diamond anvil cell (DAC) is a well established method to vary the intermolecular distances on a 10 per cent scale, typically between 0 and 10GPa. The intramolecular distances show a smaller variation. Measuring the linewidths at different pressures at low or even at several temperatures would yield experimental data which have to be reproduced by the theoretical potentials. That means not only the anharmonic lineshifts but also the linewidths have to be predicted by the same potential.

In this paper we report our recent pressure dependent measurements of Raman linewidths in single crystals of CS_2 at 20K. Carbon disulphide crystallizes with space group D_{2h}^{18} (C_{mca}) having two molecules per primitive unit cell which occupy sites of C_{2h} symmetry. Group theory predicts four Raman active lattice modes (A_g, B_{1g}, B_{2g}, B_{3g}), from which only three have been observed, and a split v_1 mode (A_g, B_{3g}). Furthermore Fermi resonance occurs between v_1 and the overtone of the Raman forbidden v_2 mode resulting in a $v_2(\mathbf{k})+v_2(-\mathbf{k})$ band and two bound states which are peaked at the upper edge of the band (v^+) and 150cm^{-1} lower (v^-). A schematic phonon spectrum is depicted in Fig. 1.

We recorded spectra of the fundamental modes in high resolution (<0.01 cm^{-1}) at 20K and p<3 GPa.

Frontiers of High-Pressure Research, Edited by H.D. Hochheimer and
R.D. Etters, Plenum Press, New York, 1991

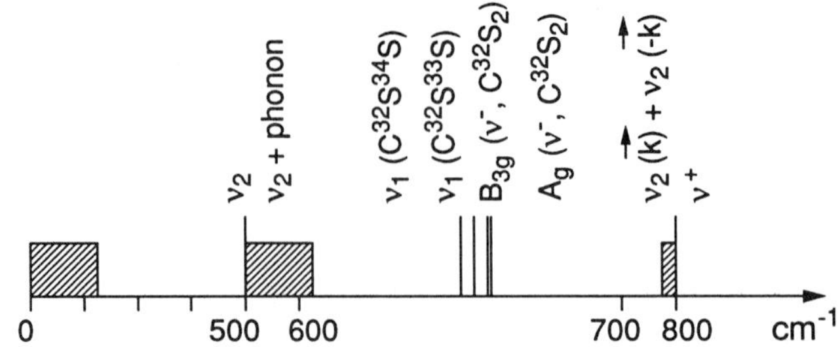

Fig. 1   Schematic of the vibrational spectrum of CS_2.

Experimental

Three single crystals of good optical quality were grown in the DAC at room temperature from commercial grade CS_2 without further purification (Crystal III was made of a glass distilled CS_2 [Aldrich product number 34,227-0] which contains a reduced amount of organic impurities.).

The sample size was 150μm in diameter and 80μm in thickness. Both decreased with increasing pressure, especially during crystal growth. The DAC was mounted in a closed cycle cryostat which allows long term cooling and measurements. Crystals I and II were cooled down during 6 hours from room temperature to 20K; 100K was reached after only 2 hours. The third crystal was cooled at a rate of approx. 12K/h so that the sample was at its lowest temperature of 20K after about 24 hours. The temperature was measured by two calibrated silicon diodes mounted at the top and at the bottom of the DAC. In addition Stokes/antistokes spectra were recorded to obtain an independent temperature estimate of the sample. No discrepancy was found between these two methods (ΔT<3K). Pressure was increased only at room temperature to reduce the generation of stress in the sample.

Raman spectra were excited by an argon ion laser running on the 488nm line in single mode at a power of 100mW. The power density over the sample area should not be too high (typically 100mW at a focus diameter of 50μm) because CS_2 tends to decompose into carbon and sulfur at pressures above approx. 2.5GPa at 20K. Raman spectra were obtained by a standard high resolution setup consisting of a pressure tuned Fabry-Perot interferometer as the high resolving element and a triple monochromator as the pre-selector[9]. The FPI was equipped with spacers of thickness 3.0mm and 2.0mm

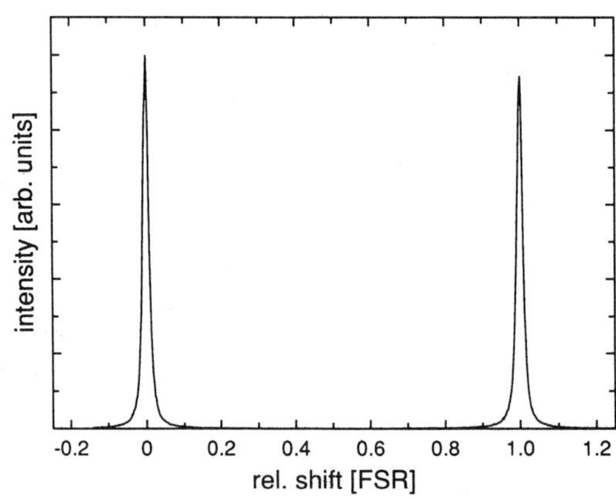

Fig. 2.   Instrumental function for a free spectral range (FSR) of 1.7cm^{-1}.

corresponding to a free spectral range of 1.7cm^{-1} and 2.5cm^{-1} resulting in a bandpass of typically 0.04cm^{-1} respectively. The width of the bandpass can be determined with a precision of 0.001cm^{-1}. Linewidths smaller than the bandpass can be measured easily if a line-profile is assumed; in our cases the interferograms were deconvoluted by a Lorentzian. A typical interferogram of the laser is shown in Fig. 2.

Theoretical Background

Finite Raman (and of course infrared) linewidths are a consequence of the presence of anharmonic terms in the expansion of the crystal potential. In the harmonic approximation the lifetime of phonons is infinitely large and the corresponding line profile is a delta peak. Anharmonic terms couple different modes to such an extent that depopulation and dephasing can occur.

The expansion of the potential can be written according to[1,2] as:

$$V = V_2 + V_3 + V_4 + \dots$$

$$= \frac{1}{2!} \sum_{lm} C_{lm} Q_l Q_m + \frac{1}{3!} \sum_{lmn} C_{lmn} Q_l Q_m Q_n + \frac{1}{4!} \sum_{lmno} C_{lmno} Q_l Q_m Q_n Q_o + \dots$$

where

$$C_{lm\dots s} = \left(\frac{\partial^s V}{\partial Q_l \partial Q_m \dots \partial Q_s} \right)_0 \delta(\mathbf{k}_l + \mathbf{k}_m + \dots + \mathbf{k}_s).$$

Introduction of phonon creation and annihilation operators leads to coefficients B which are related to C by

$$B_{lm\dots s} = \frac{1}{n!} \left(\frac{\hbar^n}{2^n \omega_l \omega_m \dots \omega_s} \right)^{\frac{1}{2}} C_{lm\dots s}.$$

Application of the thermal Green's function method results in corrections to the line position and contributions to the linewidths which are the sum of three and four phonon up and down processes and four phonon pure dephasing processes.

As an example the linewidth evolution with temperature for a three phonon down process is given by

$$\Gamma_l^{(3d)} = \frac{36 \pi}{\hbar^2} \sum_{m,n} |B_{lmn}|^2 (n_m + n_n + 1) \, \delta(\omega_l - \omega_m - \omega_n)$$

where m and n are composite indices comprehensive of the phonon branch label j and of the wavevector label k, and ω_m and ω_n are the frequencies of the phonons, in which ω_l is decaying.

To complete the picture the three phonon up process assumes the form:

$$\Gamma_l^{(3u)} = \frac{72 \pi}{\hbar^2} \sum_{m,n} |B_{lmn}|^2 (n_m - n_n) \, \delta(\omega_l + \omega_m - \omega_n)$$

where ω_l forms by fusion with ω_m a new phonon of frequency ω_n.

In the low temperature case ($T \to 0$) the residual linewidth of the three phonon down process is:

$$\Gamma_l^{(3d)} = \frac{36\,\pi}{\hbar^2} \sum_{m,n} |B_{lmn}|^2 \, \delta(\omega_l - \omega_m - \omega_n)$$

whereas the linewidth of the up process vanishes because no populated thermal bath is available. For the same reason pure dephasing is inactive at low temperatures.

The considerations above are only valid for an ideal anharmonic crystal. Real crystals, in contrast, are characterized by the presence of impurities (very often isotopes like in CS_2) and defects, which lead to energy trapping and dephasing. These processes are active even at low temperatures and increase the residual linewidth. However such processes are much more difficult to describe mathematically so that they will not be discussed here.

Formulas for the linewidth contribution of a process at different temperatures are available if constant volume over the whole temperature range is assumed. Unfortunately no such simple equations exist to our knowledge for the case of low temperature and varying pressure. In this case the coefficients B have to be calculated for every configuration corresponding to the applied pressure.

In addition to linewidth measurements we also investigated the splitting of the ν^--mode. This splitting originates from the interaction of two CS_2 molecules on different sites and can be calculated based on gas phase data like molecular polarizability and molecular quadrupole moment.

Results and Discussion

The pressure dependence of the linewidths of several modes and the splitting of the ν^- mode were measured in the pressure range between 0.44GPa and 2.4GPa.

Interferograms were obtained from the second lowest lying libration. The lowest lying one was too week for HRRS investigation and could therefore not be observed. In Fig. 3 the pressure dependent linewidth data are extracted from the interferograms. The fairly high bandwidths are mainly due to the fact that this libration is not the lowest one and three further external modes (two translational and one librational) possess lower frequency so that three phonon down processes can occur even at low temperatures. At higher pressures (above 1.5GPa) we observed an asymmetric lineshape. This may be a sign of the presence of the fourth libration which has not yet been doubtlessly observed in Raman spectroscopy. Since very sharp lines are not to be expected ($0.01cm^{-1}$-$0.1cm^{-1}$) the librational modes are not well suited for testing the crystal quality.

Apart from visual inspection the crystal quality was checked by measuring in the region of the Raman forbidden ν_2 mode. In a bad crystal (e.g. obtained by increasing the pressure at low temperature) this mode becomes visible in Raman spectra because the symmetry is lost due to strain and stress or crystal imperfections. A crystal of good quality shows no ν_2 line but a broad slightly structured vibron-phonon combination band[10]. In our crystals we observed no ν_2 line but only this band which becomes broader with increasing pressure as expected.

A series of conventional spectra of the ν^- mode is presented in Fig. 4a. The three peaks can be assigned to the several isotopes. The strongest

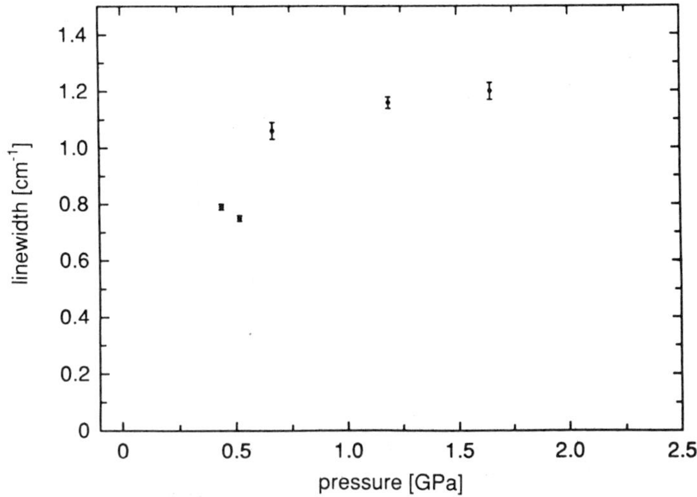

Fig. 3.   Linewidth of the second lowest lying libration versus pressure at 20K.

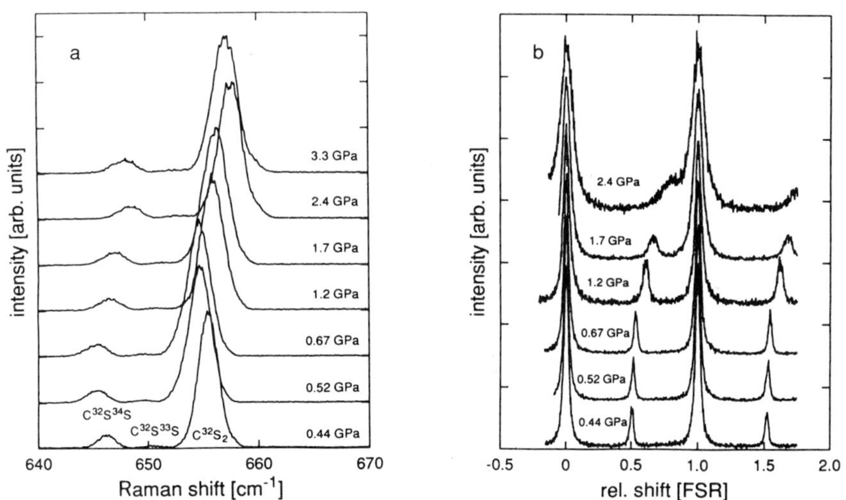

Fig. 4.   Conventional (a) and high resolution spectra (b) of the ν^- mode for several pressures at 20K.

peak ($C^{32}S_2$) and the isotope $C^{32}S^{34}S$ were measured with high resolution. Interferograms of the main peak are shown in Fig. 4b. The linewidth data of the two components A_g and B_{3g} are summarized in Fig. 5 and 6 respectively (for easier comparison of the bandwidth data the main diagram is limited to 0.35 cm^{-1}. The insert contains all datapoints.). The B_{3g} component is up to 1.5GPa narrower than the A_g mode. This can be simply understood because B_{3g} lies 1–2 cm^{-1} lower than A_g. The zero pressure linewidth[8] and our lowest pressure data are in good agreement what we interpret as a further hint for suitable crystal quality. Above 1.5 GPa the lower lying mode gets broader which is probably caused by the onset of decomposition of CS_2 in carbon and sulfur by laser irradiation.

The splitting of v^- is shown in Fig. 7. Extrapolation of our data to zero pressure results in the same splitting as was observed in Ref. 8.

The linewidths of the isotope which are summarized in Fig. 8 are slightly larger than of the two components of v^-. This was not expected because the isotope lies approx. 10 cm^{-1} lower than the doublet. Relaxation processes should hardly occur.

Finally we present the results of the v^+ mode which is a resonant state at the upper edge of the two-phonon band. On the low frequency side the slope of the line is much steeper than on the other side. Nevertheless we fitted the interferograms to a lorentzian lineshape and obtained the bandwidth data of Fig. 9. Our lowest pressure data are again in good agreement with the ambient pressure data[8]. This mode is broader than the low energy component of the Fermi dyad because it can decay via the two-phonon band (Fig. 1). Theory predicts also two components of the v^+ but we found no hint for a doublet structure even under pressure.

The diagrams of the linewidths $\Gamma(p)$ (Figs. 8 and 9) contain some outlying points at the highest pressure values. These data were obtained from measurements in the pressure region where decomposition was observed.

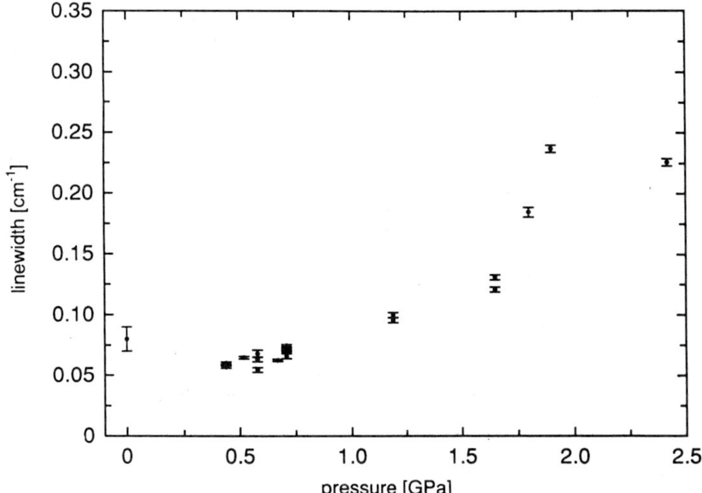

Fig. 5.   Linewidth of the A_g component of the v^- mode versus pressure at 20K.

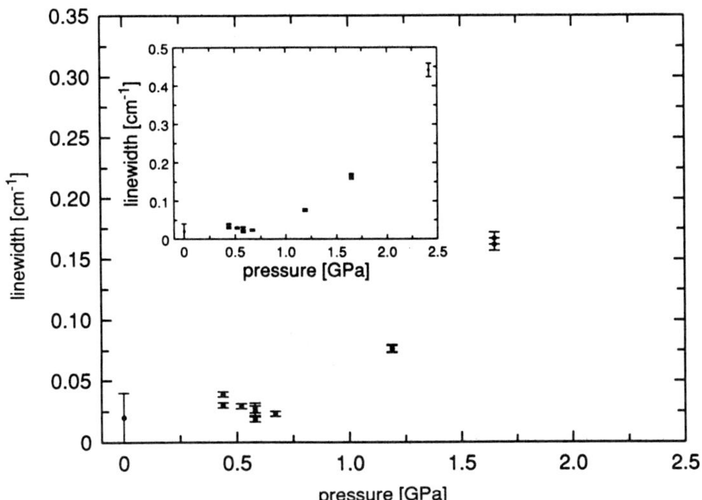

Fig. 6.   Linewidth of the b_{3g} component of the v^- mode versus pressure at 20K.

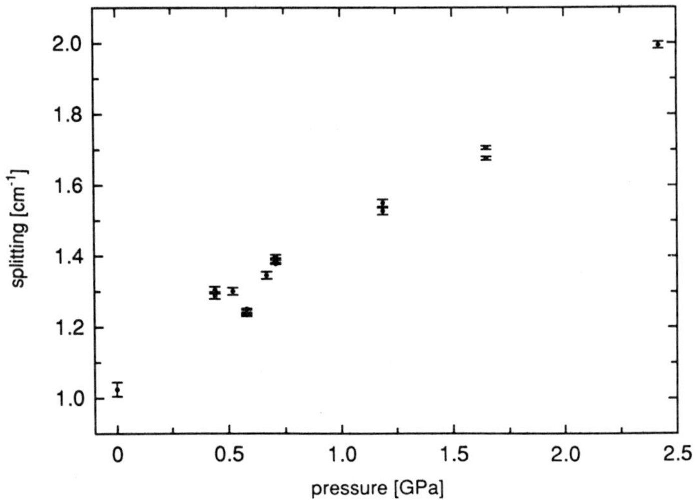

Fig. 7. A_g-B_{3g} splitting of v^- versus pressure at 20K.

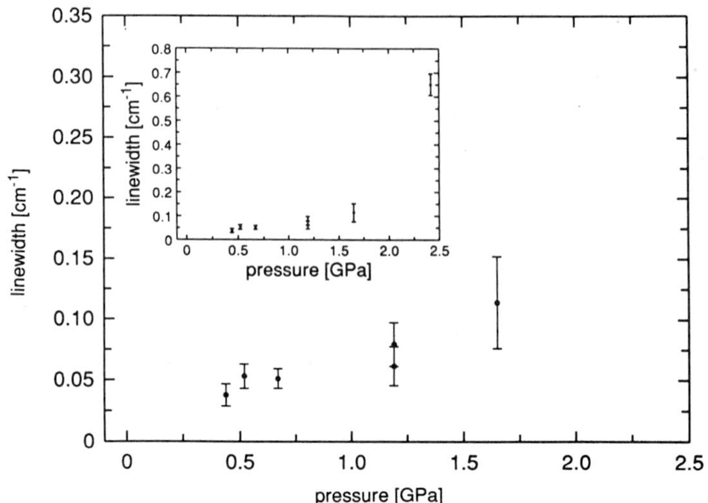

Fig. 8.   Linewidth of the isotope $C^{32}S^{34}S$ versus pressure at 20K.

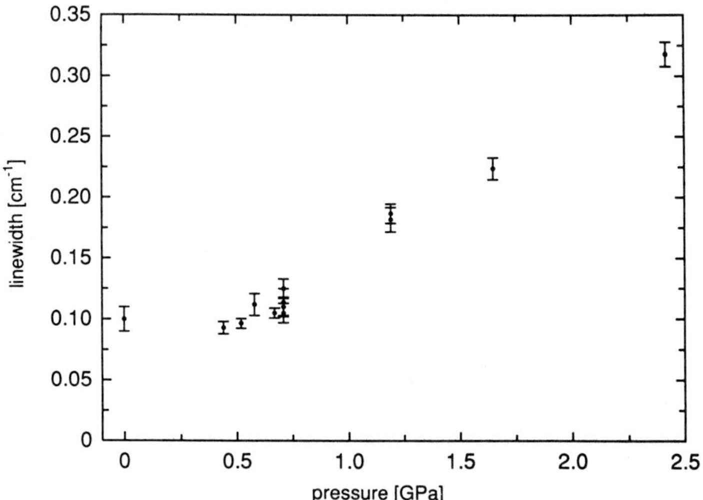

Fig. 9.   Linewidth of the ν^+ mode versus pressure at 20K.

Summary

In this project we have shown in the case of CS_2 - which should be
considered only as one example of molecular crystals - that measuring very
narrow linewidths is also possible at high pressure as is the case at low
temperature. We are able to grow single crystals of sufficiently good
quality; we can increase the pressure without damaging the crystal at
least up to 3GPa.

Due to the lack of theoretical data it is hard to interpret our results. The presence of isotopic impurities leads to further complications
in a mathematical description. To avoid this, future experiments should be
executed on isotopically pure CS_2 such that the decay and dephasing channels caused by isotopes are not available. In principle this should lead
to much smaller linewidths ($0.001 cm^{-1}$) for the two components of v^- [8].

Pressure dependent FTIR investigations at low temperature would be of
great interest because the low lying translations at about 66 cm^{-1} could be
measured where we expect much narrower linewidths than observed in Raman
spectroscopy.

Nevertheless all Raman experiments are restricted to a maximum pressure of 2-3 GPa because pressure induced photochemistry occurs.

To overcome this problem experiments will be executed on CO_2 next by
us at high pressure because there is no photochemistry by laser and/or
high pressure to be expected.

References

1.   S. Califano, V. Schettino, N. Neto: *Lattice Dynamics of Molecular
     Crystals*, (Springer Verlag Berlin, Heidelberg, New York 1981).
2.   S. Califano, V. Schettino: Int. Rev. Phys. Chem. 7, 19, (1988).
3.   R. Ouillon, C. Turc, J. P. Lemaistre, P. Ranson: J. Chem. Phys. 93,
     3005 (1990).
4.   P. Ranson, R. Ouillon, S. Califano: J. Raman Spectrosc. 17, 155,
     (1986).
     P. Ranson, R. Ouillon, S. Califano: J. Chem. Phys. 89, 3592, (1988).
     V. K. Jindal, R. Righini, S. Califano: Phys. Rev. B: 38, 4259,
     (1988).
5.   P. Ranson, R. Ouillon, S. Califano: Chem. Phys. 86, 115, (1984).
6.   R. Ouillon, P. Ranson, S. Califano: Chem. Phys. 91, 119, (1984).
7.   R. Torre, R. Righini, L. Angeloni, S. Califano: J. Chem. Phys. 93,
     2967, (1990).
8.   L. Angeloni, R. Righini, P. R. Salvi, V. Schettino: Chem. Phys. Lett.
     154, 432, (1989).
     G. Cardini, P. R. Salvi, V. Schettino: Chem. Phys. 117, 341,(1987).
9.   A. S. Pine, P. E. Tannenwald: Phys. Rev. 178, 1424, (1969).
10. F. Bolduan, H. D. Hochheimer, H. J. Jodl: J. Chem. Phys. 84, 6997,
     (1986).

CALCULATED HIGH PRESSURE PROPERTIES OF SOLIDS

COMPOSED OF NON-CENTROSYMMETRIC MOLECULES

B. Kuchta and R. D. Etters

Physics Department
Colorado State University
Fort Collins, Colo.

INTRODUCTION

The study of structural phase transitions and other thermodynamic
properties of solids is an active area of experimental and theoretical
research. Essential to this progress are techniques that now make measurements at 100 GPa and at relatively high temperatures routinely available. Advances in calculational skills have also been important. In this
report we discuss recent progress in understanding solid N_2O and CO, which
are particularly interesting because the molecules have no center of
symmetry.

Interest in condensed phases composed of simple linear molecules is a
natural evolution from the monatomic rare gas solids that have been ex-
haustively studied. Because they have orientational, sometimes magnetic,
as well as translational degrees of freedom, the phase diagrams of these
molecular materials are generally complex and various similarities and
differences in the behavior of H_2, N_2, O_2, F_2, CO, CO_2, N_2O, and other
linear systems has been noted.[1] Among these the pairs (N_2, CO) and (CO_2,
N_2O) are particularly interesting because the molecules in each separate
pair are isoelectronic and have identical mass. Thus, it is not surprising that N_2 and CO have very similar properties, as is evident by comparing the phase diagrams on Fig. 1. Note that their molar volumes at zero
pressure and low temperature are nearly identical, as is the melting
curve. The quantitative differences are understood to be primarily a consequence of the larger quadrupole moment in CO which, among other things,
favors the cubic Pa3 phase observed in both solids[2,3] at low temperatures.
However, there are some pronounced quantitative differences. Carbon mon-
oxide, unlike N_2, forms a polymer at relatively low pressures,[4] and it
does not exhibit the tetragonal γ-N_2 phase, which has never been satisfactorily predicted. Finally, because CO is not centro-symmetric, the
relative end-to-end ordering of the molecules is important. For example,
complete order gives the $P2_13$ structure in the ground state whereas random
disorder leads to the Pa3 structure. While it is not settled, there is
increasing evidence that this latter case obtains.[5]

The isoelectronic pair CO_2 and N_2O also have very similar phase diagrams. Until recently there was no evidence of any solid-solid transition
from the cubic Pa3 phase in either of these systems. Because N_2O is not
centro-symmetric, the same uncertainty in the end-to-end order/disorder

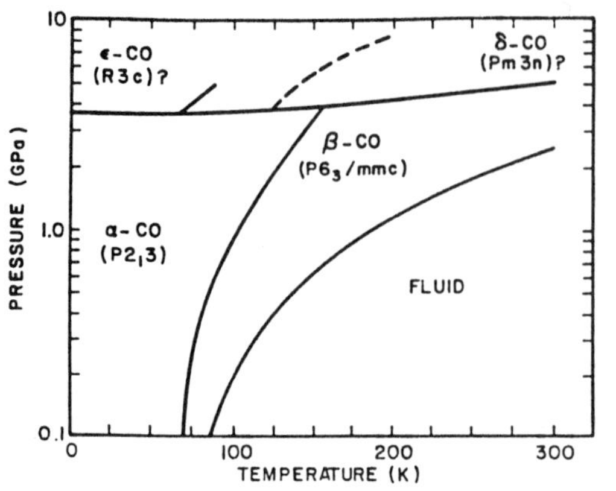 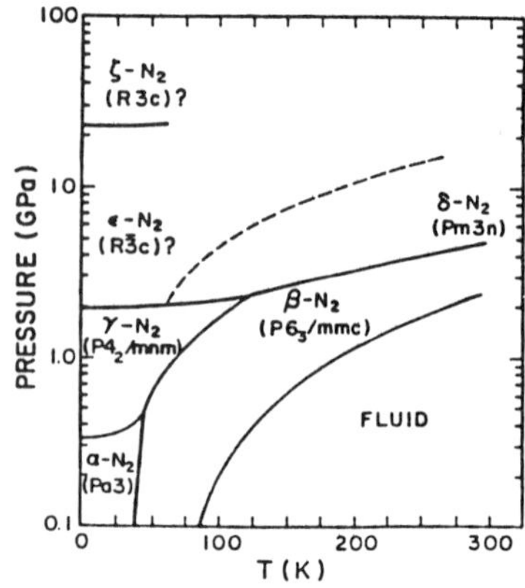

Fig. 1. The phase diagrams of CO and N_2.

persists in the low temperature solid,[5] as in CO. Evidence, including
recent x-ray diffraction measurements, indicates that there is random disorder in the solid at low temperatures.[5,6] Raman scattering measurements[7]
have recently shown what appears to be a phase transition in CO_2 and N_2O
at high pressures, although the results were insufficient to identify the
new phases. Calculations[8] and x-ray diffraction[6] have now identified a
transition in N_2O at about P = 4.8 GPa, into the orthorhombic Cmca phase.
The transition pressure is nearly independent of temperature over the
range 0 ≤ T ≤ 300 K. Calculations[9] have identified the transition in CO_2
to also be into the CmCa phase at somewhat higher pressures. This transition is yet to be confirmed experimentally.

THE POTENTIAL

Central to obtaining quantitatively accurate predictions for the
properties of condensed phases is a reliable representation of the interaction potential. While it is desirable to have this information from _ab-
initio_ results, it is seldom available or is not reliable. Thus, a combination of _ab-initio_ and experimental information is required, as it is in
this case. Following is a brief description of that process for N_2O and
CO. Details have been published elsewhere.[8]

Three force centers have been located on each N_2O molecule near the
nuclear sites and the overlap-dispersion interaction between sites on
different molecules is given in terms of an exp-6 functional form. In
addition there are large electric multipole interactions. These were
determined by first calculating the charge distribution for an individual
molecule. From this the four leading multipole moments were found and
three charges were located along each molecular axis to exactly reproduce
them. The multipole potential energy was then found by summing the
coulomb interaction between charges on different molecules. This sum and
that of the overlap-dispersion interactions gives the total potential
energy of the system.

It was found that the intermolecular potential must accurately reflect the shape of the molecular charge distribution, shown on Fig. 2, to
obtain reliable results at high pressures. Note the depression near the
midpoint of the molecule that becomes more pronounced near the longitudinal axis. This feature is apparent in Fig. 3 where the potential between N-N, N'-N', and O-O sites on two different molecules are displayed.
N' refers to the site near the central nitrogen atom, and similarly for

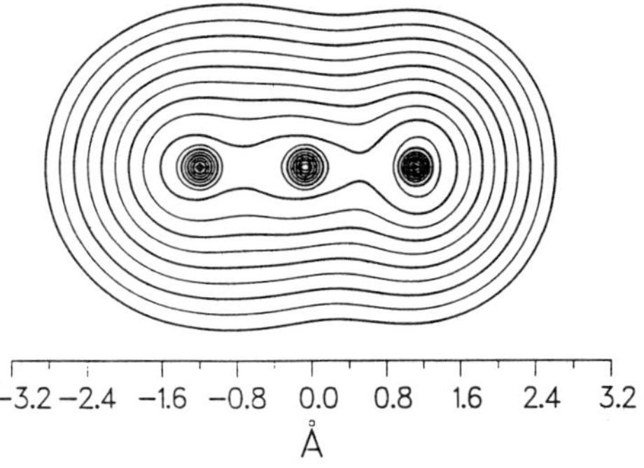

−3.2 −2.4 −1.6 −0.8 0.0 0.8 1.6 2.4 3.2

Å

Fig. 2. The calculated charge distribution of the N_2O molecule.

the others. Note that the repulsion core for the central N′-N′ interaction is much shorter in range than the outside ones. As will be discussed, this feature is essential to a description of the high pressure
properties. The charge distribution of CO_2 is very similar to that of
N_2O. Therefore, if our arguments are correct, the features of the potential should be similar to N_2O. A previously established representation[9]
of the CO_2 pair potential shows that this is the case. Because of the
success achieved in predicting the properties of CO_2, we find this agreement gratifying.

Because of the great similarity in the CO and N_2 phase diagrams, and
because their differences are believed to be primarily due to the larger
quadrupole moment of the former, it was natural for us to suppose that the
N_2-N_2 pair potential, with the electric multipoles replaced by those of

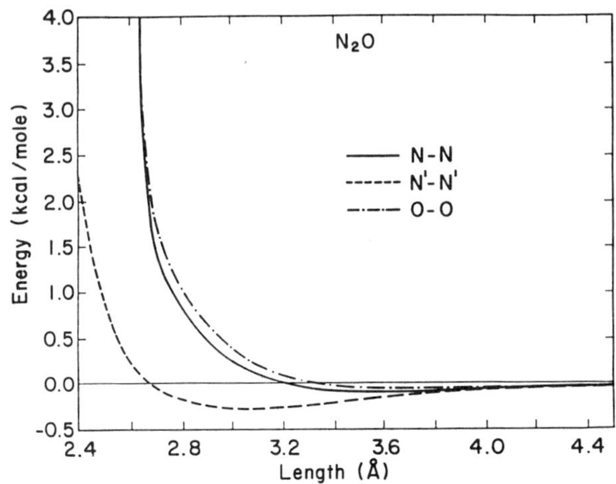

Fig. 3. The dashed, solid, and dot-dashed curves
are the N′-N′, N-N, and O-O site-site
potentials for N_2O, respectively.

CO, would be a good representation of CO-CO potential. This notion was
fortified in a study of CO adlayers on graphite.[10] Applying this representation to bulk CO shows that the virial coefficients, the cubic Pa3
structure, molar volume, sublimation energy, and the zone center libron
frequency, at zero pressure and in the limit of zero temperature, all
agree very well with experiment.[3,11] Unfortunately, calculations performed
by applying pressure at low temperature predict that the cubic phase
remains thermodynamically stable and does not transform into the rhombo-
hedral $R\bar{3}c$ structure that is observed[4] at P ~ 3.7 GPa. Numerous adjust-
ments to the short ranged part of the potential and locations of the force
centers failed to improve matters. In retrospect this result is not surprising since, while the transition from γ-N_2 to the $R\bar{3}c$ structure has
been correctly predicted[12] in N_2, the cubic phase is predicted as thermodynamically stable with respect to either of them. However, by introducing a three-center potential, where the repulsive region reflects the
shape of the CO charge distribution, substantial improvements were noted.
In addition, by reducing the quadrupole moment by 10% and the octapole
moment by 60%, the calculated Pa3 to $R\bar{3}c$ transition pressure is near the
observed value. Further investigation of this matter is in progress.

METHOD

At zero Kelvin the structures that minimize the enthalpy are determined at each pressure using Powell's energy optimization technique.[13] At
finite temperatures, structures and thermodynamic averages were calculated
using Monte Carlo constant pressure (NPT) or constant volume, (NVT) ensem-
bles, where N is the number of molecules in the Monte Carlo cell. N = 32
and 108 were used in these calculations. The quantities (P, V, T) are the
pressure, volume, and temperature. Lattice sums were taken out to 12 Å
with continuum corrections beyond. Averages are taken over approximately
10^4 steps, after "quasi" equilibrium is reached. Dynamical modes of
excitation are calculated using standard lattice dynamics.

The constant pressure Monte Carlo method with deformable, periodic
boundary conditions is a powerful technique for determining phase transitions since the volume fluctuations, common to phase changes, are allowed.
Moreover, the capability of the boundary conditions to deform from those
appropriate for one phase into another makes it possible to predict transformations without _a-priori_ bias, and to continuously monitor the systems
behavior throughout the transition region. Unfortunately, this procedure
does not always work for practical reasons. Potential barriers and other
sources of hysteresis, and difficulties in constructing a unit cell sufficiently flexible to match all possible boundary conditions, are problems.
Consequently another method has been developed that makes it possible to
accurately calculate the Gibbs free energy G of any phase at any temperature. The method employs an artificial control parameter λ to provide a
reversible path from one phase to another. First, we construct

$$U_i(\lambda) = \lambda U_i + (1 - \lambda)U_m \tag{1}$$

where $0 \leq \lambda \leq 1$ is a control parameter and U_m is the potential energy of
a model system of non-interacting local harmonic oscillators, three translational and two orientational. U_i is the energy of the physical system,
either cubic or orthorhombic for N_2O. Monte Carlo calculations are conducted on the system defined by Eq. (1) using an (N, V, T) ensemble, where
the Einstein oscillators of the model system are localized on the lattice
sites of the physical system at each volume V. The following exact
thermodynamic relationships apply:

$$dF(\lambda)/d\lambda = \langle dU_i(\lambda)/d\lambda \rangle = \langle U_i - U_m \rangle_\lambda \tag{2}$$

$$F_i(V) = F_m(V) + \int_0^1 \langle dU_i(\lambda)/d\lambda \rangle \, d\lambda \tag{3}$$

$$F_i(V) = F_i(V_0) - \int_{V_0}^V P_i(V) \, dV \tag{4}$$

$$\Delta G_i(P) = \Delta F(P) + P\Delta V(P) \tag{5}$$

where the brackets represent thermodynamic averages with respect to $U_i(\lambda)$; $F_i(V)$ and $F_m(V)$ are the Helmholtz free energies for the physical and model systems respectively. The subscript on the bracket of Eq. (2) is a reminder that the average depends on λ even though the argument does not. By solving Eqs. (2-4) for any two phases, Eq. (5) gives the difference in Gibbs free energy between them. Clearly the transition occurs when

$$\Delta G(P) = 0 \; . \tag{6}$$

Since the absolute free energy of the model system can be determined analytically, so can that of the physical system. This is a valuable feature.

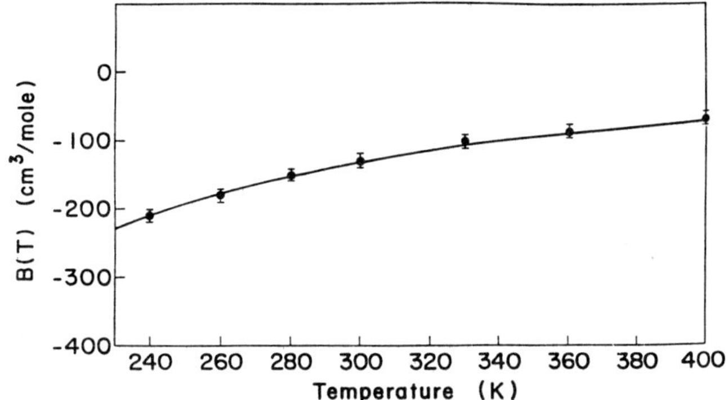

Fig. 4. The solid line gives the measured Ref. (11) second virial coefficients of N_2O and the flags represent the experimental uncertainty. The solid line represents the calculated values.

RESULTS

Figures 4-7 show a comparison of the calculated and measured values of the second virial coefficients, the pressure-volume relation, the lattice parameters, and the zone center libron frequencies of N_2O. A phase transition into the Cmca phase is predicted at $P_t = 4.75$ GPa in the limit of zero temperature. The volume change on transition is predicted to be $\Delta V \approx 0.25$ cm^3/mole. Utilizing Eqs. (2-6), Fig. 8 shows that the calculated phase transition at room temperature occurs at $P_t \approx 4.85$ GPa.

For CO, the results are preliminary. The calculated second virial coefficients are compared with experiment[11] on Fig. 9. Moreover, a transition into the observed $R\bar{3}c$ phase is calculated to occur between

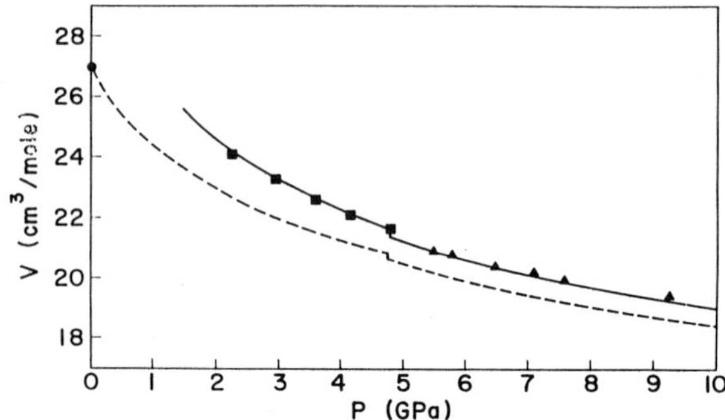

Fig. 5. The circle shows the measured (Ref. 14) molar
volume of N_2O in the limit T, P = 0, and the
squares and triangles are measured (Ref. 6)
values at room temperature for the Pa3 and Cmca
phases, respectively. The dashed and solid
lines are the calculated P-V curves at zero and
300 K, respectively. The statistical uncertainty in the calculated values are negligibly
small on this scale.

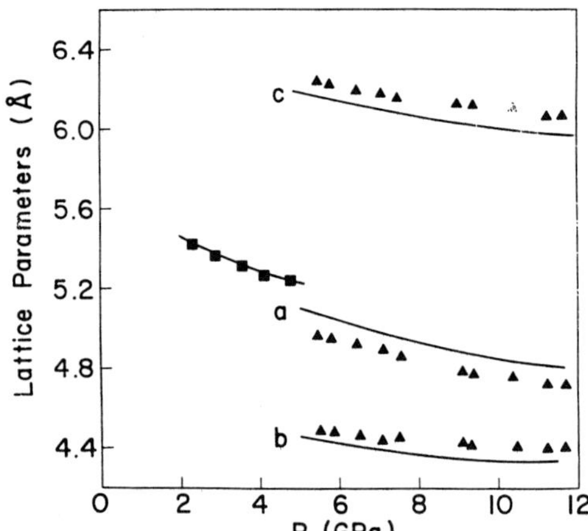

Fig. 6. The squares are the experimental
(Ref. 6) values of the lattice parameter of N_2O in the cubic Pa3 phase, and
the triangles are the lattice parameters of the Cmca phase. The solid
lines represent calculated values.

3 and 4 GPa. P-V results and libron frequencies also appear to be close
to measured values. Work is still in progress.

SUMMARY

We have stressed the importance of requiring that the short ranged
part of the interaction between molecules reflects the shape of the inner
contours of the charge distribution. This is particularly true at high
pressures where the charge clouds substantially overlap. It is not surprising then that the fidelity of the potential representation in this
regard may strongly influence the predicted stability of various high
pressure phases. This is primarily a packing consideration. We have
found that the use of three rather than two center representations of the
site-site potential facilitate this fidelity.

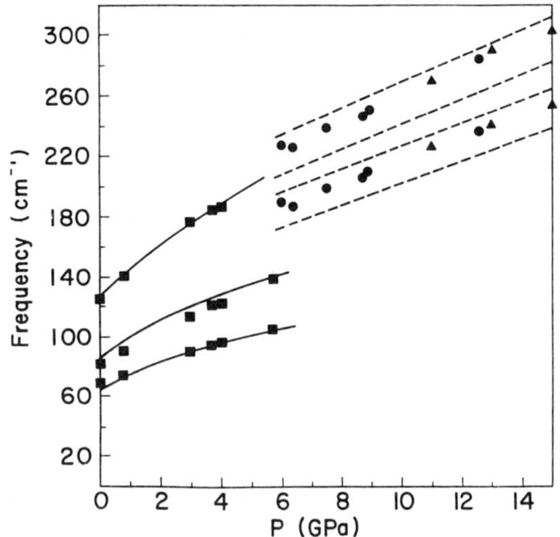

Fig. 7. The squares are the observed (Ref. 7)
         libron frequencies of N_2O in the Pa3
         phase at 20 K, and the triangles and
         circles are measured values in the high
         pressure phase at 20 and 300 K, respec-
         tively. The solid and dashed lines give
         the calculated libron frequencies in the
         Pa3 and Cmca phases, respectively.

Our calculations[8] on N_2O reveal not only the orthorhombic structure
of the phase transition but also the transition pressure and its temperature dependence. Here theory and experiment are in concert. All other
comparisons are also quite satisfactory. Calculations were based upon an
end-to-end randomly disordered structure which appears in both theory and
experiment to be at least metastable with respect to the ordered system.

For CO we again find that a three center potential representation is
critical in establishing the high pressure transition to the $R\bar{3}c$ phase.

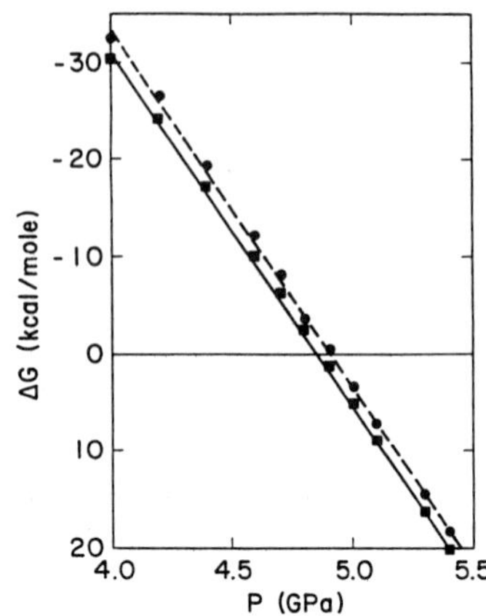

Fig. 8. The circles show the calculated difference in Gibbs
free energy between the cubic
and orthorhombic phases of
N_2O at 300 K, using the harmonic model reference system
described in the text. The
squares are the results of a
slightly different calculation not described here.

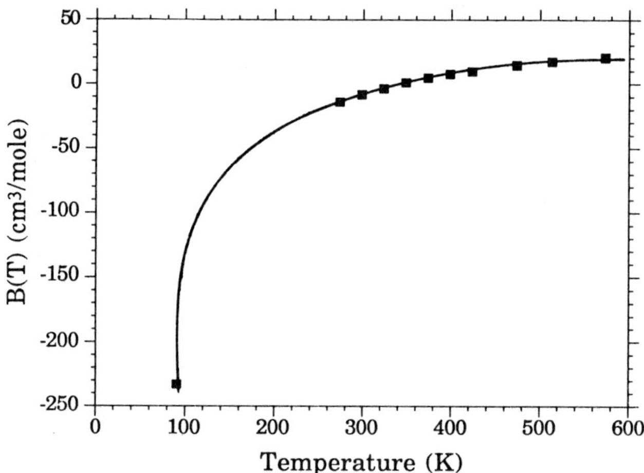

Fig. 9. The solid line gives the calculated
values of the second virial coefficients in CO, and the points are
experimental (Ref. 11) values.

All indications are that most other calculated properties are in reasonably good agreement with experiment. Still these results are preliminary and more work is in progress.

ACKNOWLEDGEMENTS

This work was supported by Department of Energy Grant No. DE-FG02-86ER45238.

REFERENCES

1. R. D. Etters, "Simple molecular systems at very high density,"
    A. Polian et al., eds., Plenum Press, New York (1989).
2. J. A. Venables and C. A. English, Acta Cryst. B $\underline{30}$, 929 (1974).
3. A. Anderson, T. Sun, M. Donkersloot, Can. J. Phys. $\underline{48}$, 2665 (1970);
    B. O. Hall, and H. M. James, Phys. Rev. B $\underline{13}$, 3590 (1976).
4. A. Katz, D. Schiferl, R. L. Mills, J. Phys. Chem. $\underline{88}$, 3176 (1984).
5. K. Nary, P. Kuhns, and M. Conradi, Phys. Rev. B $\underline{26}$, 3370 (1982).
6. R. L. Mills, B. Olinger, D. Cromer, and R. LeSar, J. Chem. Phys. $\underline{95}$,
    (1991).
7. H. Olijnyk, H. Daufer, M. Rubly, H. Jodl, and H. Hochheimer, J. Chem
    Phys. $\underline{93}$, 45 (1990).
8. B. Kuchta and R. D. Etters, J. Chem. Phys. $\underline{95}$, (1991).
9. B. Kuchta and R. D. Etters, Phys. Rev. B $\underline{38}$, 6265 (1988).
10. J. Belak, K. Kobashi, and R. D. Etters, Surf. Sci. $\underline{161}$, 390 (1985).
11. J. Dymond and E. Smith, <u>The Virial Coefficients of Gases</u>, Clarendon,
    Oxford (1969).
12. V. Chandrasekharan, R. D. Etters, and K. Kobashi, Phys. Rev. B $\underline{28}$,
    1095 (1983).
13. W. Press, B. Flannery, S. Teukolsky, and W. Vetterling, <u>Numerical
    Recipes</u>, Cambridge, New York (1986).
14. I. Krupskii, A. Prokhvatilov, and A. Erenburg, Sov. J. Low Temp.
    Phys. $\underline{6}$, 569 (1981).

ELASTIC PROPERTIES OF RARE GAS-SOLID

Alain Polian

Physique des Milieux Condensés
CNRS - URA 782
Université Pierre et Marie Curie, T 13 E 4
4 Place Jussieu
F-75252 Paris Cedex 05

Abstract

Brillouin scattering results from the interaction of light with thermal excitations (acoustic phonons in a crystal) of a material, or, from a classical point of view, with density waves. The physical phenomenon involved in Brillouin scattering is the same as in Raman scattering, but there are three substantial differences: First of all the frequency shifts are of the order of 1 cm^{-1} in the former; Second, there is <u>always</u> at least one allowed Brillouin mode; finally due to its nature, Brillouin scattering do not need long range order, and therefore is well suited for fluids and amorphous materials studies.

These characteristics have made Brillouin scattering a powerful technic in the study of rare gases under pressure: at ambient temperature, rare gases crystallize in the face centered cubic structure (except helium which structure was recently found to be hexagonal) and are therefore Raman and infrared inactive.

Experimental results will be presented on rare gases and rare gas mixtures in the fluid phase, like He-Ne and He-H$_2$ in relation with recent measurements of the frequency of global oscillations of Jupiter.

I. Introduction

Brillouin scattering is the result of the interaction of light with thermal excitations in a material, like acoustical phonons in a crystal. The result of the interaction is a shift in the energy of the scattered light, and therefore the measurement of this frequency shift gives information on the phonons of the material.

The theory of the Brillouin scattering has been extensively developed in the literature[1-4]. Therefore, only the main ideas will be developed here. In the interaction process between the photon and the thermal excitation, the energy and momentum are conserved.

$$h\,\Omega = \pm h\,(\omega_i - \omega_s) \tag{1}$$

$$\vec{q} = \pm(\vec{k_i} - \vec{k_s}) \tag{2}$$

where ω and k are the wave number and the wave vector of the photon, the subscripts i and s refer to the incident and scattered beams, and Ω and q are the wave number and the wave vector of the phonon; the + sign corresponds to the creation of a phonon (Stokes), and the minus to the annihilation (anti-Stokes). It should be noted here that the equations of conservation hold for Brillouin scattering as well as for Raman scattering because the physical processes are identical. On the opposite, the experimental point of view, the differences are large, coming from the difference in the energies involved in the processes: the energy of the optical phonons giving rise to Raman scattering are of the order of 10 to 1000 cm^{-1} (1.2 to 120 meV), whereas the Brillouin energy shift is of the order of 1 cm^{-1}. Therefore a much more dispersive apparatus has to be used to discriminate the Brillouin scattered light from the elastic Rayleigh scattering as for Raman scattering where grating spectrometer are very well suited. Such a dispersive apparatus is the Fabry-Pérot (FP) interferometer. Recent advances in Fabry-Pérot spectroscopy have been recently reviewed[5] and will be only briefly described here.

A Fabry-Pérot consists basically into two partially transmitting plane mirrors, separated by a distance d. A parallel beam of light is incident along the mirror axis. Due to interferences between the two parallel mirrors, the light is only transmitted when

$$d = m\,\frac{\lambda}{2} \tag{3}$$

where λ is the wavelength of the light, and m an integer. Therefore, the wavelength of the transmitted light is selected by varying the distance d between the mirrors.

Two parameters describe the quality of a FP: the first one is the contrast (C), defined by the ratio of the maximum intensity transmission to the minimum intensity transmission, and then qualify the ability of the set-up to measure signals small in comparison with the Rayleigh scattering. The second one is the finesse (F), defined by the ratio of the distance between two successive transmission peaks by the full width at half maximum of a transmission peak, and provides the capability of measuring peaks close to the Rayleigh peak. The most important parameter is the contrast. For a good quality FP, one get typically contrasts of the order of few hundreds which is far from to be sufficient to measure Brillouin scattering in most materials. The solution to increase considerably the contrast is the multipass FP which was introduced by Sandercock[6] and yields contrasts above 10^{10}.

One measurement is not sufficient to determine the wave number of a phonon: it is only determined modulo one "free spectral range" (FSR) i.e. a wave number equal to the distance between two successive transmission peaks of the FP. The solution was again given by Sandercock[7], with the tandem FP interferometer, where two interferometers with different spacings are scanned simultaneously.

II. Brillouin scattering

Brillouin scattering provides information about the acoustical branches of the dispersion curves of the material under study. The measured frequency shift of the radiation is equal to that of the phonon under consideration (Eq. 1), and its wave vector is deduced from Eq. 2, so the sound velocity may be calculated by:

$$v = \Omega\,/\,q \tag{4}$$

This relation may also be written:

$$\Delta\sigma(\text{cm}^{-1}) = \frac{2nv}{\lambda c}\sin\frac{\theta}{2} \tag{5}$$

where $\Delta\sigma$ is the measured frequency shift (in cm^{-1}), n the refractive index at λ, the wavelength of the exciting light, c the speed of light in the vacuum and θ the angle between k_i and k_s. The sound velocity is related to the elastic constants by:

$$X = \rho v^2 \tag{6}$$

where X is a combination of elastic constants depending on the direction of propagation of the phonon with respect to the crystallographic axis, and ρ the density. For example, in a cubic crystal, like every rare gas crystal, a longitudinal phonon propagating along the [1 1 0] direction is related to $X = (C_{11} + C_{12} + 2C_{44})/2$ and the two non degenerate transverse phonons are related to $X = (C_{11} - C_{12})/2$ and $X = C_{44}$. Therefore, by correctly choosing the geometry of the experiment, it is possible to determine the complete set of elastic constants.

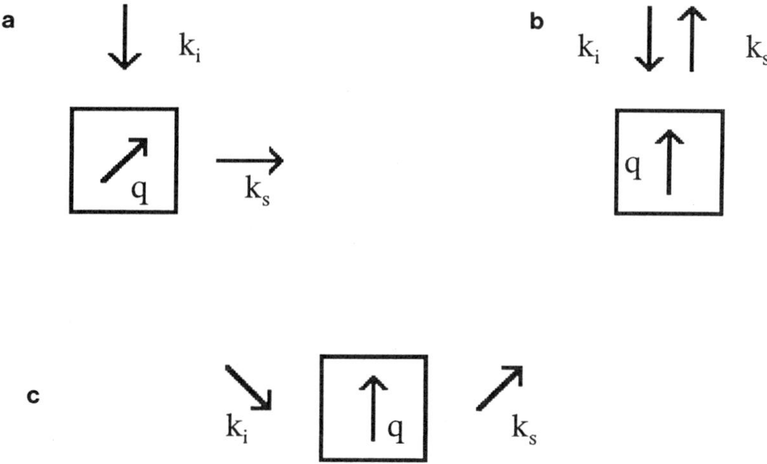

Figure 1.   *Most commonly used geometries in Brillouin scattering. (a) 90° scattering. (b) Backscattering geometry. (c) Platelet geometry.*

Fig. 1 shows scattering geometries commonly used in order to determine the elastic constants. In Fig. 1(a), the modulus of the phonon wave vector is equal to $nk\sqrt{2}$ where $k \approx k_i \approx k_s$. In Fig. 1(b), $|q| = 2nk$. Fig. 1(c) shows an interesting geometry for Brillouin scattering under pressure (provided that the ad-hoc cell is available), because the measurement of the refractive index is no more required. Indeed, in the crystal the modulus of the wave vector of light is equal to nk. Due to the incidence angle, $|q| = 2nk\,\sin(r)$, where r is the angle of the refracted photon inside the crystal with respect to the normal to the surface. Using the Descartes law of refraction, one obtain $|q| = 2k\,\sin(i)$ where i is the known angle of incidence.

It is clear that, if one knows the relative orientation of k_i and k_s with respect to the crystallographic axes, one knows also that of q. The knowledge of the selections rules[2] allows then to deduce the related combination of elastic constants.

III. Brillouin scattering in the diamond anvil cell

With a classical diamond anvil cell (DAC), i.e. with only optical accesses along the axis of the diamond, it is mainly possible to use the backscattering geometry, even if one experiment was performed in the platelet geometry[8]. Unfortunately, in this geometry, the selection rules favor the longitudinal mode, so most of the experiments performed in this geometry give only information on these phonons. Another problem which occurs is due to the diamond anvils which have also acoustical phonons, and the difference of volume between the anvils and the sample make the intensity of the Brillouin of the former much more intense than that of the latter.

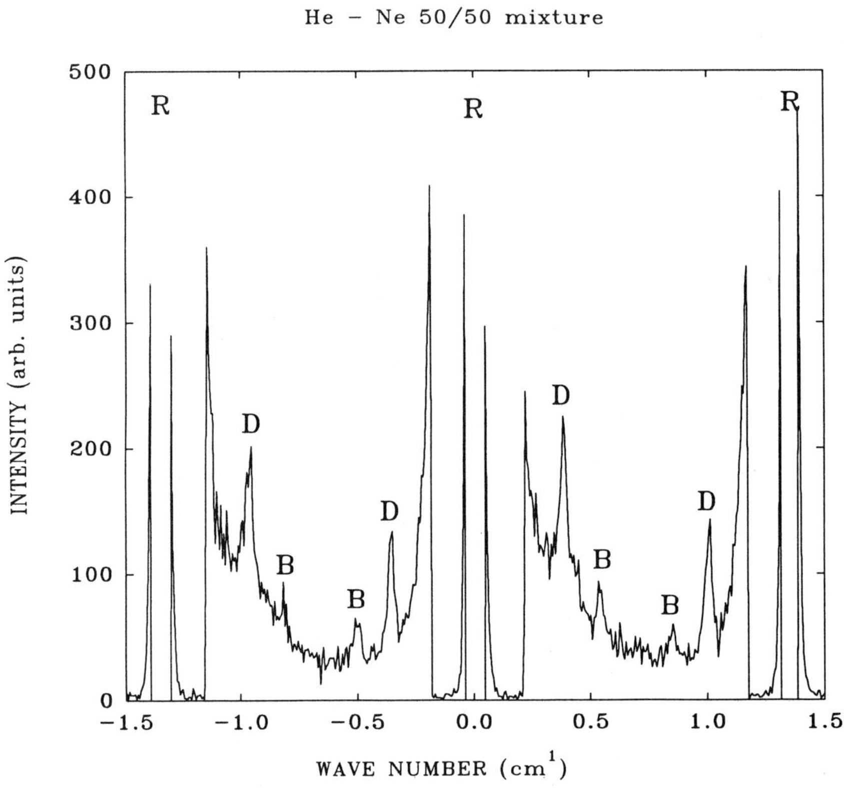

Figure 2. *Brillouin scattering of the He-Ne 50-50 mixture at 3.95 GPa (fluid phase). Three interference order of the FP are shown (0, -1.3, 1.3 cm⁻¹). The peaks labelled "D" come from the diamond anvils, the peaks "B" represent the signal due to the sample, and "R" the elastically scattered Rayleigh lines. Every laser line (Rayleigh) produces Stokes and anti-Stokes scattering, this explains why the position of the Brillouin peaks is symmetrical with respect to the middle of the interval between two Rayleigh lines.*

In Fig. 2, an example of the problem due to the diamonds is shown: the sample is a 50-50 mixture of He and Ne at 3.95 GPa. The FSR was 1.3264 cm⁻¹. With this FSR, the very intense longitudinal mode of the diamond in the [100] direction (5.2 cm⁻¹) is hidden by the Rayleigh lines, but due to small disorientation of the anvils a transverse mode of the anvils is observed. The problem of the longitudinal modes of the diamonds is avoided by using a tandem FP.

The use of "swiss cheese" DAC[9] allows to measure in the platelet geometry, and then gives information on the transverse modes. Table I summarize the Brillouin experiments performed up to now in a DAC with the studied material, the geometry and the maximum pressure.

Table I - *Materials studied to date by Brillouin scattering in a DAC.*

MATERIAL	SCATTERING GEOMETRY	MAX. PRESSURE (GPa)	Ref.
NaCl	angle	4	10
H_2, D_2	angle	20	11-13
a-SiO_2	angle	14	14,15
a-SiO_2	back.	17	16,17
GaS	back.	18	18,19
Mg_2SiO_4	back.	4	20
H_2O, D_2O	back.	74	21-23
a-B_2O_3	back.	16	24
a-GeO_2	back.	20	24
a-As_2S_3	back.	4	24
TiO_2	back.	14	25
N_2	back.	20	26
Ar	back.	35	27,28
HF	angle	12	29
He	back.	20	30
CH_4	back.	32	31
CS_2	angle	7	32
Kr	back.	33	33
$BaTiO_3$	back. & angle	11 & 8	34,35
$SrTiO_3$	back. & angle	36 & 9	36-38
H_2S	back. & angle	12	39
NH_3	back.	20	40
C_6H_6	back.	30	41
CH_3OH	angle	8.4	42
ZnO	angle	4	43
Xe	back.	31	44

As can be seen in the table, the maximum reached pressure is obtained, in average, in a backscattering geometry.

Once the spectrum is recorded, the sound velocity should be computed, which means that (Eq. 5) the refractive index at the wavelength of the laser and at the desired density should be measured or computed. The measurement of the pressure dependence of the refractive index of compounds which are solid at ambient is not easy. Various techniques have been proposed and applied in few cases, for H_2 and D_2[11,12], GaS[19], xenon[45], argon[46], and helium[47]. In case where n is not measured with pressure, a good approximation is given by the Clausius-Mossotti equation, provided the equation of state is known.

Once the sound velocity is determined, Eq. 6 gives the effective elastic modulus corresponding to the direction of propagation of the measured phonon. In order to determine the involved linear combination of elastic constants, one has to know the crystallographic orientations of the crystal inside the DAC. This may be done when the sample is solid[10, 18-20, 34-38] at ambient conditions. When the sample is loaded as a fluid in the DAC, there is no control on the crystal orientations, and the determination of individual elastic constants is not straightforward. This is the case of the rare gas solids, and it will be discussed in the next section.

IV. Experimental results on rare gas solids

A - Argon

Brillouin scattering has been measured in argon up 34 GPa[28] in backscattering. Using the measured index of refraction[46] and equation of state[47,50], the effective elastic moduli could be computed (Fig. 3).

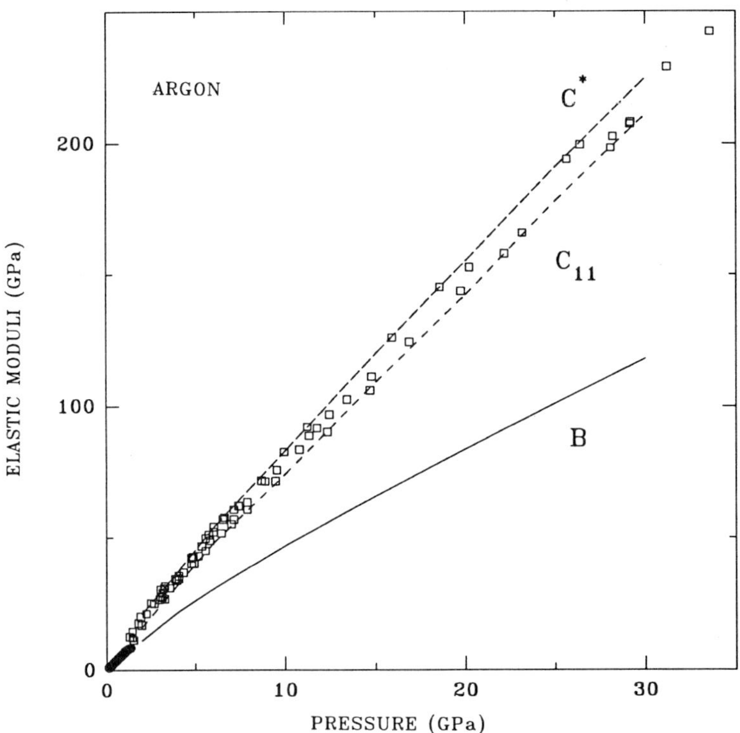

Figure 3. *Elastic moduli of argon vs pressure[28]. Full circles: fluid. Empty squares: solid. The two dashed lines represent the envelope curves from which C_{11} and C^* were determined. Solid line: isothermal bulk modulus determined by x-ray measurements[47-49].*

In the solid phase, argon is elastically anisotropic. Making the hypothesis that in different loading all the orientations were probed (and indeed, even a pressure increase could change the orientation of the crystal in the cell), it was possible to determine the three elastic constants: in argon, the elastic modulus corresponding to the largest longitudinal sound velocity is $C^* = (C_{11} + 2C_{12} + 4C_{44})/3$ (propagation along [111]), and the smallest one is

186

C_{11}(propagation along [100]). So the envelop curves shown in Fig. 3 by dashed lines were identified with C^* and C_{11}. Moreover, the bulk modulus $B = (C_{11} + 2C_{12})/3$ was measured by x-ray diffraction[48-50] (see Fig. 3).

Comparisons of theoretical calculations with (i) the equation of state, and (ii) the elastic moduli showed that both sets of results could not be fitted simultaneously using pair potentials. At high density, it is necessary to use three body forces at least.

B - Krypton

A similar analysis has been performed on krypton[33], with the difference that the refractive index was not measured, but calculated from that of xenon[45] obtained in a one electron model by:

$$\frac{n^2 - 1}{n^2 + 2} = 3.304 \left(\frac{5.4}{E_b^2 - E^2} + 2.5 \; 10^{-4} \right) \rho \tag{7}$$

where

$$E_b = 15.46 + 0.381 \, (0.5234 \, \rho - 1)^{1.8} \tag{8}$$

and E is the photon energy at which the refractive index is calculated. E and E_b are in eV and ρ in g/cm^3. The equation of state has been measured by x-ray diffraction[33]. The results are shown in Fig. 4.

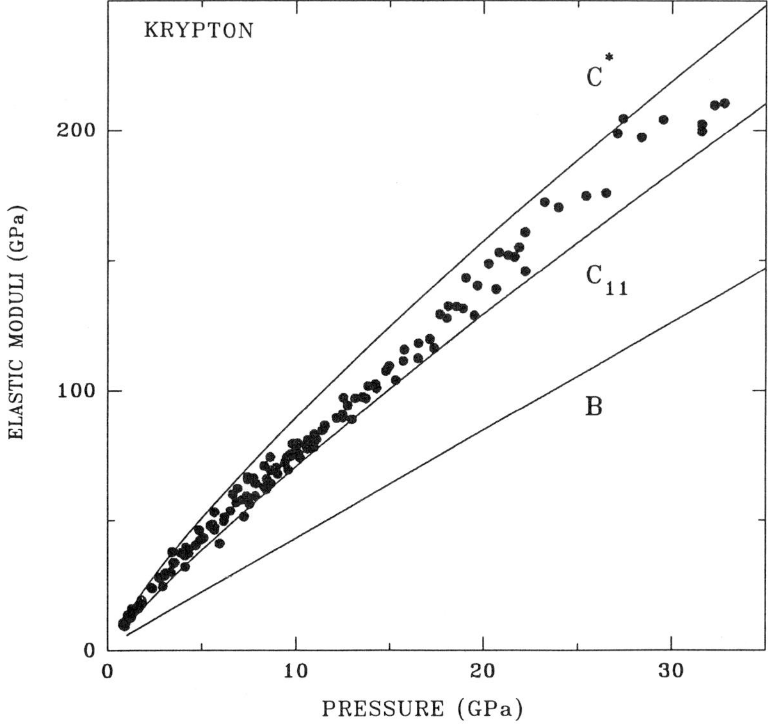

Figure 4. *Elastic moduli of krypton vs pressure.*

C - Xenon

Xenon, the heaviest of the stable rare gases, has been studied by Brillouin scattering up to 30 GPa. The fcc structure is not stable in xenon: a transition to an intermediate close-packed phase has been detected by x-ray diffraction at 14 GPa, and another one to an hcp structure at 74 GPa[51]. Unfortunately, due to the elastic anisotropy of the fcc phase, the transition at 14 GPa is hardly detected by Brillouin scattering. In Fig. 5, is shown the

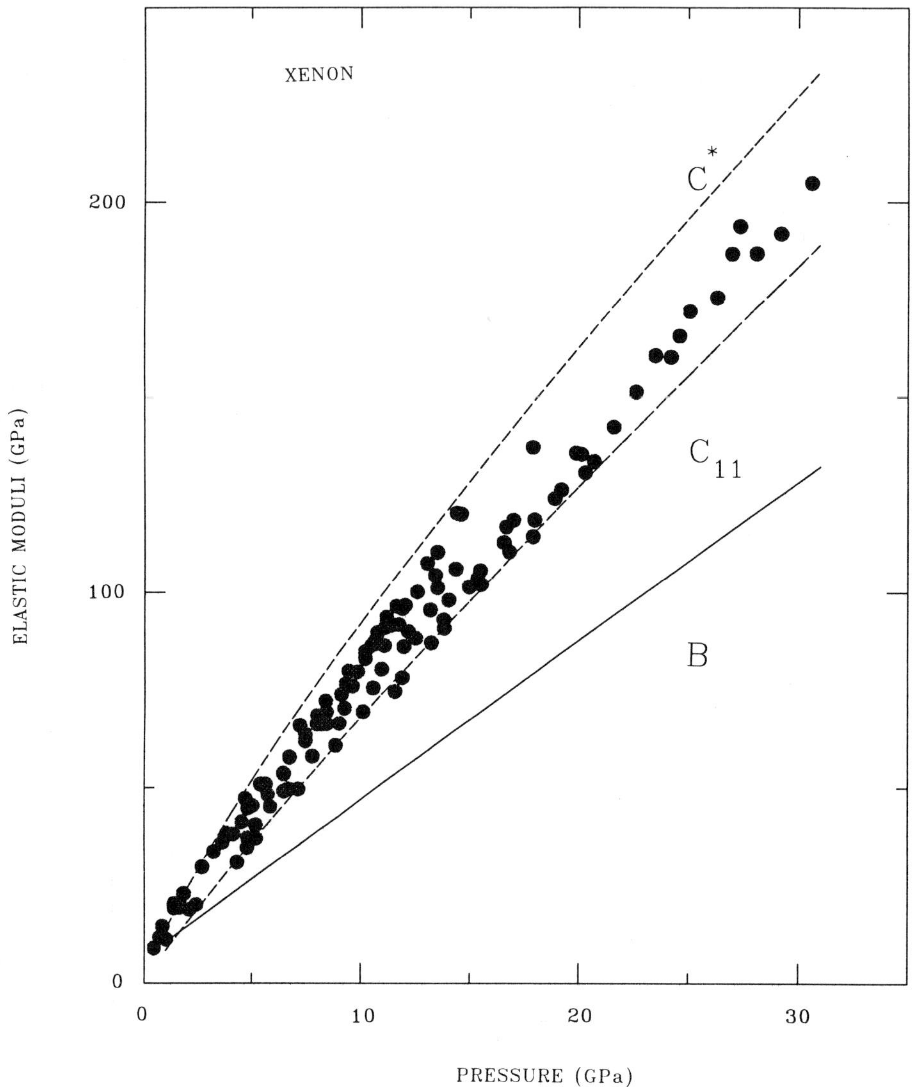

Figure 5. *Elastic moduli of xenon vs pressure.*

pressure dependence of the elastic moduli of xenon[44]. The refractive index has been determined[45] by the index matching technique between the xenon and a series of crystals with known index, these data being fitted in a one electron model with an equation similar to Eq. 7. The equation of state has been obtained by x-ray diffraction[52].

D - Helium

Helium is the lightest of rare gases, and also that which has the most complicated (to date) phase diagram. At ambient temperature, the fusion pressure lies at 11.4 GPa[53]. In contrast with the other rare gases, it crystallizes in the hcp structure[54] and not in the fcc one. So, in principe it could be possible to study solid helium by Raman scattering. Unfortunately, the polarizability of helium is so small, that even in a classical large volume high pressure cell[55], the intensity of the peak was of the order of 3 counts/s, and then such an experiment is not possible in a DAC and Brillouin is the only spectroscopic usable technique. Brillouin scattering has been performed mainly in the fluid[30] (Fig. 6).   In a fluid, there is only one

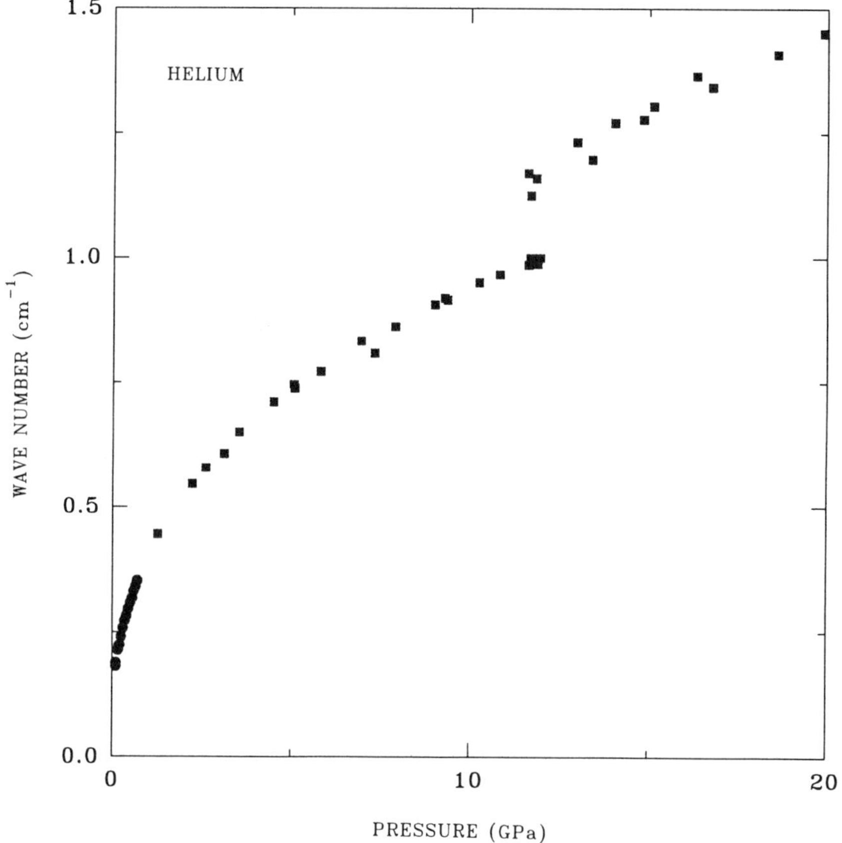

Figure 6. *Pressure dependence of the Brillouin shift of helium. The circles were measured using a classical large volume pressure cell, and the squares in a DAC. The fluid-solid transition is clearly seen at 11.4 GPa.*

elastic constant, which is equal to the bulk modulus. The elastic isotropy of the fluid is clearly seen in Fig. 6, where the dispersion of the measured points is small in contrast with the situation in the solid. Using a model for the refractive index (at that time not measured) and for γ, the heat capacity ratio, the first experimental room temperature equation of state (EOS) for fluid helium was deduced. More recently, the refractive index was measured, and using new hypotheses for γ, a new EOS of fluid helium was determined[47].

E - Mixtures

The study of simple fluids mixtures under pressure is interesting because it allows to test theories of mixtures. In that respect, the study of the sound velocity is of prime interest, because it tests very precisely the interaction potentials. From another point of view, the knowledge of the pressure dependence of the sound velocity of the He-H$_2$ mixture is of first interest: a recent observation of the global oscillations (pressure modes) of Jupiter[56,57] have shown a discrepancy of the order of 15% for the frequency of the observed modes with the standard model. The origin of that departure should be determined: does it come from the standard model, i.e. the composition of the planet, from the fact that there may be a singularity in the sound velocity for the jovian composition of the He-H$_2$ mixture (~ 20-80), or from the uncertainties in the observed values.

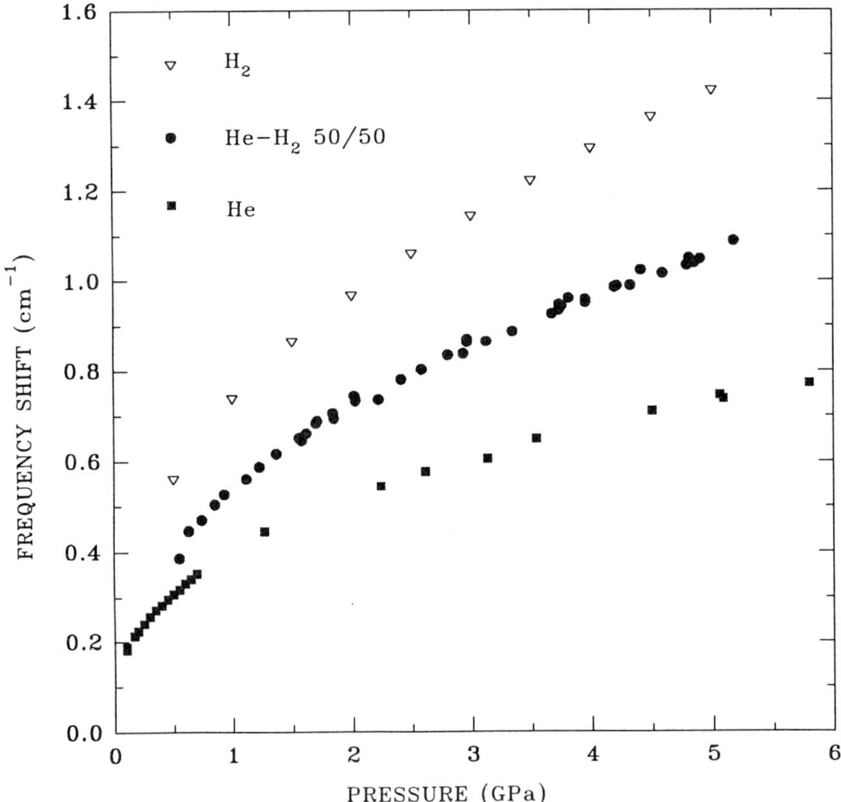

Figure 7. *Pressure dependence of the frequency shift for helium[30] (full squares), hydrogen[11] (open triangles) and an equimolar mixture of helium and hydrogen (full circles).*

As a first step, an equimolar mixture of helium and hydrogen has been studied in the homogennous fluid phase, i.e. below 5 GPa. As it can been seen in Fig. 7, the frequency of the acoustical mode of the mixture lies between those of the pure constituents. The comparison of the sound velocities deduced from these measurements with calculations based (i) on an ideal mixture, and (ii) on the theory of an effective fluid[58] shows that, below 1.5 GPa, the mixture is ideal, but above this pressure, the experimental sound velocity is larger than the ideal one. On the contrary, in the effective fluid model, the calculated sound velocity is systematically higher than the experimental one. Improvements are needed for a better agreement with experimental results.

Nevertheless, a point deduced from this preliminary study is that, for an equimolar mixture, there is no singularity in the composition dependence of the sound velocity.

V. Conclusion

In this paper, we have reviewed the elastic properties of solid rare gases and of fluid mixtures. It should be emphasized that Brillouin scattering is one of the few experimental techniques which provide information on the behavior of these materials under pressure, and is complementary (and indeed need the results of) to x-ray scattering and refractive index measurements. It was shown that, due to their elastic anisotropy and provided that the bulk modulus is known, it is possible to determine the complete set of elastic constants using the backscattering geometry. Moreover, the results obtained in argon have shown that the determination of the EOS alone is not sufficient to refine the interatomic potentials: it is possible to determine a pair potential which fits either the equation of state, **or** the elastic constants, but not both quantities together: three body contributions are needed. It has also been shown that the use of Brillouin scattering open the possibility to determine equation of state of fluids, and therefore of glasses and amorphous materials.

The comparison of direct measurements of the vibrational properties of giant planets, which were recently obtained, with Brillouin scattering measurements of mixtures in the same conditions (pressure, composition, and later temperature) gives the possibility to test directly the planetary models. In particular, preliminary results obtained on equimolar He-H$_2$ mixture already show the need of an improvement of the theory of mixtures and will be extended with composition equal to the cosmological abundance (80% H$_2$, 20% He).

ACKNOWLEDGMENTS

This work was supported by the Commissariat à l'Energie Atomique under grant DAM/CEB.3/N°1617, and by the Institut National des Sciences de l'Univers under grant N° 83-070978.

REFERENCES

1. M. Born and K. Huang, *Dynamical Theory of Crystal Lattices* (Oxford Univ. Press, London, 1954)
2. H. Z. Cummins and P. E. Schoen, in *Laser Handbook*, edited by F. T. Arecchi and E. O. Schulz-Dubois (North Holland, Amsterdam, 1972), p. 1029
3. W. Hayes and R. Loudon, *Scattering of Light by Crystals* (Wiley, New York, 1978)
4. J. G. Dil, Rep. Prog. Phys. **45**, 285 (1982)
5. F. Nizzoli and J. R. Sandercock, in *Dynamical Properties od Solids*, edited by G. K. Horton and A. A. Maradudin (North Holland, Amsterdam, 1990), p. 281
6. J. R. Sandercock, in *Light Scattering in Solids*, edited by M. Balkanski (Flammarion, Paris, 1971), p. 9
7. J. R. Sandercock, in *Light Scattering in Solids* III, edited by M. Cardona and G. Güntherodt (Springer, Berlin, 1982), p. 173

8. S. A. Lee, D. A. Pinnick, S. M. Lindsay and R. Hanson, Phys. Rev. **B 34**, 2799 (1986)

9. W. A. Bassett, D. R. Wilburn, J. A. Hrubec and E. M. Brody, in *High Pressure Science and Technology*, edited by K. D. Timmerhaus and M. S. Barber, (Plenum, New York, 1979), Vol. 2, P. 75

10. C .H. Whitefield, E. M. Brody, and W. A. Bassett, Rev. Sci. Instrum. **47**, 942 (1976)

11. H. Shimizu, E. M. Brody, H. K. Mao and P. M. Bell, Phys. Rev. Lett. **47**, 128 (1981)

12. H. Shimizu, E. M. Brody, H. K. Mao and P. M. Bell,in *High Pressure Research in Geophysics*, edited by S. Akimoto and M. H. Manghnani (Center Acad. Publ. Japan, Tokyo, 1982), p. 135

13. H. Shimizu, T. Kumazawa, E. M. Brody, H. K. Mao and P. M. Bell, Jpn. J. Appl. Phys. **22**, 52 (1983)

14. J. Schroeder, K. J. Dunn and F. P. Bundy, in *Proc. of the VIIth AIRAPT conference*, edited by C. M. Backman, T. Johanisson and L. Tegner, Uppsala 1981, p. 259

15. J. Schroeder, T. G. Bilodeau and X. S. Zhao, High Pres. Res. **4**, 531 (1990)

16. M. Grimsditch, Phys. Rev. Lett. **52**, 1312 (1984)

17. M. Grimsditch, Phys. Rev. **B 34**, 4372 (1986)

18. A. Polian, J. M. Besson, M. Grimsditch and H. Vogt, Appl. Phys. Lett. **38**, 334 (1981)

19. A. Polian, J. M. Besson, M. Grimsditch and H. Vogt, Phys. Rev. **B 25**, 2767 (1982)

20. H. Shimizu, W. A. Bassett and E. M. Brody, J. Appl. Phys. **53**, 620 (1982)

21. A. Polian and M. Grimsditch, Phys. Rev. **B 27**, 6409 (1983)

22. A. Polian and M. Grimsditch, Phys. Rev. Lett. **52**, 1312 (1984)

23. A. Polian and M. Grimsditch, Phys. Rev. **B 34**, 6362 (1984)

24. M. Grimsitch, R. Bhadra and Y. Meng, Phys. Rev. **B 38**, 7836 (1988)

25. A. Polian and M. Grimsditch,J. Physique Lett (Paris) **45**, L1131 (1984)

26. M. Grimsditch, Phys. Rev. **B 32**, 514 (1985)

27. A. Polian and M. Grimsditch, Physica **139&140 B**, 187 (1986)

28. M. Grimsditch, P. Loubeyre and A. Polian, Phys. Rev. **B 33**, 7192 (1986)

29. S. A. Lee, D. A. Pinnick, S. M. Lindsay and R. Hanson, Phys. Rev. **B 34**, 2799 (1986)

30. A. Polian and M. Grimsditch, Europhys. Lett. **2**, 849 (1986)

31. P. Hebert, A. Polian, P. Loubeyre and R. Le Toullec, Phys. Rev. **B 36**, 9196 (1987)

32. H. Shimizu, S. Sasaki and T. Ishidate, J. Chem. Phys. **86**, 7189 (1987)

33. A. Polian and M. Grimsditch, J. M. Besson and W. A. Grosshans, Phys. Rev. **B 39**, 1332 (1989)

34. M. Fischer and A. Polian, Phase Trans. **9**, 205 (1987)

35. T. Ishidate and S. Sasaki, Phys. Rev. Lett. **62**, 67 (1989)

36. M. Fischer and A. Polian, in *High Pressure Geosciences and Materials Synthesis*, edited by H. Vollstädt, (Akademie Verlag, Berlin, 1988) p. 156

37. M. Fischer, B. Bonello, A. Polian and J.M. Leger, in *Perovskite: A Structure of great Interest in Geophysics and Materials Science*, A.G.U. Monographs **45**, 125 (1989)

38. T. Ishidate, S. Sasaki and K. Inoue, High Pres. Res. **1**, 53 (1988)

39. H. Shimizu and S. Sasaki, to be published.

40. M. Gauthier, Ph. Pruzan, J. C. Chervin and A. Polian, Solid State Commun. **68**, 149 (1988)

41. M. Gauthier, A. Polian, J. C. Chervin and Ph. Pruzan, High Pres. Res. to be publlished

42. S. A. Lee, A. Anderson, S. M. Lindsay and R. C. Hanson, High Pres. Res. **3**, 230 (1990)

43. T. Ishidate, K. Yamamoto, K. Inoue and M Shibuya, in *Solid State Physics under Pressure: Recent Advance with Anvil Devices*, edited by S. Minomura (KTK Scientific Publishers, Tokyo, 1985), p. 99

44. A. Polian, I. Zouboulis and M. Grimsditch, to be published

45. J. P. Itié and R. Le Toullec, J. Physique (Paris) Colloque, **45**, C8-53 (1984)

46 M. Grimsditch, R. Le Toullec, A. Polian and M. Gauthier, J. Appl. Phys. **60**, 3479 (1986)

47. R. Le Toullec, P. Loubeyre and J. P. Pinceaux, Phys. Rev. **B 40**, 2368 (1989)

48. R. M. Hazen, H. K. Mao, L. W. Finger and P. M. Bell, Carnegie Inst. Washington Yearb. **79**, 348 (1980)

49. G. Zou, H. K. Mao, and P. M. Bell, Carnegie Inst. Washington Yearb. **81**, 392 (1982)

50. J. Xu, K. K. Mao and P. M. Bell, High Temp. High Pressures, **16**, 495 (1984)

51. A.P. Jephcoat, H. K. Mao, L. W. Finger, D. E. Cox, R. J. Hemley and C. S. Zha, Phys. Rev. Lett. **59**, 2670 (1987)

52. K. Asaumi, Phys. Rev. **B 29**, 7026 (1984)

53. J. M. Besson and J. P. Pinceaux, Science **206**, 1073 (1979)

54. H. K. Mao, R. J. Hemley, Y. Wu, A. P. Jephcoat, L. W. Finger, C. S. Zha and W. A. Bassett, Phys. Rev. Lett. **60**, 2649 (1988)

55. G. H. Watson and W. B. Daniels, Phys. Rev. **B 31**, 4705 (1985)

56. F.X. Schmider, B. Mosser and E. Fossat, Astron. Astrophys. to be published

57. B. Mosser, F.X. Schmider, Ph. Delache and D. Gautier, Astron. Astrophys., to be published

58. P. Loubeyre et al, to be published

PRESSURE AND SUBSTITUTIONAL DISORDER IN MOLECULAR CRYSTALS

K.H. Michel

Department of Physics
University of Antwerp UIA
2610 Wilrijk, Belgium

INTRODUCTION

The study of mixed crystals containing molecules or molecular ions in high concentration is of interest in statistical mechanics and solid state physics. We consider systems $(MY)_x (MX)_{1-x}$. Here M is a metal counter ion, X a halogen ion while Y is a molecular ion with orientational degrees of freedom. In particular we have in mind the mixed alkali cyanides-alkali halides. Then M stands for Na, K,Rb; X is a halogen Cl, Br, I; Y denotes the CN^- ion [1]. One could also think of the corresponding nitrites, where Y stands for NO_2. (We do not know whether mixed alkali nitrites-alkali halide crystals with x > 0.1 have been grown.) These mixed crystals present rich phase diagrams as a function of substitutional disorder [1-3].

The pure compounds (x = 1) exhibit at high T an orientationally disordered (plastic) phase, the average lattice structure has cubic symmetry Fm3m. At lower temperature there occurs a transition to a phase with lower symmetry, the structural change is accompanied by orientational ordering. Examples are the ferroelastic transitions in MCN [4] and in KNO_2, $RbNO_2$ [5]. In the high T phase, these structural changes are announced by a drastic softening of the elastic constants c_{44} with decreasing temperature. The ferroelastic transition is in general of first order. In principle one could argue that this is a trivial statement, since renormalization group calculations for an Ising model on a compressible cubic lattice show that the transition becomes always first order [6]. However the temperature interval $t = (T_1 - T_c)/T_c$ between the actual first order transition temperature T_1 and the virtual second order transition temperature T_c is of the order $(c_{44}/3c_{11})^{+7}$ for a cubic lattice. For the alkali cyanides, this would

Frontiers of High-Pressure Research, Edited by H.D. Hochheimer and
R.D. Etters, Plenum Press, New York, 1991

correspond to 10^{-9}. Experimentally one obtains $t \approx 10^{-1}$ in KCN. We therefore conclude that in MCN and related compounds, the renormalization group arguments are not sufficient to explain the strong first order character of the phase transition. We will study anharmonic cubic couplings between molecular reorientations and acoustic lattice displacements, and investigate which mechanism is relevant for the first order nature of the phase transition. This brings us also to the study of the pressure dependence of the phase transition and of related physical properties in these substances [7,8].

In mixed crystals, substitutional disorder strongly affects the phase diagram. For instance in $M(CN)_x X_{1-x}$, one finds that the low T structure is orthorhombic, rhomboedral or monoclinic [1-3] for $x_c < x \leqslant 1$. For $x < x_c$, where x_c depends on the nature of M and X, the average cubic structure remains stable. This was first discovered in $K(CN)_x Cl_{1-x}$ [9]. At low T, the orientational disorder is frozen. This is the orientational glass state, the existence of which is revealed by the appearance of an elastic central peak in the elastic neutron scattering law [10]. $K(CN)_x Br_{1-x}$, $x \approx 0.5$ is an example of a quadrupolar glass, while $(Rb)_{1-x}(NH_4)_x H_2 PO_4$ is a dipolar glass [11]. For a recent review on orientational glasses, see Ref.(12). In quadrupolar glasses, the orientational freezing is accompanied by a freezing of static strain fields [13, 14, 15]. The strain fields which are a consequence of substitutional disorder, play the role of "internal pressure" [16]. It is therefore instructive to compare the role of substitutional disorder and of hydrostatic pressure on the phase diagram and on physical properties of mixed crystal. Such a study is the aim of the present paper.

2. MODEL

We consider a mixed crystal $M(Y)_x X_{1-x}$ in the high temperature fcc phase (NaCl structure). At random lattice sites, the halogen ions X are replaced by molecular ions Y. In particular, we take $Y = CN^-$. For our purposes, we can neglect the electric dipole of CN^- and approximate CN^- by symmetric dumbbells (quadrupoles). The orientational degrees of freedom are described by means of symmetry adapted functions, which take into account the symmetry of the molecule and of the lattice [17]. We will restrict ourselves here to the three symmetry adapted functions $Y_\alpha(\vec{\Omega})$:

$$Y_1(\Omega) = n\hat{x}\hat{y}, \quad Y_2(\Omega) = n\hat{y}\hat{z}, \quad Y_3(\Omega) = n\hat{z}\hat{x} \quad , \tag{2.1}$$

$n = (15/4\pi)^{1/2}$. The polar angles $(\Theta,\varphi) \equiv \Omega$ fix the orientation of the linear molecule, with $\hat{x} = \sin\Theta\cos\varphi$, $\hat{y} = \sin\Theta\sin\varphi$, $\hat{z} = \cos\Theta$. These

symmetry adapted functions transform according to the T_{2g} representation
of the cubic group. For additional details, see Ref. (18). In general the
index α accounts for the ℓ,Γ,τ, where ℓ is the angular momentum quantum
number, Γ labels the irreducible representation of the site group and τ
labels the columns (rows) of the representation. The elastic degrees of
freedom are formulated in terms of acoustic lattice displacements $\vec{s}(\vec{q})$,
where $\vec{q}$ is the wave vector.

Here and in the following, substitutional disorder between Y and X
sites is taken into account by the occupation variable $\sigma(\vec{n})$ with value
+1 in presence of Y and value 0 in presence of an X ion at lattice site
$X(\vec{n})$. We therefore write for the orientational coordinate

$$Y_\alpha(\vec{q}) = \frac{1}{\sqrt{N}} \sum_{\vec{n}} \sigma(\vec{n})\, Y_\alpha(\Omega(\vec{n}))\, \exp(-i\vec{q}.\vec{X}(\vec{n})) \ . \tag{2.2}$$

Here N is the total number of unit cells. We recall that the substituional
disorder is quenched.

The total interaction potential $M(Y)_x X_{1-x}$ is given by

$$V = V^{TT} + V^{RR} + V^R + V^{TR} + V^{RRT} + V^{SR} \ . \tag{2.3}$$

Here the harmonic acoustic-phonon potential is given by

$$V^{TT} = \frac{1}{2} \sum_{\vec{q}} M_{ij}(\vec{q}) s_i(\vec{q}) s_j(\vec{q}) \quad , \tag{2.4}$$

$M(\vec{k})$ stands for the corresponding bare dynamical matrix. The elastic
quadrupole-quadrupole interaction among neighbouring CN ions [19,20] is
given by

$$V^{RR} = \frac{1}{2} \sum_{\vec{q}} J_{\alpha\beta}(\vec{q}) Y_\alpha(\vec{q}) Y_\beta(\vec{q}) \quad . \tag{2.5}$$

An explicite form of $J(\vec{q})$ is given in appendix. The term V^R in (2.3)
refers to the crystal field potential in absence of translation-rotation
(T-R) coupling. The term V^{TR} accounts for the bilinear T-R coupling:

$$V^{TR} = \sum_{\vec{q}} \hat{v}_{\alpha i}(\vec{q}) Y_\alpha(\vec{q}) s_i(\vec{q}) \quad . \tag{2.6}$$

The coupling matrix $\hat{v}$ is specified in appendix. In the alkalicyanides-
halides, the bilinear coupling leads to a lattice mediated interaction

$$C_{\alpha\beta}(\vec{q}) = \hat{v}_{\alpha i}(\vec{q})\, M_{ij}^{-1}(\vec{q})\, \hat{v}_{j\beta}^*(\vec{q}) \tag{2.7}$$

between quadrupoles. This "elastic dipole"-type interaction dominates
over the electric quadrupole interaction in KCN and in NaCN and drives the
ferroelastic phase transition at the Brillouin-zone center.
In case of a compressible lattice, the interaction J has to be expanded in

terms of lattice displacements. We thereby obtain

$$V^{RRT} = \frac{1}{2\sqrt{N}} \sum_{\vec{p}\,\vec{q}} J'_{\alpha\beta i}(\vec{p},\vec{q}) \; Y_\alpha(\vec{p}+\vec{q}) \; Y_\beta(\vec{p}) s_i(\vec{q}) \quad . \qquad (2.8)$$

The coefficients $J'_{\alpha\beta i}$ are specified in appendix. The term V^{SR} in Eq.(2.3) accounts for the interaction of residual molecules Y with the static random strain fields which are generated by the substitutional halogen ions [13]

$$V^{SR} = \sum_{\vec{q}} Y_\alpha(\vec{q}) h_\alpha(-q, \sigma) . \qquad (2.9)$$

Here h_α is a static random field for which we assume a Gaussian distribution. We have $\bar{h}_\alpha = 0$ and

$$\overline{h_\alpha h_\beta} = x\,(1-x) H \,\delta_{\alpha\beta} \quad , \qquad (2.10)$$

where H is a measure of the square of the field strength. The overbar stands for a configurational average. A potential with contributions (2.3) was introduced in Ref. 21. in order to study the structural phase transitions and polymorphism in mixed crystals $M(Y)_x X_{1-x}$ with Y = CN. We show in appendix that the interaction coefficients $\hat{v}$, M, J and J' depend on the lattice constants and therefore on hydrostatic pressure. We remark that in KCN, the cubic lattice constant 2a = 6.5 $\overset{o}{A}$ and in NaCN 2a=5.9 $\overset{o}{A}$.We therefrom expect that NaCN at atmospheric pressure has a behaviour similar to KCN under higher hydrostatic pressure. This is indeed the case as we will see later.

3. FREE ENERGY

In the previous section, we have derived the model interaction V for a system of interacting molecules, randomly distributed on a compressible lattice in the presence of static random strain fields. In Refs. 13 and 21 we have calculated the free energy by taking the virtual crystal approximation in treating the effective orientational interaction. In contradiction with conventional spin glass theory [22] we neglect the random bond aspects of the molecular interaction. This approximation is probably not unreasonable in the present system, because the "elastic dipole" interaction (2.7) varies on a length scale much larger than the average distance between neighbouring CN molecular ions. On the other hand we treat the effect of the X ions as static random fields, which couple to the orientational degrees of freedom of the molecules. As has been shown for an Ising system, the static random field problem can be treated within molecular field theory [23]. We will here adopt the same point of view. Following then Ref. 21, we obtain the free energy

$$F = F_R + F_I + F_S \tag{3.1}$$

Here F_R is the orientational single particle free energy in the average crystal field. More interesting are the contributions F_I from interaction and F_S from entropy. We quote

$$F_I = \Sigma \ \{\tfrac{1}{2} M_{ij}(\vec{q}) s_i^{e\dagger}(\vec{q}) s_j^{e}(\vec{q}) + \hat{v}_{\alpha i}(\vec{q}) \ Y_\alpha^{e\dagger}(\vec{q}) s_i^{e}(\vec{q}) +$$

$$\tfrac{1}{2} \left[c^s \delta_{\alpha\beta} + J_{\alpha\beta}(\vec{q}) \right] Y_\alpha^{e\dagger}(\vec{q}) \ Y_\beta^{e}(\vec{q}) + \frac{1}{2\sqrt{N}} J'_{\alpha\beta i}(\vec{p},\vec{q}) Y_\alpha^{e\dagger}(\vec{p}+\vec{q}) Y_\beta^{e}(\vec{p}) s_i^{e}(\vec{q}) \} .$$

$$\tag{3.2}$$

Here the summation sign runs over wave vectors $\vec{p}$ and $\vec{q}$ and over repeated indices. $Y_\alpha^{e}(\vec{q})$ and $s_i^{e}(\vec{q})$ denote instantaneous expectation values. C° is the lattice mediated orientational self interaction;

$$c^s = \frac{1}{N} \ \underset{\vec{q}}{\Sigma} \ C_{\alpha\alpha}(\vec{q}) \quad , \qquad \alpha \text{ fixed} \quad , \tag{3.3}$$

is independent of α. The entropy contribution reads

$$F_S = \frac{1}{2} \frac{T}{y} \Sigma \ Y_\alpha^{e\dagger}(\vec{q}) \ Y_\alpha^{e}(\vec{q}) + F_S^{(3)} + F_S^{(4)} + F_S^{(6)} \ . \tag{3.4}$$

Here

$$y = x \ \alpha_1 / X(2) \ , \qquad X(2) = \langle Y_1^2 \rangle_\circ^{-1} \quad , \tag{3.5}$$

where $X(2)$ is a single particle average and α_1 depends on the random field [13]. $F_S^{(n)}$, $n = 3, 4, 6$ are of third, fourth and sixth order in Y^e respectively.

For a given configuration of orientations $Y^e(\vec{q})$, we determine the displacements by minimizing F_I with respect to $s_i^{e}(\vec{q})$, thereby obtaining

$$s_i^{e}(\vec{q}) = -M_{ij}^{-1}(\vec{q}) \ \{v_{j1}^{\tau}(-\vec{q}) Y_1^{e}(\vec{q}) + \frac{1}{2\sqrt{N}} \ \underset{\vec{p}}{\Sigma} \ J'_{\alpha\beta j}(\vec{p},-\vec{q}) Y_\alpha^{e\dagger}(\vec{p}-\vec{q}) Y_\beta^{e}(\vec{p})\} \ . \tag{3.6}$$

Eliminating s_i^{e}, we rewrite F_I as

$$F_I = \Sigma \ \{\tfrac{1}{2} L_{\alpha\beta}(\vec{q}) Y_\alpha^{e\dagger}(\vec{q}) Y_\beta^{e}(\vec{q}) - \frac{1}{8N} I_{\alpha\beta\gamma\delta}(\vec{p},\vec{k};\vec{q}) Y_\alpha^{e\dagger}(\vec{p}+\vec{q}) Y_\beta^{e}(\vec{p}) Y_\gamma^{e\dagger}(\vec{k}-\vec{q}) Y_\delta^{e}(\vec{k})$$

$$- \frac{1}{2\sqrt{N}} K_{\alpha\beta\gamma}(\vec{p},\vec{q}) Y_\alpha^{e\dagger}(\vec{p}+\vec{q}) Y_\beta^{e}(\vec{p}) Y_\gamma^{e}(\vec{q})\} \quad , \tag{3.7}$$

where

$$L_{\alpha\beta}(\vec{q}) = J_{\alpha\beta}(\vec{q}) - C_{\alpha\beta}(\vec{q}) + \delta_{\alpha\beta} c^s \ , \tag{3.8a}$$

$$I_{\alpha\beta\gamma\delta}(\vec{p},\vec{k};\vec{q}) = J'_{\alpha\beta i}(\vec{p},\vec{q}) \, M^{-1}_{ij}(\vec{q}) \, J'_{\gamma\delta j}(\vec{k},-\vec{q}) \quad , \tag{3.8b}$$

$$K_{\alpha\beta\gamma}(\vec{p},q) = J'_{\alpha\beta i}(\vec{p},\vec{q}) \, M^{-1}_{ij}(\vec{q}) v^{\tau}_{\gamma j}(-\vec{q}) \quad . \tag{3.8c}$$

The stability of a thermodynamic state (ordered or disordered phase) results from the competition between interaction and entropy in F, Eq. (3.1). The coefficients of the second, third and fourth order terms in Y^e occuring in F_I and in F_S respectively have a different concentration dependence and a different dependence on lattice constant 2a. Therefore we expect that by changing the concentration of substitutional disorder or by applying high pressure, we can drastically influence the phase transition behaviour. From Eqs. (3.1) and (3.3), (3.6), we obtain for the collective orientational susceptibility

$$\chi_{\alpha\beta}(\vec{q}) = \frac{\langle Y^{\dagger}_{\alpha}(\vec{q}) Y_{\beta}(q) \rangle}{T} = \chi_o \left[1 + \chi_o L(\vec{q}) \right]^{-1}_{\alpha\beta} \quad , \tag{3.9a}$$

where

$$\chi_o = y/T \tag{3.9b}$$

is the single particle orientational susceptibility. If the phase transition is of second order, $\chi(\vec{q})$ diverges at a temperature $T_c = y\lambda$, where λ is the largest eigenvalue of $|L(\vec{q}_o)|$. Here $\vec{q}_o$ is the wave vector which maximizes L. The orientational susceptibility χ is related to the inverse D of the displacement-displacement susceptibility by [24]

$$D_{ij}(\vec{q}) = \left[1 + \hat{v}^{\dagger}(\vec{q}) \chi(\vec{q}) \hat{v}(\vec{q}) M^{-1}(\vec{q}) \right]^{-1}_{ik} M_{kj}(\vec{q}) \quad . \tag{3.10}$$

This relation allows us to calculate the renormalized elastic constants.

4. DISCUSSION

We consider the ferroelastic transition from the orientationally disordered cubic phase to the orthorhombic phase. Such a transition is found experimentally in KCN, NaCN and in the region near $0.95 < x < 1$ in $M(CN)_x X_{1-x}$.[2, 3, 4] One order parameter component of T_{2g} symmetry condenses with wave vector $\vec{q}_o$ at the Brillouin zone center[25]. We take $\vec{q}_o = (q_x, o, o)$, with $q_x \to 0$. The order parameter is $Y^e_1(\vec{q}_o) \equiv \sqrt{N}\,\eta$, with $Y^e_{\alpha} = 0$ for $\alpha = 2, 3$. The free energy (3.1), with (3.4) and (3.7), then simplifies since third order terms in F_I and F_S are zero. Using the explicit expressions of the interactions from the appendix, we get

$$(1/N)F = A\eta^2 + C\eta^4 + D\eta^6 \quad , \tag{4.1}$$

with

$$A = \frac{1}{2}\left[L_{11}(q_o) + Ty^{-1}\right] \quad , \tag{4.2}$$

$$C = -C_I + C_S \quad , \tag{4.3a}$$

$$C_S = \frac{T\,a_2}{4!x^3\alpha_1^4}\left[3X^2(2) - X(4,2)\right] , \tag{4.3b}$$

$$C_I = \frac{1}{8}\,J'_{11x}(o,\vec{q}_o)\,M_{xx}^{-1}(\vec{q}_o)\,J'^{*}_{11x}(o,\vec{q}_o) \quad . \tag{4.3c}$$

The coefficient C_S follows from the calculation of $F_S^{(4)}$ (see Ref. 21). We have defined

$$X(4,2) = \langle Y_1^4\rangle_o\, X^4(2) \quad , \tag{4.4}$$

with $X(2)$ given by (3.5), and where $\langle\ \rangle_o$ denotes again a crystal field average. The coefficients α_1 and a_2 are close to unity for x close to 1. For an Ising system, molecular field theory yields $C_S = T/12$. Here C_S is larger, due to the factor $X^2(2)$. $X(2)$ is of the order of $10^{-1}.$ In addition, C_S increases with decreasing x due to substitutional disorder. We have not calculated explicitly D_S, which is the coefficient of the order parameter in $F_S^{(6)}$. All we need is that $D_S > 0$. For an Ising system, one gets $D_S = T/30$ by molecular field theory. Applying Landau theory to the free energy (4.1), we find a first order or a second order phase transition, depending on whether the coefficient C is negative or positive respectively. The condition

$$C_I > C_S \tag{4.5}$$

apparently holds in $M(CN)_x X_{1-x}$, M = K or Na for $0.95 < x \leqslant 1$. The first order transition occurs at a temperature

$$T_1 = T_c + \frac{C^2 y}{2D_S} \tag{4.6}$$

with $T_c = y\left|L_{11}(\vec{q}_o \to 0)\right|$. The discontinuity of the order parameter at T_1 is given by

$$\Delta\eta = \left[\frac{C}{2D_S}\right]^{1/2} \tag{4.7}$$

The static orientational susceptibility does not diverge but has the maximum value $2D_S C^{-2}$ at T_1.

The elastic constant c_{44} is obtained by means of Eqs. (3.10) and (3.9a).
Using the explicite form of M and $\hat{v}$ as given in appendix, we obtain at T_1

$$c_{44} = c^\circ_{44} \left[1 + \frac{2\,B^2}{ac^\circ_{44}} \frac{2D_s}{C^2} \right]^{-1} .$$
(4.8)

As long as $\Delta\eta \neq 0$, $c_{44} > 0$ at T_1. Experimentally, this effect is more
pronounced in NaCN [26] than in KCN [27]. A small admixture of substitutional
X ions not only decreases T_1, but also diminishes the first order character
of the phase transition. This experimental fact is well established[29,1,2,28].
With increasing substitutional disorder, the phase transition becomes more
second order like. We attribute this effect to the different concentration
dependence of the terms $C_I \sim x^\circ$ and $C_S \sim x^{-3}$. With decreasing x, the
inequality (4.5) can be reversed. The neighbourhood of a second order
phase transition is apparently a favorable condition for the occurence of
the orientational glass state [30, 31].

Applied external pressure has the opposite effect of substitutional disorder.
The inequality $C_I > C_S$ gets more stabilized and therefore the phase transition has a more pronounced first order character [7,8]. An increase of
pressure leads to a decrease of the lattice contant 2a and therefore C_I
increases. On the other hand, the entropy term C_S is much less affected
by pressure or equivalently lattice fluctuations. T_c, the virtual second
order transition temperature is almost independent of pressure. This is
probably due to the fact that T_c results from a competition between
attractive Coulomb forces and repulsive Born–Mayer forces which are affected
in opposite way by pressure, such that the net effect compensates. The
increase of T_1 with pressure is clearly due to the increase of C_I which
contributes to the second term on the right hand side of (4.6).

While the present considerations agree qualitatively well with experiment, quantitative agreement is more problematic. We first estimate C_S.
Taking $a_2 \approx 1$, $\alpha_1 \approx 1$ near x = 1, taking numerical results $\langle Y^4 \rangle_\circ = 0.013$,
$\langle Y^2 \rangle_\circ = 0.082$, we get

$$C_S = \frac{T}{0.16} \approx 10^3 [K]$$
(4.9)

for T = 160 [K]. On the other hand, we have estimated C_I, taking into account
the compressible electric quadrupole interaction

$$C_I = \frac{(5\pi Q^2)^2}{a^{10}_2{}^3} \left(\frac{19}{10}\right)^2 x \frac{1}{4a^3 c^\circ_{11}} = 10.9 \text{ K}$$
(4.10)

with Q = (0.72) 4.64 x 10^{-10} esu Å^2, a = 3.25 Å, c°_{11} = 5.12 x 10^{11} dyn/cm^2
for KCN.

Obviously this interaction is too small. We have also calculated the corresponding contribution to C_I due to the repulsive Born-Mayer forces between neighbouring CN^- ions. Also that contribution is too small. The only interaction which we can think of to be relevant, is the coupling of orientations to quadratic lattice displacements:[32]

$$V^{TTR} = \sum_{\vec{k}\,\vec{p}\,\vec{h}} w_{\alpha ij}(\vec{h},\vec{k},\vec{p})\, Y_\alpha(\vec{h}) s_i(\vec{k}) s_j(\vec{p}) \quad , \tag{4.11}$$

where

$$w_{\alpha ij}(\vec{h},\vec{k},\vec{p}) - \frac{4}{m\sqrt{N}} \sum_\kappa \{ v^{(2)}_{\alpha ij}(\kappa)\, \sin\left(\frac{\vec{k}\cdot\vec{t}(\kappa)}{2}\right) \cdot$$

$$\cdot \sin\left(\frac{\vec{p}\cdot\vec{\tau}(\kappa)}{2}\right)\cos\left(\frac{\vec{h}\cdot\vec{\tau}(\kappa)}{2}\right)\, \delta(\vec{h}+\vec{k}+\vec{p}) \quad . \tag{4.12}$$

Here $\kappa = 1-6$ runs over the six nearest neighbour M^+ ions surrounding a CN^- ion. The coefficients $V^{(2)}_{\alpha ij}(\kappa)$ are quoted in appendix. We see that only terms with $i \neq j$ are coupled. In particular for $\alpha = 1$, $Y^e_1(\vec{q}_o) \equiv \sqrt{N}\eta$, $\vec{q}_o = (q_x,\, o,\, o)$ and for a wave propagating in x-direction, we get (for fixed q_x)

$$V^{TTR} = \frac{4a^2}{m}\,\tilde{K}\, q_x s_x(q_x) q_x s_y(-q_x)\eta \quad , \tag{4.13}$$

which corresponds to the coupling of orientational order to lattice stretching and shear modes. We replace the shear wave by an orientational wave by using the bilinear coupling:

$$q_x s_y(q_x) = -\frac{i\,B\,\sqrt{m}}{c^o_{44}a^2}\,\sqrt{N}\eta \quad . \tag{4.14}$$

Subsequently, we minimize $(V^{TT} + V^{TTR})$ with respect to the compressional wave:

$$q_x s_x(q_x) = -\frac{2\,\tilde{K}}{a\,c^o_{11}}\, q_s s_y(q_x)\eta \quad . \tag{4.15}$$

Eliminating $(q_x s_x)$ from $V^{TT} + V^{TTR}$ and using (4.14), we finally obtain an additional contribution to $F^{(4)}_I$:

$$F^{(4)TTR}_I = -C^{TTR}_I \eta^4 \quad , \tag{4.16}$$

where $C^{TTR}_I = 4\tilde{K}^2 B^2/a^3 c^o_{11}(c^o_{44})^2$. We have calculated this quantity by using the numerical estimates $K = 937\ [K/\mathring{A}^2]$, $B = 1824\ [K/\mathring{A}]$, $c^o_{44} = 0.45$ and $c^o_{11} = 5.12\ 10^{11}\ dyn/cm^2$ for KCN. We then find $C^{TTR}_I = 860\ [K]$. This value is much more satisfactory and comes close to our estimate for C_S. Given the numerical uncertainties, in particular in c^o_{44}, c^o_{11} which are probably overestimated here, it is highly plausible that $C_S \gtrsim C_I^{TTR}$, in KCN. F. Lüty has been suggested that near T_C, KCN acts as a "bistable borderline case" between the ferroelastic NaCN and the antiferroelastic RbCN.

We have also estimated C_I^{TTR} for NaCN. Taking $a = 2.95$ Å, $B = 2500$ $[K/Å]$, $c_{44}^{\circ} = 0.75$ and $c_{11} = 5.7$, $\tilde{K} = 1300$ we obtain $C^{TTR} = 1273$ $[K]$. Summarizing our results we comme to the conclusion that the anharmonic interaction V^{TTR} and not the conventional "compressible interaction"-type mechanism V^{RRT} is the dominant factor which determines the nature of the phase transition and the pressure dependence in $M(CN)_x X_{1-x}$. We expect that this mechanism is important in general in crystals which contain counterions. The role of anharmonicities has been emphasized before on experimental grounds in analyzing pressure effects in MCN [8].

APPENDIX

Here we specify the interaction matrices which have been used in the previous considerations. The bare dynamical matrix has the elements

$$M_{11} = [\, q_x^2 c_{11}^{\circ} + (q_y^2 + q_z^2)\, c_{44}^{\circ}]\, \rho^{-1} \quad , \tag{A.1}$$

$$M_{12} = M_{21} = [\, q_x q_y\, (c_{12}^{\circ} + c_{44}^{\circ})]\, \rho^{-1} \quad , \tag{A.2}$$

etc. by permutation of indices. $\rho = m/2a^3$ is the mass density. The bilinear T-R coupling reads

$$\hat{v}(\vec{q}) = -\frac{i2Ba}{\sqrt{m}}
\begin{bmatrix}
q_y & q_x & o \\
o & q_z & q_y \\
q_z & o & q_x
\end{bmatrix}
\tag{A.3}$$

in the long wavelength limit. Here B is a material constant which is calculated from the microscopic potentials: Born-Mayer, Coulomb and van der Waals. m is the total mass per primitive cell. The electric quadru-pole interaction reads for $\vec{q} \to o$:

$$J(\vec{q}) = \frac{Q^2 \pi}{a^5 \sqrt{2}} \{J^{+(yz)} + J^{-(yz)} + J^{+(xy)} + J^{-(xy)} +$$

$$J^{+(zx)} + J^{-(zx)}\} \quad . \tag{A.4}$$

Here Q is the quadrupole moment of the CN$^-$ ion.

The matrices J^{+}_{-} are given by

$$J^{+}_{-}(yz) = \begin{bmatrix} -\frac{3}{5} & o & \mp 1 \\ o & \frac{19}{10} & o \\ \mp 1 & o & -\frac{3}{5} \end{bmatrix} \quad , \tag{A.5}$$

the others follow by cyclic permutation. The change of the quadrupole
interaction by acoustic lattice displacements is described by the coupling

$$J'_{\alpha\beta i}(\vec{p},q) = J'^{(xy)}_{\alpha\beta i}(\vec{p},\vec{q}) + J'^{(yz)}_{\alpha\beta i}(\vec{p},\vec{q}) + J'^{(z,x)}_{\alpha\beta i}(\vec{p},\vec{q}) \quad ,$$

where

$$J'^{(xy)}_{\alpha\beta z}(\vec{p},\vec{q}) = o$$

$$J'^{(xy)}_{\alpha\beta x}(\vec{p},\vec{q}) = \frac{-i\ 5\pi Q^2}{\sqrt{m}\ a^5(\sqrt{2})^3} \{(q_y+q_x)J^{+(xy)}_{\alpha\beta} + (q_x-q_y)J^{-(xy)}_{\alpha\beta}\} \quad , \tag{A.6}$$

for $\vec{p} \to o$. The other elements follow by cyclic permutation.

The anharmonic TTR-coupling (4.11) follows from the expansion of the
bilinear T-R coupling in terms of lattice displacements. The result (4.12)
is derived in Ref. 32. See also Ref. 18. In coefficients $v^{(2)}_{1ij}(\kappa)$,
the index 1 refers to $\alpha = 1$, T_{2g}. The index κ specifies the location
of the six M atoms at the corner of the octahedron surrounding a CN molecule. One has

$$v^{(2)}_1(1,o,o) = \begin{bmatrix} o & \tilde{K} & o \\ \tilde{K} & o & o \\ o & o & o \end{bmatrix} \quad ,$$

where (1,o,o) refers to position (a,o,o) of M.

$$v^{(2)}_1(o,1,o) = \begin{bmatrix} o & \tilde{K} & o \\ \tilde{K} & o & o \\ o & o & o \end{bmatrix} \quad ,$$

$$v^{(2)}_1(o,o,1) = \begin{bmatrix} o & \tilde{H} & o \\ \tilde{H} & o & o \\ o & o & o \end{bmatrix} \quad .$$

Here $\tilde{H}$ and $\tilde{K}$ are calculated numerically [18].

REFERENCES

1. F. Lüty, in <u>Defects in Insulating Crystals</u>, edited by V.M. Tuchkevich and K.K. Shvarts (Springer-Verlag, Heidelberg, 1981), p. 69.

2. K. Knorr and A. Loidl, Phys. Rev. B <u>31</u>, 5387 (1985).

3. J.M. Rowe, J. Bouillot, J.J. Rush and F. Lüty, Physica B + C <u>136 B</u>, 498 (1986).

4. J.M. Rowe, J.J. Rush, and E.Prince, J. Chem. Phys. <u>66</u>, 5147 (1977). J. Ortiz-Lopez and F.Lüty, Phys. Rev. B <u>37</u>, 5452 (1988).

5. P.W. Richter and C.W.F.T. Pistorius, J. Solid State Chemistry <u>5</u>, 276 (1972); S. Hirotsu, M. Miyamoto and I. Yamamoto, Jap. J.Appl. Phys. <u>20</u>, L 917 (1981).

6. D.J. Bergman and B.I. Halperin, Phys. Rev. B <u>13</u>, 2145 (1976); A.I. Larkin and S.A. Pikin, Zh. Eksp. Teor. Fiz. <u>56</u>, 1664 (1969) Sov. Phys. JEP <u>29</u>, 891 (1969) .

7. W. Dultz and H. Krause, Phys. Rev. B <u>18</u>, 394 (1978); W. Dultz, H.H. Otto, H. Krause and J.L. Buevoz, ibid. <u>24</u>, 1287 (1981).

8. H.D. Hochheimer, W.F. Love and C.T. Walker, Phys. Rev. Lett. <u>38</u>, 832 (1977).

9. M. Julian and F. Lüty, Ferroelectrics <u>16</u>, 201 (1977); D. Durand and F. Lüty, ibidem 205 (1977).

10. J.M. Rowe, J.J. Rush, D.J. Hinks and S. Susman, Phys. Rev. Lett. <u>43</u>, 1158 (1979); K.H. Michel and J.M. Rowe, Phys. Rev. B <u>22</u>, 1417 (1980).

11. E. Courtens, Jpn. J. Appl. Phys. <u>24</u>, Suppl. 24-2, 70 (1985).

12. U.T. Höchli, K. Knorr and A. Loidl, Adv. Physics <u>39</u>, 405 (1990).

13. K.H. Michel, Phys. Rev. Lett. <u>57</u>, 2188 (1986); Phys. Rev. B <u>35</u>, 1405 (1987).

14. L.J. Lewis and M.L. Klein, Phys. Rev. Lett. <u>57</u>? 2698 (1986).

15. J.O. Fossum and C.W. Garland, Phys. Rev. Lett. <u>60</u>, 592 (1988).

16. T. Lüty and R. Fouret, J. Chem. Phys. <u>90</u>, 5696 (1989); T.H.K. Barron, T. G. Gibbons and R.W. Munn, J. Phys. C: Solid State Phys. <u>4</u>, 2805 (1971).

17. W. Press and A. Hüller, Acta Cryst. A <u>29</u>, 252 (1973); M. Yvinec and R.M. Pick, J. Physique <u>41</u>, 1045, 1053 (198).

18. K.H. Michel and J.M. Rowe, Phys. Rev. B <u>32</u>, 5818 (1985).

19. M.L. Klein and I.R. McDonald, Chem. Phys. Lett. <u>78</u>, 383 (1981); R.M. Lynden-Bell, I.R. McDonald and M.L. Klein, Mol. Phys. <u>48</u>, 1093 (1983).

20. S.D. Mahanti and D. Sahu, Phys. Rev. Lett. <u>48</u>, 936 (1982); D. Sahu and S.D. Mahanti, Phys. Rev. B <u>26</u>, 298 (1982).

21. K.H. Michel and T. Theuns, Phys. Rev. B <u>40</u>, 5761 (1989).

22. For a review, see K. Binder and A.P. Young, Revs. Mod. Physics <u>58</u>, 801 (1986).

23. T. Schneider and E.Pytte, Phys. Rev. B $\underline{15}$, 1519 (1977); A. Aharony,
    ibid. $\underline{18}$, 3318 (1978).

24. K.H. Michel and J. Naudts, Phys. Rev. Lett. $\underline{39}$, 212 (1977); eidem
    J. Chem. Phys. $\underline{67}$, 547 (1977).

25. K. Parlinski, Z. Phys. B $\underline{56}$, 51 (1984).

26. S. Haussühl, J. Eckstein, K. Recker and F. Walbrafen, Acta Cryst. A $\underline{33}$,
    847 (1977).

27. S. Haussühl,Solid State Commun. $\underline{13}$, 147 (1973).

28. W. Rehwald, J.R. Sandercock and M. Rossinelli, phys. stat. sol. (a)
    $\underline{42}$, 699 (1977).

29. J. Ortiz-Lopez and F. Lüty, Phys. Rev. B $\underline{37}$, 5461 (1988).

30. K.H. Michel, Z. Phys. B $\underline{68}$, 259 (1987).

31. A.P. Mayer and R.A. Cowley, J. Phys. C $\underline{21}$, 4835 (1988).

32. K.H. Michel, Z. Phys. B $\underline{61}$, 45 (1985).

STIMULATED BRILLOUIN GAIN SPECTROSCOPY AT HIGH PRESSURES

J. S. Friedman, B. L. Bracewell, H. D. Hochheimer, C. Y. She

Department of Physics
Colorado State University
Fort Collins, CO 80523

I. INTRODUCTION

The use of spontaneous Brillouin spectroscopy with a Fabry-Perot interferometer is a well established, powerful technique for studying thermally excited sound waves in liquids and solids[1-4]. There are well known problems associated with the Fabry-Perot interferometer, however. Its limited finesse (resolution), low contrast, nonlinearities of the scanning system, and the absence of an internal standard for frequency calibration all contribute to a shortfall in the precision and accuracy of the experimental data.

In order to perform rigorous tests of theoretical predictions, and to better determine model parameters, one needs measurements accurate enough to allow comparison with data obtained from ultrasonic or neutron scattering studies. Hence, considerable efforts have been made to overcome the limitations of the conventional spectrometer. Pioneering work was done by R. J. Sandercock[5-8], who developed a tandem multipass interferometer. Further improvements concerning precision measurements have been reported by Sussner and Vacher[9]. They modulated a reference beam at a microwave frequency corresponding exactly to the shift of the Brillouin line and obtained the Brillouin frequency shift by direct comparison. This technique allowed them to determine the absolute hypersound velocity with a 10^{-4} relative uncertainty, and to improve the accuracy of a Brillouin experiment (usually 10^{-3} accuracy on an absolute scale[5,10-12]) by more than an order of magnitude.

Vacher et al.[13] also constructed a Brillouin spectrometer with a frequency resolution ($\nu/\Delta\nu$) of 10^8 and a contrast of more than 10^{10} by arranging a triple-pass plane and a confocal Fabry-Perot interferometer in tandem. Using this arrangement, one can achieve a resolution of 2 to 50 MHz depending on the confocal Fabry-Perot interferometer, but the operation of such a system is very tedious and one loses 50% of the intensity in the confocal interferometer. Thus the use of their technique has limited application.

Although spontaneous Brillouin scattering has been used extensively to study liquids and solids, only a few measurements on liquids[14-19] and solids[20-29] under pressure have been reported. This is due both to the extreme experimental demands which must be met in order to combine the two techniques and to the limitations of spontaneous Brillouin scattering mentioned above.

The work of Nelson and coworkers[30-32] has shown that the time-domain impulsive stimulated light scattering technique can be exploited to study sound waves in liquids and solids. Since the time-domain technique is insensitive to vibrations, which cause enormous problems for a conventional Brillouin scattering experiment, the impulsive stimulated Brillouin scattering technique would be appropriate for studies at high pressures[33].

Another coherent light scattering technique is stimulated Brillouin gain (SBG) spectroscopy. The phenomenon of stimulated Brillouin scattering was first observed by Chiao et al.[34] in the mid 1960's, but it is only in the last decade that coherent Brillouin scattering as a high resolution spectroscopic tool has been developed[35-38]. In SBG spectroscopy, two laser beams (pump and probe) of slightly different frequencies are overlapped in the sample. The pump and probe fields selectively excite a normal mode of density fluctuations in the sample, with the wave vector and frequency of the normal mode determined by momentum and energy conservation. When the frequency difference between the beams equals the Brillouin frequency, energy is transferred from the higher frequency beam to the lower frequency one. Thus the normal modes of the medium can be obtained by scanning the frequency of one beam and monitoring the gain and/or loss in the probe beam intensity.

There are several advantages of coherent spectroscopy over spontaneous light scattering spectroscopy[37]. First, the elastically scattered light is highly suppressed because both gain and loss of coherent elastic scattering occur when pump and probe beams are at the same frequency and therefore cancel each other. Second, the wave vector of the medium under investigation is well defined and is specified by the wave vectors of the pump and probe beams. Third, the spectral resolution can be much higher since

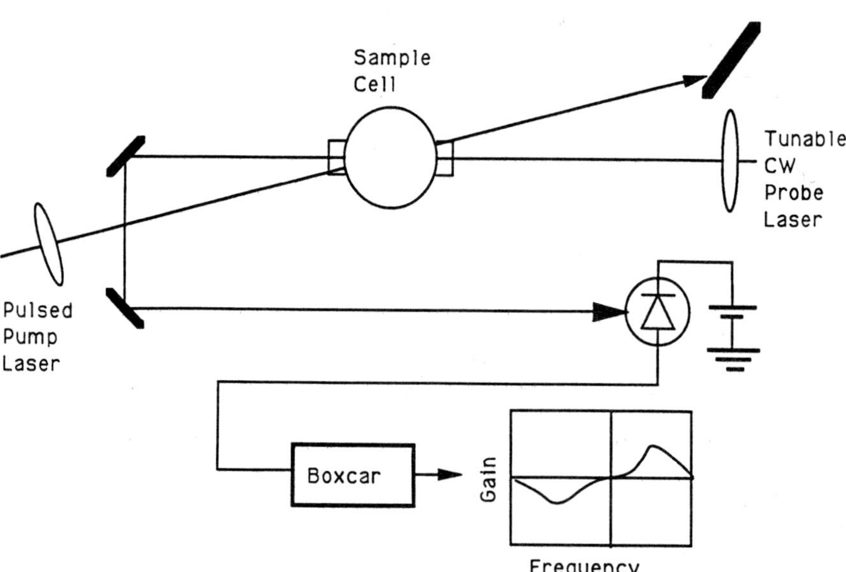

Fig. 1. Experimental setup for SBG spectroscopy. The sample cell is described in Fig. 3. As the probe laser is scanned about the pump frequency, the SBG spectrum is obtained.

it is limited by the laser linewidths instead of the instrumental bandwidth
of the Fabry-Perot interferometer.

Since the effect of stray light is absent in a coherent scattering
spectrum, the minimum detectable Brillouin shift is limited only by the
laser linewidths and by noise introduced due to sources other than stray
light scattering. In our previous pulsed laser experiments[36] the resolution
was limited by the pump laser linewidth to 100 MHz, which is no better than
the results obtained by spontaneous scattering[10-12]. Using a cw pump laser
instead of a pulsed laser we have demonstrated that the resolution can be
improved to be better than 4 MHz[36,37]. Assuming a free spectral range (FSR)
of 15 GHz, this is comparable to a Fabry-Perot interferometer with a finesse of about 3500. (A FSR of 15 GHz is appropriate, for example, in the
study of CS_2, a sample with a very strong SBG spectrum.) CW stimulated Brillouin gain (SBG) spectroscopy with this resolution was first applied to
study SF_6 gas at 10.5 atm[37] and has recently been used to measure Brillouin
shifts in simple liquids at room temperature and atmospheric pressure[39].

These advantages motivated us to combine SBG spectroscopy with high
pressure techniques. Here we report the first room temperature SBG spectroscopic study at high pressure. In order to show that SBG spectra at
high pressure can be obtained from weaker scatterers we chose ethanol instead of CS_2 as a sample.

II. EXPERIMENTAL

The experimental setup for SBG spectroscopy is shown in Fig. 1. The
pump laser is a Hänsch style pulsed dye laser pumped by a Q-Switched frequency doubled Nd-YAG laser with a 10 Hz repetition rate. It operates at a
center wavelength of about 585 nm and has a bandwidth of less than 0.5 GHz.
It is crossed at a near counter-propagating angle by a cw single-mode dye
laser which acts as the probe shown in Fig. 1. The probe laser, which has
a bandwidth of about 2 MHz, can be scanned over a range of 30 GHz around
the pump frequency. The probe laser is passed through a spatial filter to
reduce stray scattered pump light from the high pressure cell before it
illuminates a fast photodiode. A signal is seen as a pulse of gain or loss
in the probe beam when the pump laser fires. It is detected by a gated

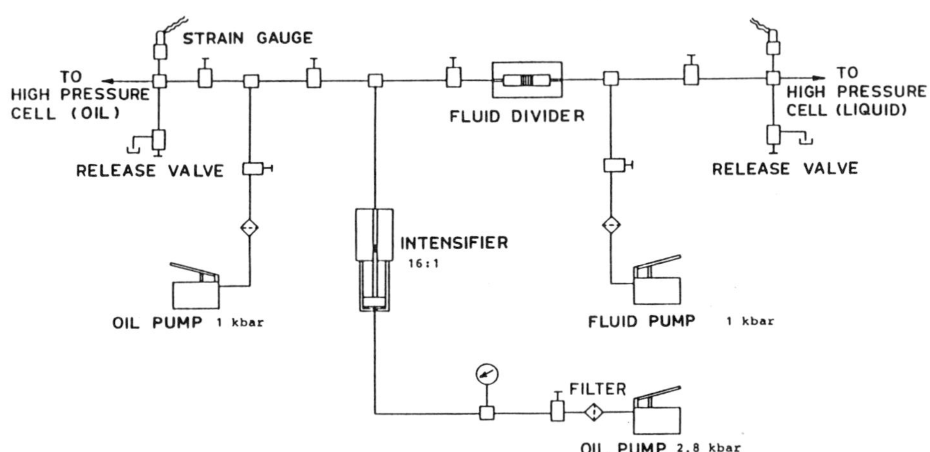

Fig. 2. High pressure generating system. The fluid divider separates the
primary pressure medium (oil) from the working medium (ethanol).

integrator and boxcar averager and recorded by a computer. As the probe
laser is scanned the SBG spectrum is obtained.

The high pressure generating system is shown in Fig. 2. The crucial
part in this system is the fluid divider which ensures that the sample
liquid is not contaminated by oil, the primary pressure medium. The divider is basically a 1:1 intensifier with a very simple packing arrangement.
It can be easily disassembled, cleaned and used again with another liquid
as sample or pressure medium. Figure 3 shows the high pressure cell. For
light scattering experiments, three of the ports are fitted with sapphire
windows and the fourth is plugged. For the experiment reported here, however, only two of the ports are fitted as entrance and exit windows for
the pump and probe beams and the other two are blocked. The sapphire windows have a clear aperture of 10mm and are sealed against the high pressure
plugs as described by Hochheimer, et al.[40].

III. RESULTS AND DISCUSSION

Figure 4 shows room temperature SBG spectra for ethanol at different
pressures. Data were taken in the pressure range between 0 and 4 kbar.

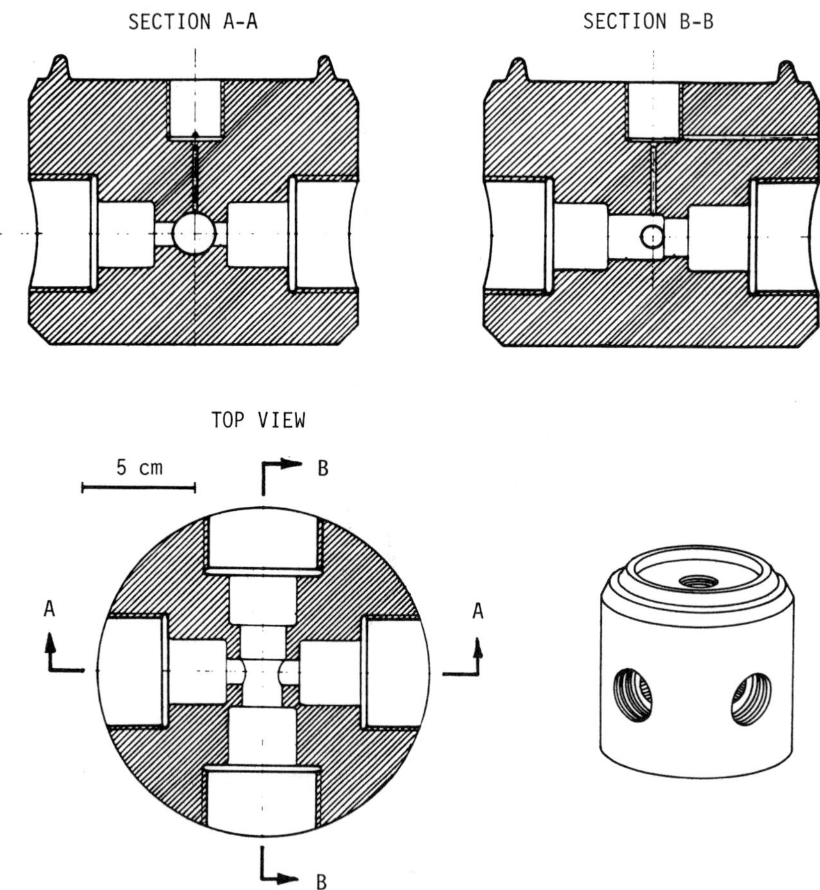

Fig. 3. High pressure cell. Only two of the four ports are fitted with
sapphire windows for SBG spectroscopy. The other two ports are
plugged.

The scattering angle of 178° in the liquid is defined by the crossing angle
of the laser beams in the liquid under study. The beams are weakly focussed with 50 cm focal length lenses. The resulting spectra show a clear
shift to higher frequencies as pressure is increased. The fluctuations
near the unshifted frequency are not random noise. Rather, they are due to
the frequency structure of the pump laser, which has three modes in a bandwidth of roughly 0.5 GHz. This pulsed laser is far from the ideal pump
laser for such an experiment. The fact that reasonably good spectra can be
obtained from a weak scattering medium with a marginal pump laser suggests
that considerable improvements in signal-to-noise as well as spectral resolution can be expected for future studies of liquids and amorphous solids
under high pressures.

In Fig. 5 the Brillouin frequency versus pressure is plotted. The
resulting pressure shift is 1.4 GHz/kbar. We have compared our results
with the work of Polian et al.[23]. Judging from their graph, their measured
shift is 1.1 GHz/kbar in a mixture of ethanol and methanol, a value which
agrees very well with ours for pure ethanol. This is not surprising because
the densities of both alcohols differ by no more than 2% up to 20 kbar.

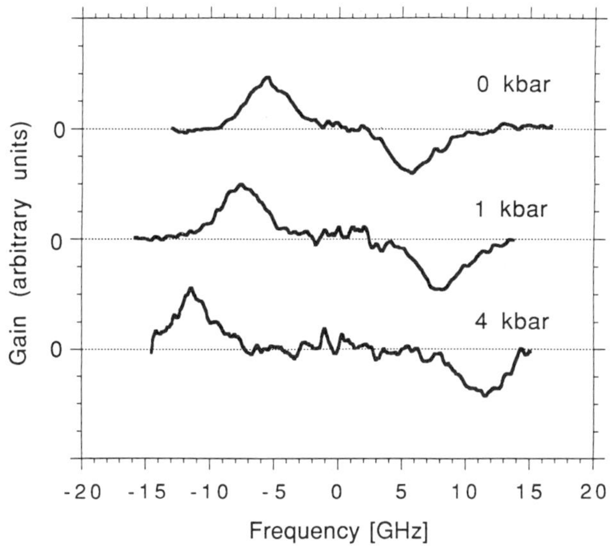

Fig. 4. Room temperature SBG spectra for ethanol at 0, 1
         and 4 kbar. The fluctuations near the unshifted
         frequency are due to the mode structure of the pump
         laser.

IV. SUMMARY

We have shown that we can combine high resolution SBG spectroscopy
with high pressure techniques. These combined techniques will allow us to
study the pressure dependence of structural relaxation, translation-vibra-
tion coupling, and motion in viscoelastic liquids. Even more interesting

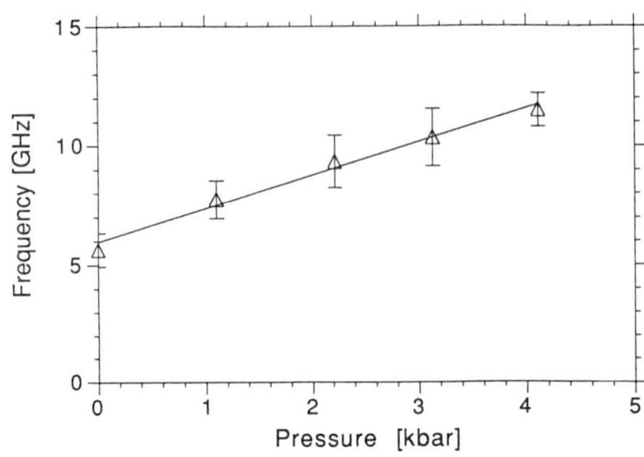

Fig. 5. Brillouin frequency versus pressure for
         ethanol. The frequency shift is 1.4 GHz/kbar.

will be studies of the glass transition for viscoelastic solids such as
high molecular weight polymers, and studies of amorphous solids with a resolution far better than achievable with spontaneous Brillouin scattering
using a Fabry-Perot interferometer.

V. ACKNOWLEDGEMENTS

We would like to acknowledge discussions with the late Professor Ian. L.
Spain whose interests in laser probing of condensed matter under high pressure prompted the initiation of this experiment. We would also like to
acknowledge the help of S. Y. Tang for his advice concerning SBG experiments and for his expertise with the cw single-mode dye laser.

REFERENCES

1. B. J. Berne and R. Pecora, Dynamic Light Scattering, Wiley, New York
   (1976)
2. G. Benedek, J. B. Lastovka, K. Fritsch, T. Greytak, J. Opt. Soc. Am.
   54, 1284 (1964)
3. R. Y. Chiao, B. P. Stoicheff, J. Opt. Soc. Am. 54, 1286 (1964)
4. G. B. Benedek, K. Fritsch, Phys. Rev. 149, 647 (1966)
5. J. R. Sandercock, Opt. Commun. 2, 76 (1970)
6. J. R. Sandercock in: Proc. 2nd Int. Conf. on Light Scattering in Solids, ed. M. Balkanski, Flammarion, Paris (1971), p. 9
7. J. R. Sandercock, R.C.A. Rev. 36, 89 (1975)
8. J. R. Sandercock, Solid State Commun. 26, 5218 (1976)
9. H. Sussner, R. Vacher, Appl. Optics 18, 3815 (1979)
10. R. Vacher, L. Boyer, M. Boissier, Phys. Rev. B6, 674 (1972)
11. W. S. Gornall, B. P. Stoicheff, Solid State Commun.8, 1529 (1970)
12. M. Bush, J. C. Toledano, J. Torres, Opt. Commun. 10, 273 (1974)
13. R. Vacher, H. Sussner, M. von Schickfus, Rev. Sci. Instrum. 51, 288
    (1980)
14. J. H. Stith, L. M. Peterson, D. H. Rank, T. A. Wiggins, J. Acoust. Soc.
    Am. 55, 785 (1974)
15. M. Sedlacek, Z. Naturforsch. 29a, 1622 (1974)
16. M. Sedlacek, A. Asenbaum, Phys. Lett. 50A, 245 (1974)
17. F. D. Medina, D. C. O'Shea, J. Chem. Phys. 66, 1940 (1977)

18. A. Asenbaum, H. D. Hochheimer, J. Chem. Phys. 74, 1 (1981)
19. A. Asenbaum, H. D. Hochheimer, Z. Naturforsch. 38a, 980 (1983)
20. C. H. Whitfield, E. M. Brody, Rev. Sci. Instrum. 47, 942 (1976)
21. H. D. Hochheimer, W. F. Love, C. T. Walker, Phys. Rev. Lett. 38, 832 (1977)
22. H. D. Hochheimer, W. F. Love, C. T. Walker in: High Pressure and Low Temperature Physics, ed. C. W. Chu and J. A. Woolam, Plenum, New York (1978) p. 299
23. J. Schroeder, K. J. Dunn, F. P. Bundy in: Proc. 8th AIRAPT Conf., eds. C. M. Backman, T. Johannisson, and L. Tegner, Arkitektkopia, Uppsala (1982) p. 259
24. A. Polian, J. M. Besson, M. Grimsditch, H. Vogt, Phys. Rev. B25, 2767 (1982)
25. A. Polian, M. Grimsditch, Phys. Rev. B27, 6409 (1983)
26. E. M. Brody, H. Shimizu, H. K. Mao, P. M. Bell, W. A. Bassett, J. Appl. Phys. 52, 3583 (1981)
27. S. A. Lee, D. A. Pinnick, S. M. Lindsay, R. C. Hanson, Phys. Rev. B34, 2799 (1986)
28. K. Ströbner, W. Henkel, H. D. Hochheimer, M. Cardona, Solid State Commun. 47, 567 (1983)
29. A. Asenbaum, O. Blaschko, H. D. Hochheimer, Phys. Rev. B34, 1968 (1986)
30. Y. X. Yan, K. A. Nelson, J. Chem. Phys. 87, 6240, 6267 (1987)
31. Y. X. Yan, L. T. Cheng, K. A. Nelson, J. Chem. Phys. 88, 6477 (1988)
32. S. M. Silence, S. R. Goates, K. A. Nelson, Chem. Phys. 149, 233 (1990)
33. Y. X. Yan, private communications
34. R. Y. Chiao, C. H. Townes, B. P. Stoicheff, Phys. Rev. Lett. 12, 592 (1964)
35. A. G. Jacobson, Y. R. Shen, Appl. Phys. Lett. 34, 464 (1979)
36. C. Y. She, G. C. Herring, H. Moosmüller, S. A. Lee, Phys. Rev. Lett. 51, 1648 (1983)
37. S. Y. Tang, C. Y. She, S. A. Lee, Opt. Lett. 12, 870 (1987)
38. G. W. Faris, L. E. Jusinski, M. J. Dyer, W. K. Bischel, A. P. Hickman, Opt. Lett. 15, 703 (1990)
39. K. Ratanaphruks, W. T. Grubbs and R. A. MacPhail, Chem. Phys. Letters (in press)
40. H. D. Hochheimer, M. L. Shand, J. E. Potts, R. C. Hanson, C. T. Walker, Phys. Rev. B14, 4630 (1976)

PROTON NMR CHEMICAL SHIFTS IN ORGANIC LIQUIDS MEASURED AT HIGH PRESSURE USING THE DIAMOND ANVIL CELL

K.E. Halvorson*, D.P. Raffaelle**, G.H. Wolf*, and R.F. Marzke**

*Department of Chemistry, **Department of Physics
Arizona State University, Tempe AZ 85282, USA

ABSTRACT

Chemically-shifted proton NMR lines of organic liquids, including glycerol, methanol, benzyl alcohol, and 1,2-propanediol, have been resolved for the first time in the diamond anvil cell (DAC) at pressures up to and beyond 10 kbar. The change of the chemical shift difference between two closely-spaced groups of lines, one arising from the hydroxyl protons and other arising from protons in the backbone, was measured as a function of pressure in both glycerol and 1,2-propanediol. Deshielding of the hydroxyl protons relative to the backbone increased approximately linearly with pressure at the rate of 0.06 ppm per kilobar for glycerol, and 0.04 ppm per kilobar for 1,2-propanediol. These experiments prove that line broadening arising from a spatially inhomogeneous magnetic environment surrounding the sample in the DAC need not preclude NMR studies at moderately high resolution (1 ppm). This opens new regions of very high pressure for the study of liquids by NMR.

INTRODUCTION

The study of matter at high pressure by NMR presents formidable challenges, and the pressures achieved in most work to date have been limited to values below 10 kilobars. NMR spectrometers generally utilize either piston-cylinder or Bridgeman anvil devices, which can produce maximum sample pressures ranging from only tens of kilobars for the former to approximately 100 kilobars for the latter (Vaughn et al., 1971). For many years much higher pressures have been available for optical studies using the diamond anvil cell (DAC), a device that is capable of reaching megabar (100 GPa) pressures, but only recently has it been adapted for NMR (Lee et al., 1987). The key to its successful application was the realization that the metal gasket of the DAC need not strongly shield the sample from the radio frequency magnetic field required for NMR. The very small size of DAC samples makes detection difficult, but good signal-to-noise ratios can be obtained with the use of standard signal averaging techniques, at least for NMR frequencies in the vicinity of 400 MHz.

In their paper, Lee et al. (1987) reported measurements of unresolved proton spin-lattice, and spin-spin relaxation times, T_1 and T_2, in a 4:1 methanol to ethanol solution up to 54 kilobar. Although reasonable signal to noise values were obtained, they indicated that high resolution measurements of chemical shifts would probably be precluded by the large, inhomogeneous width of the proton NMR line they observed, which presumably was caused by the differing magnetic susceptibilities of the materials surrounding the samples in the DAC. We have recently found, however, that it is possible to obtain NMR spectra from organic liquids in a DAC with resolution sufficient to observe well-separated, chemically shifted lines arising from inequivalent groups of protons. We began our high-pressure NMR investigation with glycerol since it has a well known pressure-induced glass transition that has been extensively studied by numerous techniques. In addition, several nuclear magnetic relaxation investigations have been performed on glycerol with temperature and pressure as variables (Diehl et al., 1990; Kuhns and Conradi, 1982; Wolfe and Jonas, 1979; Fiorito and Meister, 1972). Once it became obvious that moderate resolution was possible (1 ppm), we expanded our investigation to include other alcohols. We were also interested in studying the relaxation properties of liquids, from T_1 and T_2 measurements, as functions of pressure through the glass transition. Our findings for these quantities will be the subject of a later report (Raffaelle et al., 1991).

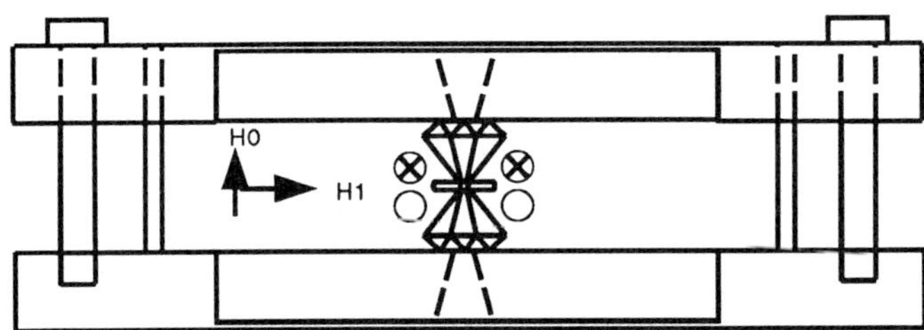

Figure 1. Schematic of the DAC

EXPERIMENTAL

Figure 1 shows a diagram of the modified Merrill-Bassett cell used for the NMR measurements. The cell was constructed of hardened beryllium-copper alloy No. 25. The gasket used was also made from hardened Be-Cu. Typical gasket thickness was 0.013 inches, with a 500 micron diameter hole drilled in the center. The gasket was not preindented, in order to maximize the sample volume. For most measurements two large type II diamonds were used, one with an 800 micron culet, the other with a 1000 micron culet. On some measurements smaller type I diamonds having 500 micron culets were employed, with 300 micron holes drilled in the gaskets. The diamonds were mounted using pressed indium, which was removed after pressure was achieved. The cell was then placed in appropriate solvents and ultrasonically cleaned. This resulted in a cell that was free of protons, except for those inside the sample chamber. Cell pressure was measured using the ruby fluorescence technique (Barnett et al., 1973). The NMR probe was built around a simple double-loop coil, tuned and matched to 400.1 MHz using two ceramic trimmer capacitors. The probe and cell were designed for use on a conventional Bruker AM-400 spectrometer in the NMR Facility of the Chemistry department at Arizona State University. No modifications to the spectrometer, aside from the probe itself, were required for obtaining the data.

RESULTS

A typical high resolution spectrum of neat glycerol, at atmospheric pressure in a standard NMR tube, is shown in the bottom trace of figure 2. This spectrum was taken with the magnet deuterium-locked and shimmed. In all spectra the chemical shifts were not referenced to tetramethyl silane (TMS), but to a line arising from within the molecule. The peaks near zero ppm are assigned to the protons attached to the carbon frame of the molecule: the small peak downfield (on the left) of this group is assigned to the single proton attached to the central carbon and the other two peaks represent the outer protons. The two peaks around 1.7 ppm are assigned to the hydroxyl protons: the small downfield peak of this group represents the central hydroxyl proton, while the large peak is assigned to the end hydroxyl protons. The top trace of figure 2 shows the NMR spectrum of neat glycerol obtained in a DAC at atmospheric pressure. In the DAC spectrum the backbone and hydroxyl proton bands are clearly separated, although not resolved into their constituent peaks. This spectrum was taken with the magnet locked but not shimmed. In the course of our study we could detect no difference between locked and unlocked spectra in the DAC.

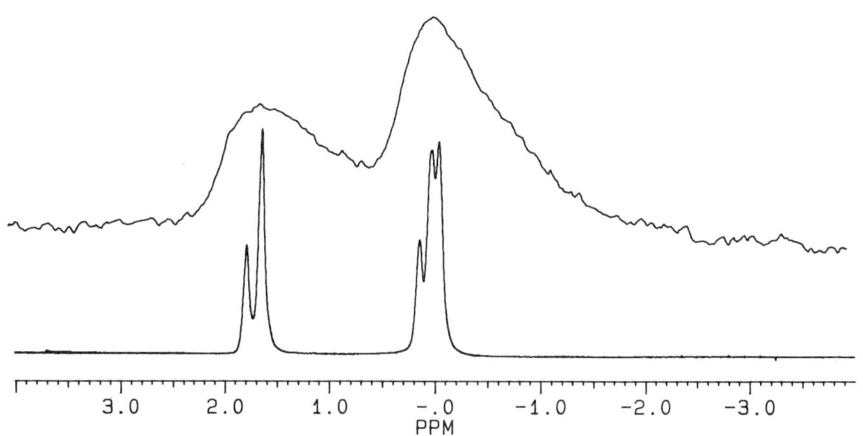

Figure 2. Glycerol NMR spectra at one atmosphere. Top trace inside DAC; bottom trace high resolution spectrum in a normal NMR sample tube.

We have obtained similar resolution at higher pressures in glycerol and in three other organic liquids, as shown in figures 3-6. The two glycerol peaks of figure 3 are well-separated at 6.4 kbar, but broaden beyond the limits of resolution at pressures above 10 kbar. Figure 4 shows a spectrum of 1,2-propanediol taken at 6.5 kbar. Three bands are evident. The band at zero ppm is assigned to the methyl protons, the middle band near 2.4 ppm belongs to the other backbone protons, and the most downfield band near 4.5 ppm is assigned to the hydroxyl protons. The three bands of 1,2-propanediol could still be resolved up to pressures of 12 kbar when the cell decompressed disastrously.

A typical methanol spectrum taken at 11 kbar appears in figure 5. The methyl proton band is centered near -0.5 ppm. The band downfield near 1.5 ppm is attributed to the hydroxyl proton. The chemical shift separation of the hydroxyl and methyl peaks is similar to those seen in the other organic liquids. We made no attempt to account for the

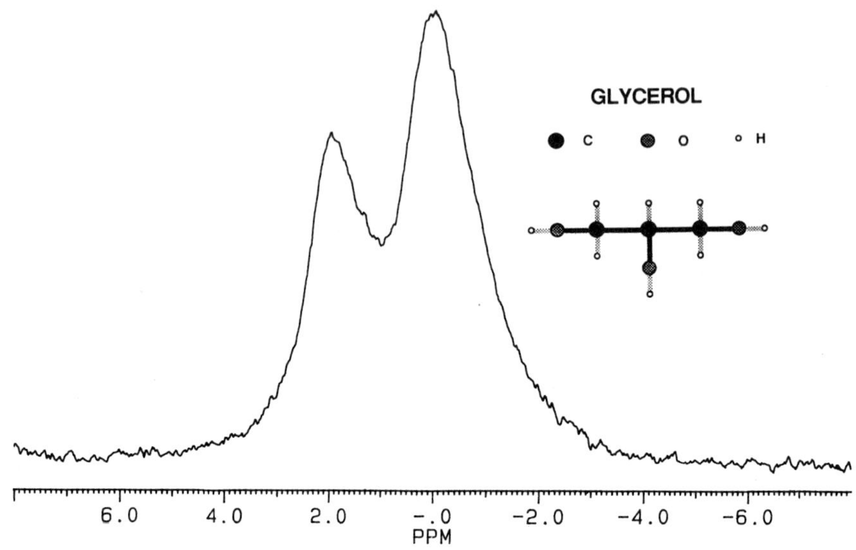

Figure 3. NMR spectrum of glycerol in a DAC at 6.4 kbar.

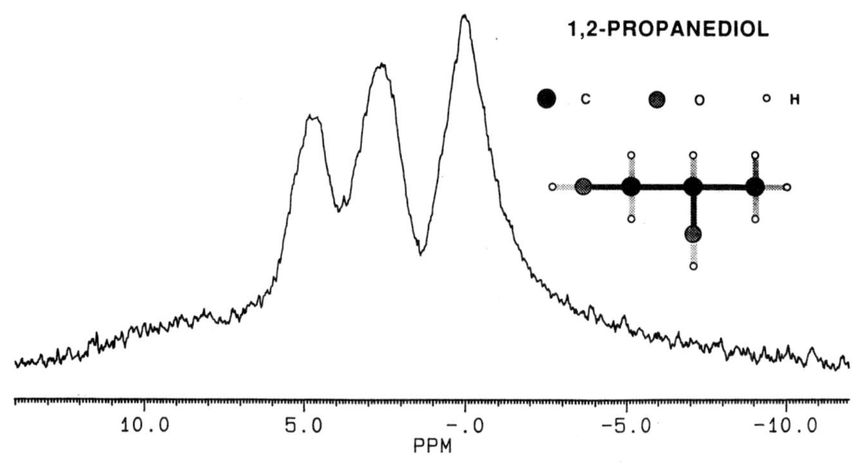

Figure 4. NMR spectrum of 1,2-propanediol in a DAC at 6.5 kbar.

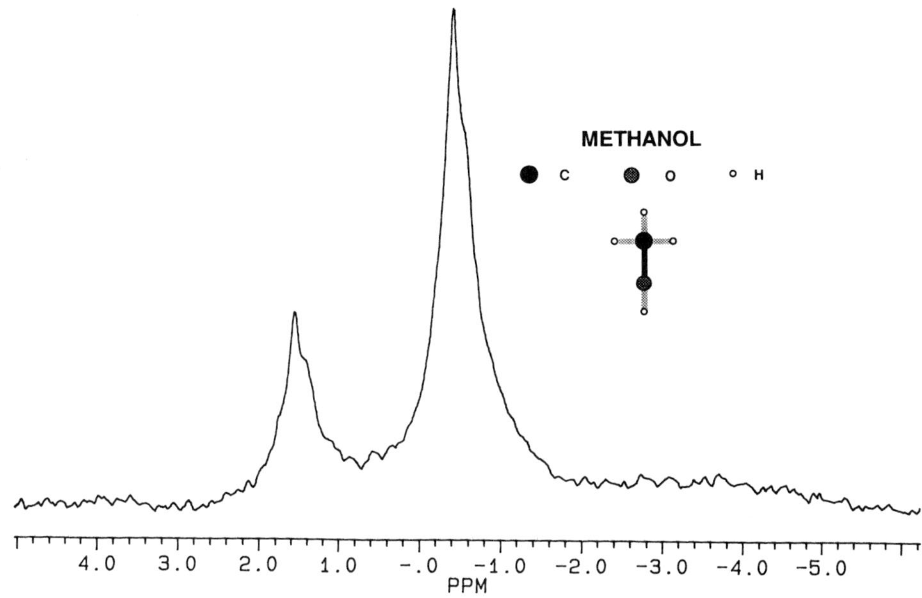

Figure 5. NMR spectrum of methanol in a DAC at 11 kbar.

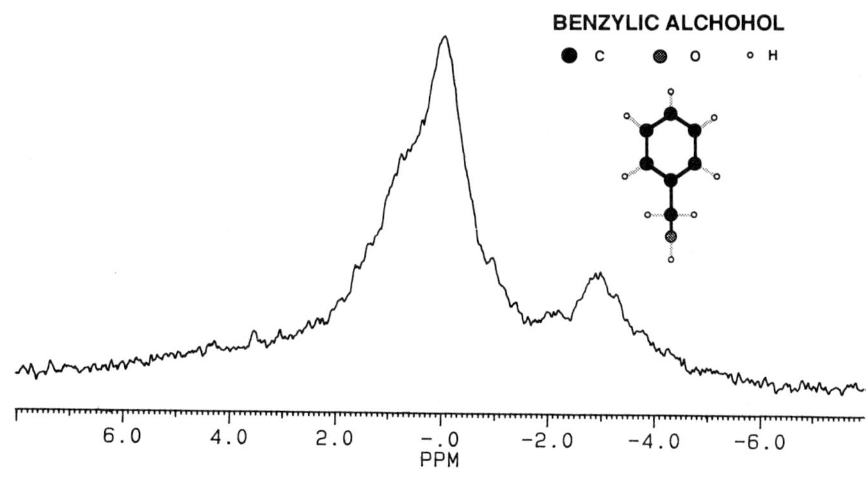

Figure 6. NMR spectrum of benzyl alcohol in a DAC at 5.5 kbar.

small shoulders upfield of each band. The spectrum for benzyl alcohol in figure 6 was taken at 5.5 kbar. The large band at zero ppm is attributed to the ring protons, while the shoulder on this band near 1 ppm is attributed to the hydroxyl protons. The small upfield band near -3 ppm is attributed to the methene protons. The quality of the benzyl alcohol used for this spectrum is questionable and may contain water.

In the course of developing the technique for performing NMR in a DAC we noted that the difference between chemical shifts of the hydroxyl protons and the backbone protons increased with increasing pressure. This indicated a deshielding of the hydroxyl protons, relative to the backbone. In figure 7 this difference for glycerol is plotted against pressure. As noted previously, above 10 kbar it becomes extremely difficult to resolve the peaks. When the data are fitted to a linear expression, a slope of 0.06 ± 0.01 ppm/kbar is obtained (approximately 24 Hz/kbar at 400.1 MHz). 1,2-propanediol also exhibited a similar pressure dependence in its hydroxyl peak. Figure 8 shows the pressure dependence of the chemical shifts of both the hydroxyl protons (squares) and the non-methyl backbone protons (circles). The backbone proton chemical shift does not change appreciably with increasing pressure, while the hydroxyl proton peak shifts considerably, at the rate of about 0.04 ppm/kbar.

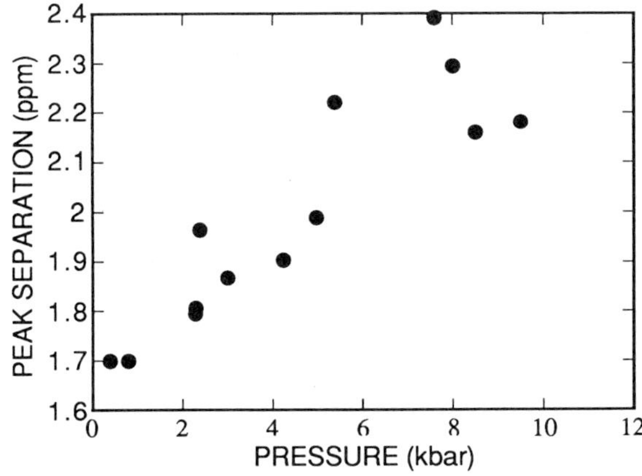

Figure 7. Glycerol chemical shift difference between the hydroxyl and backbone protons as a function of pressure.

DISCUSSION

The above results show that it is possible to obtain moderate resolution in NMR spectra taken in a DAC. As discussed by Lee et al. (1987), the inhomogeneity of the static magnetic field cannot be responsible for the limit in resolution. Instead, at low pressure, resolution is limited by the differing susceptibilities of the materials in the neighborhood of the sample. At higher pressures the limits of resolution are found to be in general agreement with measurements of the spin-spin relaxation time T_2, as will be discussed in a later publication (Raffaelle et al., 1991).

Improvements in resolution at low pressures may be possible, with better probe and cell designs using lower susceptibility materials. Accuracy could also be improved by computer fitting to the observed lineshapes, especially at high pressures.

We now consider briefly the origins of the observed dependences of hydroxyl proton chemical shifts upon pressure, and begin by citing the effects of changing other experimental variables, such as temperature and concentration. Figure 9 gives the variation in chemical shift of the backbone and hydroxyl protons in neat glycerol with temperature. These data were taken with the external field locked on a signal from an ampoule of deuterium oxide centered in a normal NMR sample tube. It is evident that the chemical shift of the backbone protons does not appreciably change with increasing temperature. However, the hydroxyl proton chemical shift shows a marked, nearly linear dependence upon temperature with a negative slope. This same behavior has been observed in the hydrogen bonded liquids methanol and ethanol (Arnold and Packard, 1951). Comparing figures 7, 8, and 9, we see that for these systems increasing pressure has roughly the same effect as decreasing temperature.

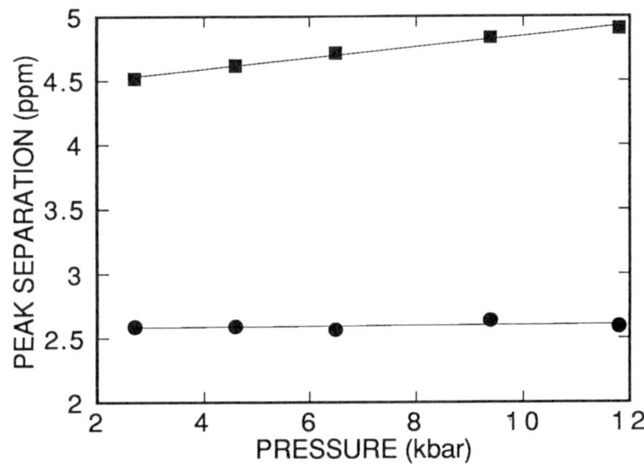

Figure 8. Measured difference in the chemical shifts of
1,2-propanediol relative to the methyl protons as a
function of temperature. Squares represent hydroxyl
protons, circles represent non-methyl backbone protons.

Concentration can also affect chemical shift differences in these systems. The concentration dependence of the difference between the chemical shift of the methyl group and the hydroxyl group for methanol mixed with $CDCl_3$ is plotted in figure 10. Here it is apparent that increasing the concentration produces changes similar to those caused by increasing pressure.

There are, in general, three main mechanisms that may account for changes in chemical shift, all of which are associated with hydrogen bonding. (1) The hydrogen bond itself may change. Since the D-H...A hydrogen bonding system is mainly electrostatic in nature, the electronegative acceptor (A) site both draws the proton toward it

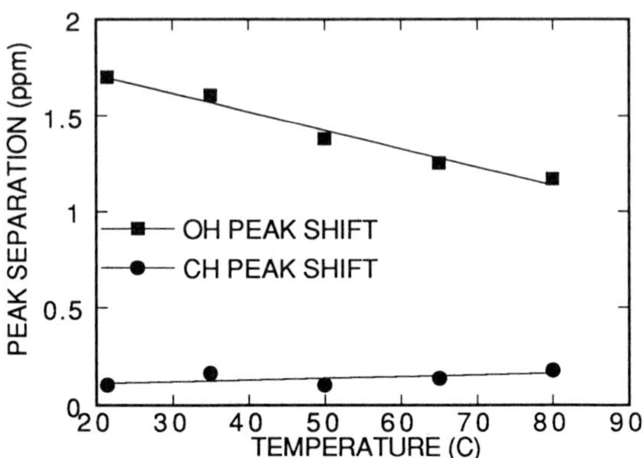

Figure 9.   Glycerol chemical shift difference between the hydroxyl
and backbone protons as a function of temperature.

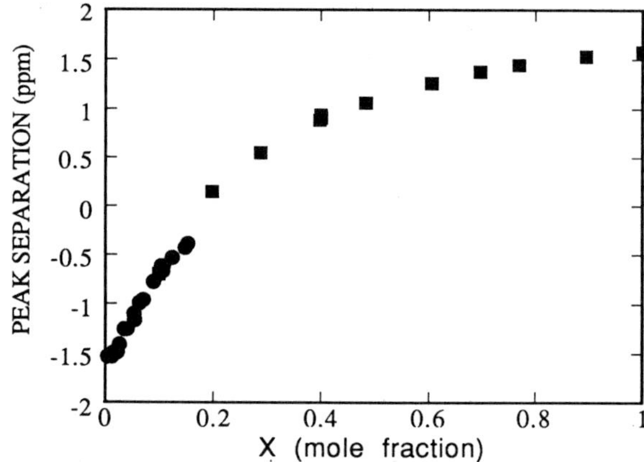

Figure 10. Shift with concentration of OH peak relative to CH_3
peak of methanol in $CDCl_3$. Squares, this study. Circles,
Cohen and Reid, 1951.

and repels the electron density of the donor-hydrogen (D-H) bond toward the electronegative D site. Pressure, for example, forces the donor and acceptor sites together. This results in the proton moving away from the donor towards the acceptor, and the bonding electron density is shifted away from the proton. This in turn causes the amount of deshielding of the proton to increase. (2) The fractions of hydrogen bonded molecules may change. Each proton's chemical shift is an average of associated and non-associated positions, since the formation of associated molecules happens at frequencies much faster than characteristic NMR measuring times. Increasing pressure will be expected to increase the fraction of molecules that are associated, thus effectively deshielding the H-bonded proton. (3) The effects of magnetic anisotropy can be present in the acceptor group (see for example, Bovey, 1988). For liquids, observed shieldings are an average over all orientations of tumbling molecules. Departure from spherical symmetry hinders electron circulation, which will give rise to magnetic anisotropy that even the rapid rotation of a tumbling molecule may not cancel out. This contribution can have either sign, although it will be positive if the principal symmetry axis of the acceptor site is directed toward the hydrogen. Increasing the pressure will slow the motion of the molecules, and may also affect the anisotropy of their magnetic susceptibilities, thus altering this contribution to the chemical shift.

Of the above three contributions to the deshielding effects of pressure, the first is probably the dominant one, while the last is probably the least effective. Our results can be understood in terms of the first two mechanisms, the distortion of the electron density about the D-H bond, and changes in the amount of association in the H-bonding molecules. protons. We find that, as expected, pressure stabilizes the hydrogen-bond network, resulting in a more viscous liquid. Although ours is not the first study of the pressure dependence of the proton chemical shift, (Yamada, 1972), we have found the largest change in chemical shift with pressure (0.06 ppm/kbar for glycerol and 0.04 ppm/kbar for 1,2-propanediol, as opposed to 0.015 ppm/kbar for a mesitylene and benzene solution).

Our measurements of the changes of chemical shifts with temperature and with concentration are in general agreement with those reported in similar studies. Regarding the effects of temperature, in 1951 both Liddel et al. (1951) and Arnold et al. (1951) reported variations in absolute chemical shift of ethyl alcohol. They found that increasing temperature increased the amount of shielding at the hydroxyl proton. This is expected, since increasing the temperature decreases the fraction of molecules that are hydrogen bonded (second mechanism), and may decrease the actual strength of the hydrogen bond itself (first mechanism). In either event, the result is a shorter OH bond on average, and an increase in the amount of shielding at the proton, as shown in Figure 9.

Similarly, when the concentration is reduced, the overall hydrogen bonding network is broken up with solvent molecules inserted between the hydrogen-bonded molecules. This was first observed by Arnold et al. (1951), and accounted for by Cohen et al, (1956). Cohen explained the variation of proton chemical shift using the first two of the above mentioned mechanisms. The charge density of the "unshared pair" of electrons in an oxygen acceptor site as it approaches a proton causes an O-H bond to become even more polar than in the isolated molecule. This site then is more likely to form hydrogen bonds with neighboring molecules. Increasing concentration thus leads to the formation of dimers, and higher oligamers. On the other hand, decreasing the concentration inserts solvent molecules, which results in weaker, more extended hydrogen bonds and again an increase in shielding at the hydroxyl protons, as seen in Figure 10.

CONCLUSION

We have demonstrated moderate resolution NMR spectroscopy at high pressure using a diamond anvil cell with a variety of organic liquids. Two resolved peaks were recorded for glycerol at pressures up to 10 kilobar, three resolved peaks for 1,2-propanediol up to 13 kilobar, two peaks and a shoulder for benzyl alcohol at 5.5 kilobar, and two peaks for methanol at 6.7 kilobar. We have also demonstrated the effects of pressure upon the chemical shifts of the hydroxyl protons in both glycerol (0.06 ppm/kbar) and 1,2-propanediol (0.04 ppm/kbar). This has opened a new area of research regarding the hydrogen bond, and has established the limiting factors involved in performing moderate resolution NMR spectroscopy of liquids at extremely high pressures.

ACKNOWLEDGMENTS

We would like to gratefully acknowledge the help of R. A. Nieman of the NMR Facility in the Department of Chemistry at Arizona State University.

REFERENCES

Abragam, A., 1961, "The Principles of Nuclear Magnetism", Oxford University Press, Oxford.
Arnold, J. T., and Packard, M. E., 1951, J. Chem. Phys., 19:1608.
Barnett, J. D., Block, S., and Piermarini, G. J., 1973, Rev. Sci. Intr., 44:1.
Bovey, F. A., 1988, "Nuclear Magnetic Resonance Spectroscopy", Academic Press Inc., San Diego.
Cohen, A. D., and Reid, C., 1956, J. Chem. Phys., 25:790.
Conradi, M. S., 1990, Private Communication.
Diehl, R. M., Fujara, F., and Sillescu, H., 1990, Europhys. Lett., 13:257.
Fiorito, R. B., and Meister, R., 1972, J. Chem. Phys., 56:4605.
Jonas, J., 1990, High Pressure Research, 4:573.
Kuhns, P. L., and Conradi, M. S., 1982, J. Chem. Phys., 77:1771.
Lee, S. H., Luszczynski, K., Norberg, R. E., and Conradi, M. S., 1987, Rev. Sci. Instr., 58:415.
Liddel, U., and Ramsey, N. F., 1951, J. Chem. Phys., 19:1608.
Raffaelle, D. P., Halvorson, K. E., Marzke, R. F., and Wolf G., 1991, to be submitted.
Vaughn, R. W., Lai, C. F., and Ellemen, D. D., 1971, Rev. Sci. Instr., 42:626.
Wolfe, M., and Jonas, J., 1979, J. Chem. Phys., 71:(8):3252.
Yamada, H., 1972, Chem. Lett., 747.

THE APPLICATION OF THE MÖSSBAUER EFFECT FOR PROBING ELECTRONIC

PROPERTIES OF THE PRESSURE-INDUCED MOTT TRANSITION

Moshe P. Pasternak[1,2], R. Dean Taylor[2] and Raymond Jeanloz[3]

[1]School of Physics and Astronomy, Tel Aviv University, 69978 Tel Aviv, ISRAEL. [2]Physics Division, Los Alamos National Laboratory, Los Alamos, NM 87545. [3]Department of Geology and Geophysics, University of California, Berkeley, CA 94720.

INTRODUCTION

The problem of the Mott insulator[1] and its transition into a *metallic* state (the Mott transition) is considered to be one of the most serious challenges to the prevailing concepts of solid state physics. At present it remains an unsolved problem. The subject of Mott insulators began in 1937 when De Boer and Verwey[2] presented their experimental results on the electrical conductivity of transition-metal (TM) oxides (the oxides of Ni, Co, Mn and Fe). The fact that the majority of these oxides were *insulators* did not fit the conventional Bloch-Wilson band picture. Assuming the compounds were highly ionic would imply partially filled $3d$ bands and therefore be *metallic*! In discussion that followed Peierls suggested that the Coulomb repulsion was responsible for the $3d$-electron localization. The TM-oxides such as NiO, CoO and MnO are classic examples of Mott insulators (MI). The phenomenological aspects of a MI can be described as follows: It is an *antiferromagnetic* insulator whose local moments persist unchanged above T_N .

Both the Néel transition and the spin-wave spectrum can be described with an effective spin Hamiltonian of the Heisenberg form. The optical band is large - of the order of 2-4 ev (40-80 times larger than that cor-

Frontiers of High-Pressure Research, Edited by H.D. Hochheimer and
R.D. Etters, Plenum Press, New York, 1991

responding to a typical T_N) and independent of T. Within the gap, optical spectra exhibit the excitons predicted by crystal-field theory. Any magnetic insulator - a material whose local moments and energy gap persist above the spin-ordering temperature - can be classified as a Mott insulator. Mott insulators constitute about half of all binary TM compounds, and the majority of the TM -oxides, -halides and -chalcogenides are MI. Magnetism and the band structure of MI are strongly related.

Though the experimental evidence and the phenomenology of MI are quite rich[3], information on the insulator-metal transition in MI, the so-called Mott transition[4] (MT), is scarce. A MT may be induced either by alloying or by applying external pressure[5]. However, in contrast to classical semiconductors, doping of narrow-band insulators results in *severe electronic* and *structural disorder* in the MI under study. Thus, the only "clean" way to investigate this intricate insulator-metal transition is by applied pressure.

The *concurrence* of an isostructural insulator-metal transition and of *d*-electron delocalization are inherent features of a Mott transition. Delocalization implies a transition from the magnetic to a non-magnetic state. Therefore, to probe experimentally the pressure-induced MT one requires high-pressure techniques for X-ray diffraction, resistivity and magnetism. Such measurements on NiI_2 were recently published by Pasternak et al.[6]. The purpose of this paper is to present the unique features of the *transferred hyperfine interaction* (THI) measured by Mössbauer spectroscopy (MS) in diamond-anvil cells (DAC) for studying: 1) - the *magnetic* properties, on a local or atomic scale, of the Mott insulator as a function of temperature and pressure and 2) - the mechanisms that lead to metallization at the Mott transition. As will be shown, the THI is unique and perhaps the only available method for studying *magnetic structure* at very high pressures. We are currently studying a series of structurally analogous TM-iodides using ^{129}I MS, X-ray diffraction and electrical conductivity methods. In this paper we present results carried out in the antiferromagnetic-insulators NiI_2 and CoI_2 emphasizing the application of ^{129}I MS.

The iodides of Ni and Co are layered compounds that crystallize respectively in the CdI_2 (space group $R\bar{3}m$) and $Cd(OH)_2$ ($C\bar{3}m$) structures[7]. They consist of alternating layers of weakly interacting *I-TM-I* (see Fig. 1) and can be described as a hexagonal lattice with three NiI_2 units or

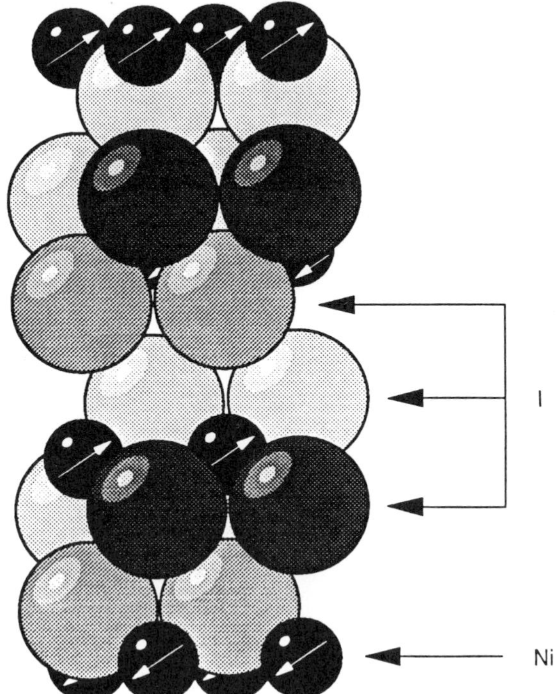

Fig. 1 - Spin structure of NiI_2 and CoI_2 compatible with the Mössbauer and neutron diffraction results. The ferromagnetic sublattice is on the *ab*-plane and the spins are tilted at 55 ± 5° with respect to the *c*-axis.

one CoI_2 unit per unit cell. They order antiferromagnetically[8] with T_N = 75 K for NiI_2 and 12 K for CoI_2. Each iodine is located at the top of a tri-pyramid of three equivalent nearest neighbor metal ions (see Fig. 1). Along the *c*-axis the I is axially symmetric (C_{3v}). In the spin-ordered state, the TM^{2+} ions located within a sublattice plane (*ab*-plane) are *ferromagnetically* ordered, and the hyperfine field H_{hy} at the iodine nucleus at T = 0 (H_{hy}^{sat}) is proportional to the vector sum of the TM^{2+} moments. Thus, H_{hy} serves as a local probe for the sublattice magnetization and its structure at each temperature and pressure. The magnetic structure of both NiI_2 and CoI_2 at ambient pressure have been determined by neutron diffraction[9]. Ambient pressure ^{129}I Mössbauer studies[10,11] of NiI_2 agree with the neutron diffraction picture, namely a ferromagnetically ordered sublattice with Ni moments canted at 55° with respect to the *c*-axis. For CoI_2, Friedt et al.[11] suggest an antiferromagnetic structure with the moments aligned parallel within the planes. Neutron studies suggest a helix structure of type 1 with a spiral angle of 45° in the *ab* plane.

EXPERIMENTAL

Sample Preparation. The samples were prepared in milligram amounts by direct metal-vapor reaction of high purity Ni or Co with stoichiometric amount of elemental I_2 enriched with ^{129}I. The latter was obtained by thermal decomposition of $Pd^{129}I_2$ at 350°C. The final reaction took place at approximately 700°C in evacuated quartz tubes. Due to their hygroscopic nature, the samples had to be loaded into the DAC under a N_2 atmosphere. The DAC used was a modified Merrill-Bassett type with a $Ta_{0.9}W_{0.1}$ gasket that also served as an efficient γ-ray (E_γ = 27.8 keV) collimator[12]. Typical cavity sizes were 250-μ diameter by 40-μ thickness resulting in ^{129}I absorber thickness of 1 - 2 mg/cm^2. Argon was loaded as the pressurizing medium. Pressures were measured by the ruby fluorescence method[13]. In general, pressure uncertainties did not exceed ± 0.4 GPa.

X-ray Diffraction and *Conductivity Measurements.* Samples with natural isotopic abundance were prepared for for X-ray diffraction and conductivity studies. X-ray diffraction was carried out at 300 K by the angle-dispersive method using a Mo $K\alpha$ monochromatic radiation. The relative resistance measurements (R(P,T)) were obtained using a quasi-four-point method[6].

Mössbauer Effect Measurements. The single line 27.8-keV γ-ray source was in the form of $Mg_3^{129m}TeO_6$ ($T_{1/2}$ = 33 d) prepared by thermal neutron irradiation of $Mg_3^{128}TeO_6$. Typical source diameters were 2-3 mm. Measurements were carried out at variable temperatures (4 < T < 320 K) and at pressures in the range of 0 - 32 GPa. At selected pressures, absorption spectra were recorded as a function of temperature. From these measurements the pressure dependence of the saturation hyperfine field (H_{hy}^{sat}), the Néel temperature (T_N) and the isomer shift (IS) were deduced. In the paramagnetic state the spectrum was fitted assuming an axially symmetric electric field gradient (*efg*) using the following Hamiltonian:

$$H = e^2 q_{zz} Q / [4I(2I-1)] \, [3\hat{I}_z^2 - I(I+1)] \tag{1}$$

acting at the ground (I = 7/2) and excited (I* = 5/2) states of ^{129}I.

At temperatures below T_N the experimental data were least-squares fitted using the following Hamiltonian[14] to calculate the transition energies and intensities:

$$H = \mu H_{hy} (\hat{I}_Z/I) + \{\tfrac{1}{2}e^2 q_{zz}Q \ (3\cos^2\theta - 1)/[4I(2I-1)]\} \ \{3\hat{I}_z^2 - I(I+1)\} \qquad (2)$$

where μ is the nuclear moment, $\hat{I}_Z$ and $\hat{I}_z^2$ are spin operators in the systems of reference of H_{hy} and the *efg* (eq_{zz}), respectively, $e^2 q_{zz}Q$ is the quadrupole coupling, and θ is the angle formed between H_{hy} and eq_{zz}, the latter being along the *c*-axis (see Fig. 2). In fitting the spectra at T < T_N, the quadrupole coupling was fixed to that measured at T > T_N, enabling us to determine θ, the relative orientation of the sublattice magnetization as a function of pressure.

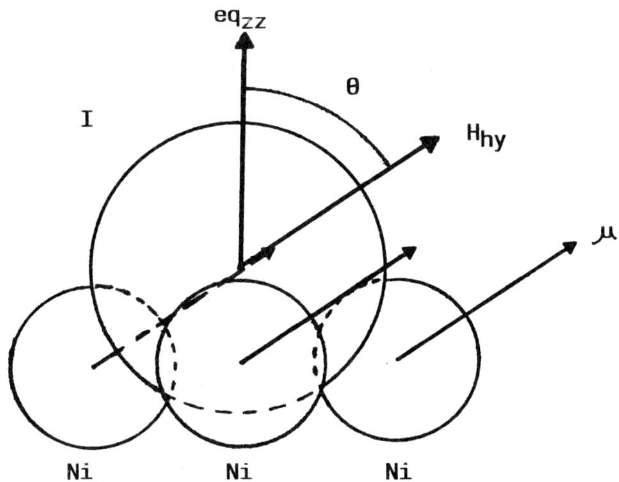

Fig. 2 - Principal axes and direction of the *efg* and H_{hy} at the iodine nucleus. The three TM cations are equally distanced from the I. Below T_N the ordered TM moments polarize the 5*sp* orbitals inducing a H_{hy} at the iodine nucleus. AT T > T_N the spins are randomly oriented resulting in H_{hy} = 0. Due to the axial symmetry, the asymmetry-parameter η of the *efg* vanishes.

RESULTS

Mössbauer Effect Results

<u>NiI$_2$</u>. Typical MS spectra of Ni^{129}I$_2$ at various temperatures measured at 6.7 GPa are shown in Fig. 3. We note the following features: 1) The intensities of hyperfine splitting are symmetric. This feature is a con-

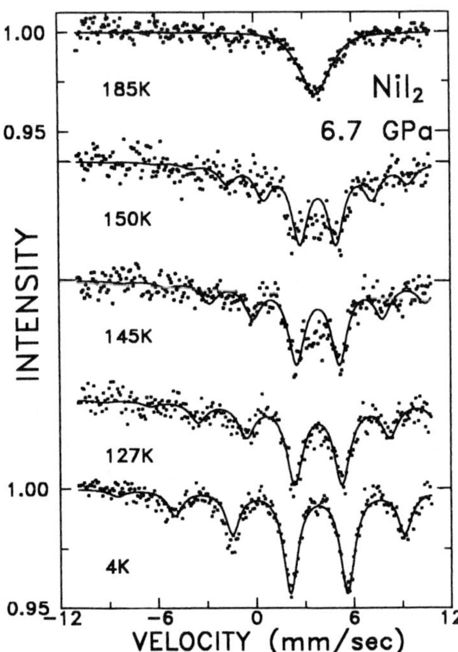

Fig. 3 - Ni^{129}I$_2$ Mössbauer spectra recorded at 6.7 GPa and various temperatures. The solid line is a theoretical spectrum obtained from the least-squares fit using the spin-Hamiltonians of Eqs. 2 and 3.

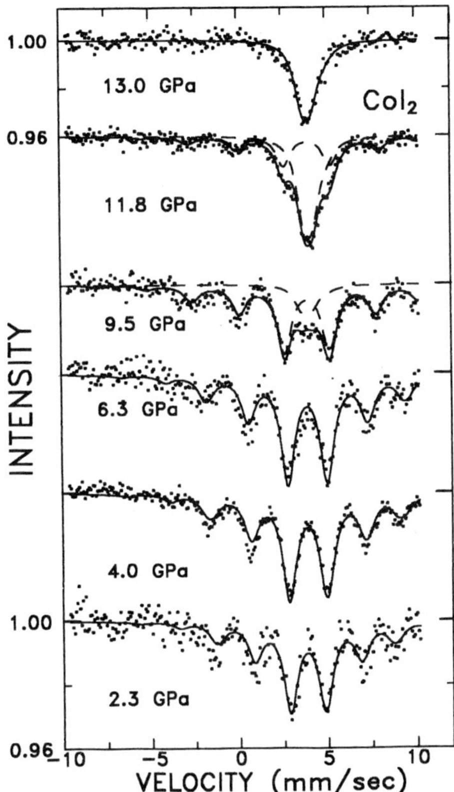

Fig. 4 - The Mössbauer spectra of $Co^{129}I_2$ at 4K at various pressures. Note the coexistence of two sites at P = 9.5 and 11.8 GPa, corresponding to a magnetic and quadrupolar (non-magnetic) interactions. The fit was carried out assuming such interactions with free relative intensities of the subspectra. At 13 GPa only the quadrupole interaction is present.

sequence of the value of $\theta \cong 55°$ (see Eq. 2). In this special case the quadrupole contribution vanishes ($3 \cos^2\theta = 1$), and the interaction is purely magnetic. To the highest pressure where the the material is still magnetic, no change in θ was observed. 2) The hyperfine field decreases with increasing T. Plotting $H_{hy}(T)$ at each pressure allows us to determine $T_N(P)$ from an extrapolation of $H_{hy}(T)$ to zero. The spectrum measured at 185 K corresponds to the paramagnetic state and was fitted with a pure quadrupole interaction.

$\underline{CoI_2}$. Mössbauer spectra of $Co^{129}I_2$ measured at 4 K for different pressures are shown in Fig. 4 from which the saturation hyperfine fields $H_{hy}^{sat}(P)$ were determined. As can be seen, a nonmagnetic site appears for P > 9 GPa, increasing in abundance with increasing pressure and reaching unity at P $\approx$ 12.5 GPa. The pressure dependences of IS, H_{hy}^{sat} and T_N for NiI_2 and CoI_2 are shown in Figs. 5, 6 and 7. The relative abundance of the non-magnetic component appearing above 9 GPa is depicted in Fig. 8.

X-ray Diffraction and Conductivity Results

$\underline{NiI_2}$. The pressure dependence of the normalized volume, V/V(0) measured at 300 K is shown in Fig. 9. The solid curve is a least-squares fit to the Birch-Murnaghan equation which yields values for the bulk modulus $K_o = 27.7 \pm 0.9$ GPa and its pressure derivative $K_o' = 4.8 \pm 0.2$. Measurements[6] of R(P,T) show the following features: 1) For P $\leq$ 17 GPa, logR decreases linearly with $\Delta V/V$ ($\Delta V = V(0) - V(P)$) and near 19 GPa drops rapidly and levels off for P > 19 GPa. The temperature dependent of R below 19 GPa (defined as the critical pressure P_c) are characteristic of nonmetallic systems; above 19 GPa R(T) shows metallic behavior.

$\underline{CoI_2}$. Preliminary X-ray diffraction and R(P,T) measurements[15] in CoI_2 clearly show as in NiI_2, that the transition to the nonmagnetic state at P_c ($P_c = 12.5$ GPa) is isostructural and that R(T) for P > P_c increases with T, i. e., the CoI_2 system also becomes metallic.

DISCUSSION

The Pressure Dependence of the THI

The hyperfine fields observed at the I nucleus in the magnetic phase are due to a spin polarization of the electrons of the originally diamagnetic ligand. The unpaired spin-density arises from covalency effects,

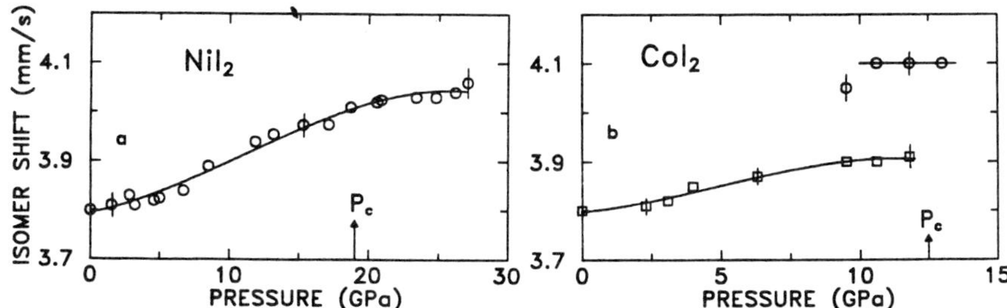

Fig. 5 - The pressure dependence of the IS for NiI_2 (a) and CoI_2 (b). The monotonic increase in IS with pressure reflects the increase in the electron density $|\psi_s(0)|^2$ at the nucleus with decreasing volume. The differential increase in IS with pressure is similar for both the Ni and Co compound. The discontinuity in CoI_2 reflects different $5p$-holes per iodine ion when crossing from a nonmetal to a metal state. Such an abrupt transition is not observed in NiI_2. The curves are a guide to the eye.

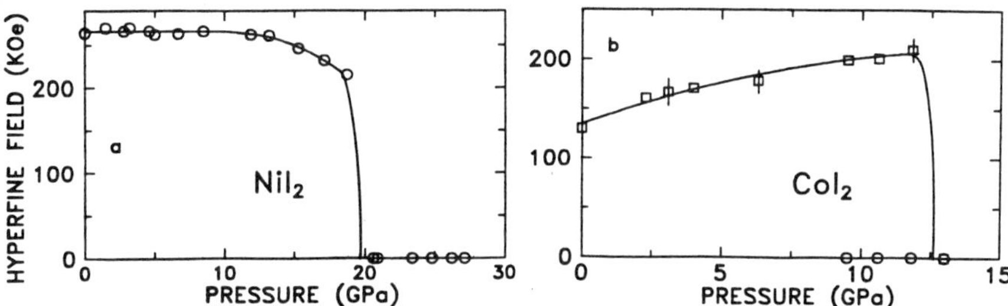

Fig. 6 - The pressure dependence of the saturation hyperfine field, H_{hy}^{sat} for NiI_2 (a) and CoI_2 (b). In NiI_2 H_{hy}^{sat} remains constant up to $\approx$ 13 GPa (0.7 P_c), decreases approximately by 8% at $P \cong P_c$ and collapses to zero at P_c. In CoI_2 we observe a monotonic increase in H_{hy}^{sat} and the coexistence of magnetic and non-magnetic phases in $9.5 < P < P_c$. The curves are a guide to the eye.

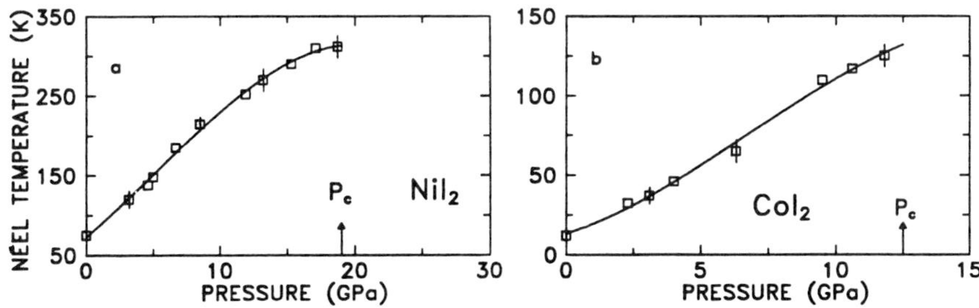

Fig. 7 - The pressure dependence of T_N for NiI_2 (a) and CoI_2 (b). Whereas in NiI_2 T_N increases fourfold in the pressure range of 0 - 19 GPa, in CoI_2 the increase in tenfold in 0 - 13 GPa. The solid line curves are a guide to the eye.

235

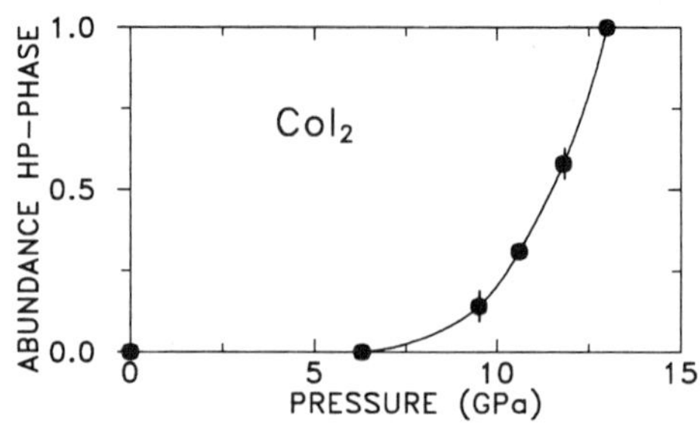

Fig. 8 - The pressure dependence of the relative abundance of the non-magnetic phase in CoI$_2$ in the 9 - 13 GPa pressure range. The coexistence of these two phases is not observed in NiI$_2$. The solid line through the experimental points is a guide to the eye.

i.e., effects associated with the electron transfer between the magnetic ions and their diamagnetic surroundings[16]. The transfer of spin-density from metal to ligand site is described using a molecular orbital scheme for the cation and considering independently the three orthogonal bonds of iodine to its nearest neighbors; the transferred interaction is thus calculated along each of the bonds and the individual contributions are summed up[17]. Such an approach is justified by the localized bonding occurring in these covalent compounds as opposed to conducting systems. Bonding of iodine involves the $5s$ and $5p$ electrons.

The antibonding molecular orbitals which contain the unpaired spin-density are written in the case of an octahedral TM complex:

$$\psi_e^a = N_e[d_e - \lambda_s\chi_{5s} - \lambda_\sigma\chi_{5p\sigma}],$$

$$\psi_t^a = N_t[d_t - \lambda_\pi\chi_{5p\pi}],$$

(3)

where χ_{5s}, $\chi_{5p\sigma}$ and $\chi_{5p\pi}$ are appropriate linear combinations of the ligand atomic orbitals, d_e and d_t are the e_g and t_{2g} metal orbitals, N_e and N_t are normalization constants (the overlap integrals are neglected) and λ_s, λ_σ and λ_π represent the mixing coefficients. For d orbitals that are singly occupied the spin-densities in the s, σ and π orbitals are deduced from Eq. 3[18]:

$$f_s = \tfrac{1}{3} N_e^2 \lambda_s^2, \qquad f_\sigma = \tfrac{1}{3} N_e^2 \lambda_\sigma^2, \qquad f_\pi = \tfrac{1}{3} N_t^2 \lambda_\pi^2.$$

(4)

In the case of a singlet orbital ground state of the metal ion, e.g., for electron configuration $3d^8$ in Ni^{2+}, it can be shown[18,19] that for a parallel arrangement of the three nearest moments, the transferred hyperfine field H_{hy} is due the *isotropic* Fermi contact term with components:

$$H_{hy}^{i} = -8\pi \, \mu_B \, \rho(0) \, f_s \, <S_i>/S \qquad \text{with } i = x, y, z. \tag{5}$$

where μ_B is the Bohr magneton, $\rho(0)$ is the 5s electron density at the nuclear site, S is the spin of the metal ions and $<S_i>$ are the time average values of S_i. This average value is related to the magnetic moment.

In the case of Co^{2+} ($3d^7$ configuration) the ground state is orbitally degenerate in a cubic crystal-field; spin-orbit coupling has to be taken in account to obtain the ground state wave functions and thus to describe properly the spin-transfer. This problem has been discussed in the literature[20]. The three components of H_{hy} can be written as:

$$H_{hy}^{i} = -[16\pi \, \mu_B \, \rho(0) \, (\tfrac{5}{9}) \, f_s + 4\mu_B \, <r^{-3}> \, \tfrac{14}{45} \, f_\pi] \, <S_i>/S \tag{6}$$

where $<r^{-3}>$ is an average for the 5p-orbit.

The pressure dependences of H_{hy}^{sat} for NiI_2 and CoI_2 are depicted in Fig. 6. It is clear that within the experimental error $H_{hy}^{sat}(Ni)$ remains constant to $P = 0.8 \, P_c$, slightly decreases near P_c and vanishes abruptly at and above P_c. On the other hand, $H_{hy}^{sat}(Co)$ *increases* with increasing pressure. The difference in the $H_{hy}^{sat}(P)$ for CoI_2 is due to the contribution of the second term in Eq. 6. Whereas we expect that the spin density (the product $\rho(0) \cdot f_s$) and the TM moment $<S>$ will not change appreciably with pressure, the second term in Eq. 6 is strictly *volume* dependent and should increase with pressure. However, at the critical pressure the d-orbitals delocalize, and the magnetic moments (and therefore H_{hy}^{sat}) vanishes in both cases.

The Mott Transition.

The simplest, nearest neighbor tight-binding model Hamiltonian to describe a Mott insulator was formulated by Hubbard[21]:

$$H = - t \sum_{ij\sigma} a_{i+j}^{\dagger} \, a_{i\sigma} + U \sum_{i} \hat{n}_{i\uparrow} \, \hat{n}_{i\downarrow} \tag{7}$$

where U ("the Hubbard U") is the energy spacing between the the "upper" (empty) and "lower" (filled) d-band and t is the hopping parameter[1]. For simple lattices, the ordinary bandwith can be expressed as $W = 2zt$ where z is the nearest neighbor coordination number. $U > W$ denotes a ground state

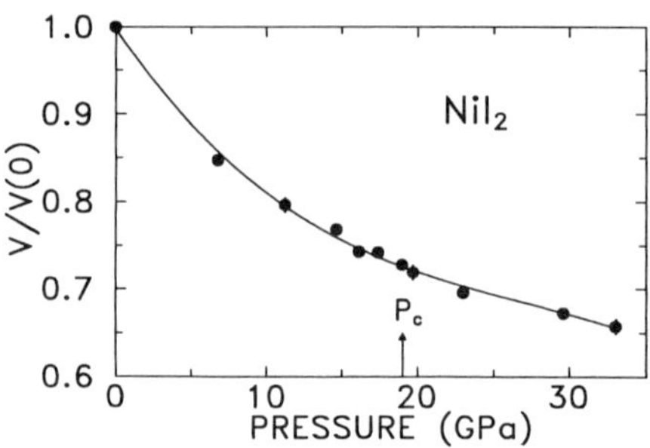

Fig. 9 - The normalized volume vs pressure of NiI$_2$ at 300 K. The solid line is a least-squares fit to the Birch-Murnaghan equation.

MI whereas for $W > U$ the system is in a metallic state. With pressure increase W/U increases, reaching a point where a 1st-order phase transition occurs; the system goes from a correlated to an uncorrelated d-state. The d-electrons delocalize resulting in metallization and a tran-sition to a nonmagnetic state. In principle such a transition should result in a lattice contraction.

From the pressure dependence of H_{hy}^{sat} (Fig. 6) one clearly sees the abrupt transition at P_c concurring with a metallic state has been confirmed[6] for NiI$_2$ by $R(T)$ measurements and shown to be in the vicinity of P_c. For CoI$_2$ preliminary results[15] also suggest metallization at $P \geq P_c$ (13 GPa). It is clear however that confirmation of the MT by magnetic measurements (namely through MS) is superior. The collapse of the magnetic state is rather dramatic whereas the determination of P_c via electrical conductivity is rather sluggish.

It has been shown for both Co and Ni iodide that the MT is isostructural. Can we detect a lattice contraction? As shown in Fig. 9, if such a contraction occurs in NiI$_2$ it is less than 1%. X-ray diffraction evidence for contraction in CoI$_2$ is not yet available. The pressure dependence of the IS in NiI$_2$ (Fig. 5a) near P_c shows no evidence for lattice contraction within the experimental error. However, for CoI$_2$ (Fig. 5b), the positive jump in IS (ref. 21) indicates a substantial decrease of the lattice volume in the metallic state. Careful X-ray diffraction studies are now in progress to confirm this suggestion. Another explanation for the increase in IS ($\Delta IS(P_c) = 0.20 \pm 0.04$ mm/s) could be that it is a result of *transfer* of electrons from the I $5p$-band into the Co "upper" $3d$-band at P_c thus increasing the number of I p-holes (h_p). It has been shown that the IS of $5sp$ elements can appoximated by a linear function of h_p[22,23]. The IS(hp_p) with respect to Mg_3TeO_6 at ambient pressure can be expressed as:

$$IS = 1.27\ h_p + 3.58\ mm/s. \tag{8}$$

At P_c, the change $\Delta IS(CoI_2)$ corresponds to the creation of 0.16 p-holes. Thus, the onset of a metallic state is accompanied by a transfer of $0.32e^-$ from the iodine $5p$ band into the cobalt $3d$-band. This *charge-transfer* (CT) mechanism can be expressed in the form: $d_i^n \rightarrow d_i^{n+1}\ \underline{L}$ where $\underline{L}$ is a hole in the anion valence-band. CT excitations have been proposed by Zaanen et al.[24] as the main excitations within the gap of a MI. In NiI_2 the $3d$ intra-band overlap seems to be the major factor responsible for the MT.

We define t_c/t_0 as the normalized *critical* hopping parameter ($t_c = t(P_c)$, $t_0 \equiv t(0)$). To estimate the values of t_c/t_0 we use the following expression[1] for T_N, valid in the MI regime:

$$T_N = (4k_B)^{-1}(2zt/U) \tag{9}$$

Assuming to a first approximation that U is pressure independent, and using values of T_N at $P = 0$ and P_c (see Fig. 7), we find $t_c/t_o = 2$ and 3 for NiI_2 and CoI_2, respectively.

CONCLUSION

It has been shown that high-pressure Mössbauer spectroscopy is an unique tool for investigating the Mott insulators and its various aspects of magnetism and structure. Information obtained from the MS features, combined with electrical conductivity and X-ray diffraction measurements, provide for the first time a global picture of the various parameters involved in the pressure-induced Mott transition. As demonstrated, the THI provides an accurate probe for determining the critical pressure where the system becomes a normal metal. Furthermore, the method allows one to measure coexistence of electronic states; nonmagnetic and magnetic sites in a rather narrow pressure range. The IS, a parameter unique to the MS, provides important information concerning the role of particular bands involved in the metallization process. Further studies are now in progress with other structurally analogous TMI_2 compounds to further elucidate this exquisite and important transition process in solid state physics categorized as *Mott insulators.*

REFERENCES

1. H. Brandow, *Int. J. of Quant. Chem.*, *Symp.* No. 10:417 (1976).

2. J.H. de Boer and E.J.W. Verwey, *Proc. Roy. Soc.* (London) 49:59 (1937).

3. J.A. Wilson, in: *The Metallic and Nonmetallic States of Matter*, P.P. Edwards and C.N.R. Rao, eds., Taylor and Francis, London (1985), p215

4. N.F. Mott, *Adv. Phys.* 21:785 (1972).

5. The cuprate high temperature superconductors are examples of metals that have gone through a MT by alloying of the MI CuO.

6. M.P. Pasternak, R.D. Taylor, A. Chen, C. Meade, L.M. Falicov, A. Giesekus, R. Jeanloz and P.Y. Yu, *Phys. Rev. Lett.* 65:790 (1990).

7. R.W.G. Wyckoff, *Crystal Structures*, Vol. I, Interscience, New York (1963).

8. L.G.Van Uitert, H.J. Williams, R.D. Sherwood and J.J. Rubin, *J. Appl. Phys.* 36:1029 (1965).

9. S.R. Kuindersma, J.P. Sanchez and C. Haas, *Physica* 111B:231, (1981).

10. M. Pasternak, S. Bupkshpan and T. Sonnino, *Solid State Comm.* 16:871 (1975).

11. J.M. Friedt, J.P. Sanchez and G.K. Shenoy, *J. Chem. Phys.* 65:5093 (1976).

12. M.P. Pasternak and R.D. Taylor, *Hyperfine Interact.* 47:415 (1989); R.D. Taylor and M.P. Pasternak, *loc. cit.* 53:159 (1990).

13. A. Jayaraman, *Rev. Mod. Phys.* 55:65 (1983).

14. This Hamiltonian is for a combined quadrupole-magnetic interaction, appropriate for an axially symmetric *efg* and for $\mu H_{hy} \gg e^2 qQ/4I(2I-1)$.

15. E. Sterer and M.P. Pasternak, private communication.

16. G.A. Sawatzky and F. Van der Woude, *J. Physique.* (Paris), *Colloque C6*, 35:47 (1974).

17. F. Keffer, T. Oguchi, W. O'Sullivan and Y. Yamashita, *Phys. Rev.* 115:1553 (1959).

18. A. Abragam and B. Bleaney, *Electron Paramagnetic Resonance of Transition Ions*, Clarendon Press, Oxford, p.761 (1970).

19. J. Owen and J.H.M. Thorley, *Rep. Prog. Phys.* 29:675 (1966).

20. W. Low and M. Weger, *Phys. Rev.* 118:1119 (1960).

21. J. Hubbard, *Proc. Roy. Soc.* (London), A276:238 (1963); A281:401 (1964).

22. S.L. Ruby and G.K. Shenoy, <u>in</u> *Mössbauer Isomer Shifts*, G.K. Shenoy and F.E. Wagner, ed., North Holland, Amsterdam (1978), pp 617.

23. M. Van der Heyden, M.P. Pasternak and G. Langouche, *J. Phys. Chem. Solids*, 46:1221 (1985).

24. J. Zaanen, G.A. Sawatzky and J.W. Allen, *Phys. Rev. Lett.* 55:418 (1985).

KINETICS OF HIGH PRESSURE PHASE TRANSITIONS IN THE

DIAMOND ANVIL CELL

Bernd Lorenz and Ingo Orgzall

Institute of High Pressure Research
Telegrafenberg, Potsdam, O-1585
Germany

INTRODUCTION

Phase transitions under high pressure have attracted widespread interest in different fields of natural sciences, e.g. in connection with high pressure materials synthesis, phase transitions in geological materials or fundamental investigations of solids under pressure. For *in situ* investigations of small volume samples within a broad spectrum of physical methods the diamond anvil cell (DAC) technique has been successfully developed over the last decade (Jayaraman, 1986). Besides the measurement of physical properties in thermodynamic equilibrium the DAC also allows the investigation of dynamic or time dependent phenomena if the appropriate experimental methods are applied.

The kinetics of the phase changes in various materials under pressure is of basic importance for understanding the physical and chemical processes which lead to the formation of high pressure phases with special properties. The knowledge of the physical mechanisms of phase transformations is a preposition for controlling the synthesis of new materials. On the other hand, the investigation of the phase transition kinetics and its slowing down near the equilibrium point is useful in order to determine the transition lines in the phase diagram as accurate as possible (Merkau and Holzapfel, 1986; Krüger et al., 1990). This method is of special significance for those systems where the high pressure phase remains metastable at zero pressure. However, the extrapolation of the equilibrium pressure from time resolved measurements at the phase transformation requires a physical understanding and the modelling of the kinetic processes.

Commonly, the kinetics of phase changes under high pressure is studied by energy or angle dispersive X-ray methods. The resolution of time dependent effects presupposes the high intensity of synchrotron radiation. From the relative intensities of the diffraction peaks of the high and low pressure phases the time dependent degree of transformation x(t) is calculated and compared with the thermodynamic theory of nucleation and growth (Christian, 1965). By this method the kinetics of structural phase transitions has been investigated for various solids in the DAC (Brar and Schloessin, 1981; Merkau and Holzapfel, 1986; Krüger et al., 1990) as well as in other high pressure devices (Hamaya and Akimoto, 1981; Hamaya et al., 1986). The essential problems of the diffraction methods lie in the interpretation of the measured transformation degree x(t) and its fit to the equation

$$x(t) = 1 - \exp\left[- K \ t^n \right] \tag{1}$$

Frontiers of High-Pressure Research, Edited by H.D. Hochheimer and
R.D. Etters, Plenum Press, New York, 1991

without knowing further characteristic quantities like nucleation rates or growth velocities. Equation (1) is valid for fixed thermodynamic conditions (pressure, temperature) and the value of the exponent n ($1 \leq n \leq 4$) depends on the growth kinetics (Christian, 1965). In the pressure chamber of the DAC, however, it is well known that the pressure changes during a phase transition due to the volume change of the sample (Hirsch and Holzapfel, 1983) and other effects related to the special preparation technique (Däßler, 1991). Due to this pressure change the validity of (1) and the physical meanings of the parameters K and n are questionable.

In order to get a better insight into the kinetics of high pressure phase transitions in the DAC a method has been developed (Däßler, 1988) which allows the experimental determination of nucleation rates, cluster distributions and other characteristic quantities of the nucleation and growth process. This method is based on the observation of the growing high pressure phase with a microscope and an image processing system. As an example the precipitation of the cubic B2-phase of KCl from a solution at a pressure of about 2 GPa has been studied (Däßler, 1991; Lorenz et al., 1991). The precipitation of the solid phase is governed by a nucleation and growth process with changing pressure. Thereby, after raising the force of the press, an initial pressure increase due to plastic deformations of the gasket material is followed by a decrease depending on the transition to the denser phase. Because of the strong pressure change during the transition an interpretation of the experimental data for x(t) by equation (1) is ruled out.

It is the aim of this paper to present a new method which is suitable to interpret the phase transition kinetics under changing pressure and even more complicated conditions. The procedure makes use of the experimentally determined nucleation rate I(t) and the transformation degree x(t) as an input for a numerical simulation of the transition. As an example, for one of the precipitation experiments mentioned above the measured functions I(t) and x(t) are shown in Fig. 1. Note that the nucleation rate as function of time exhibits a distinct maximum reflecting the variation of pressure and the concentration change in the solution during the kinetic process.

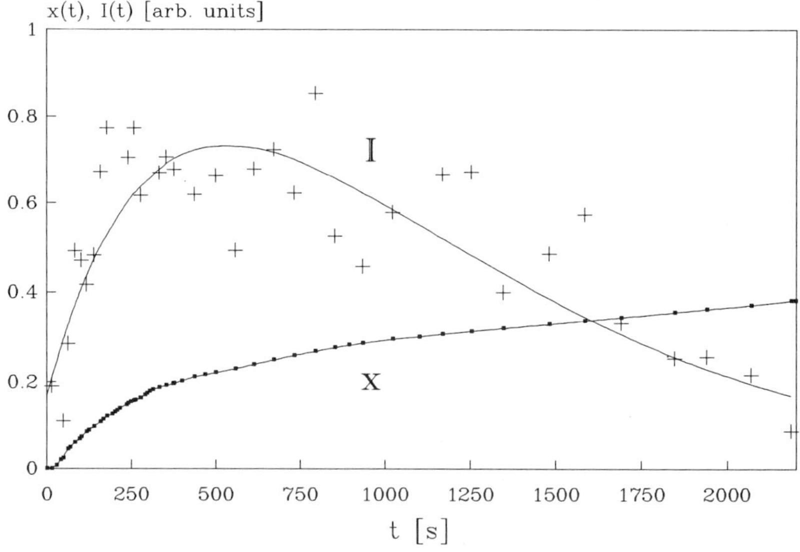

Fig. 1.   Nucleation rate I(t) and precipitation degree x(t) as evaluated from the digitized microscope images for the precipitation of B2-phase KCl from the solution at about 2 GPa (Däßler, 1991).

The theoretical approach to the understanding of the experimental data and the simulation method are developed in the following section. The results are compared with the experiment of Fig. 1 and discussed in the final section.

THEORETICAL APPROACH

For a deeper understanding of the kinetics of phase transformations under high pressure we have to consider the classical theory of nucleation and growth in more detail. The two basic processes are the creation of nuclei of the critical size and the growth of these nuclei with linear velocity v. Both processes strongly depend on the thermodynamic parameters and on the underlying physical mechanism (Christian, 1965). In the case of homogeneous nucleation due to statistical fluctuations the rate I only depends on pressure and temperature. Although the qualitative behaviour of I(t) in Fig. 1 can be explained by the pressure change in the chamber a quantitative description by the theory of homogeneous nucleation is questionable since the microscopic mechanism of nucleation is not clear. As to the growth velocity of the nuclei there are two basic mechanisms which are commonly discussed:

(i) diffusion limited growth

If the growth of crystal faces is controlled by diffusion of atoms to the interface, the growth velocity of a given nucleus decreases with time,

$$v_d \; \propto \; t^{-0.5} \; . \tag{2}$$

The diffusion limited process results in a crystal size dependent growth velocity.

(ii) interface controlled growth

A qualitative different time dependence of v is observed for growth processes where diffusion is fast and the incorporation of atoms into the crystal face is the rate limiting step. Then the growth velocity is constant,

$$v_i \; = \; \text{const.} \; , \tag{3}$$

and the growing crystals show smooth facets.

Equations (2) and (3) are only valid if the pressure and temperature are fixed during the phase transition which is obviously not the case in the DAC experiments. Taking into account this implicit time dependence the growth velocity v at time t for nuclei which have been created at time t_1 can be written in a more general form, $v(t,t_1) = v_o(t) \, f(t-t_1)$. $v_o(t)$ represents the implicit time dependence due to the pressure and concentration changes and $f(t-t_1)$ depends on the growth mechanism,

$$f(t-t_1) \; = \; \left\{ \begin{array}{ll} c \, (t-t_1)^{-0.5} & \text{for case (i)} \\ 1 & \text{for case (ii)} \end{array} \right\} \; .$$

c is a constant of dimension $(\text{time})^{-0.5}$. In terms of the nucleation rate I(t) and the growth velocity $v(t,t_1)$ the degree of precipitation x(t) is expressed by (Christian, 1965)

$$\ln\,[\,1\text{-}x(t)\,] = -\,\alpha \int_0^t I(t_1) \left[\int_{t_1}^t v(t_2,t_1)\,dt_2 \right]^2 dt_1 \; . \tag{4}$$

α denotes a constant of order unity depending on the shape of the growing crystals. Note that equation (4) has been written down for a two dimensional system. This is reasonable since in the DAC experiments as described above a two dimensional projection over a finite thickness of the sample is observed. Although the nucleation rate I(t) and the transformation degree x(t) are deduced from the experiments (Fig. 1) equation (4) is hard to solve since neither the function $v(t_2,t_1)$ nor the growth mechanism are known. In order to resolve this puzzle we introduce a nonlinear transformation of time scale by $t = t(\tau)$. The function $t(\tau)$ is arbitrary and is chosen in an appropriate manner later on. The main idea of this transformation is to compensate the implicit time dependence of the growth velocity expressed by $v_o(t)$ in the new scale τ in order to simulate the nucleation and growth process using numerical procedures developed earlier (Orgzall and Lorenz, 1988; Lorenz, 1989).

In the transformed time scale τ equation (4) reads

$$\ln [\ 1\text{-}x(\tau)\ ] = -\ \alpha \int_0^\tau I(\tau_1)\ w(\tau_1)\ \left[\int_{\tau_1}^\tau v_o(\tau_2)\ f(\tau_2\text{-}\tau_1)\ w(\tau_2)\ d\tau_2 \right]^2 d\tau_1\ ,\quad (5)$$

with $w(\tau) = dt/d\tau$, the derivative of time t with respect to τ. Introducing renormalized variables by $\hat{I}(\tau) = I(\tau)\ w(\tau)$ and choosing the time transformation $t(\tau)$ such that $v_o(\tau)\ w(\tau) = \hat{v}_o = \text{constant}$, i.e. $w(\tau) = \hat{v}_o/v_o(\tau)$, we get

$$\ln [\ 1\text{-}x(\tau)\ ] = -\ \alpha\ \hat{v}_o^2\ \beta^2 \int_o^\tau \hat{I}(\tau_1)\ (\tau\text{-}\tau_1)^m\ d\tau_1\ ,\quad (6)$$

where $m = 1, \beta = 2c$ and $m = 2, \beta = 1$ for the growth processes (i) and (ii), respectively. The problem now consists in the calculation of $\hat{I}(\tau_1)$ from the measured functions I(t) and x(t), shown in Fig. 1. At first, using the inverse function t(x) the nucleation rate I is expressed as function of x. Then, by definition, we get

$$\hat{I} = I(x)\ \frac{dt}{d\tau} = I(x)\ \frac{dt}{dx}\ \frac{dx}{d\tau}\ .\quad (7)$$

Inserting (7) into (6) results in an integral equation which can be solved numerically with the experimental input functions I(x) and dt/dx and the initial conditions $t(\tau\text{=}0) = 0$ and $(dt/d\tau)_{\tau=0} = \text{const}$. The solutions for $\hat{I}(\tau)$ and $x(\tau)$ from (6) are shown in Fig. 2 for the diffusion limited growth process (m=1). Obviously, the time scale τ is stretched appreciably at the beginning of the phase transition so that the function $\hat{I}(\tau)$ exhibits a distinct maximum in the intermediate stage.

With the time scaling $\tau(t)$ the nucleation and growth problem under the changing pressure conditions in the DAC is mapped to the much simpler problem of nucleation and growth where the external conditions are fixed and the effective nucleation rate is given by $\hat{I}(\tau)$. However, it is difficult to determine the basic growth mechanism and the exponent m in (6). In a simplified model calculation (for constant nucleation rate) it has been shown that the structure of the growing phase qualitatively depends on the underlying growth mechanism (Orgzall and Lorenz, 1991). Therefore, we compare the cluster-size distribution as determined from the experiments with the corresponding quantity calculated for diffusion as well as interface limited growth. In the first step the nucleation rate $\hat{I}(\tau)$ ist determined by solving equation (6) for both cases (i) and (ii). In the second step the nucleation and growth process is numerically simulated with the frequency of nucleation $\hat{I}$ and either size dependent (i) or constant (ii) growth velocities and the structures of the growing phases are evaluated.

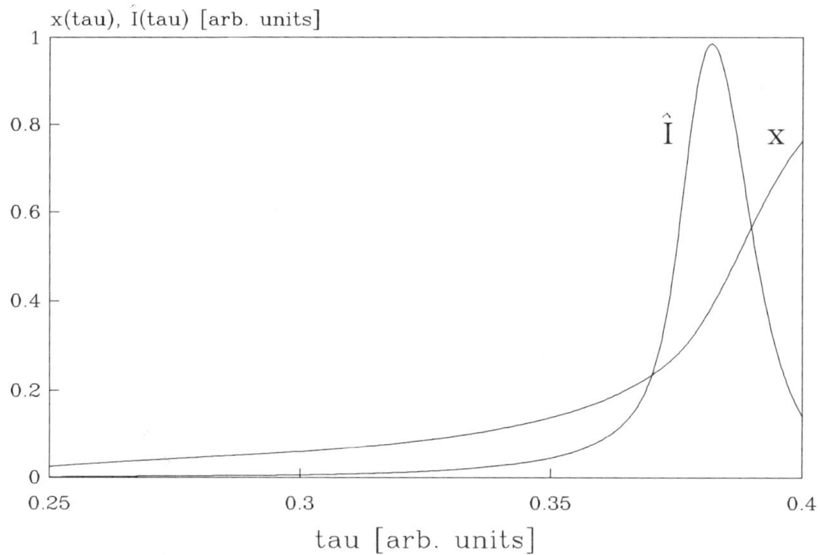

Fig. 2.   Transformation degree $x(\tau)$ and nucleation rate $\hat{I}(\tau)$ as calculated from equation (6) for diffusion limited growth.

CONCLUSIONS

During the nucleation and growth process different nuclei impinge to one another forming larger and larger clusters of grains. The size distribution of these clusters has been shown to be a characteristic function which strongly depends on the physical mechanism of nucleation and growth (Orgzall and Lorenz, 1988; 1991). The experimental and calculated cluster-size distributions (CSD) as function of an effective diameter d (defined as the square root of the cluster area) are shown for three different stages of the phase transition (x=0.06, 0.2, and 0.4) in Figs. 3a to 3c. Thereby, the variable d is scaled with respect to its average value and the CSD's are normalized to one. The experimental distribution (E) exhibits a characteristic maximum near the average cluster size and becomes zero for small d. With increasing x(t) the shape of the CSD is preserved and no qualitative changes can be detected. This behaviour of the experimental size distribution is in good agreement with the calculated one for the case of diffusion limited growth (D). On the other hand, the corresponding distribution function for interface limited growth (I) decreases gradually with increasing size and, contrary to the experimental data, tends to a finite value for small cluster sizes. From the comparison of the measured CSD with the results calculated for the two growth models (i) and (ii) it becomes obvious that in the present DAC experiment the growth of nuclei is limited by diffusion of atoms to the crystal surface. Note that this conclusion which follows from the experimental and theoretical evaluation of the microstructure of the growing phase cannot be obtained from the time dependence of the degree of precipitation x(t).

The procedures and conclusions outlined in the present paper represent a first step for a better understanding and interpretation of the phase transition kinetics in the diamond anvil cell. However, the success of the method depends on the experimental ability of measuring the microstructure (cluster-size distributions etc.) of the growing high pressure phase. If the basic physical mechanisms of nucleation and growth have been clarified a thermodynamic model taking into account the dependences of the nucleation rate and the growth velocity on e.g. pressure and concentration (Christian, 1965) can be used for further investigations.

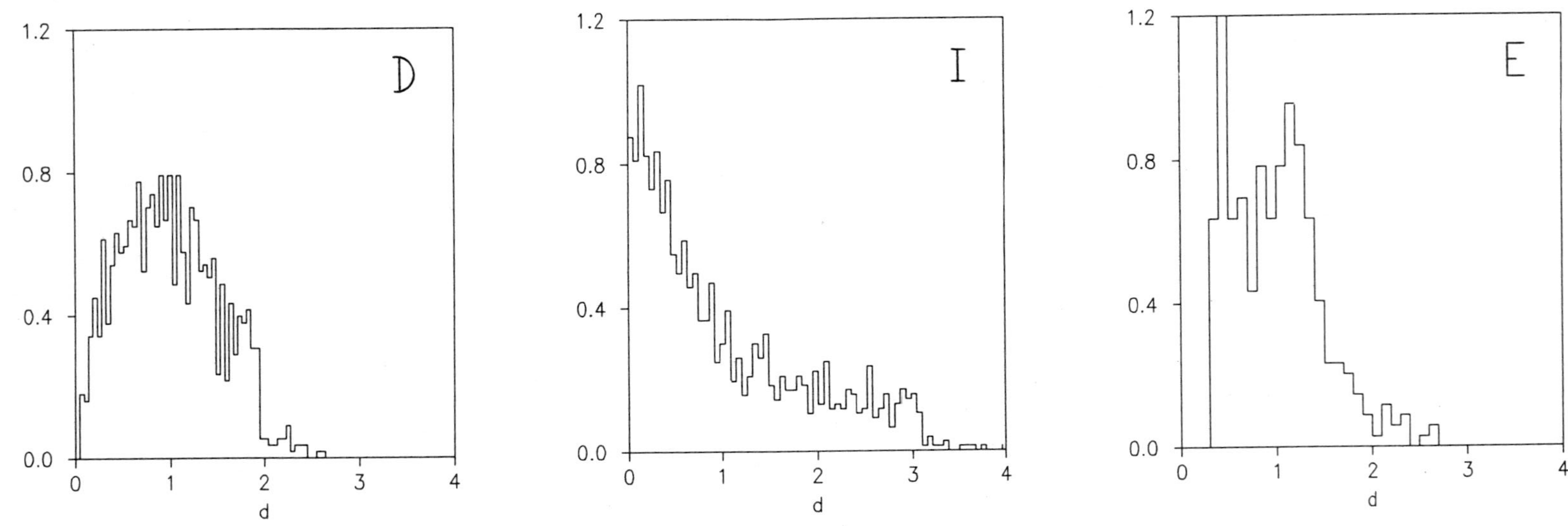

Fig. 3a. Cluster-size distributions in the early stage of the transition (x=0.06) as calculated from the numerical simulation of diffusion (D) and interface (I) limited growth processes in comparison with the experimental result (E).

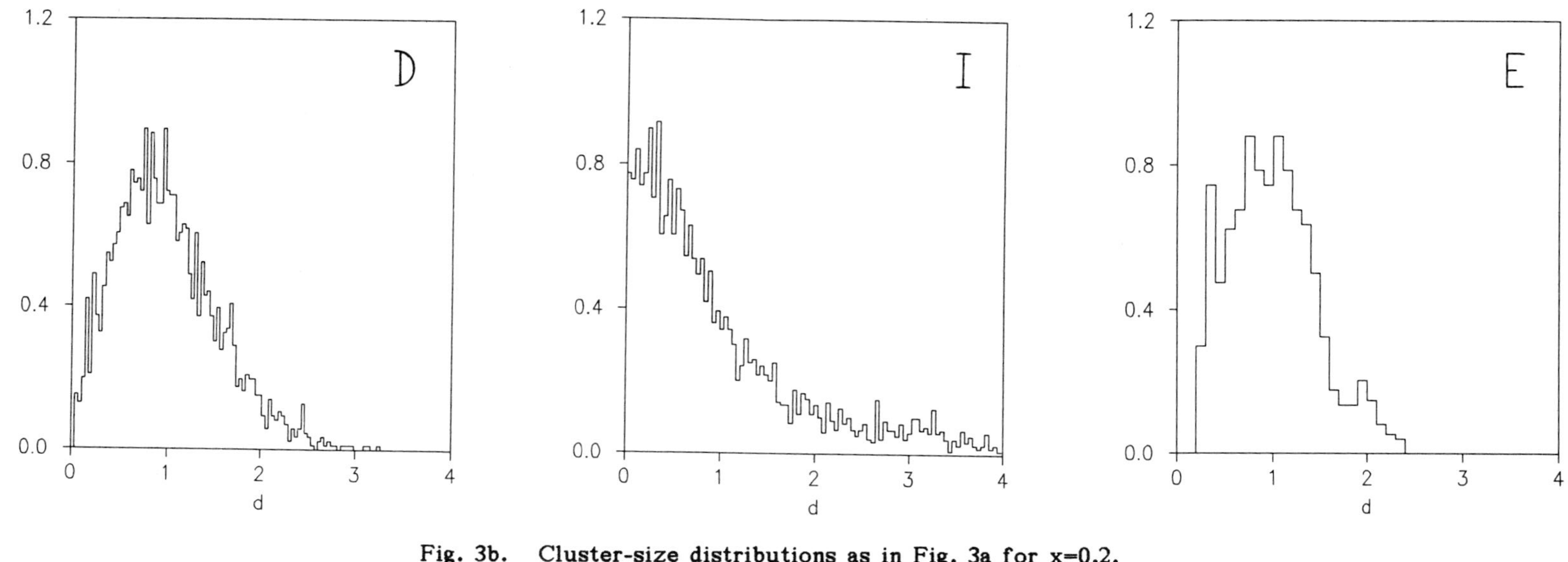

Fig. 3b.   Cluster-size distributions as in Fig. 3a for x=0.2.

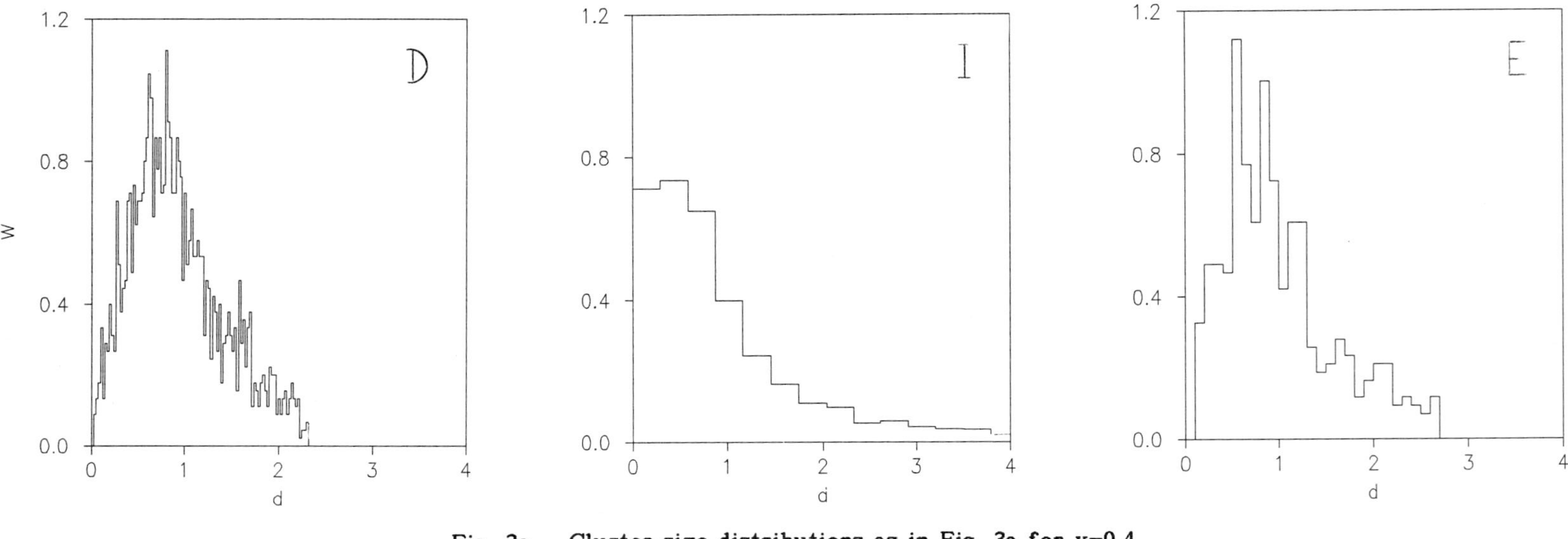

Fig. 3c.   Cluster-size distributions as in Fig. 3a for x=0.4.

REFERENCES

Brar, N. S., and Schloessin, H. H., 1981, The kinetics of the GeO_2 (alpha-quartz) $\rightarrow$ (rutile) transformation under high pressure, High Temp. - High Press., 13:347.

Christian, J. W., 1965, "The Theory of Transformations in Metals and Alloys", Pergamon Press, Oxford.

Däßler, R., 1988, An optical method for measurements of the kinetics of pressure induced phase transformations in the diamond anvil cell, High Temp. - High Press., 20:661.

Däßler, R., 1991, Kinetics of pressure-induced nucleation and growth process in the diamond anvil cell, High Temp. - High Press., in press.

Hamaya, N., and Akimoto, S., 1981, Kinetics of pressure-induced phase transformation in KCl at room temperature, High Temp. - High Press., 13:347.

Hamaya, N., Yamada, Y., Axe, J. D., Belanger, D. P., and Shapiro, S. M., 1986, Neutron scattering study of the nucleation and growth process at the pressure induced first-order phase transformation of RbI, Phys. Rev B, 33:7770.

Hirsch, K. R., and Holzapfel, W. B., 1983, A realtime ruby luminescence spectrometer for pressure determinations, J. Phys. E: Sci. Instrum., 16: 412.

Jayaraman, A., 1986, Ultrahigh pressures, Rev. Sci. Instrum., 57:1013.

Krüger, T., Merkau, B., Grosshans, W. A., and Holzapfel, W. B., 1990, Kinetics and systematics of structural phase transitions in the regular lanthanide metals under pressure, High Press. Res., 2:193.

Lorenz, B., 1989, Simulation of grain-size distributions in nucleation and growth processes, Acta metall., 37:2689.

Lorenz, B., Orgzall, I., and Däßler, R., 1991, Theoretical and experimental investigation of the precipitation kinetics of B2-phase KCl from the solution under high pressure, High Press. Res., in press.

Merkau, B., and Holzapfel, W. B., 1986, Kinetics of the dhcp-fcc phase transformation in La under pressure, Physica B, 139&140:301.

Orgzall, I., and Lorenz, B., 1988, Computer simulation of cluster-size distributions in nucleation and growth processes, Acta metall., 36:627.

Orgzall, I., and Lorenz, B., 1991, The cluster-size distribution in nucleation and growth processes: Diffusion controlled vs. interface limited growth, to be published.

HIGHLIGHTS OF THE ROUND TABLE DISCUSSION

ON MOLECULAR CRYSTALS AT ULTRA HIGH PRESSURE

N. W. Ashcroft

Laboratory of Atomic and Solid State Physics
Cornell University, Clark Hall
Ithaca, NY, U.S.A.

Panel: N. W. Ashcroft, Chairman, Cornell University, Ithaca, NY, U.S.A.
H. J. Jodl, Universität Kaiserslautern, Germany
B. Kuchta, Colorado State University, Fort Collins, CO, U.S.A.
K. H. Michel, Universiteit Instelling, Antwerpen, Belgium
A. Polian, Universite Pierre et Marie Curie, Paris, France
I. Silvera, Harvard University, Cambridge, MA, U.S.A.

This panel focused its attention on the future prospects of research
at high pressure for mainly molecular systems. Each panelist was asked to
provide a working definition for the concept of "high pressure" as might
possible be guided by their own particular interests and to present state-
ments on what they considered to be the key experimental and theoretical
issues to be faced in the immediate and perhaps more distant future.

Dr. Jodl opening the proceedings by noting that for many molecular
systems the energy can be delivered by compressional work still remains a
relatively small fraction of interesting bond energies. Thus, the choice
of system is important, if the effect on pressure is to be significant.
Nevertheless (and here there were several suggestions from the audience)
there remain many possibilities even within such constraints, among them
mixed systems and disordered systems ("matrix isolated systems" to use Dr.
Jodl's term) because a foreign environment for a given molecule can often
enhance the pressure effect. Such environments can also lead to quite
different local symmetries, and then via the corresponding changes in
oscillator strengths, to more effective probes. Impurities then become
microscopic "sensors" of the changes brought about by the high pressure
environment, though as was pointed out in discussion, it is important to
separate out the effects being sought from those owing their origin to
homogenous broadening.

Dr. Kuchta turned to more calculational aspects remarking (to the
general amusement of the audience) that so far as computers are concerned
the concept of ultra-high pressure is somewhat irrelevant since the computer does not "feel it." Dr. Kuchta made the important point that underlying much of the high pressure physics is the aim to understand the properties of matter, and particularly molecular systems, in terms of plausible interactions. For complex systems these still have to be regarded as
"effective" interactions, or potentials, rather than fundamental, and this

obviously involves physical modeling. Under conditions of high pressure,
there are changes in electronic structure, (the ultimate example being the
case where chemical reaction or electron transfer takes place) and these
influence vary much the viewpoint that should be taken on the nature of
effective interactions. In the case of the metallic state of matter, the
existence of iternant electrons, which by definition are sensitive to a
volume boundary condition, state dependence of pair potentials is always
present. It is here that very effective use can be made of modern computational methods since they enable a rapid transcription from proposed
intermolecular potentials to physical observables within the framework of
models proposed for particular high pressure environments. There was a
suggestion from the audience that in the quest for more complete understanding of the behavior of effective interactions with pressure, more
consideration should be given to the halogens since they possess the capability of providing a bridge between the molecular (insulating) and metallic states. Iodine is here a demonstrated case in point.

Dr. Michel agreed that emphasis needs to be placed on the modeling
issue, so far as our understanding of effective interactions is concerned,
and that in such an exercise the molecular systems, because of their relative simplicity play an important role. However, he also emphasized the
fact that these interactions are in large part electronic in origin, and
that in consequence it is implicit that in approaching the modeling issue,
considerable thought has to be given to basic electronic structure and its
progression with pressure or density. A specific example is the case of
charge transfer systems (for instance potassium cyanide) where the progressive redistribution of electronic charge alters the underlying electrostatic problem to the point where the multipole character, in particular the quadrupole component, can play a very significant role in the
ensuing sequence of phase transitions. This structural theme was taken up
by members of the audience who pointed out that temperature dependence is
also an important issue, and that there is a very serious need for additional efforts on the instrumentation side in order to exploit the emerg-
ing possibilities for combining high pressure with high temperatures.
This is particularly true in the geophysical context.

Dr. Polian emphasized the point that the meaning to be attached to
the term "high pressure" or "ultra-pressure" depends very much on what it
is one wishes to measure. Thus if the intent is to study the way in molecular levels can be "tuned" by the effects of density increase of their
environment, and here an example might be benzene, then relatively large
changes can be induced by quite modest pressures (a point that has been
strongly emphasized in the past by, for example, Drickamer). Dr. Polian
also reiterated the earlier observations that impurity atoms can be very
effective probes of high pressure states of condensed matter, and especially molecular insulators, citing the example of hydrogen in dense helium. Here the probes are mainly optical in nature, including the systematic tracking of Raman and infra-red active modes. In such studies the
prospects of making effective use of temperature should also be acknowledged, for example through the use of laser heating. From the audience
came the additional suggestion that studies using this mode of heating
might be linked to dynamical techniques which achieve heating during shock
conditions, and are often relatively close to equilibrium states.

Our last panelist, Dr. Silvera, took the position that in order to
keep a field active, including the field of molecular systems at high
pressure, it is necessary to keep a steady stream of lively ideas flowing,
and to resists the temptation to measure systems at high pressure "just
for the sake of measurement." As an example of a lively area, he cited
the enduring problem of the metal-insulator transition, where in the context of high pressure physics, the controlling parameter is the thermody-

namic density. One of the most challenging systems in which to study such
a phenomenon, is also in a sense of simplest: dense hydrogen. Dr.
Silvera suggested that this subject is still in need of critical experiments, particularly in elucidating the possible onset of a metal-insulator
transition. He suggested that the optical signatures of such a transition
are currently far from reliable. He also drew attention to the need for
further theoretical study of the paired metallic phase, and stated that
the subject of superconductivity in molecular (metallic) hydrogen now
needs to be addressed further. Since hydrogen at high pressures continues
to exist in rotational states, Dr. Silvera stressed that the need for a
theory of the electronic structure of such phases is now with us, a point
that has also been made in the past. But going beyond hydrogen, other
systems of a more imaginative kind might easily include metallic polymers
and other quasi one-dimensional systems, possibly new classes of super
hard materials (or even materials that become super hard in the high pressure range) and new magnetic materials. And surely the system of most
recent world-wide fascination, namely condensed $C60$ molecules (or "fullerite") is a material about which high pressure techniques will have much to
say. For example, the high pressure equation of state will be quite revealing in the debate on the nature of the inter-$C60$ bonding issue, and
the transport and optical properties at high pressure will give considerable insight into the excited state electronic structure, and the mainte-
nance of an overall energy gap. Finally, Dr. Silvera drew attention to
the "impending limit" of diamond cells implied by the eventual closing of
the optical windows, and also to the need for constant development of
reliable pressure scales wholly centered on diamond cells (and hence independent of other source, such as diffraction probes).

    The last point prompted a response from the audience, namely that it
is crucially important for many experiments to have a quite detailed understanding of the actual conditions prevailing in a diamond cell, for
example local states of strain, thermal gradients, pressure gradients and
the like. These are not always revealed by local probes. On a related
theme, the point was also made that problems of interpretation often ensue
when there is over-reliance on a single technique; put in other terms,
cross-linking of different experiments on a single system but within the
diamond anvil cell, would be highly desirable. An example was offered,
namely the possibility of carrying out Brillouin scattering on highly
compressed hydrogen which would aid in the direct determination of the
acoustic phonon structure, so far inferred only from the extrapolated
equation of state. This information, in conjunctions with the known progression of the vibron structure, could assist markedly in refining the
predictions for the onset of vibrational instability (and the transition
to the monatomic phase).

    The roundtable concluded on a note of relatively high optimism!

ENHANCED STABILITY OF HETEROSTRUCTURES UNDER PRESSURE

B.A. Weinstein, L.J. Cui, U.D. Venkateswaran,* and F.A. Chambers[1]
Department of Physics, 239 Fronczak Hall
SUNY at Buffalo
NY 14260 U.S.A.
[1]Amoco Technology Co., P.O. Box 3011
Naperville, Il 60566

I. INTRODUCTION

Epitaxial heterostructures configured as single or repeated layers, and, more recently, as islands, have become ubiquitous within semiconductor physics because of the diverse phenomena and applications made possible through their tailored growth.[1] The practical limits of heterostructure tailoring are defined by a variety of instabilities related to bulk, interface, and local bonding properties, and the study of these instabilities offers fundamental insight into the competition between mechanical and chemical forces at each structural level. Despite the importance of such issues, most experimental and theoretical work on heterostructure stability has been limited to external environments compatible with growth, and, hence, has not considered the effects of extreme hydrostatic pressure.[2–4] Although this is understandable in terms of current interest trends, it overlooks the nature of pressure as a thermodynamic parameter capable of shifting heterostructures into regions of phase space which ordinarily are either inaccessible, or only attained in chemically different materials systems.

A case in point is provided by the authors' recent studies of the high pressure phase stability of AlAs/GaAs multilayers and superlattices (SLs).[5,6] In this materials system it was found that zincblende (zb) AlAs could be strongly overpressed above its bulk boundary for transition to the rocksalt (B1) structure. Indeed, in the regime $12.3 < P < 17.2$ GPa, between the bulk AlAs and GaAs transitions, we establish that AlAs/GaAs SLs enter a metastable phase. This phase is completely analogous to, for example, the 1 atm. as-grown state of MnSe/ZnSe SLs containing thin layers of zb MnSe, instead of the normal B1 form.[7] Essentially, a pressure of 1 atm. represents substantial overpressing for zb MnSe since its zb→B1 transition is regarded to occur at negative pressure.[8] Several other II-VI and I-VII semiconductors fall into this category. The present review amply illustrates the advantages of pressure experiments over chemical modifications for studying the physics of interface-altered stability in semiconductor heterostructures. The interested reader will find greater details in Refs. 5 and 9 for the AlAs/GaAs, and the ZnSe/GaAs systems, respectively.

Frontiers of High-Pressure Research, Edited by H.D. Hochheimer and
R.D. Etters, Plenum Press, New York, 1991

II. EXPERIMENTS

High pressure phase transitions in bulk-like AlAs (BL1), GaAs (BL2) epitaxial films, and six GaAs/AlAs superlattices (SLs 1–6), grown by molecular beam epitaxy or metal-organic chemical vapor deposition were studied using Raman scattering. The sample characteristics are given in Table I. BL samples were thinned to a total thickness of ~ 20 μm, and SL samples were made 'free-standing' by selectively etching away the entire GaAs substrate. A cleaved sample of ~ 50 μm $\times$ 70 μm was loaded in the diamond anvil cell (DAC) along with several tiny chips of ruby for pressure calibration. Often the GaAs LO(Γ) phonon peak frequency was used for secondary pressure calibration. The pressures deduced from Raman LO-phonon and ruby R-line fluorescence were found to agree within $\pm$ 0.4 GPa. Above 9.5 GPa, the DAC was annealed at ~ 70–$75°$C for about an hour to minimize pressure gradients. The Raman experiments were performed at room temperature using a double monochromator with either photomultiplier or intensified Si-diode array detectors and computer controlled data acquisition and processing. Kr$^+$ or Ar$^+$ laser lines ($\leq$ 30 mW) were used for excitation, and a custom-built microprobe focussed the beam to ~ 10 μm.

TABLE I. Summary of sample characteristics and observed pressures for forward $\alpha \to \beta$ and reverse $\beta \to \alpha$ transitions. P_1^t and P_2^t denote the first (AlAs) and second (GaAs) thresholds on increasing pressure. R^t denotes the pressure at which transparency returns on decompression.

Samples	GaAs/AlAs Layer widths	Periods	Growth Method	Transition Pressures (GPa)		
				P_1^t	P_2^t	R^t
BL1	0.1/2 μm	1	MBE	12.3±0.4	17.2±0.4	6.0±1.5
BL2	8/0.0 μm	1	MBE	—	17.3±0.4	9.0±1.5
SL1	160/290Å	60	MOCVD	13.8±0.4	16.8±0.8	8.2±1.5
SL2	225/75Å	73	MBE	14.7±0.4	17.4±0.8	9.2±1.5
SL3	200/50Å	80	MBE	16.0±0.4	17.4±0.8	8.8±1.5
SL4	50/25Å	250	MBE	17.3±0.4	17.3±0.4	6.0±1.5
SL5	50/20Å	100	MBE	17.6±0.4	17.6±0.4	6.4±1.5
SL6	25/60Å	200	MOCVD	17.0±0.4	17.0±0.4	11.5±1.5

III. RESULTS

Visual microscopic observations provide a convenient means for detecting the change from the transparent zb α-phase to the opaque high-pressure β-phase. When combined with Raman measurements, several interesting features of the $\alpha - \beta$ transition kinetics can be deduced. As we shall see, the studies suggest many interconnections as well as unique characteristics of the SL phase transitions compared to the analogous bulk transitions.[10]

A. Phase transitions in bulk AlAs and GaAs

$\alpha - \beta$ transitions in bulk samples serve as a benchmark for the superlattice results to be discussed in Sec. III B. Figure 1 shows typical Raman spectra for AlAs (sample BL1) and GaAs (sample BL2) epitaxial films recorded in the back-scattering geometry. The transparent or opaque visual appearance of each sample is represented by the shading or its absence in the rectangular inset sketches and the position of the laser spot is denoted by the solid circle.

The dominant feature at 1 bar is the zone-center longitudinal optic [LO(Γ)] phonon. At higher pressures, due to the increase of the band-gap energy, the sample becomes progressively more transparent to visible radiation. Concurrent with this, the transverse optic [TO(Γ)] peak appears with increasing intensity, arising from allowed forward scattering of reflected laser light from the specimens' back surface. The transition to the β-phase is detected both by the onset of visual opacity and by the disappearance of the zb phonon peaks in the Raman spectrum. As can

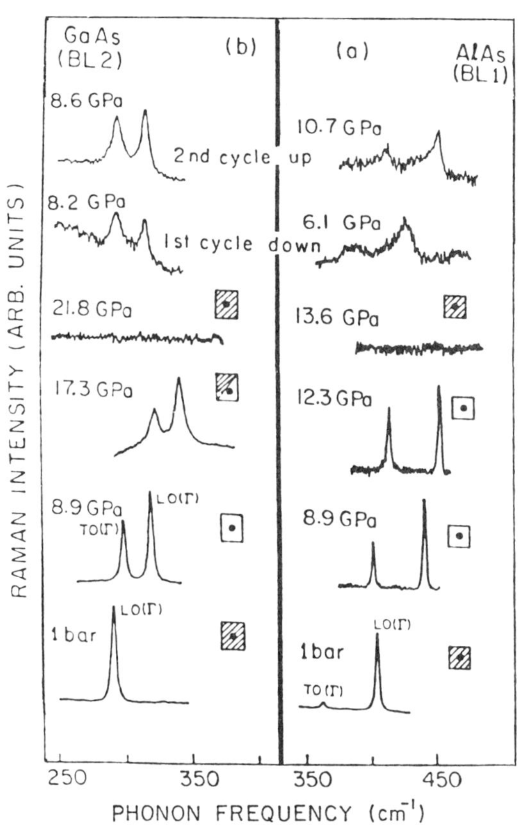

Fig.1. *Raman spectra of bulk-like AlAs (a) and GaAs (b) epitaxial films at various pressures. Visual specimen appearance and laser location are indicated by the shading and solid circles in the rectangular inset sketches. Top two traces show reversal to the zincblende phase on decompression and subsequent recompression. Spectra during 1st pressure cycle were excited with 647.1 nm, and the reversal spectra with 568.2 nm.*

be seen from Fig. 1(a) for AlAs, and Fig. 1(b) for GaAs, the $\alpha - \beta$ change in these semiconductors occurs at 12.4 GPa and 17.3 GPa, respectively. The transformation is observed to take place very rapidly in AlAs, going to completion in a few tenths of a second, whereas the transition is found to be rather sluggish ($\sim$ mins) in GaAs. Thus, it has been possible to retain BL2 in a partially transformed state, while a mixed $\alpha - \beta$ phase has not been observed in BL1. We always find that the Raman spectra from transparent regions exhibit the standard zb signatures. No Raman signal could be detected in the opaque high pressure phases of either bulk AlAs or GaAs.

The reversal back to the α-phase after cycling through the $\alpha - \beta$ transition is also studied in detail. The top two traces in Figs. 1(a) and 1(b) show the reappearance of the zb phonons during the first down-cycle and the second up-cycle of pressure. Notice that the second-cycle spectra are more intense compared to their first down-cycle counterparts, suggesting the sluggish nature of the reversal. Reduced transparency and speckled appearance of the samples are sometimes associated with reversal. In addition, there is considerable hysteresis in the reverse threshold, $\sim$ 6 GPa for AlAs and $\sim$ 8 GPa for GaAs. Another interesting observation is that the α-phase reverses only if the maximum pressure (P_{max}) reached is below 20 GPa. If P_{max} exceeds 20 GPa, GaAs returns to 1 bar in an opaque metastable phase.[10] There is no evidence for a metastable phase in AlAs.

The Raman spectra observed after reversal are weaker and broader than before transformation, but they grow in strength and sharpen as the pressure is reduced to near 1 bar. Whenever possible, specimens are retrieved from the DAC after pressure cycling, to investigate their 1 bar phases. We note three main differences between the 1 bar spectra after reversal compared to the unpressurized zb phase. These differences are: (i) the zb reversed LO(Γ) peak lies $\sim$ 3 cm^{-1} below the original 1 bar LO(Γ) frequency, (ii) the peaks in retrieved samples show at least a factor of two asymmetric broadening on the low energy side, and (iii) the TO(Γ) intensity becomes comparable to LO(Γ) after cycling, while the TO(Γ) peak was forbidden before. The presence of the TO(Γ) phonon and the asymmetric broadening are indicative of increased disorder. Consistent with this, the negative frequency shift of the LO(Γ) peak is clear evidence for microcrystallite formation. Based on results from ion-implantation studies on GaAs [11], and using a spatial correlation model, we find that the microcrystallite diameter is $\sim$ 65 Å in GaAs and $\sim$ 175 Å in AlAs samples undergoing the $\beta - \alpha$ reversal. Knowledge of the microcrystalline grain size, and of the hysteresis in the transition threshold leads us to an estimate of the $\alpha - \beta$ surface energy, as will be discussed below.

B. Phase transitions in superlattices

As in the bulk case, visual observations in combination with Raman data yield a great deal of information concerning the $\alpha - \beta$ transitions in GaAs/AlAs superlattices. Our studies on the six different SLs listed in Table I provide evidence that SL transitions occur either separately in each layer-type (SLs 1–3), or simultaneously in both layers (SLs 4–6). It is convenient to group SLs into "thick-layer" and "thin-layer" categories, respectively, according to this behavior. Optical microscopic observations on the "thick-layer" SLs show transitions by rapid (< 0.1 sec) step-wise darkening, while the "thin-layer" specimens transform by continuous expansion of opaque nucleation regions. Concurrent with the onset of opacity, the Raman results show the disappearance of only the zb-AlAs phonons in the "thick-layer" SLs, whereas both the GaAs and AlAs zb-phonons disappear in the "thin-layer" samples.

A typical example of the behavior in a thick layer SL is presented in Fig. 2 for sample SL2. At 1 bar [Fig. 2(a)], and also during the first cycle of increasing pressure when the sample gained visual transparency [Fig. 2(b)], the zb phase $LO(\Gamma)$ and $TO(\Gamma)$ phonons of both the GaAs and AlAs layers are observed. Near 14.4 GPa (when opacity first sets in), the SL specimen is held in a partially transformed state, and Raman spectra are measured from both its opaque and transparent portions, [Figs. 2(c) and (d)]. It is clear that above 14.4 GPa the AlAs zb-phonons disappear, while the GaAs zb-phonons persist to much higher pressure. Thus there are two distinct and layer-specific transitions in this SL sample. The AlAs layers transform first at $P_1^t=14.4$ GPa which is 2 GPa higher than the bulk AlAs $\alpha - \beta$ transition threshold; the GaAs layers are found to transform at $P_2^t=17.4$ GPa, which is close to the bulk GaAs threshold of 17.3 ± 0.4 GPa. Upon decreasing the pressure, we observe the return of visual transparency and a corresponding reappearance of the zb phonons. Figure 2(e) shows the reversal in the AlAs layers after cycling SL2 through P_1^t but not P_2^t. A similar reversal is also observed in the GaAs layers as long as $P_{max} < 20$ GPa. When $P_{max} > 20$ GPa, the zb phonons do not return, and the pressure cycled specimen, when retrieved at 1 bar, exhibits a four peak Raman spectrum similar to that observed for the metastable phase in bulk GaAs.

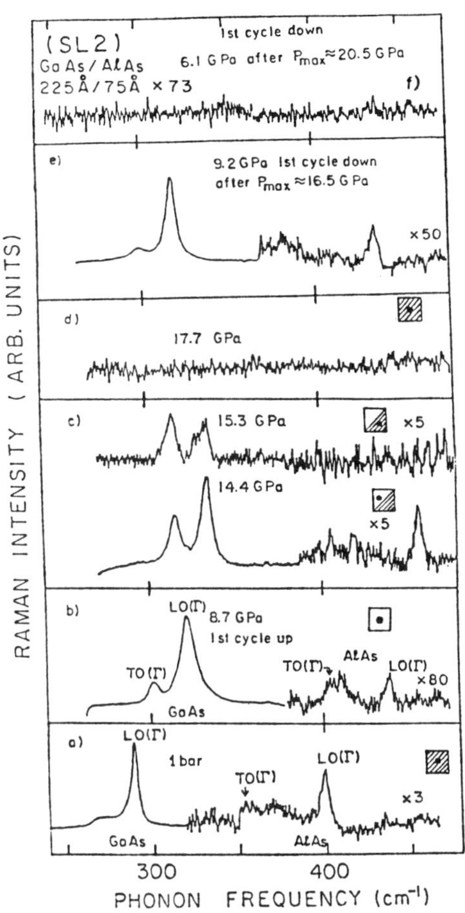

Fig.2. *Raman spectra showing separate $\alpha - \beta$ transitions in the "thick-layer" SL2 sample. AlAs peaks vanish at the onset of opacity at P_1^t (c), but GaAs peaks persist until P_2^t (d). (e) and (f) show reverse transition when $P_{max} = 16.5$ GPa, and its absence when $P_{max} = 20.5$ GPa, respectively.*

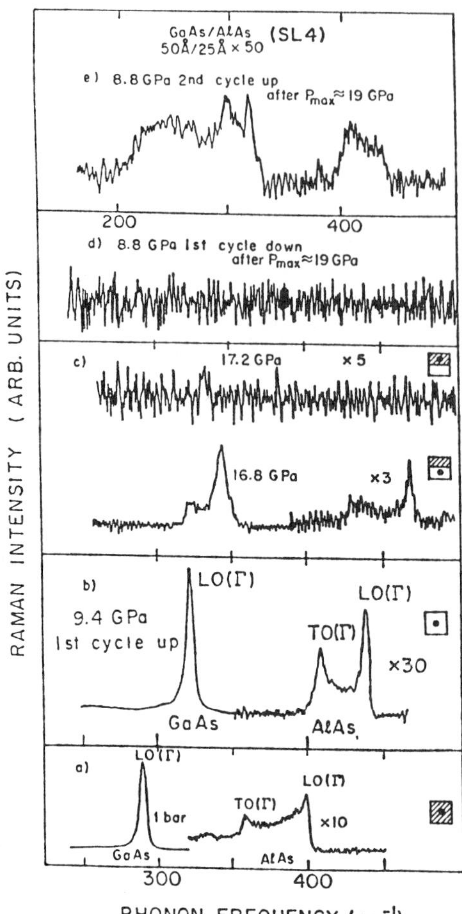

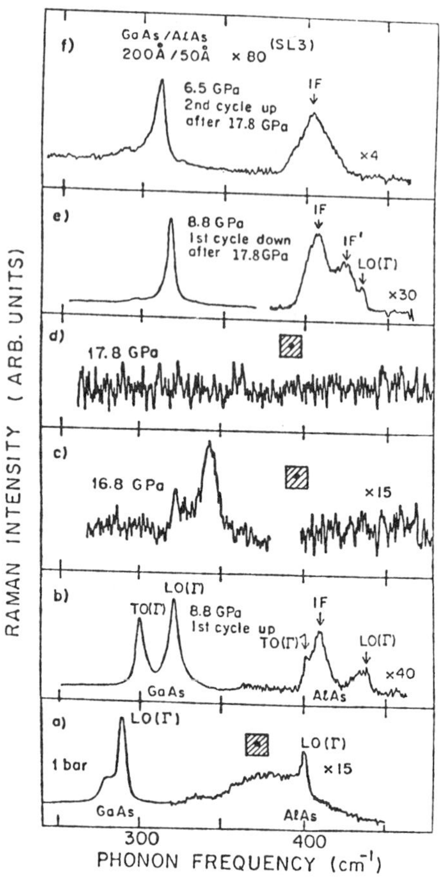

Fig.3. *Raman data showing simultaneous AlAs and GaAs $\alpha - \beta$ phase changes in the "thin-layer" SL4 sample. Zincblende peaks of both AlAs and GaAs disappear at 17.2 GPa (c). Reversal to the zincblende phase and transition-induced disorder are shown in (e).*

Fig.4. *Sequential pressure Raman study of SL3. (a) 1 bar spectrum; (b) 8.8 GPa pretransition data. Note AlAs-like interface mode (IF); (c) AlAs peaks vanish at P_1^t; (d) GaAs peaks vanish at P_2^t; (e) Spectra at 8.8 GPa on 1st cycle down. Note, additional AlAs-interface mode (IF '); (f) 2nd cycle up at 6.5 GPa. IF broadening indicates increased interface disorder.*

A typical phase transition and its reversal for one of the "thin-layer" SLs is illustrated in Fig. 3 for SL4. The zb phonons in both the GaAs and AlAs layers are observed up to 16.8 GPa. Above 17 GPa these phonon peaks all disappear _simultaneously_ in the opaque regions of the specimen [see Fig. 3(c)]. The reverse transition in this sample is sluggish, taking $\sim$ 20 hours for completion, after which the zb phase contains considerable disorder as portrayed in the second pressure-cycle spectra shown in Fig. 3(e).

A further illustration of transition-induced interface disorder is displayed in Fig. 4 for SL3. A pronounced AlAs-like interface mode [labeled IF in Fig. 4(b)] is observed at 8.8 GPa prior to any phase transition. After separate $\alpha - \beta$ transitions in the AlAs and GaAs layers [Figs. 4(c) and (d)], a second interface peak labeled IF' appears [Fig. 4 (e)]. Further pressure cycling then leads to a triangular shaped AlAs optic phonon band [see Fig. 4(f)] indicating increased interface disorder during the $\beta \rightarrow \alpha$ reversal. Evidence for heterointerface disordering is found in all the SL samples except SL2.

The observed threshold pressures for AlAs (P_1^t) and GaAs (P_2^t) are listed in Table I for all the SL and BL samples studied. It can be seen that in the "thick-layer" specimens SL1–3, the AlAs layers transform at a pressure substantially above the bulk AlAs threshold found in BL1. This implies that the AlAs layers in the SLs are _overpressed_ beyond their bulk stability limit. However, the GaAs layers always transform close to the bulk GaAs threshold indicating that there is _no_ complementary underpressing of GaAs. For the "thin-layer" samples, we find that $P_1^t \simeq P_2^t$ and, further that this single transition pressure agrees with the bulk GaAs threshold within experimental accuracy (± 0.4 GPa). These results are summarized in Fig. 5, where we plot the transition thresholds P_1^t (crosses) and P_2^t (squares) as a function of the

AlAs layer thickness. Bulk thresholds are denoted by arrows. We have also included a datum at 0.06 Å^{-1} from Ref. 12. Briefly stated, this figure shows that the amount of AlAs overpressing increases with decreasing AlAs layer thickness, until for sufficiently thin layers, a single transition of both AlAs and GaAs occurs near the bulk GaAs transition threshold. Hence, the effective stability of GaAs/AlAs SLs against hydrostatic compression is controlled by GaAs, and such enhanced stability leads to possible interesting applications for mismatched or metastable heterostructures belonging to other materials systems.

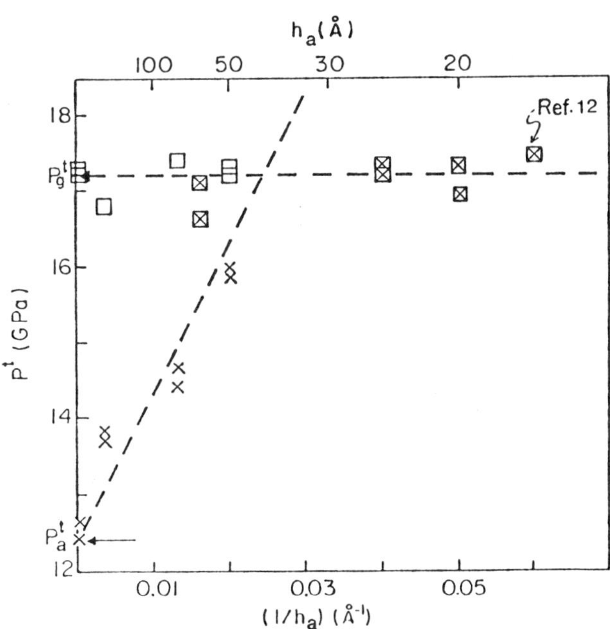

Fig.5. _Measured $\alpha - \beta$ transition pressures P_1^t for AlAs (crosses) and P_2^t for GaAs (squares) in SLs 1–6. Bulk AlAs (P_a^t) and GaAs (P_g^t) thresholds are marked by horizontal arrows. Data at 0.06 Å^{-1} is after Ref. 12. Dashed lines show difference between "thick-layer" SLs (sloped line) and "thin-layer" SLs (horizontal line)._

IV. THEORY and DISCUSSION

The $\alpha \rightarrow \beta$ transitions in GaAs and AlAs are first order phase changes with $\sim$ 18% volume decrease, and the reverse transitions are accompanied by considerable hysteresis. As in most semiconductors, these transitions proceed *via* a nucleation and growth mechanism. The growth of a 6-fold coordinated β-embryo within a 4-fold coordinated α-matrix can proceed only when the radius of the β-embryo exceeds a certain critical value r*. For homogeneous bulk materials, it can be shown using simple thermodynamic energy considerations that [13],

$$r^* \simeq -\frac{2\sigma_0^{\beta\alpha}}{(P-P_0)(\frac{\Delta V}{V})^{\beta\alpha}},$$ (1)

where $\sigma_0^{\beta\alpha}$ is the surface energy density of the β/α *homo*interface, P_0 is the true equilibrium pressure of the transition, and $(\frac{\Delta V}{V})^{\beta\alpha}$ is the fractional volume change during the phase transition. Eq. (1) applies equally (same $\sigma_0^{\beta\alpha}$) to both the forward $\alpha \rightarrow \beta$ and the reverse $\beta \rightarrow \alpha$ transitions. Since there is hysteresis, it is reasonable to replace $(P - P_0)$ by $(P^t - R^t)/2$, where P^t and R^t are the observed forward and reverse transition pressures. We can now calculate $\sigma_0^{\beta\alpha}$ from our estimates of the microcrystallite size and the observed hysteresis. We find that $\sigma_0^{\beta\alpha} \sim 0.05 - 0.08$ eV/Å^2 in GaAs and $\sim 0.04 - 0.15$ eV/Å^2 in AlAs.

Because of the chemical and structural similarity of AlAs and GaAs, the kinetic boundary between the β-nucleus and α-matrix considered above should not differ appreciably from the static β-AlAs/α-GaAs *hetero*interfaces encountered in super-lattice transitions. We have considered in detail the role and importance of the static epitaxial interfaces in treating the phase transitions in SLs. A detailed thermody-namic formulation taking into account heterointerface dislocation and strain energies is given elsewhere.[5, 14] This treatment leads to the α-phase overpressing of material 'a' according to,

$$-\frac{2(\sigma^{\beta\alpha} - \sigma^{\alpha\alpha})}{(\frac{\Delta V}{V})^{\beta\alpha}h_a^{\alpha\alpha}} = P_1^t - P_a^t,$$ (2)

where the σ's denote interface-energy densities, and h_a is the layer thickness cor-responding to material 'a' (AlAs in AlAs/GaAs SLs). Notice that the amount of overpressing is proportional to $(1/h_a)$ as observed for the "thick-layer" SLs in our experiments (see Fig. 5).

Consider now the second SL transformation at P_2^t involving material 'g' (GaAs layers). The α-phase AlAs overpressing requires a complementary underpressing of the α-phase of GaAs, given by,

$$\frac{2(\sigma^{\beta\beta} - \sigma^{\beta\alpha})}{(\frac{\Delta V}{V})^{\beta\alpha}h_g^{\alpha\beta}} = P_g^t - P_2^t.$$ (3)

However, underpressing in the GaAs layers is not observed in our experiments. This can be explained if, based on the evidence for post-transition interface degradation in Figs. 3(e) and 4(f), we adopt a disordered-interface (DI) model. Here, we reason that the first $\alpha/\alpha \to \beta/\alpha$ transition at P_1^t produces so many dislocations that it is not possible for the interface to reorder during the subsequent $\beta/\alpha \to \beta/\beta$ transition at P_2^t. The interface energies are then dominated by approximately the same density of dislocations for both the β/α and β/β interfaces. Hence, we get $\sigma^{\beta\beta} \approx \sigma^{\beta\alpha}$, and since $\sigma^{\alpha\alpha} \sim 0$ in the lattice-matched GaAs/AlAs system, Eqs. (2) and (3) predict AlAs overpressing _without_ GaAs underpressing, in agreement with our experimental observations.

A third case to consider is the extreme situation when the layer thicknesses approach atomic dimensions. For atomic monolayers the expected stability is that of the equivalent homogeneous alloy (EA). This EA limit should be reached when the SL period is smaller than the minimum nucleation radius given by Eq. (1). It is then possible for a β-nucleus to encompass both layer-types, yielding a transition pressure determined by the layer-thickness-weighted average of the bulk transition thresholds, $i.e.$,

$$\bar{P}^t = \frac{h_a^{\alpha\alpha} P_a^t + h_g^{\alpha\alpha} P_g^t}{h_a^{\alpha\alpha} + h_g^{\alpha\alpha}}. \tag{4}$$

In fact, we find no evidence of EA behavior in our results, even for the thinnest-layer SL studied. Rather, our observations of interface disorder in SLs undergoing reversal favor the DI model. Evidently, the EA limit requires SL periods thin enough that several are included within a single nucleation radius.

Now let us compare our experimental results with existing microscopic calculations. Based on the overpressing predicted by Eq.(2), the initial positive slope of P_1^t vs $1/h_a$ gives an empirical value for the change in the interface energy density $(\sigma^{\beta\alpha} - \sigma^{\alpha\alpha})$ at the $\alpha/\alpha \to \beta/\alpha$ transition. Neglecting $\sigma^{\alpha\alpha}$ in comparison with $\sigma^{\beta\alpha}$ (justified for the lattice-matched GaAs/AlAs system [15]), the slope in Fig. 5 yields, $\sigma^{\beta\alpha} = 0.12 \pm 0.02$ eV/Å^2. Using the Matthews and Blakeslee treatment of misfit dislocations,[2] this value of $\sigma^{\beta\alpha}$ corresponds to a dislocation spacing of ~ 28 Å or a dislocation density of 3.6×10^6 cm^{-1}, which is quite reasonable considering the large volume change ($\sim 18\%$) accompanying the $\alpha \to \beta$ transition.

Total energy calculations by Martin [16] have suggested that a stable coherently-strained pseudomorphic β/α interface is possible for sufficiently thin layers in GaAs/AlAs SLs. For the obtained value of $\sigma^{\beta\alpha} = 0.12$ eV/Å^2, this strained-layer structure should become stable in the range 50 Å$< h_a <$ 90 Å. However, our Raman measurements on SLs 2–6 for $P > P_1^t$ fail to show any abrupt shift in the GaAs zb-phonon frequencies, indicating that no coherent biaxial strain is present. We conclude that the DI limit is the proper picture, and that during the first $\alpha - \beta$ transition in the AlAs layers enough dislocations are created to relieve any homogeneous strain.

The observed overpressing of AlAs layers beyond the bulk $\alpha - \beta$ transition threshold implies that for $P_1^t < P < P_2^t$, AlAs/GaAs superlattices can exist in a metastable state. Similar metastable SL phases should be possible in several other two-component systems. For example, a number of $A_N B_{8-N}$ semiconductors with the rock-salt structure at 1 atm. are thought to 'undergo' the $\alpha - \beta$ transition at negative pressures.[8] With the right choice of zb substrate, it might be possible to 'overpress' the normally unstable zb phase of these materials into the 1 atm. pressure regime, thereby allowing the growth of several novel systems. Recent *ab-initio* theoretical work further supports this possibility.[4]

V. SUMMARY

A systematic Raman study of the zb(α) to high pressure (β) phase transitions in bulk-like epitaxial films of GaAs and AlAs and GaAs/AlAs superlattices has been discussed. The transition thresholds are $P_a^t = 12.4 \pm 0.4$ GPa for bulk AlAs and $P_g^t = 17.3 \pm 0.4$ GPa for bulk GaAs. In SLs, the $\alpha - \beta$ transitions may occur either separately in each constituent layer (thick layer SLs) or simultaneously in both layers (thin layer SLs). From the slope of P_1^t *vs* $1/h_a$ in the former case, an empirical value for the energy/area of the β-AlAs/α-GaAs interface is determined and found to be 0.12 ± 0.02 eV/Å^2. The forward $\alpha - \beta$ transition produces a large number of dislocations at the heterointerfaces which survive even when the pressure is decreased through transition reversal. Both the thick- and the thin-layer SLs exhibit overpressing of zb AlAs without any underpressing of zb GaAs. We suggest that SL stability is governed by the constituent with the higher $\alpha - \beta$ transition threshold *viz.*, GaAs in AlAs/GaAs SLs. This property leads to interesting possibilities for new heterostructure systems.

ACKNOWLEDGEMENTS

The authors are grateful to R.D. Burnham for the growth of several OMCVD samples and acknowledge with pleasure his collaboration in this study. Financial support for this work was provided by ONR contract No. N00014-89-J-1797, and by a grant from the Xerox Webster Research Center.

VI. REFERENCES

* Present address: Department of Physics, Oakland University, Rochester, MI 48309.

[1] See for example, Semiconductors and Semimetals, Vols. 24, 25, and 32 (Academic Press, New York); L.L. Chang, J. Vac. Sci. Technol. **B1**, 120 (1983); P. Grambow, E. Vasiliadou, T. Demel, K. Kern, D. Heitmann, and K. Ploog, Microelectronic Eng., **11**, 47 (1990).

[2] j.W. Matthews and A.E. Blakeslee, J. Cryst. Growth **27**, 118 (1974); **29**, 273 (1975); Epitaxial Growth, edited by J.W. Matthews (Academic Press, New York, 1975).

[3] J.H. Van der Merwe, J. Woltersdorf, and W.A. Jesser, Mats. Sci. Eng. **81**, 1 (1986).

[4] A.A. Mbaye, D.M. Wood, and A. Zunger, Phys. Rev. **B37**, 3008 (1988); D.M. Wood and A. Zunger, Phys. Rev. Lett. **61**, 1501 (1988); S. Froyen, S.-H. Wei, and A. Zunger, Phys. Rev. **B38**, 10124 (1988).

[5] L.J. Cui, U. D. Venkateswaran, B. A. Weinstein, and F.A. Chambers, in *The Physics of Semiconductors*, Vol. 2, (Proceedings of the 20th International Conference on the Physics of Semiconductors) edited by E. Anastassakis and J.D. Joannapoulos (World Scientific, Singapore, 1991), p. 953; L.J. Cui, U.D. Venkateswaran, B.A. Weinstein, and F.A. Chambers, Semicond. Sci. Technol.

6, 469 (1991); L.J. Cui, U.D. Venkateswaran, B.A. Weinstein, and F.A. Chambers, to be published.

[6] B.A. Weinstein, S.K. Hark, R.D. Burnham, and R.M. Martin, Phys. Rev. Lett. **58**, 781 (1987); B.A. Weinstein, S.K. Hark, and R.D. Burnham, in *The Physics of Semiconductors*, Vol. 1 (Proceedings of the 18th International Conference on the Physics of Semiconductors) edited by O. Engström (World Scientific, Singapore, 1987), p. 707.

[7] L.A. Kolodziejski, R.L. Gunshor, N. Otsuka, B.P. Gu, Y. Hefetz, and A.V. Nurmiko, Appl. Phys. Lett. **48**, 1482 (1986); W. Giriat and J.K. Furdyna in *Semiconductors and Semimetals* Vol. 25 edited by J.K. Furdyna and J. Kossut (Academic Press, New York, 1988), Chap. 1.

[8] A. Navrotsky and J.C. Phillips, Phys Rev. **B11**, 1583 (1975).

[9] L.J. Cui, U.D. Venkateswaran, B.A. Weinstein, and B.T. Jonker, to be published.

[10] U.D. Venkateswaran, L.J. Cui, B.A. Weinstein, and F.A. Chambers, Phys. Rev. **B43**, 1875 (1991).

[11] K.K. Tiong, P.M. Amirtharaj, F.A. Pollak, and D.E. Aspnes, Appl. Phys. Lett. **44**, 122 (1984); M. Holtz, R. Zallen, O. Brafman, and S. Matteson, Phys. Rev. **B37**, 4609 (1988); R. Zallen, M. Holtz, A.E. Geissberger, R.A. Sadler, W. Paul, and M-L. Théye, J. Noncrystalline Solids **114**, 795 (1989).

[12] M.Holtz, K. Syassen, and K. Ploog, Phys. Rev. **B40**, 2988 (1989).

[13] R.E. Hanneman, M.D. Banus, and H.C. Gatos, J. Phys. Chem. Solids **25**, 293 (1964).

[14] B.A. Weinstein, Semicon. Sci. Technol. **4**, 283 (1989).

[15] R.G. Dandrea, J.E. Bernard, S.-H. Wei, and A. Zunger, Phys. Rev. Lett. **64**, 36 (1990); S.-H. Wei and A. Zunger, Phys. Rev. Lett. **61**, 1505 (1988); D.M. Wood, S.-H. Wei, and A. Zunger, Phys. Rev. **B37**, 1342 (1988).

[16] R.M. Martin in *The Physics of Semiconductors*, Vol. 1 (Proceedings of the 18th International Conference on the Physics of Semiconductors) edited by O. Engström (World Scientific, Singapore, 1987), p. 639.

INFLUENCE OF PSEUDOMORPHIC CONSTRAINTS ON THE PRESSURE-RESPONSE OF SEMICONDUCTOR HETEROSTRUCTURES

L. J. Cui, U. D. Venkateswaran*, B. A. Weinstein
Physics Dept., 239 Fronczak Hall, SUNY at Buffalo, NY 14260

B.T. Jonker
Naval Research Laboratory, Washington DC 20375-5000

F.A. Chambers
Amoco Technology Co., P. O. Box 3011, Naperville IL 60566

I. INTRODUCTION

Hydrostatic pressure (P) is an important tool for the study of semiconductor heterostructures. It can tune their electronic energy bands, and as a consequence, has been used to study the band offset at heterointerfaces,[1] the possibility of tunable quantum well lasers,[2] and the indirect-gap related DX-defects.[3] The phase stability of heterostructures can be studied under high pressure more conveniently than by modifying chemical composition, and novel superpressing phenomena have been observed.[4] Phase stability is discussed elsewhere in this proceedings.[5] External hydrostatic pressure can also affect the mechanical stability of heterostructures by tuning lattice mismatch.[6] This tuning is of interest and importance for strained-layer heterostructures now finding wide device applications, since lattice-mismatch strain can strongly influence electronic properties.[7]

The present account reviews work [6,8] on hydrostatic pressure tuning of lattice-mismatch generated internal strain in two-component epitaxial films and multilayers. Explicit calculations of internal strain and its effect on the first order Raman frequencies are performed for samples grown along the [001] and [111] directions. The calculated results are then tested comprehensively by a series of pressure-Raman measurements on two heterostructure systems— –GaAs/AlAs, and ZnSe/GaAs. The contrast between these two systems is instructive since the former is composed of elastically similar materials, but the latter of materials that are elastically dissimilar. The experimental results for the ZnSe/GaAs system lead us to propose a new method for enhancing heterostructure growth, and this is also discussed.

Frontiers of High-Pressure Research, Edited by H.D. Hochheimer and
R.D. Etters, Plenum Press, New York, 1991

II. CALCULATION OF PHONON PRESSURE-SHIFT IN COHERENT EPITAXIAL LAYERS

A theory originally developed to treat the effects of applied uniaxial stress[9] can be used to calculate the phonon shift in coherent epitaxial layers under applied <u>hydrostatic</u> pressure. In this theory, one expands the effective phonon spring constants linearly in the strain components, and then solves the resulting secular equation as in degenerate perturbation theory to the first order in the strain. For zincblende-type heterostructures with film plane oriented along the [001] axis the relation between the induced phonon-shift $\Delta\omega$ and the strain components ϵ_{ij} in the standard cubic coordinate system is [9,10]

$$
\begin{aligned}
\left(\frac{\Delta\omega}{\omega_0}\right)_{LO} &= \frac{1}{2}\tilde{K}_{12}(\epsilon_{11} + \epsilon_{22}) + \frac{1}{2}\tilde{K}_{11}\epsilon_{33}, \\
\left(\frac{\Delta\omega}{\omega_0}\right)_{TO} &= \frac{1}{4}(\tilde{K}_{11} + \tilde{K}_{12})(\epsilon_{11} + \epsilon_{22}) + \frac{1}{2}\tilde{K}_{12}\epsilon_{33}.
\end{aligned}
\tag{1}
$$

For heterostructures grown with a [111]-oriented film plane the analogous expressions are,

$$
\begin{aligned}
\left(\frac{\Delta\omega}{\omega_0}\right)_{LO} &= \frac{1}{6}(\tilde{K}_{11} + 2\tilde{K}_{12} - 2\tilde{K}_{44})(\epsilon_{11}^* + \epsilon_{22}^*) + \frac{1}{6}(\tilde{K}_{11} + 2\tilde{K}_{12} + 4\tilde{K}_{44})\epsilon_{33}^*, \\
\left(\frac{\Delta\omega}{\omega_0}\right)_{TO} &= \frac{1}{6}(\tilde{K}_{11} + 2\tilde{K}_{12} + \tilde{K}_{44})(\epsilon_{11}^* + \epsilon_{22}^*) + \frac{1}{6}(\tilde{K}_{11} + 2\tilde{K}_{12} - 2\tilde{K}_{44})\epsilon_{33}^*,
\end{aligned}
\tag{2}
$$

where the ϵ_{ij}^* are rotated strain components refered to the axes $[1\bar{1}0]$, $[11\bar{2}]$ and $[111]$. Here the $\tilde{K}_{ij}$ denote first-order strain derivatives of the $LO(\Gamma)$ and $TO(\Gamma)$ spring constants in a given material. These coefficients are known for many bulk semiconductors from various uniaxial and hydrostatic stress experiments.(See table in Ref. 10.)

The strains in each layer of a heterostructure can be derived from the conditions of pseudomorphism and mechanical equilibrium. In this paper we have considered [001]- and [111]- grown epitaxial films and two-component superlattices only, but the calculation can be generalized to any growth direction or number of constituents.

II-a) [001]-oriented epitaxial systems

In a two-component superlattice (SL) system the strain tensor (refered to cubic coordinates) can be written for each SL constituent labeled by the superscripts 'a' or 'b' as,

$$
\epsilon_{ij}^{a,b} = M_{ij}^{a,b} + \Pi_{ij}^{a,b}
$$

Here the $M_{ij}^{a,b}$ are the initial as-grown misfit strains, and the $\Pi_{ij}^{a,b}$ are induced strains proportional to P. Since hydrostatic pressure preserves symmetry, we have $\epsilon_{ij}^{a,b} = 0$ for $i \neq j$, and $\epsilon_{11}^{a,b} = \epsilon_{22}^{a,b} \neq \epsilon_{33}^{a,b}$. For [001]-growth the strain tensor is biaxial, and it remains so at all pressures.

If $f_0 \equiv \frac{a_0^a - a_0^b}{\bar{a}_0}$ is the customary as-grown lattice misfit at P=0, the constraints of pseudomorphism at all pressures require

$$M^b_{11} - M^a_{11} = f_0; \quad \Pi^b_{11} = \Pi^a_{11}, \tag{3}$$

and mechanical equilibrium gives

$$(\partial U/\partial M^{a,b}_{ij})_{\Pi^{a,b}_{ij}=0} = 0; \quad (\partial U/\partial \Pi^{a,b}_{ij})_{M^{a,b}_{ij}} = -P. \tag{4}$$

Here U is the volume energy density of the entire SL, given by

$$U = (h^a u^a + h^b u^b)/(h^a + h^b), \tag{5}$$

where h^a and h^b are the SL layer thicknesses, and u^a and u^b are the elastic energy densities of each material, given in terms of the elastic compliance constants C_{ij}, by the standard expressions.[11]

The theory of pseudomorphic growth [12] requires the a and b layers to come to mechanical equilibrium individually along the growth direction, but jointly along the directions in the SL plane. Equations (3) and (4) give for the initial misfit strains M_{ij} the well known results, [12]

$$\begin{aligned}
M^a_{11} &= M^a_{22} = -\frac{h^b G^b}{h^a G^a + h^b G^b} f_0, \\
M^b_{11} &= M^b_{22} = \frac{h^a G^a}{h^a G^a + h^b G^b} f_0, \\
M^{a,b}_{33} &= -\Gamma^{a,b} M^{a,b}_{11},
\end{aligned} \tag{6}$$

with

$$\Gamma^{a,b} = 2C^{a,b}_{12}/C^{a,b}_{11}; \quad G^{a,b} = \left(C^{a,b}_{11} + C^{a,b}_{12} - \Gamma^{a,b} C^{a,b}_{12}\right).$$

The pressure-induced strain components are found to be

$$\begin{aligned}
\Pi^a_{11} &= \Pi^b_{11} = -P\frac{h^a D^a + h^b D^b}{h^a G^a + h^b G^b}, \\
\Pi^{a,b}_{33} &= -\frac{P}{C^{a,b}_{11}} - \Gamma^{a,b}\Pi^{a,b}_{11},
\end{aligned} \tag{7}$$

with $D^{a,b} = (C^{a,b}_{11} - C^{a,b}_{12})/C^{a,b}_{11}$.

II-b) [111]-oriented epitaxial systems

In this case the internal strain is again biaxial due to symmetry. To solve the problem, the plane-parallel ($\epsilon^*_{11} = \epsilon^*_{22}$) and plane-normal ($\epsilon^*_{33}$) strain components must be rotated into the standard cubic coordinate system, and after rotation the strain tensor acquires non-zero off-diagonal elements. Hence the elastic compliance constant C_{44} is required, and the energies $u^{a,b}$ have an extra term. Nevertheless, the constraints of pseudomorphism and mechanical equilibrium are the same as for the [001]-growth case. We then find, after rotating back to the starred coordinates, that the final expressions for the strain components M^*_{ij} and Π^*_{ij} are the same as Eqs. (6) and (7), except that the elastic compliances C_{11} and C_{12} are replaced by new constants C^*_{11} and C^*_{12} defined by,

$$C_{11}^* = (C_{11} + 2C_{12} + 4C_{44})/3, \quad C_{12}^* = (C_{11} + 2C_{12} - 2C_{44})/3. \tag{8}$$

This leads to $\Gamma^* = 2(C_{11} + 2C_{12} - 2C_{44})/(C_{11} + 2C_{12} + 4C_{44})$. Equations (6), (7) and (8) generalize results previously obtained for the P=0 case.[10]

II-c) Bulk and epilayer limits

Now, let us consider two limiting cases. If $h^b/h^a \to \infty$ the strain components reduce to $\epsilon_{11}^b(\epsilon_{11}^{*b}) = \epsilon_{33}^b(\epsilon_{33}^{*b}) = -\frac{P}{3B_0^b}$ for both growth directions, where $B_0^b = \frac{1}{3}(C_{11}^b + 2C_{12}^b)$ is the standard bulk modulus of material 'b'. This limit corresponds to material 'b' becoming bulk-like. On the other hand, if $h^b/h^a \to 0$, the strain components in material 'b' become,

$$\epsilon_{11}^b = f_0 - \frac{1}{3}\frac{P}{B_0^a},$$

$$\epsilon_{33}^b = -\Gamma\{f_0 + \frac{P}{3}(\frac{1}{B_0^b} - \frac{1}{B_0^a})\} - \frac{P}{3B_0^b}, \tag{9}$$

for [001]-orientation, and, with the appropriate change Γ to Γ^*, the same expressions apply for ϵ_{11}^{*b} and ϵ_{33}^{*b} in the [111]-orientation case. Here, again, B_0^b and B_0^a represent the bulk moduli of each material. To first order, the Eq. (9) expression in curly brackets is the pressure dependent misfit f(P) for an epilayer of material 'b' on a substrate of material 'a' in either growth orientation. If the misfit should vanish under pressure, the epilayer will be perfectly matched to the substrate, and the pressure P_m satisfing $f(P_m)=0$ is defined as the matching pressure. This will be important to our later discussion.

Once the strain components are known, the pressure-shift of the zone-center optical phonons in a heterostructure can be obtained straight forwardly from Eqs.(1) and (2) for [001]- and [111]-growth directions, respectively. For the two limiting cases discussed above, we have calculated the P-shifts of the first order Raman modes allowed in backscattering. The results for some common epilayer/substrate combinations are shown in Table I. In this table the nomenclature has been changed from 'a', 'b' to the more descriptive 'e' (epilayer) 's' (substrate), and bulk in order to clarify the meaning of the various columns. The quoted pressure-coefficients always apply to the phonons of material 'e'. In the second, third and fourth columns, $(\partial\omega/\partial P)^e$ indicates the P-shift for an epilayer of material 'e' on a substrate of material 's'; in the fifth and sixth columns $(\partial\omega/\partial P)^{bulk}$ indicates the P-shift for the <u>same</u> material (i.e., the 'e'-material) in its <u>bulk</u> form. The last column gives the calculated pressures for lattice matching between the epilayer and substrate materials. In the numerical calculation, the values of the C's are taken from Ref. [13].

According to the above, the calculated difference between the phonon P-coefficients of the epilayer and bulk forms of a given material depends on the difference in elastic constants between this material and its substrate. Additionally, there is a weak dependence on the orientation. For the GaAs/AlAs system, the two constituents have such similar elastic properties (with bulk moduli 755 kbar and 781 kbar for GaAs and AlAs, respectively [13]), that an externally applied pressure does not generate significant internal biaxial strain. Indeed, according to Table I, the P-

TABLE I. Calculated linear pressure coefficients for the backscattering-allowed optical phonon frequencies in some common epilayer(e)/substrate(s) systems. Columns 2-4 pertain to the 'e'-material coherently bonded to the given substrate for [001]-growth and [111]-growth; columns 5 and 6 are for the same material in bulk form. All pressure coefficients have the units $cm^{-1}/kbar$. The last column gives the calculated lattice-matching pressure P_m in kbar for each e/s system; asterisks mark the possible candidates for mechanically buffered growth.

| Samples | [001]-growth | [111]-growth | | bulk form of 'e' | | |
e/s	$(\frac{\partial\omega}{\partial P})^e_{LO}$	$(\frac{\partial\omega}{\partial P})^e_{LO}$	$(\frac{\partial\omega}{\partial P})^e_{TO}$	$(\frac{\partial\omega}{\partial P})^{bulk}_{LO}$	$(\frac{\partial\omega}{\partial P})^{bulk}_{TO}$	P_m
GaAs/AlAs	0.469	0.471	0.483	0.477	0.493	−82.7
ZnSe/GaAs*	0.330	0.351	0.426	0.378	0.482	22.8
GaAs/Si	0.424	0.436	0.424	0.477	0.493	399.2
Ge/Si	0.396	0.416	0.379	0.447	0.447	397.9
InAs/Si	0.351	0.384	0.338	0.416	0.431	538.9
GaP/Si*	0.465	0.474	0.369	0.491	0.391	102.9
InAs/GaAs	0.383	0.400	0.384	0.416	0.431	676.9
GaP/GaAs	0.540	0.522	0.433	0.491	0.391	556.4
GaAs/GaP	0.442	0.450	0.448	0.477	0.493	556.4
InAs/InP	0.389	0.402	0.393	0.416	0.431	380.7
AlSb/InAs	0.563	0.567	0.457	0.574	0.467	668.1
AlSb/GaSb	0.590	0.585	0.482	0.574	0.467	−221.1
InAs/GaSb	0.431	0.423	0.452	0.416	0.431	128.4
InAs/ZnTe*	0.451	0.433	0.481	0.416	0.431	67.0
GaSb/ZnTe*	0.525	0.515	0.566	0.503	0.530	20.2

coefficients of the GaAs LO(Γ) frequency in an epilayer of GaAs on AlAs and in bulk GaAs differ by only 1.7% (for [001]-growth).

In contrast, for the ZnSe/GaAs system the bulk modulus of ZnSe (595 kbar) is significantly smaller than that of GaAs. As a result the corresponding difference between the P-coefficients of the bulk and epilayer forms of ZnSe is 13.6%. This corresponds to a significant frequency difference of 5 cm^{-1} in 100 kbar, which is easily detectable by Raman scattering. For other materials systems the calculated differences between bulk and epilayer pressure-coefficients range from 1%–17%. (See Table I.)

III. EXPERIMENTAL RESULTS FOR GaAs/AlAs SLs AND EPITAXIAL ZnSe FILMS ON GaAs

III-a) Techniques

GaAs/AlAs SLs, bulk-like GaAs and AlAs films, ZnSe epilayers, and melt-grown bulk ZnSe are studied by high pressure Raman scattering in this work. Hydrostatic pressures are generated using a ruby-calibrated diamond-anvil cell (DAC) with 4:1 methanol:ethanol as the pressure medium. Except for the melt-grown ZnSe sample, all the specimens have been grown by molecular beam epitaxy (MBE) on [001]-oriented GaAs substrates. The configuration and layer-dimensions of each sample are given below in the keys of the various figures in which the samples are discussed. All the GaAs/AlAs SLs except one have been prepared by removing their GaAs substrates using selective etching[14] prior to measurement. The single epilayer specimens have been prepared for DAC loading by thinning their substrates to $\sim$30μm using mechanical and chemical polishing; it is important to achieve the <u>final</u> thickness using chemical polishing because mechanical methods can introduce dislocations that destroy the pseudomorphism, especially for strained ZnSe epilayers. In our ZnSe experiments, a chip of bulk (i.e., melt-grown) ZnSe is always loaded into the DAC alongside of each ZnSe epilayer specimen in order to improve the accuracy of comparison to bulk behavior.

The Raman measurements are performed at room temperature in the backscattering geometry using a double monochromator with microscope foreoptics[14] and an intensified Si-diode array detector. Kr^{+} and Ar^{+} laser lines are used as Raman excitation sources. The laser power is kept below 30mW, which is less than any observed laser heating levels.

III-b) The GaAs/AlAs system

To test the predicted minimal effect of pseudomorphic constraints in the GaAs/AlAs system, we have recorded over 300 measurements at various pressures on six different SLs, as well as on bulk-like GaAs and AlAs films; the layer dimensions range from a few μm to 20Å.

The full compilations of this data are shown in Figs. 1a) and 1b) for the first-order Raman phonons of GaAs and AlAs, respectively. The solid curves are quadratic least-square fits to the data. The presence of the normally forbidden TO(Γ) scattering is probably due to allowed <u>forward</u> scattering of reflected laser light from the film's back surface.[8] The experimental uncertainty of the plotted points for all modes, except the TO(Γ) of AlAs, is $\sim \pm 0.6$ cm^{-1} (about half the width of the symbol size). This overall uncertainty is due to pressure measurement errors, pressure gradients above 100 kbar, and occasional sample bending (the latter especially for free-standing SLs) caused by contact with the gasket or ruby flakes. The TO(Γ) peak of AlAs is observed only in three of the SL samples, and the data for this mode is a factor of two less certain because of weaker scattering intensity.

As can be seen in Fig. 1, for each mode the points from different samples fall along the same curve within the statistical accuracy of our measurements. The minimal deviation from this behavior for the AlAs TO(Γ) mode is attributed to the greater uncertainty in its frequency and is not regarded as significant; otherwise the AlAs LO(Γ) would have shown similar deviations. Hence, in the GaAs/AlAs system, any layer-thickness dependent trends that might reflect the effect of pseudomorphic constraints on phonon pressure response, as predicted above, are too small to be measured.

III-c) Experimental results for ZnSe epilayers on GaAs.

Two epilayers with very different ZnSe layer-thicknesses are studied here. The ZnSe thickness is 775Å for one epilayer– – below the <u>measured</u> critical thickness h$_c$ $\sim$1500Å;[15] in contrast, the ZnSe layer is 2μm thick, about ten times greater than h$_c$, for the other epilayer sample. X-ray and low temperature reflectivity measurements[16] are used to check for pseudomorphism in each specimen, and the data show as expected that only the thinner ZnSe epilayer is pseudomorphic.

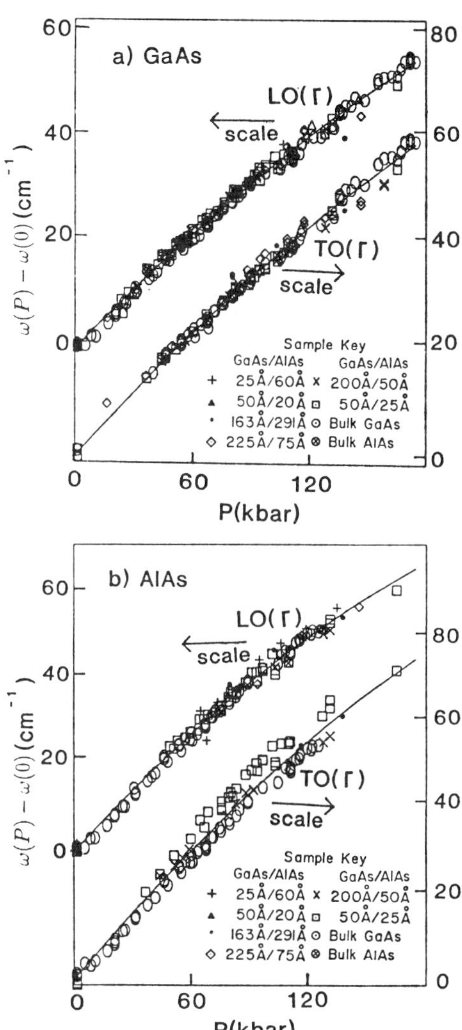

Figure 1. Pressure-shifts of the GaAs a) and AlAs b) long-wavelength optical phonons in eight different SL and bulk samples. Within experimental error, there are no layer-thickness dependent trends in the pressure-shifts.

A comparison between the LO(Γ) Raman spectra of the 775Å epilayer and the co-loaded melt-grown (bulk) ZnSe specimen is shown in Fig.2. At 1 bar the epilayer LO(Γ) frequency is higher than in bulk ZnSe, indicating that the 775Å epilayer is biaxially compressed by its GaAs substrate. Under applied pressure the LO(Γ) peaks of both specimens shift to higher frequency in the normal fashion. However, the shift in the epilayer is smaller than in the bulk specimen, so that the frequency difference between the two peaks reverses in sign at $\sim$21 kbar (see traces (b) and (c)). This reversal indicates that the biaxial strain in the 775Å epilayer changes from compressive to tensile at $\sim$21 kbar.

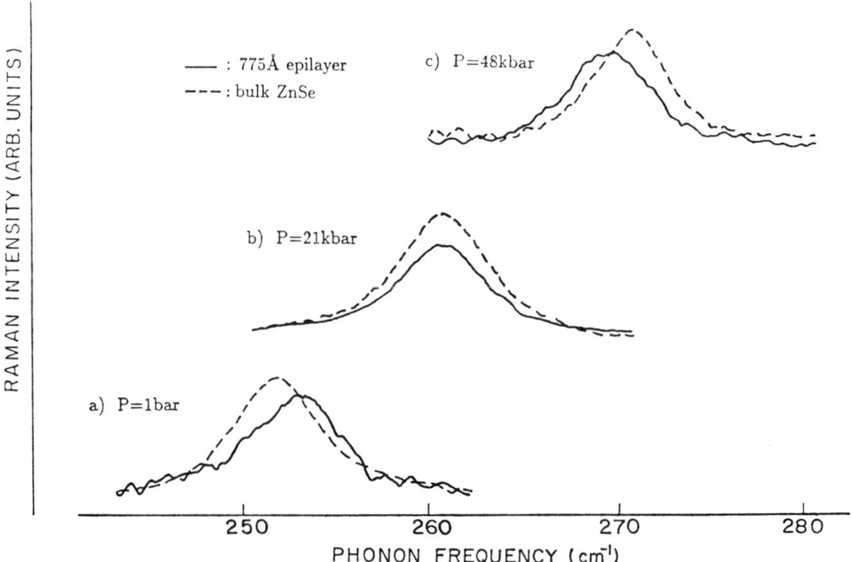

Figure 2. Comparison between the LO(Γ) peaks of the 775Å ZnSe epilayer (solid spectra) and that of bulk (i.e., melt-grown) ZnSe (dashed spectra) at three different pressures. The relative shift between the two peaks reverses at 21±3 kbar.

In Fig. 3 we plot, for each epilayer sample, the difference ($\Delta\omega^{e-b}$) between the epilayer LO(Γ) frequency and that of bulk (i.e., melt-grown) ZnSe as a function of pressure. Here, the solid line is <u>not</u> a linear least-square fit, but is the dependence <u>calculated</u> according to the theory outlined in section II. (The horizontal dashed line merely marks the zero position.)

This plot clearly shows that the pressure response of phonons in the pseudomorphic 775Å epilayer is different from that found in bulk ZnSe. The LO(Γ) frequency difference becomes zero at 21±3 kbar, which is in good agreement with the predicted matching pressure, P_m=22.8kbar, in the ZnSe/GaAs system (see Table I). At this pressure the substrate-generated biaxial strain vanishes, and the epilayer's elastic response should be the same as for bulk ZnSe. For pressures above P_m, the 775Å epilayer data continue to fall along the same line up to 60kbar––indicating that pseudomorphism is preserved even when the (now) tensile strain becomes almost twice as large as the initial compressive strain.

In contrast, (see triangles in Fig. 3) the LO(Γ) pressure shift for the non-pseudomorphic 2μm ZnSe film exhibits no significant difference from that of bulk ZnSe. Hence, pressure-Raman measurments can easily distinguish the presence of pseudomorphic from non-pseudomorphic interfaces in the ZnSe/GaAs system, and the use of this technique for interface characterization in other strained-layer systems should be pursued.

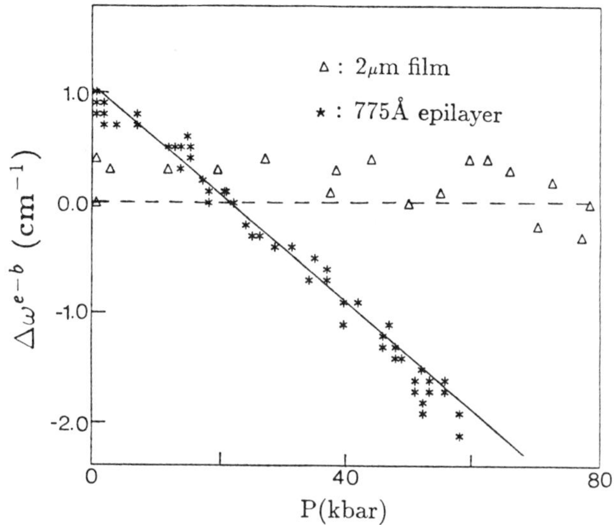

Figure 3. LO(Γ) frequency difference $\Delta\omega^{e-b}$ for each ZnSe epilayer with respect to a co-loaded bulk (melt-grown) ZnSe specimen. For the 775Å epilayer, this difference obeys the calculated response (solid line). For the 2μm film, $\Delta\omega^{e-b}$ is pressure independent within the experimental error. Dashed line marks the abscissa.

IV. DISCUSSION AND CONCLUSIONS

It has been demonstrated that, because the GaAs/AlAs system is composed of elastically similar materials, the effects of pseudomorphic constraints are too small to significantly alter the phonon pressure response from that measured in its bulk constituents. However, we find that in the ZnSe/GaAs system—— composed of elastically dissimilar materials—— the reverse is true. Furthermore, in the ZnSe/GaAs case it is quite apparent from Fig.3 that the lattice mismatch can be tuned by appreciable amounts using applied hydrostatic pressure. It is of interest to ask, what is the critical misfit $f_c(P)$ for which dislocations are expected to form in the 775Å epilayer studied here?

At 1 bar, where $|f_0|=0.27\%$, the empirical critical thickness for a ZnSe epilayer on GaAs is ~1500Å.[15] We can estimate the critical misfit by simply scaling f_0 according to the models of either Matthews and Blakeslee,[17] or People and Bean;

277

[18][19] this procedure gives f_c=0.46% and f_c=0.35%, reached at 61kbar and 52 kbar, respectively. Accordingly, one expects that the magnitude of $\Delta\omega^{e-b}$ in Fig.3 will begin to decrease as strain relieving interface dislocations are generated at the highest pressures attained in our experiments. Our data do not show this tendency at all; in fact, if anything, $\Delta\omega^{e-b}$ exhibits a slight superlinear behavior in the opposite direction. Hence, we find that the 775Å ZnSe epilayer can be compressed into a <u>mechanically metastable</u> state. This result can be attributed to kinetic effects.[20] The thermal barriers to formation and motion of interface dislocations, created during layer-by-layer growth at 320°C,[16] are apparently too high to be surmounted at room temperature.

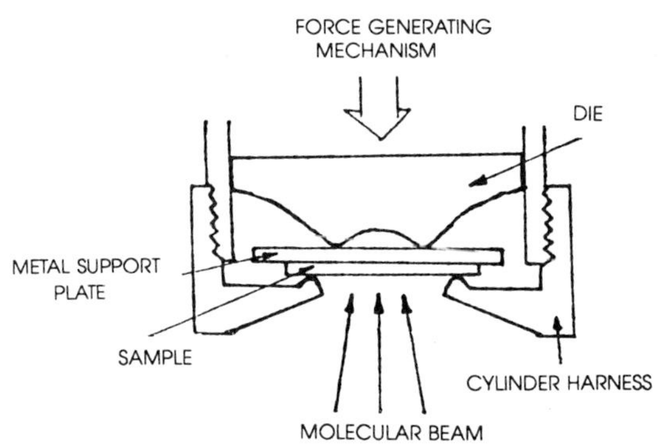

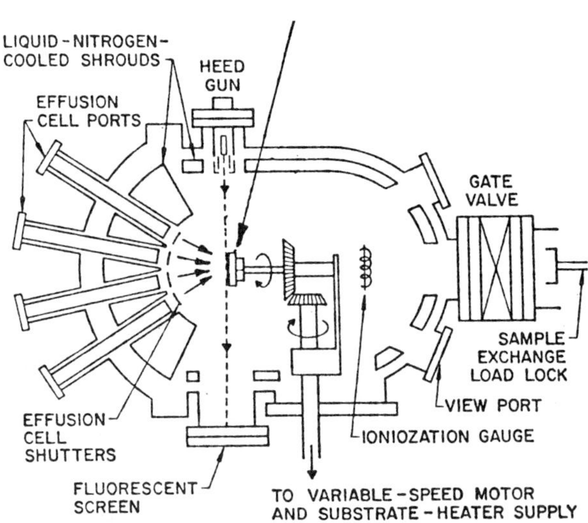

Figure 4. Schematic of a possible MBE arrangement with substrate placed in a bending harness for "mechanically buffered" growth. The bending harness is enlarged inside the top frame. It is used to compensate the ambient misfit by generating a biaxial strain of the opposite sign.

The above results suggest an interesting method for enhancing the epitaxial growth of mismatched heterostructures. By applying an external biaxial strain to the substrate during MBE deposition, it should be possible to reduce or, in favorable cases, even eliminate lattice-mismatch with the material being deposited. After the growth is complete, the sample can be "quenched" to room temperature before releasing the strain, thereby exploiting any frozen-in barriers in order to hinder formation and motion of dislocations. By such a "mechanical buffering" technique, it should be possible to increase the effective critical thickness for the strained-layer system being grown. Fig. 4 shows a possible bending harness appropriate for such an application. A metal plate is used to support the substrate so that the bending force is applied first to the metal plate and then distributed uniformly over the substrate. A major practical concern is, of course, to avoid fracturing the semiconductor substrate. One may gauge the fracture limit from reports of behavior under applied uniaxial stress. For example, GaAs often fractures under a [001] uniaxial compression of 12 kbar, corresponding to a biaxial strain in the plane normal to [001] of 0.44%.[21] Since this far exceeds the ambient mismatch of 0.27% in ZnSe/GaAs, the proposed mechanical buffering method is expected to be useful for growth of this materials system.

For other heterostructures the strain needed to create lattice matching can be sufficient to bring about phase transitions in one or both of the constituents. Among the 15 epilayer/substrate combinations listed in Table I, the four combinations marked with an asterisk have the possibility of attaining complete matching before a phase transition occurs. Heterostructures composed of these materials or their alloys are, therefore, good candidates for 'mechanically buffered' growth, and further investigation of them is warranted to verify that they exhibit the expected strain behavior.

ACKNOWLEDGEMENTS

Financial support for this work was provided by ONR contract No. N00014-89-J-1797.

V. REFERENCES

* Present address: Physics Dept., Oakland Univ., Rochester, MI 48309.

[1]   U. Venkateswaran, M. Chandrasekhar, H.R. Chandrasekhar, B.A. Vojak, F.A. Chambers, and J.M. Meese, Phys. Rev. **B33**, 8416 (1986).

[2]   S.W. Kirchoefer, N. Holonyak, Jr., K. Hess, K. Meehan, D.A. Gulino, H.G. Drickamer, J.J. Coleman, and P.D. Dapkus, J. Appl. Phys. **53**, 6037 (1982).

[3]   M.F. Li, P.Y. Yu, E.R. Weber, and W. Hansen, Phys. Rev. **B36**, 4531 (1987).

[4]   B.A. Weinstein, S.K. Hark, R.D. Burnham, and R.M. Martin, Phys. Rev. Lett. **58** 781 (1986); L.J. Cui, U.D. Venkateswaran, B.A. Weinstein, and F. A. Chambers, to be published.

[5]   B.A. Weinstein, L.J. Cui, U.D. Venkateswaran, and F.A. Chambers, "Enhanced Stability of Heterostructures Under Pressure", in the present proceedings.

[6]   L.J. Cui, U.D. Venkateswaran, B.A. Weinstein, and B.T. Jonker, to be published.

[7]   See for example, A.R. Adams, "Hydrostatic Pressure Investigations of Quantum Well Optoelectronic Devices", in the present proceedings.

[8] L.J. Cui, U.D. Venkateswaran, B.A. Weinstein, and F. A. Chambers, Semicond. Sci. and Technol. **6**, 469 (1991).

[9] S. Ganesan, A.A. Maradudin, and J. Oitmaa, J. Ann. Phys. (N.Y.) **56**, 556 (1970); E.M. Anastassakis, A. Pinczuk, E. Burstein, F.H. Pollak, and M. Cardona, Solid State Commun. **8**, 133 (1970).

[10] B. Jusserand and M. Cardona, in *Light Scattering in Solids V,* Vol. **66** of *Topics in Applied Physics*, edited by M. Cardona and G. Güntherodt (Springer, New York, 1989) pp. 124–128, and 145.

[11] See for example, J.F. Nye, *Physical properties of Crystals*, (Oxford University Press, 1987).

[12] See for example, E. P. O'Reily, Semicond. Sci. Technol. **4** 121 (1989) and references therein.

[13] The C_{ij} values for all materials, except InAs, are taken from: S.S. Mitra and N.E. Massa, in *Handbook on Semiconductors*, vol. 1, edited by T.S. Moss (North Holland, New York, 1982) p.96. The values for InAs are from: R. Reifenberger, M.J. Kerk, and J. Trivisonno, J. Appl. Phys. **40** 5403 (1969).

[14] L.J. Cui, Ph.D. Thesis, Physics Dept. SUNY at Buffalo, unpublished.

[15] H. Mitsuhashi, I. Mitsuishi, M. Mizuta, and A. Kukimoto, Jap. J. Appl. Phys. **24**, L578 (1985); J. Kleiman, R.M. Park, and S.B. Qadri, J. Appl. Phys. **61**, 2067 (1987).

[16] B.T. Jonker, J.J. Krebs, S.B. Qadri, and G.A. Prinz, Appl. Phys. Lett. **50**, 848 (1987); W.C. Chou, X. Liu, and A. Petrou, private communication.

[17] J.W. Matthews and A.E. Blakeslee, J. Cryst. Growth **27**, 118 (1974); J.W. Matthews, J. Vac. Sci. Tech. **12**, 126 (1975).

[18] R. People and J.C. Bean, Appl. Phys. Lett. **47**, 322 (1985).

[19] In the force-balance model of Ref. 17, f_c varies essentially as $(\frac{1}{h})ln(\frac{h}{b})$, where h is the layer thickness and b the Burger's displacement; in the local energy-balance picture of Ref. 18, the *square* of f_c is proportional to this expression. The non-equilibrium model of Ref. 18 is probably more appropriate here because, unlike the growth process at 320°C, the room temperature condition of our experiments should hinder the formation and motion of dislocations.

[20] R. People and S.A. Jackson in *Strained Layer Superlattices: Physics,* edited by T.P. Pearsall, Semiconductors and Semimetals, Vol. 32 (Academic Press, New York, 1990) Chap. 4.

[21] B.A. Weinstein and M. Cardona, Phys. Rev. **B5**, 3120 (1972).

HYDROSTATIC PRESSURE INVESTIGATIONS OF

QUANTUM-WELL OPTOELECTRONIC DEVICES

A.R.Adams

Department of Physics
University of Surrey
Guildford, GU2 5XH, UK

Quantum wells and superlattices offer exciting new possibilities for engineering the electronic band structure of semiconductor optoelectronic devices. Hydrostatic pressure also causes changes in the band structure which in many respects mimic the effects of quantum confinement and hence allows a systematic study to be made by varying the pressure on a single device. This avoids unwanted chemical changes that can occur when the band structure is varied in a growth sequence and thus allows parameters dependent only on the band structure to be clearly identified.

Having established potential beneficial and deleterious properties of the band structure using hydrostatic pressure or uniaxial stress, it is now possible to achieve improved properties by building strain permanently into quantum-well and superlattice structures. Using growth by MBE or MOVPE, layers can be grown deliberately mismatched to the substrate crystal e.g. InGaAs grown on GaAs is subjected to biaxial compression and, providing the layer does not exceed a certain critical thickness, it is structurally stable.

Here we will review some investigations of semiconductor laser emitters and avalanche photo-detectors using hydrostatic pressure and some resulting proposals and experiments on strained layer devices.

1) LASERS

a) The Density of States

Semiconductor diode lasers are very efficient because under forward bias the energising current pumps equal numbers of electrons and holes directly into the active region. Thus electrical energy is turned into light with the minimum of intermediate stages. To reach effective population inversion, ie when stimulated emission is as likely as absorption and the semiconductor appears transparent, it is necessary to separate the quasi Fermi levels F_c and F_v by more than the band gap energy. For this to happen at a low injected carrier density and low current, one would require the density of states to be as low as possible. However, for lasing action to occur one further requires the gain in the device to overcome the losses; including the desired emission of photons from the laser. For high gain and hence high emission, we require that there is a high density of interacting electrons and holes and hence the density of states should be as high as possible. These two conflicting requirements can be optimized by attempting to concentrate the densities of states in the conduction and valence bands at that energy where the lasing transitions occur.

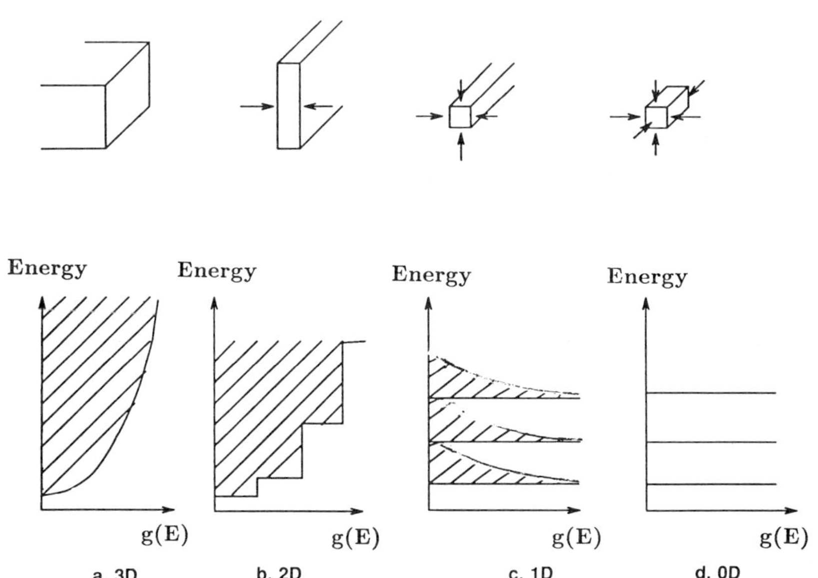

Figure 1.   The energy dependence of the density of allowed states for carriers whose freedom of movement becomes progressively more restricted by size effects.

The density of states (DOS) distribution in the conduction band for a bulk semiconductor laser is shown in Fig 1a. For carriers free to move in all three dimensions (3D) the DOS goes to zero at the band edges and in order to obtain sufficient gain for lasing action the quasi Fermi level has to be pumped high into the band. The electrons, and holes, are therefore spread over many kT in energy and can recombine non-radiatively or by spontaneous emission at energies other than that of the lasing mode. However, if the active region of the laser is a quantum well formed by the growth of a lower bandgap layer less than about 100Å thickness, the electrons in the well are only able to move in the plane of the layer (2D). Consequently the DOS is constant, increasing in a step wise manner as each new quantized level is reached as shown in Fig 1b. The distribution of injected electrons and holes ideally now takes the form illustrated by the shaded areas shown in Fig 2. The maximum electron and hole combined density of states now occurs at the band edges, although, as is illustrated in Fig 2a, since the DOS is proportional to the effective mass, m*, if m*$_e$ for electrons is much less than m*$_h$ of holes, electrons have to be pumped quite high into the conduction band before enough holes are present in the valence band for the lasing condition to be reached. Figure 1c shows the density of states distribution for carriers confined to move in one dimension (1D) and, as can be seen, in such a quantum wire the situation is even more advantageous, while in a quantum box, as illustrated in Fig 1d, the ideal density of states is achieved as in an isolated atom. However, 1D and 0D structures require extremely fine lithography and are difficult to incorporate into a diode structure and so far no improved performance has been reported in quantum-wire or quantum-box devices.

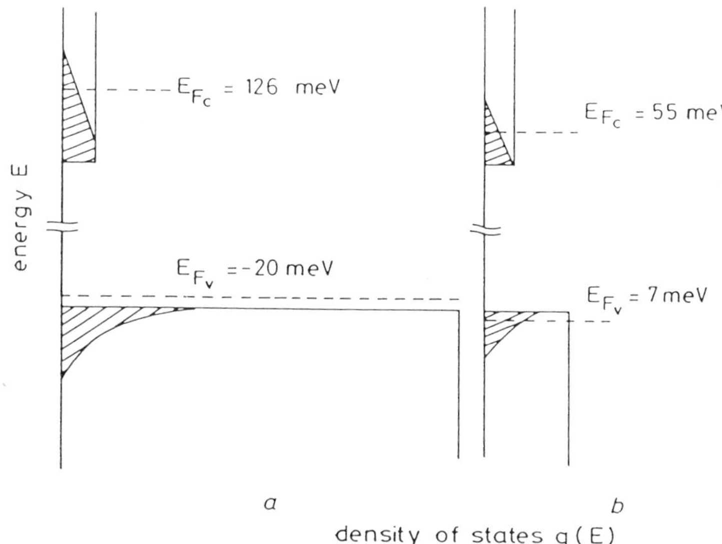

Figure 2. The density of states for a' an unstrained quantum well, b' a well in compressive strain. Also shown are the quasi Fermi levels at lasing threshold. The shaded areas indicate the distribution of electrons and holes.

<u>b) Long-wavelength Quantum-well Lasers</u>

Quantum-well lasers consisting of InGaAs wells between InGaAsP barriers have been grown by MOVPE lattice matched to InP. Of particular interest are devices operating near 1.55 μm wavelength where optical fibres have their minimum loss. Although these have improved performance over bulk devices, they suffer from the fact that their threshold current and quantum efficiency above threshold are very temperature sensitive. This is not the case in the shorter wavelength GaAs/AlGaAs quantum well lasers, and it appeared likely that the 1.55 μm devices suffer from intervalence band absorption[1] (IVBA) and Auger recombination[2] (AR) which are loss mechanisms which become worse as the operating wavelength or the temperature increase. If this is the case, the lasers should improve as they are subjected to hydrostatic pressure since the direct band gap increases at about 10 meVkb^{-1} and therefore the wavelength shortens. Such investigations have already been carried out successfully on 1.55μm bulk devices[3].

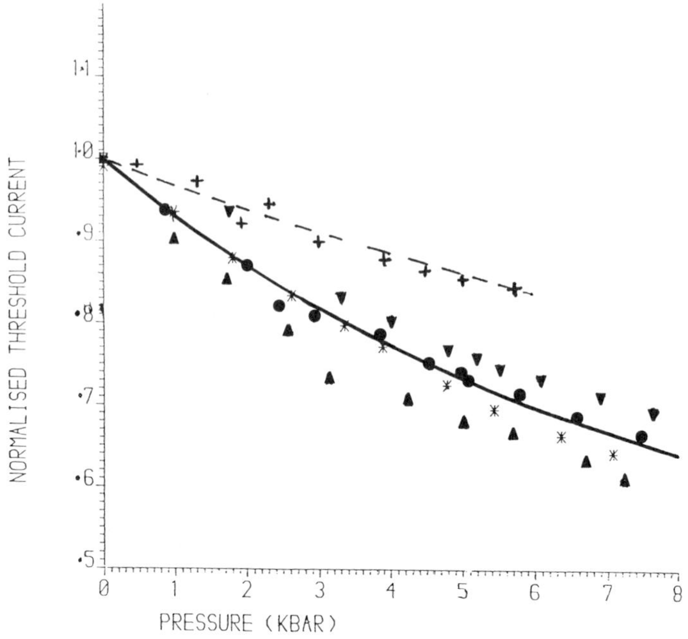

Figure 3.   The dependence on hydrostatic pressure of the threshold current of 1.55 µm InGaAs quantum well lasers (▲ ● ▼) with InGaAsP barriers, (*) with InGaAlAs barriers. The solid curve is a theoretical fit. Also shown are the results for an InGaAs well with 1.8% compressive stain (+).

The variation of threshold current with pressure is shown normalised in Figure 3 for a variety of quantum well devices. The well width was 100Å and the barrier was either InGaAsP or InGaAlAs(4). As can be seen, the threshold current decreases by more than 30% in 8kb and there is little difference between the different growths,indicating that the effect is intrinsic. As illustrated in Fig 4a, in IVBA an emitted photon is reabsorbed by lifting an electron from the spin-split valence band into a hole injected into the heavy hole band. However, at high pressure this process is reduced, as indicated by the dotted lines in Fig 4a. Since the photon energy is larger, its reabsorption occurs at a larger k (momentum) value where there are fewer holes available and therefore the probability is reduced. The process of AR, illustrated in Fig 4c, is also reduced. Here the energy and momentum of the recombining electron-hole pair are imparted to a third particle. This is an electron in Fig 4c but could be a hole. As the energy which must be dissipated increases with increasing pressure so the process becomes less likely. The full curve through the points in Fig 3 was calculated assuming that at zero pressure the IVBA and AR coefficients are 40cm^{-1} for 10^{-18} holes and 1.38 x 10^{-29}cm^{6}s^{-1} respectively[4]. These values do not seem to have been appreciably improved by comparison with those in bulk devices.

Fig 5 shows the measured variation of the quantum differential efficiency η_d with pressure[5]. As can be seen, η_d increases quite remarkably also indicating that IVBA, which degrades η_d, is indeed being removed by the application of pressure.

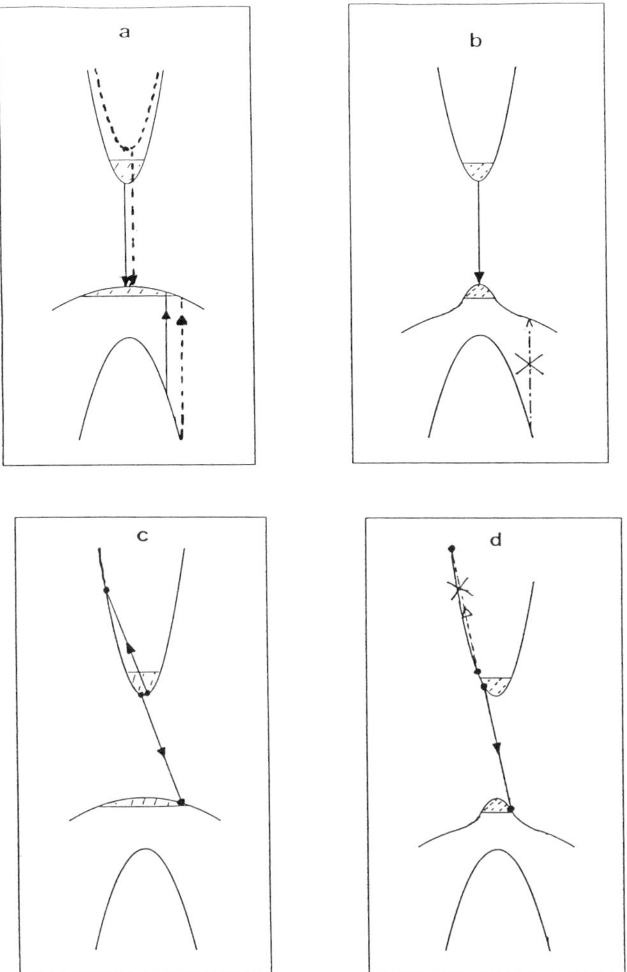

Figure 4.   Illustrates IVBA (a) and AR (c) in a quantum well laser. Their reduction or effective elimination are shown in (b) and (d) respectively. The dotted lines in (a) indicate the situation under hydrostatic pressure.

c) Strained-layer Quantum-well Lasers

Having identified the loss mechanism degrading 1.55 μm MQW lasers using pressure, it is important to consider if the devices can be improved by introducing permanent inbuilt strain. Figure 4 illustrates how this can be done[6].

If a layer is grown in a state of compressive strain (eg $In_xGa_{1-x}As$ with x>0.53 grown on InP substrates) then the valence band which is light-hole-like in the plane of the well is raised in energy with respect to the one which is heavy-hole-like in the plane. As a result, the top of the valence band can have a light-hole mass over several kT in energy[7]. This lowers the density of states and population inversion and the lasing threshold can be achieved at a low injected carrier density as discussed in section 1a. This is illustrated in both Figures 2 and 4. Fig 2 shows the distribution of electrons and holes and the energies,

E, of the quasi Fermi levels F_c and F_v. As can be seen, in the strained case E_{FV} is below the top of the valence band at lasing threshold although the total density of holes represented by the shaded area, is decreased[8]. The shaded areas in Fig 4 are simply meant to compare the spread in energy and momentum of electrons and holes attainable in the unstrained and strained cases. As can be seen, because of the light-hole cap, holes in a strained laser are much more restricted in the values of momentum they can have for the same spread in energy. But, as is also illustrated, the processes of IVBA and AR require holes with large momentum values if energy and momentum are to be conserved. Therefore, the shape of the valence band structure further reduces AR and IVBA in strained-layer lasers[6].

Lasers with 1.8% compressive strain and operating at 1.55 μm have been grown by MOVPE and investigated for their improved performance. It is found that they have a reduced threshold current, an improved quantum efficiency and are also less temperature sensitive (9). The variation with pressure of the quantum differential efficiency η_d is shown in Figure 5 (5). As can be seen η_d changes little compared with observations on similar unstrained devices. The fact that no improvement in η_d is observed as pressure

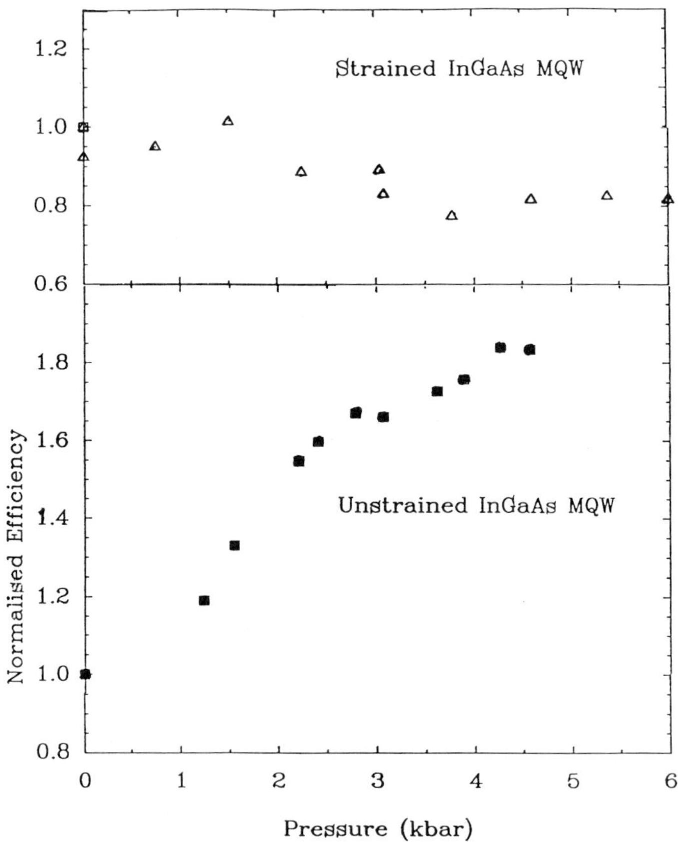

Figure 5. The variation of quantum efficiency with pressure.

is applied can be interpreted by assuming that the loss mechanism of IVBA has already been effectively removed by the inbuilt strain as described above. The variation in threshold current, I_{th}, with pressure for the strained layer device is shown in Figure 3[5]. The decrease in I_{th} is clearly less than in comparable unstrained devices. If we assume that IVBA is no longer occurring as discussed, then the decrease in I_{th} may be due to the decrease with pressure of what remains of Auger recombination. These high pressure investigations clearly support the argument that built-in strain improves the performance of 1.55 μm lasers.

2) AVALANCHE PHOTODETECTORS

a) Impact Ionisation

Impact ionisation in semiconductors is an important effect at high electric fields. Although it is usually deleterious in transistors it can be used to advantage in avalanche photodetectors (APD's). Phenomenologically it can be characterised by ionization coefficients α for electrons and β for holes, which are the reciprocal average distances between ionization events by these carriers measured in the field direction. It has been shown (1) that for optimum performance of APD's, the ratio of these quantities should be very different from unity. This is found to be the case in Si where α is measured to be much greater than β. This means that an electron generated by a photon can move through the high field region of the diode, gain sufficient kinetic energy for impact ionization to generate another electron-hole pair and the electrons can move through the diode producing more carriers in a cascade effect until they are collected at the anode. The holes are simply swept out to the cathode of the device and the electrical pulse associated with the initial photon is complete. Unfortunately, the band gap of Si is too large to use it for 1.55 μm detection. Germanium will work at 1.55 μm but it is found that here $\beta \approx \alpha$. This means that the holes generated in the ionization processes move back through the diode also generating more carriers. This leads to a much steeper increase in breakdown current with increasing voltage. This makes the device more noisy and difficult to stabilize and much less suitable for communication systems. The α/β ratio in most III-V semiconductors is unfortunately also close to unity and proposals have been made to improve the situation by engineering the band structure by introducing a superlattice into the avalanche region. However, the exact physical processes that occur in these devices where carriers are at very large energies are not well understood. We have, therefore, undertaken a series of theoretical and high pressure studies to try to identify those band-structure parameters which are of importance in determining the α/β ratios.

b) Pressure Dependence of the Breakdown Voltage

We have performed measurements of the breakdown voltage as a function of pressure to 15kb on APD's manufactured from Si[10,11], Ge [10] and GaAs[12]. The results for Si and GaAs are shown in Figs 6a and 6b. For both Si and GaAs, the breakdown voltage decreases with increasing pressure. This is, at first sight, a most unexpected result for GaAs where the direct band gap increases at about 11meVkb[-1] and an increase in breakdown voltage might have been anticipated. In order to model these results, we have applied a hard-threshold, lucky drift theory[13], in which we have included the pressure variation of the phonon mean-free path which is proportional to the phonon energy and the mass density. Calculations were then undertaken assuming the threshold energy varied

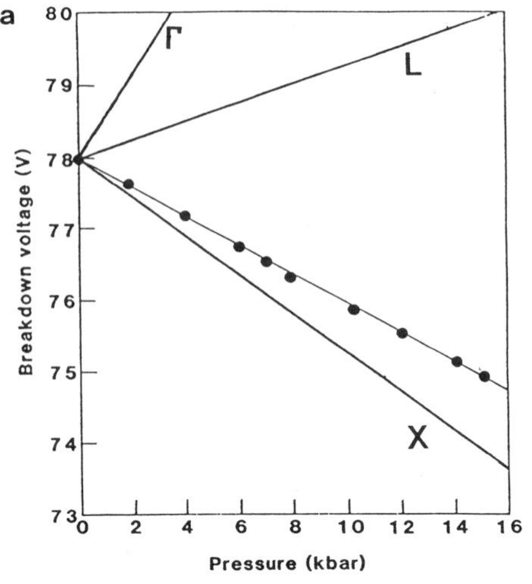
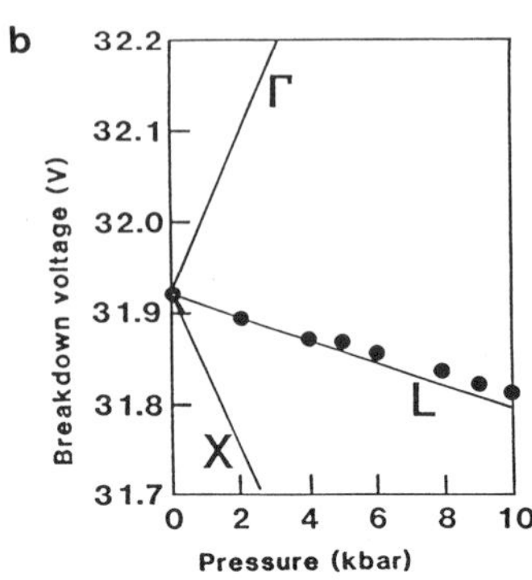

Figure 6. Pressure dependence of the breakdown voltage in a) Si and b) GaAs. Solid lines show effect of threshold energy scaling with the Γ, X and L points.

with the pressure coefficient of the Γ valley, the X valleys or the L valleys and the results are shown in Fig 6. As can be seen, the results indicate that X-related processes dominate in silicon while L-related processes dominate in GaAs, or there is some average of all Γ, L and X processes.

c) Calculations of Threshold Energies for Ionization

To try to better understand the results presented above we have calculated the threshold energies for Si from empirical pseudopotential band structures[10] using the numerical method of Anderson and Crowell[14]. The results are shown in Figure 7. We find that the ratio of the threshold energy for holes (1.71eV) to that for electrons (1.18eV) is 1.45. Since the ionization rates depend exponentially on the threshold energy, this difference is very substantial and explains the large α/β ratio in silicon. Of particular interest is the fact that the ionization energy for electrons is exactly that of the indirect band gap to the X (more exactly Δ) conduction band. This explains the observed variation with pressure shown in Fig 6a with the slight difference between experiment and theory being due possibly to some "softness" in the threshold not included in the theoretical model.

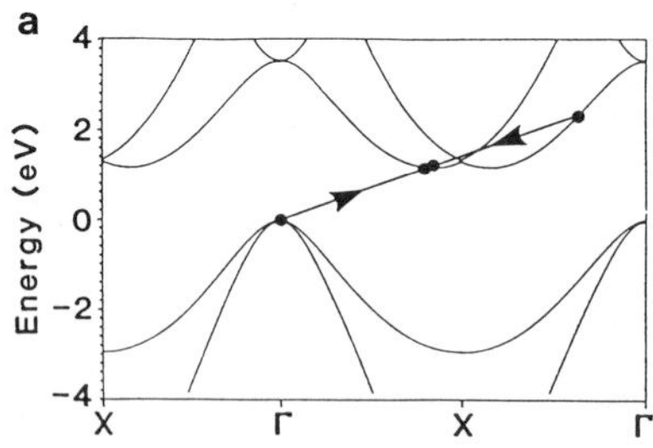
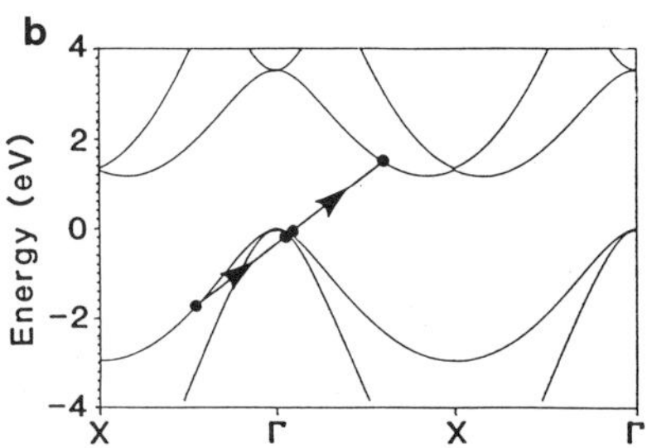

Figure 7. Lowest calculated threshold for a) electrons and b) holes in Si.

In order to understand why the α/β ratio is so different in germanium we have made similar calculations and the results are shown in Figure 8. The lowest electron initiated process is in the <111> direction with a threshold energy of 0.76 eV while the lowest hole initiated process is also in the <111> direction and has a value of 0.88 eV giving a much lower ratio of 1.16. Note that the hole initiated process shown in Figure 8b cannot occur in Si because the band-gap in the <111> direction is greater than the total spread in energy of the valence band from the Γ to the L point. This explains much of the difference in behaviour of the two materials.

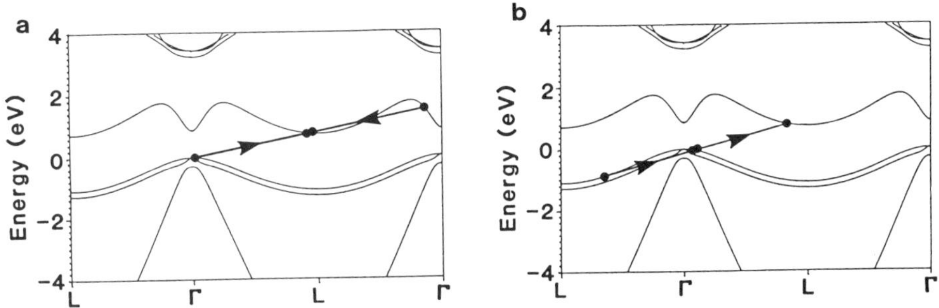

Figure 8.   Lowest calculated thresholds for a) electrons and b) holes in Ge.

d) Consequences for Strained Ge/Si Alloys

Having established a better understanding of the processes occurring from a study using hydrostatic pressure it is again interesting to ask the question, as with lasers in section 1, if improvements in device performance can be obtained using built-in strain.

Strained layer superlattices can be produced by growing $Ge_{1-x}Si_x/Si$ layers on Si substrates provided they do not exceed the critical thickness. It has been shown that the effect of strain in this material system is beneficial for the absorption characteristics of APD's[15] since it decreases the band gap. For strained layers grown on an (001) substrate the X minima will split as a result of the uniaxial component of the stress. We have seen that in Si the dominant electron-initiated ionization processes are associated with the X-minima and we can therefore expect that they will also be affected by the strain.

We have calculated the ionization thresholds for a range of relaxed and strained $Ge_{1-x}Si_x$ alloys from empirical pseudopotential band structures[16]. The results are shown in Figure 9. It can be seen that indeed the ratio of the lowest electron to hole threshold for the tetragonally distorted alloys does increase suggesting that strain may improve the α/β ratio even compared to that of silicon. These results await experimental verification.

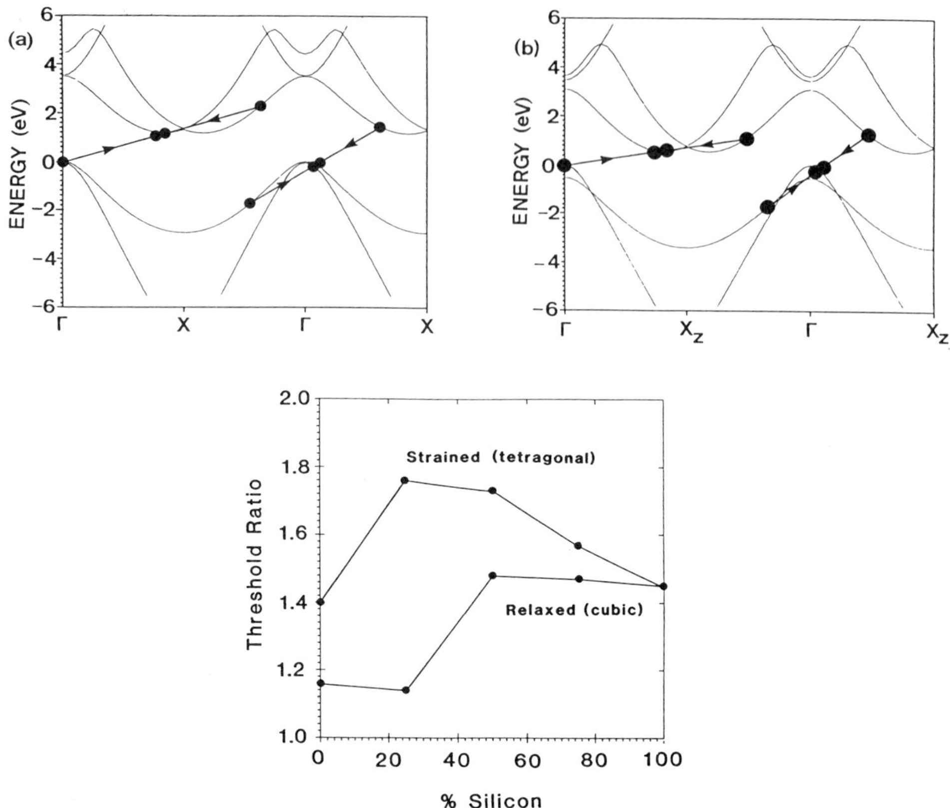

Figure 9.   Lowest calculated threshold for electrons and holes in a) relaxed Si, and b) strained Si on relaxed Ge. c) Ratio of lowest threshold energy for hole-initiated ionisation to that for electron initiated ionisation as a function of alloy composition in strained and relaxed SiGe alloys.

e) Consequences for Multiple Quantum Well Avalanche Photodiodes

It has been proposed by Chin et al[17] that the rate of impact ionization may be enhanced in a multiple quantum well by the process illustrated in Figure 10. This shows an electron which gains the conduction band offset energy as it passes from the barrier to well and hence has enough energy to initiate ionization in the well. If the conduction band offset is bigger than the valence band offset, electrons will gain from this process more than holes and so the α/β ratio will be increased. A material that has received particular attention is GaAs/AlGaAs.   However, Chin et al[17] considered only the band-edge discontinuity at the zone-centre while our high pressure measurements on GaAs described in section 2b showed that it is the L valleys that are important or a mixture of all Γ, L and

290

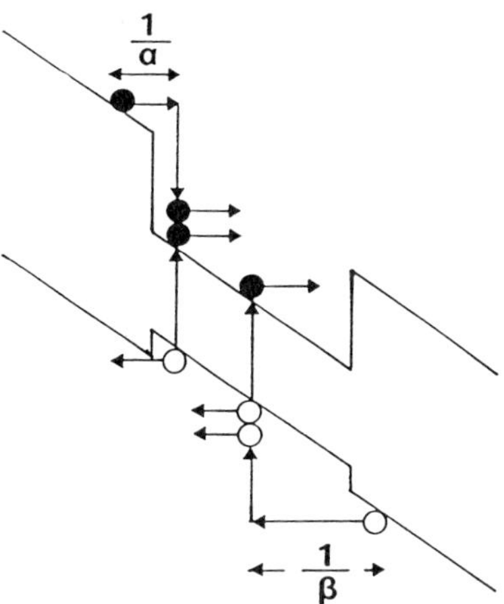

Figure 10. This illustrates how the presence of quantum wells might improve the α/β ratio if the band-offset is larger in the conduction band than in the valence band.

X valleys. To investigate the importance of this result we have undertaken Monte Carlo computer simulation of case 1 shown in Fig 11 in which both the Γ and satellite minima (assumed to be separated by 0.3 eV) have the same band offset at the well edge. This is similar to the model of Brennan[18]. We have also taken the more realistic model for GaAs/AlGaAs shown in Case II, fig 11a in which there is an offset in the Γ minima but not in the satellite minima. The calculated ionization rates as a function of electric field

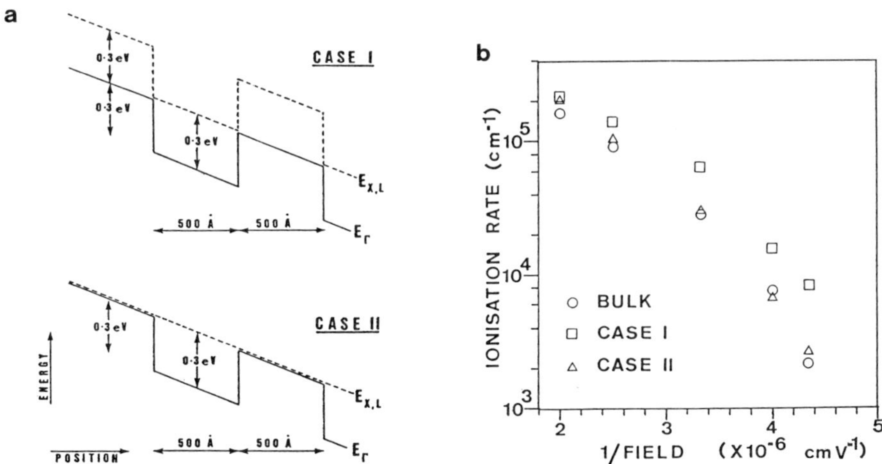

Figure 11. a) Intervalley separations and band-edge discontinuities for the two cases studied. b) Electron impact ionisation rate α against reciprocal field for the bulk well material, and for cases I and II of (a).

are shown for the two models in Fig 11b. As can be seen, while a band offset of 0.3eV throughout the Brillouin zone (case 1) leads to an increase in the ionization rate compared to GaAs, the more realistic model (case II) shows no improvement. This is basically due to the fact that at the high electric fields at which impact ionization occurs, almost all the electrons are in the satellite minima (cf. the Gunn effect) and therefore see little or no benefit as far as their energy is concerned due to the Γ band offset. Clearly this is important for consideration of multiquantum well structures and shows that we must consider materials in which there is an advantageous offset in the satellite minima or a system in which impact ionization concerns only electrons in the Γ minima.

3) CONCLUSIONS

We have considered the results of high pressure investigation of the internal physical processes occurring in semiconductor lasers and in semiconductor avalanche photodiodes. These have given considerable insight into those characteristics of the electronic band structure which have a beneficial or a deleterious effect on the devices. Making use of the fact that it is now possible to grow semiconductor layers in a state of permanent strain we have been able to propose, and in some cases test, new device structures. The results for strained layer lasers are extremely encouraging and a way forward for strained $Ge_{1-x}Si_x$ alloys is also indicated. While GaAs/AlGaAs multi-quantum well APD's may not enjoy enhancement by the mechanism which has been proposed elsewhere the problem has been more closely defined so that other materials systems may now be considered.

Acknowledgements

The author is greatly indebted to E P O'Reilly, A Ghiti, W Ring, P J Thijs, J Allam, I Czajkowski and M Silver for their assistance. He is also grateful to the Science and Engineering Research Council and to British Telecom Research for financial support.

References

1.   A R Adams, M Asada, Y Suematsu and S Arai, The temperature dependence of the efficiency and threshold current in InGaAsP lasers related to intervalence band absorption, Jpn J Appl Phys 19:L621 (1980)

2.   N K Dutta and R J Nelson, The case for Auger recombination in InGaAsP, J Appl Phys 53:74 (1982)

3.   A R Adams, K C Heasman and E P O'Reilly, "Band structure engineering in Semiconductor Microstructures", Plenum Press, 279-301 (1989)

4.   W Ring and A R Adams, Pressure dependence of the threshold current and quantum efficiency of 1.55 μm MQW lasers, to be published

5.   W Ring, A R Adams and P J A Thijs, Pressure dependence of the threshold current and quantum efficiency of 1.55 μm strained-layer lasers, to be published

6.   A R Adams, Bandstructure engineering for low-threshold high efficiency semiconductor lasers, <u>Electron Lett</u> 22:249 (1986)

7.   E P O'Reilly, Valence band engineering in strained-layr structures, <u>Semicond Sci Technol</u> 4:121 (1989)

8.   A Ghiti, E P O'Reilly and A R Adams, Improved dynamics and linewidth enhancement factor in strained-layer lasers, <u>Electron Lett</u> 25:821 (1989)

9.   P J A Thijs, L F Tiemeijer, P I Kuindersma, J J M Binsma and T Van Dongen, High performance 1.5 μm wavelength InGaAs/InGaAsP strained quantum well lasers and amplifiers, <u>IEEE J Quantum Electronics</u>, June (1991)

10.   I K Czajkowski, J Allam, M Silver, A R Adams, M A Gell, Impact ionisation thresholds in silicon and germanium under hydrostatic pressure and strain, <u>IEE Proc Pt J</u>, 137:79 (1990)

11.   J Allam, I K Czajkowski, A R Adams, M Silver, Pressure dependence of ionisation thresholds in Si, to be published

12.   J Allam, A R Adams, M A Pate, J S Roberts, Evidence for impact ionisation via X and Γ valleys in GaAs from hydrostatic pressure studies of avalanche breakdown, <u>Proc GaAs and Related Compounds</u>, 375 Jersey (1990)

13.   B K Ridley, Lucky drift mechanism for impact ionisation in semiconductors, <u>J Phys, C: Solid State Physics</u> 16:3373 (1983)

14.   C L Anderson, C R Crowell, Threshold energies for electron-hole pair production by impact ionization in semiconductors, <u>Phys Rev B</u> 5:2267 (1972)

15.   R People, Physics and applications of GeSi/Si strained layer heterostructures, <u>IEEE J Quantum Electron</u> QE-22:1696 (1986)

16.   I K Czajkowski, J Allam, A R Adams, Impact ionisation threshold in relaxed and strained Ge/Si alloys, to be published

17.   R Chin, N Holonyak Jnr, G E Stillman, J Y Tang, K Hess, Impact ionisation in multilayered heterojunction structures, <u>Electron Lett</u> 16:467 (1980)

18.   K Brennan, Theory of electron and hole impact ionisation in quantum well and staircase superlattice avalanche photodiode structures, <u>IEE Trans Electron Dev</u> ED-32:2197 (1985)

PHOTOLUMINESCENCE OF STRAINED-LAYER QUANTUM WELL STRUCTURES
UNDER HIGH HYDROSTATIC PRESSURE

V. A. Wilkinson

Strained-Layer Structures Research Group,
University of Surrey, Guildford, Surrey, GU2 5XH, UK

ABSTRACT

The photoluminescence of quantum-well structures, under high hydrostatic pressure, has been studied. An argon-loaded miniature diamond-anvil cell, which readily generates pressures in the region 0 to 200kbar, has been employed for this purpose. Structures containing strained layers are currently of great interest and are concentrated on here. High pressure techniques for determining the heterojunction band line-ups, with spectroscopic accuracy, are described. Recent results on the InGaAs/AlGaAs and GaAsSb/GaAs strained systems are discussed.

The pressure coefficients of bulk semiconductors and more recently of low dimensional structures have been reported in the literature. There is now considerable evidence that compressively strained layers exhibit pressure coefficients which are lower than expected. The influence of higher-order elastic constants and strain-dependent deformation potentials have been considered but do not adequately describe the data. This behaviour therefore remains anomalous.

INTRODUCTION

Developments in the semiconductor growth techniques, MOCVD (metalorganic chemical vapour deposition) and MBE (molecular beam epitaxy), over the last decade have made possible the growth of new structures. In particular it has been possible to grow layers only a few angstroms thick. If a semiconductor layer is grown between two layers of another semiconductor material of a higher band gap a quantum well may be formed (figure 1), where the electrons and holes are confined in a potential well. When the well is sufficiently thin ($\sim$<100Å) quantum effects are observed. A recent innovation in these

Frontiers of High-Pressure Research, Edited by H.D. Hochheimer and
R.D. Etters, Plenum Press, New York, 1991

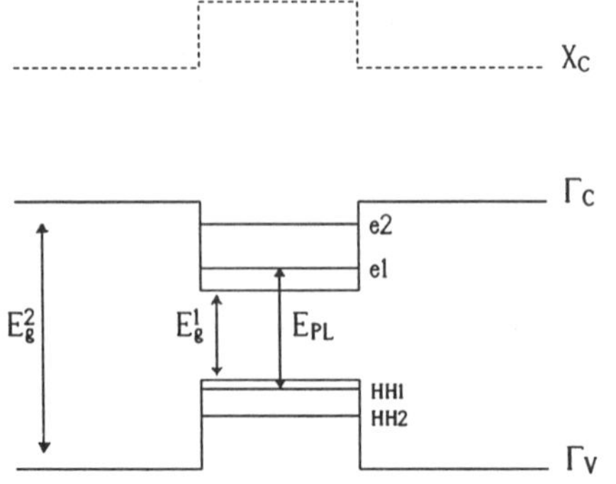

Fig. 1. A schematic diagram showing a typical band structure for an unstrained quantum well, formed by growing a thin layer of material with band gap E_g^1 between two layers of a material with band gap E_g^2, where $E_g^2 > E_g^1$. Some of the quantum confined states are indicated. The light-hole states are omitted for clarity.

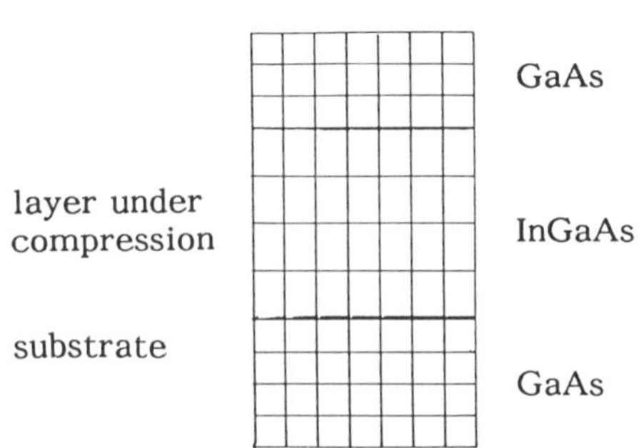

Fig. 2. A schematic representation of a structure containing a strained layer. The InGaAs/GaAs material system is given as an example of a real strained-layer system.

structures has been the growth of strained layers. In such structures a semiconductor is grown coherently on a substrate of a different natural lattice constant. If the layer is thin enough it will take on the in-plane lattice constant of the substrate. The lattice constant of the layer in the growth direction will be modified in accordance with Poisson's ratio.[1] Figure 2 illustrates the case for InGaAs grown pseudomorphically on GaAs, one of the strained systems of current interest. Adding indium to GaAs increases its relaxed lattice constant. Layers grown pseudomorphically on GaAs are therefore under biaxial compression. The incorporation of strain in the layer modifies the band structure, particularly in the valence band. Important effects are the lifting of the degeneracy of the light- and heavy-hole bands at the zone centre and the possibility of having light-hole like states highest in the valence band. Valence-band engineering in strained-layer structures has been recently reviewed by O'Reilly[1]. The importance of these effects to the operation of some devices, such as lasers, has been described by Adams.[2,3]

An area of extensive research in low dimensional structures has been the investigation of the band line-ups at heterojunctions. These line-ups have proved difficult to predict theoretically and almost equally difficult to determine experimentally. The distribution of the band-gap discontinuity between the conduction and the valence band determines the degree of confinement of the carriers and is therefore of great technological importance. Optical measurements under hydrostatic pressure have provided some of the most accurate data to date on band offsets. In the next section the behaviour of semiconductor structures under pressure is briefly discussed and a means of obtaining the valence-band offset described. In subsequent sections the experimental techniques for making optical high-pressure measurements, employing a diamond-anvil cell (DAC) are reported. Recent results from GaAs/AlGaAs quantum wells are used to illustrate the established technique for determining the band offsets in this lattice-matched system. Then data from the strained systems InGaAs/GaAs, InGaAs/AlGaAs and GaAsSb/GaAs will be presented and band alignments obtained. Finally the pressure coefficients of the PL of the strained layers mentioned above will be discussed.

SEMICONDUCTOR STRUCTURES UNDER HIGH PRESSURE

High pressure has proved a powerful diagnostic tool in the study of semiconductors. In the crystalline solid hydrostatic pressure reduces the lattice constant of the material whilst maintaining the crystal symmetry. The main influence on the band structure is illustrated in figure 3. Each of the conduction-band minima have characteristic shifts with pressure with respect to the top of the valence band ($\Gamma \sim +10$ meV/kbar, L$\sim+4$

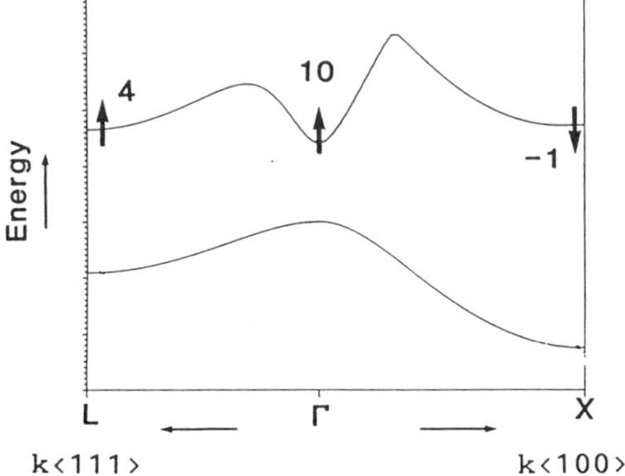

Fig. 3. A schematic diagram of the band structure of a tetrahedrally bonded semiconductor showing the characteristic pressure coefficients (labelled in the figure, in meV/kbar) of the high symmetry points with respect to the top of the valence band.

meV/kbar and X ~-1 meV/kbar).[4] These characteristic coefficients vary little across a wide range of materials and are therefore a signature of the conduction-band minimum involved in an optical transition. The application of pressure may make dramatic changes to the direct band-gap; it may alter the ordering in energy of the conduction band minima (eg turn the material from direct to indirect); it also changes the effective masses of the carriers via the k.p interactions. Warburton et al.[5] have been able to observe transitions due to each of the conduction-band minima, in a $Ga_{.85}In_{.15}Sb/GaSb$ quantum well, in photoluminescence. Over a pressure range 0-70 kbar each of the minima becomes the lowest conduction-band state in turn (Γ-L-X).

In this paper we are concerned with how the pressure induced modification to the band structure can be used to reveal information on the heterojunctions in low dimensional structures. The utility of these techniques was first demonstrated by Wolford et al.[6] and by Venkateswaren et al.[7] who studied GaAs/AlGaAs quantum wells. They found that by applying pressure to the quantum-well structure the direct Γ_6-Γ_8 PL transitions, moving at the characteristic ~10 meV/kbar, changed to a characteristic indirect transition, with pressure coefficient ~-1 meV/kbar at a certain pressure. They were able to assign this latter transition to a recombination between electrons confined in the AlGaAs barrier X-states and holes confined in the GaAs valence band: A transition which is therefore indirect in both real (Type II) and k-space. Observation of such a transition allows a spectroscopically accurate determination of the valence band offset, ΔE_v, using the following equation:

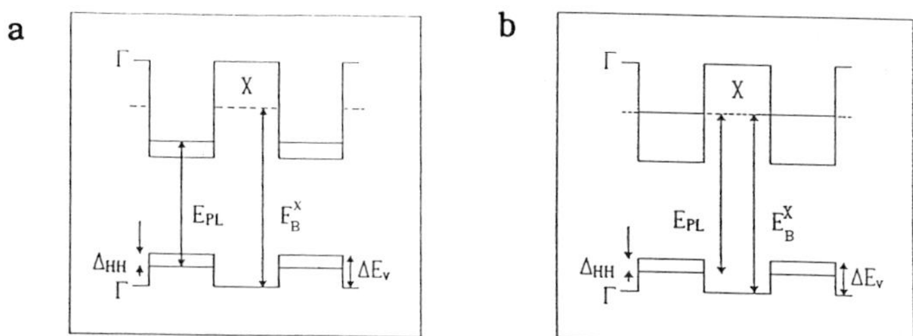

Fig. 4. The figures illustrate the two types of band alignment discussed in the text for GaAs/AlGaAs structures. (a) shows a Type I quantum-well alignment, where electrons and holes are both confined in the same material. (b) shows the situation at crossover. Any further increase in pressure will turn the quantum well Type II, in this case the electrons will be confined in the barrier X-states and the holes will be spatially separated in the well material.

$$\Delta E_v(P) = E^X_B(P) - E_{PL}(P) + \Delta_{HH}(P) - E_{EB}(P) \tag{1}$$

E^X_B is the energy of the indirect gap of the barrier material; its pressure dependence ,may be estimated from the Type II luminescence or measured in a separate experiment. E_{PL} is the measured energy of the PL transition. The exciton binding energy, E_{EB}, and the heavy hole confinement energy, Δ_{HH}, are not measured but may be calculated readily without introducing any appreciable error, especially for wide quantum wells where these energies are almost completely insensitive to small changes in well width and to errors or fluctuations in alloy composition. The data may be extrapolated back to ambient pressure to obtain the offsets there. This technique is therefore very sensitive to the band offsets, while other optical assessments which have only been able to study direct transitions in the well have proved notoriously inaccurate in determining the value of the offsets.[8] Figure 4 illustrates the quantum well alignment at ambient pressure and at some higher pressure where the structure is just turned Type II and the spatially indirect transitions are observed.

Wolford et al.[6] and Venkateswaren et al.[7] were therefore able to obtain the fractional valence-band offset, Q_v, for GaAs/Al$_y$Ga$_{1-y}$As heterojunctions which they found to be 0.31 ± 0.03 (y= 0.28) and 0.30 ± 0.04 (y = 0.33) respectively. Wolford et al. also found similar results for a sample with Al$_{0.7}$Ga$_{0.3}$As barriers ($Q_v = 0.34 \pm 0.02$), Q_v was therefore assumed to be approximately constant across the alloy range. Subsequent work on GaAs/AlAs structures, where the well and barrier widths were designed to give Type II transitions at ambient pressure, produced values of Q_v in good agreement with the high pressure work ($Q_v = 0.33\text{-}0.34$).[9]

The high-pressure technique cannot be applied directly to systems in which crossover occurs first in the well (Type I). This turns out to be the case for the strained systems which we are interested in here. A new method for determining the offsets for a solely Type I system is described and we show how this can give equally accurate values for the band offsets.

EXPERIMENTAL TECHNIQUES

The most successful device for optical, high pressure measurements has been the diamond-anvil cell (DAC). A miniature cryogenic DAC system, which has been described elsewhere,[10,11] has been used for this work. The principle of the DAC is illustrated in figure 5. The sample is contained within a steel gasket and surrounded by a pressure transmitting fluid, in this case liquid argon. In operation the diamonds are advanced by a force up to one ton and hydrostatic pressures in excess of 200 kbar may be produced. In fact work is usually done at lower pressures since most of the semiconductors have undergone a phase transition by ~100 kbar. Before loading into the DAC the samples are mechanically thinned to ~30μm thick and cleaved into ~50μm x 50μm squares. Several samples may be loaded in the cell simultaneously. A second sample may be loaded to act as a pressure gauge. Whilst much of the pressure work reported in the literature uses a ruby chip as the pressure gauge, the pressure induced shift of the ruby luminescence is in fact rather small (0.365 Å/kbar [12]). In contrast, direct-gap semiconductors emit PL which shifts at approximately 10 meV/kbar which is two orders of magnitude greater shift. While the ruby scale is accurately calibrated and highly linear, which the semiconductor gauges are not, the use of a semiconductor gauge does allow a more precise pressure measurement. Moreover, the pressure gauge is often chosen to be a material of physical significance in the experiment. For example, GaAs may be used as a pressure gauge when studying GaAs/AlGaAs or InGaAs/GaAs quantum well structures. Then the pressure shifts of the quantum-well emission relative to the well or barrier material band-gap are very accurately known. Often a conversion to pressure is not necessary, and pressure is then simply a dummy variable. Ideally the pressure gauge will be incorporated in the sample itself. For example, GaAs substrate or capping layer

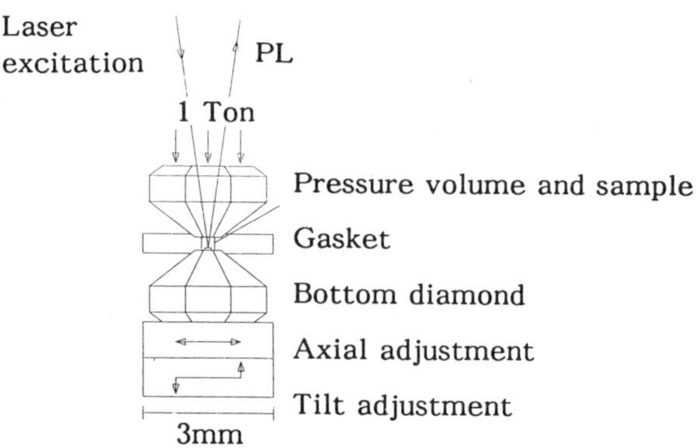

Fig. 5. A schematic diagram illustrating the principle of the diamond-anvil cell.

emission from the sample of interest may be used. Sometimes a wide GaAs quantum well is incorporated in the sample design to provide bright emission. The use of an integral pressure gauge eliminates risk of errors due to pressure inhomogeneity since the gauge and the layers under study will only be a few hundred angstroms apart.

Photoluminescence measurements are made by passing the exciting laser light through the diamond and collecting the emitted PL in the back scattering geometry as illustrated. The emitted PL is dispersed by a 1 metre Spex spectrometer and detected by either a liquid nitrogen cooled Ge detector or by a Si PIN diode depending on the spectral region of interest. Accurate focusing of the laser beam allows several samples in the DAC to be spatially resolved[13] although it is convenient if they may also be spectrally resolved. The cell is designed to fit in a standard cryostat insert (maximum diameter 19mm) so that it may be operated on almost any standard PL set-up. Measurements are made at or around liquid helium temperatures. The force to the cell is transmitted down the cryostat so that pressure changes may be made in situ and without the necessity to warm the cell up.

EXPERIMENTAL RESULTS

<u>Band Alignment of GaAs/AlGaAs Quantum Wells</u>

As discussed in the previous section, the observation of Type II transitions in the GaAs/AlGaAs material system yields a direct measure of the valence-band offset. The Type II transitions can be unambiguously assigned, once they are identified as X-related by their pressure coefficients, since the transition energies are smaller than the indirect gap of either of the constituent materials. In the rest of this section, where new strained-layer systems are considered, the correct assignment is not so clear and Type II transitions are not always observed because the Γ-X crossover may occur first in the well. Recent results from GaAs/AlGaAs quantum wells are described here to illustrate the technique where Type I/Type II conduction-band crossovers can be pressure induced. We shall then proceed to show how the technique must be modified in the case of structures which have only Type I crossovers. This new technique proves to be necessary for determining the band offsets in the strained-layer structures we are interested in.

Figure 6 shows the PL spectrum of a GaAs/Al$_{.25}$Ga$_{.75}$As sample which contained five quantum wells of varying width (14.5, 29, 58, 116 and 232Å) with 450Å wide barriers, grown on a (311)A orientated GaAs substrate. The samples were grown by MBE at the Optoelectronics Joint Research Laboratories, Japan.[14] The PL spectrum shows a sharp peak for each of the quantum wells except the widest well, whose emission is not resolved from that of the substrate. The emission energies are assigned to the direct, e1 to hh1 transition in each of the GaAs wells. On application of hydrostatic pressure each of the peaks shifts to shorter wavelength, this is shown in figure 7. In this experiment PL from the GaAs substrate was used as the pressure gauge. The plot shows how the emission peak energies from the quantum wells move relative to the GaAs band-gap energy, corresponding to pressure. As the Γ confined states in the GaAs well become coincident with the barrier X-states (figure 4(b)) the PL is strongly quenched. In a superlattice structure the Type II PL may be tracked over a considerable pressure range but in a single quantum-well sample, such as this, the PL is usually quenched beyond detection shortly after crossover. The crossover pressure may be identified by this rapid fall in intensity of the quantum well direct PL. In the figure solid lines are linear fits to the data and arrows indicate the energy at which the Γ-Γ PL is strongly quenched.

At the point of crossover the valence-band offset can be accurately determined using

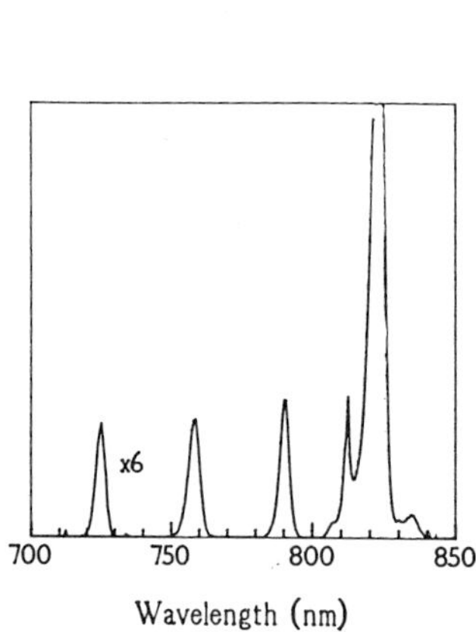

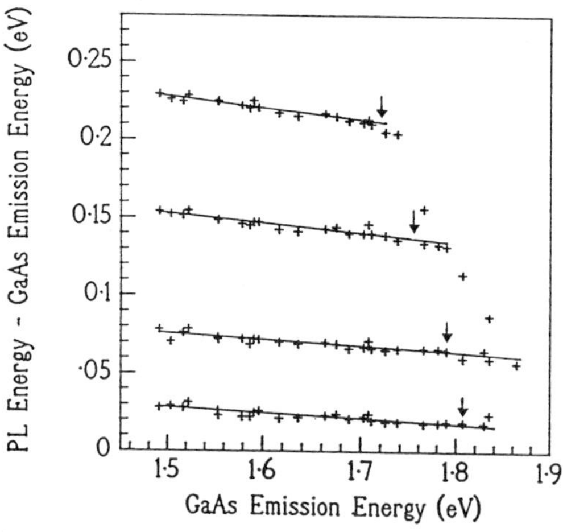

Fig. 6. The photoluminescence spectrum of the GaAs/AlGaAs structure measured at 4K and at ambient pressure.

Fig. 7. The difference between the PL peak energy and the GaAs (integral pressure gauge) PL energy is plotted against the GaAs band-gap energy for each of the wells in the GaAs/AlGaAs structure. The arrows indicate the 'cut-off' of the Γ-Γ PL intensity.

eqn. 1. The confinement energy of the heavy-hole must be calculated. This is done using the envelope-function approximation of Bastard.[15,16] Since the heavy-hole confinement energies are small, this value can be calculated without much absolute error. As previously mentioned, if the quantum wells are wide ($\sim$100Å), then confinement energies are very insensitive to small errors in layer thickness or alloy composition. Confinement energies for the AlGaAs X-states have been omitted from eqn. 1, which is correct for this single quantum well sample, but may need to be reconsidered for superlattice structures with thin barriers. Exciton binding energies have been measured for the GaAs/AlGaAs system.[17] Following the method outlined above, the fractional valence-band offset for this sample is found to be 0.35. This is in good agreement with other data in the literature for (001) orientated substrates.[6,7]

From figure 7 it is clear that the pressure coefficient of the el-hh1 transition depends

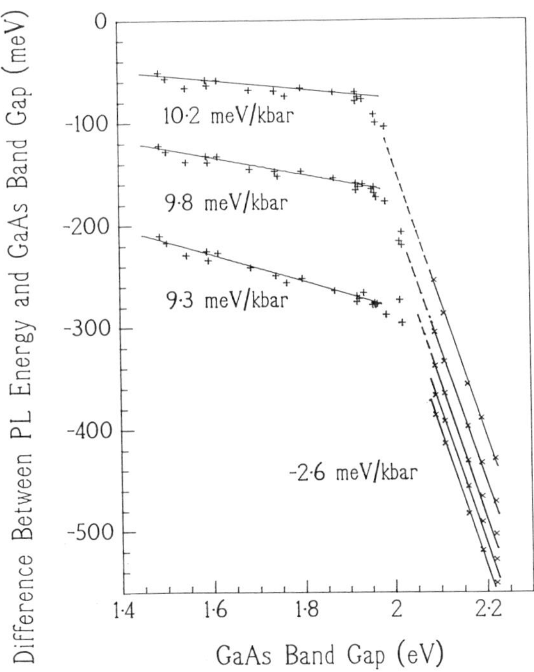

Fig. 8. The behaviour under pressure of the PL from the InGaAs structure with three quantum wells of different composition and GaAs barriers is shown. The difference between the InGaAs quantum well PL energy and the GaAs (pressure gauge) emission energy is plotted against the GaAs band gap energy. The figure shows Γ-like behaviour at low pressure turning into X-like behaviour at higher pressures. *(Reproduced from reference 22.)*

on well width. We shall return to this result later when the pressure coefficients in strained-layer quantum wells are discussed and where a quite different result is observed.

Band Alignment of InGaAs/GaAs Strained Quantum Wells

InGaAs/GaAs strained-layer structures where the InGaAs layer is under biaxial compression while the GaAs barriers (and substrate) are unstrained are currently of great interest and consequently much attention has been given to the band alignment question. Figure 8 shows a similar plot to figure 7 for a sample containing three 100Å, $In_xGa_{1-x}As$ quantum wells of varying composition (x= 0, 0.17, 0.25). In this case two pressure gauges were used, a separate piece of GaAs up to the GaAs Γ-X crossover at ~40kbar and a

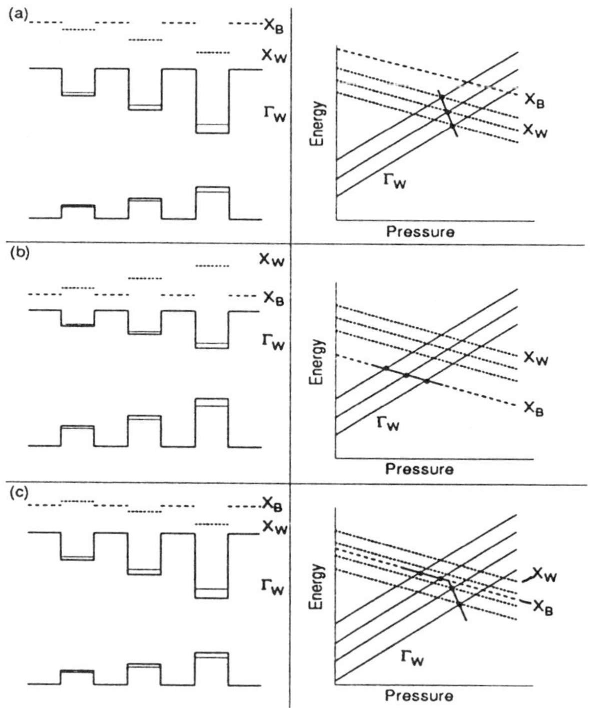

Fig. 9. The figure shows the three possible band alignments for the X minima in InGaAs/GaAs and InGaAs/AlGaAs strained-layer quantum wells. In each case three quantum wells of different InGaAs composition are shown. In configuration (a) the barrier X-minima lie higher in energy than the lowest-lying ($X_{x,y}$) well X states. Conversely in configuration (b) the barrier states are the lowest X states. In (c) the barrier X-states have been lowered with respect to the InGaAs band structure by the addition of aluminium to the barriers. To the right of each figure the behaviour of the transition energies as a function of pressure is shown.

piece of InP for higher pressures. Again the plot is against GaAs emission energy where the energy is interpolated, using known values for the pressure coefficients,[18-20] from the InP emission energy for pressures above 40kbar. At low pressures characteristic direct (ie shifting at ~10 meV/kbar) transitions are observed while at higher pressures the emissions are seen to take on the characteristics of the X-minima. In the direct regime the pressure coefficients are in fact found to drop steadily away from that of GaAs as the indium content or strain is increased; this will be discussed later. In the indirect regime the pressure coefficients are found to be -2.6 meV/kbar, which is twice that observed in GaAs.[18]

Some authors have interpreted this crossover from Γ to X-like behaviour as a Type II crossover similar to that found in GaAs/AlGaAs heterojunctions.[21] However, the intensity of the indirect transitions in these single quantum wells and the similarity of the pressures at which crossover occurs suggest that this is in fact a crossing with the lowest lying well X-minima ($X_{x,y}$). The transitions would therefore remain Type I. In the strained InGaAs band-structure the degeneracy of the X-minima is lifted by the uniaxial component of the strain, with the $X_{x,y}$ minima lower than the X_z. The relative positions of the InGaAs and GaAs X-states therefore depends on the degree of this splitting and on the band offsets. Two possible configurations are illustrated in figures 9 (a) and (b). A schematic representation of the band structure of three quantum wells, with different $In_xGa_{1-x}As$ compositions is shown. The confined Γ-states are indicated along with the lowest lying X-states in the barrier and in each of the wells. On the right of each figure a schematic illustration of the expected behaviour of the transition energies under hydrostatic pressure is shown. In the case of configuration (a) the barrier X-states lie highest in energy. As pressure is increased the Γ-Γ transition energies rise rapidly in energy until each of the confined electron ground states crosses with the $X_{x,y}$ minima in the well (Type I crossover). The presence of the barrier X-minima at higher energy is then not observed in the PL data. Note that this type of crossover leads to a quenching of consecutive wells in quick succession as the pressure is increased. In configuration (b) the barrier X-states are the lowest X-states in the heterostructure. As pressure is increased the same low pressure behaviour is observed as for (a) but the crossovers are now Type II and the well $X_{x,y}$ minima are unobserved in the experiment. This type of crossover behaviour leads to a slower rate of crossing in consecutive wells than for Type I crossings. The schematic diagrams imply that the barrier and well X-states have the same pressure coefficients and take no account of any relative movement of the two band structures under pressure. These effects need to be taken into account in the full analysis of the data.

To identify the type of crossover which occurs in InGaAs/GaAs structures, InGaAs/AlGaAs structures have been studied for comparison.[22] This work, which is discussed in the next section, confirms that in the strained InGaAs/GaAs system the pressure-induced crossovers are indeed with the well $X_{x,y}$ minima. It is not possible to obtain a reliable measure of the band offsets from purely Type I data, although an upper bound for the valence-band offset, ΔE_v, may be estimated since ΔE_v is then known to be sufficiently small not to allow a Type II crossover. In the next section we show how the band offsets for InGaAs/GaAs heterojunctions may be obtained from a study of InGaAs/AlGaAs structures.

Band Alignment of InGaAs/AlGaAs Quantum Wells

Many of the envisaged InGaAs strained-layer device structures will involve quantum wells with AlGaAs rather than GaAs barriers since this increases the depth of the quantum well and hence aids carrier confinement. It is therefore important to understand how, if at all, the addition of aluminium to the barriers modifies the band-offset ratio. The lattice constant of $Al_yGa_{1-y}As$ is almost exactly constant across the alloy range so the strain in the InGaAs is not modified by the barrier composition. In this section we report on the use of InGaAs/AlGaAs structures as a means of identifying the nature of transitions observed in the InGaAs/GaAs structures discussed above and from the data determine the fractional valence-band offset for $In_xGa_{1-x}As/GaAs$ and $In_xGa_{1-x}As/Al_yGa_{1-y}As$ heterojunctions in the composition ranges studied (x and y < 0.25).

The expected effect of adding aluminium to the barriers is to lower the AlGaAs X-minima with respect to the InGaAs band-structure in the well. It should therefore be possible to demonstrate the difference between Type I and Type II crossovers. Figure 10 shows data from three samples with different $Al_yGa_{1-y}As$ barrier compositions (y= 0, 0.10, 0.24). Each of the samples contain several quantum wells of different indium content up to a maximum of 0.25. The figure shows the energy at which crossover occurred as a function of the ambient pressure transition energy. This way of presenting the data illustrates two clear types of behaviour: A steep line for the sample with GaAs barriers and roughly parallel lines with shallower gradients for the samples with AlGaAs barriers.

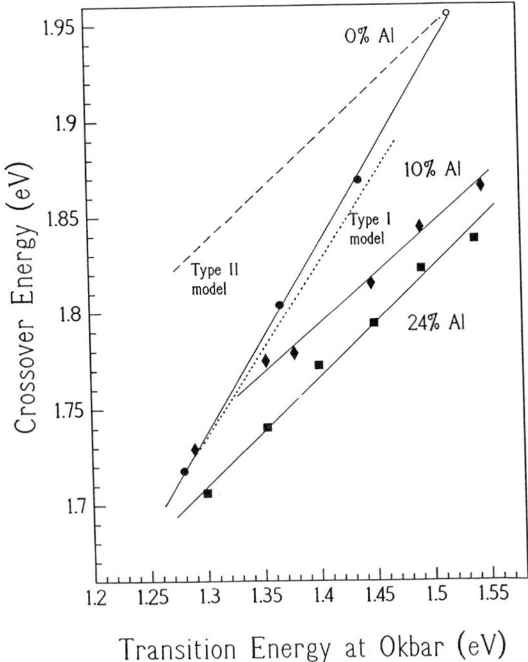

Fig. 10. The crossover energies for each of the InGaAs/GaAs and InGaAs/AlGaAs samples (solid circles, GaAs barriers, diamonds, $Al_{.10}Ga_{.90}As$ barriers, solid squares, $Al_{.24}Ga_{.76}As$ barriers) are shown as a function of the ambient pressure PL energy. The open circle indicates the bulk GaAs Γ-X crossover energy. The solid lines are least squares fits to the data. The dashed and dotted lines show the expected behaviour, for the sample with GaAs barriers, for Type II and Type I Γ-X crossovers respectively. *(Reproduced from reference 22.)*

In the figure solid lines are least squares fits to the data whilst other lines are theoretical models. The dashed and dotted lines show the predicted behaviour for a Type II and a Type I crossing respectively for a sample with GaAs barriers. The excellent agreement with the type I crossing model confirms the proposed assignment that the Γ-X crossovers observed in InGaAs/GaAs quantum wells are with the well $X_{x,y}$ states. For the samples with AlGaAs barriers the data are approximately parallel to the predicted Type II trend for GaAs barriers, confirming that this is the behaviour in these samples. In fact in the sample with $Al_{.10}Ga_{.90}As$ barriers both types of behaviour are observed. This type of behaviour is illustrated as the third configuration in figure 9. The lowering of the X-minima in the barriers means that, for some barrier compositions, these states now lie lower than some well X-states, ie for some low indium content wells, but above others with higher indium content. The predicted form of the behaviour under pressure is indicated. The wells with low indium content have their crossovers with the barrier states (Type II), whilst deeper, higher indium content wells cross Type I as in InGaAs/GaAs structures.

Whilst it is not possible to determine the band offsets from the Type I crossovers the Type II crossovers are able to reveal the fractional valence-band offset for InGaAs/GaAs heterojunctions. The change in transition energy at ambient pressure as the indium content is increased from GaAs gives, to a good first approximation, the combined change in the conduction and valence-band edges. On the other hand the change in the energy at crossover gives information on the change in the valence-band alone with indium content. Thus the gradient of the Type II data sets in figure 10 give Q_v for InGaAs/GaAs heterojunctions. Some corrections are required to this simplified picture to take account of confinement energies, the pressure dependence of the AlGaAs X-states, the pressure dependence of the offsets and exciton binding energies. These corrections have been discussed in detail elsewhere.[22] Applying the corrections to the measured gradients leads to a value for Q_v of 0.40 ±0.02 for InGaAs/GaAs quantum wells. The best fit to the data is linear, we can therefore say that Q_v is constant across the alloy range studied (x<0.25). For each of the individual InGaAs/AlGaAs quantum wells the offsets may be obtained provided that the crossover was Type II. The values of Q_v are found to be intermediate between those for GaAs/AlGaAs (0.30[6]) and InGaAs/GaAs (0.40) and to support transitivity within experimental error.[22]

Band Alignment of GaAsSb/GaAs Strained Quantum Wells

Despite the considerable current interest in InGaAs strained-layer structures the strained system GaAsSb/GaAs has received comparably little attention. It is expected that GaAsSb/GaAs quantum wells will have a large valence-band offset with holes confined in the GaAsSb. This system may therefore exhibit many of the interesting features of strained InGaAs but with greater hole confinement. Technologically this may therefore be expected to be as important as the InGaAs system. It has generally been expected that this system would display a strongly Type II band line-up with electrons confined in the GaAs. Ji et al.[23] have studied $GaAs_{1-x}Sb_x$ layers with x=0.1 using photoreflectance techniques and conclude that the band line-up is indeed strongly Type II with Q_v= 1.7. We review here work done by Prins et al., on a sample with x=0.12, who conclude from ambient pressure and high pressure photoluminescence measurements that the alignment must be Type I.[24]

The structure was grown by MBE at The University of Manchester Institute of Science and Technology (UMIST), UK. It consisted of five GaAs$_{.88}$Sb$_{.12}$ quantum wells of widths 83, 39, 28, 20 and 14Å, separated by 1000Å of GaAs. At 4K bright PL is observed from each well. Since strong PL is observed from single isolated wells this in itself indicates Type I luminescence. Figure 11 shows the behaviour under pressure of the peak emissions energies. A piece of In$_{.53}$Ga$_{.47}$As on InP has been used as the pressure gauge. The pressure coefficients of the direct transitions are indicated on the figure and will be discussed later. At around 37kbar a Γ-X crossover occurs for all wells. At higher pressure strong X-related emissions are observed and followed up to ~70kbar where the In$_{.53}$Ga$_{.47}$As pressure gauge quenches. As in the case of the InGaAs/GaAs quantum wells, this bright luminescence from a single quantum well after crossover is strong evidence for a Type I crossover. The ambient pressure transition energies and the crossover energies taken together support a small Type I offset in the conduction band, ΔE_c~25meV, for x=0.12 ($Q_v \approx .90$). This offset is in fact in good agreement with theory[25] when the strong band-gap bowing in the GaAsSb alloy system[26] is properly taken into account. Again accurate determination of the band offsets is not possible from Type I data alone. Similar work to that described for InGaAs/GaAs structures will be required to obtain Type II transitions and provide accurate quantitative data on the band offsets.

High pressure measurements by Warburton et al., on the related system Ga$_{.85}$In$_{.15}$Sb/ GaSb, show that each of the indirect minima ($X_{x,y}$ and L) have a Type I alignment in the

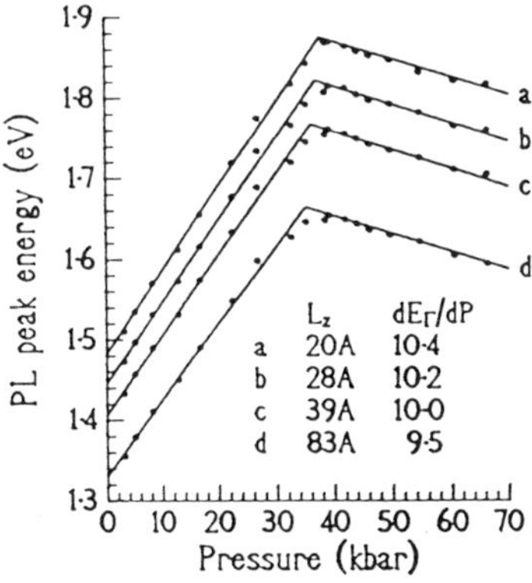

Fig. 11. The peak PL energies, for each of the GaAsSb quantum wells, are plotted against pressure. Pressure coefficients for the Γ-Γ transitions, in meV/kbar, are shown. (*Reproduced from reference 24.*)

GaInSb well.[5] In this system the first crossover is with the L minima. Again, observation of an indirect transition in a single quantum well, over a large pressure range shows that this is a Type I transition. At higher pressures an L-X conduction-band crossover is observed and the X-related luminescence followed for over 30 kbar.

PRESSURE COEFFICIENTS OF THE PHOTOLUMINESCENCE OF STRAINED-LAYER QUANTUM WELLS

The pressure coefficients of the direct Γ_6-Γ_8 photoluminescence transitions in strained layers have been measured for each of the material systems discussed above. In each case the coefficients have been surprisingly low. We have already seen that the pressure coefficients of each gap vary little across a wide range of semiconductor materials, so to find significant differences for small changes in strain or alloy composition is therefore unexpected.

Since we are always considering quantum wells in the case of strained layers, we return first to the unstrained GaAs/AlGaAs system to see what factors affect the pressure coefficient of the quantum-well transition. Using the data of figure 7, the direct pressure coefficients have been plotted as a function of well width and are shown in figure 12. A number of factors determine the pressure coefficient of the Type I

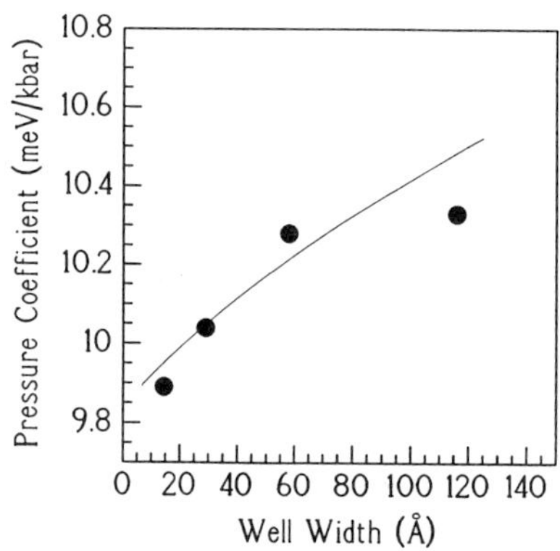

Fig. 12. The pressure coefficients of the PL of the GaAs/AlGaAs quantum wells are plotted against well width. The solid line is a guide to the eye.

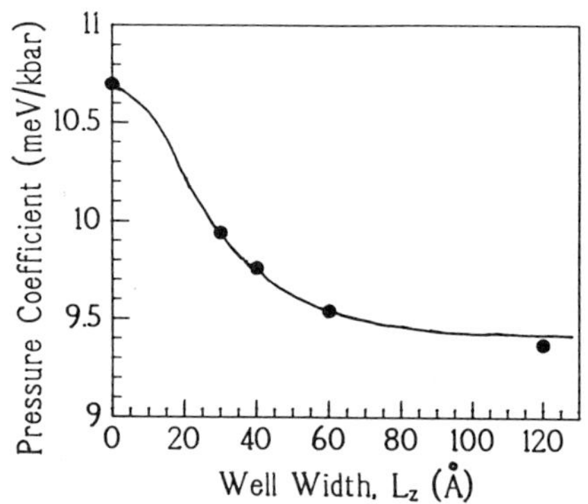

Fig. 13. The pressure coefficients, of the quantum-well transitions for the In$_{.18}$Ga$_{.82}$As/GaAs structure with wells of various widths, are shown. The solid line is a fit to the data assuming that the pressure coefficient of bulk strained In$_{.18}$Ga$_{.82}$As is 9.4 meV/kbar. *(Reproduced from reference 30.)*

310

transition. These have been discussed fully by Lefebvre et al.[16] and by Lambkin et al.[27] In the GaAs/AlGaAs system the most important factors are the coefficients of the constituent materials and the pressure dependence of the offsets. In this system the well material has a higher pressure coefficient (10.7 meV/kbar[18]) than the barrier material (9.9 meV/kbar for $Al_{.28}Ga_{.72}As$[28]). Therefore wide wells have pressure coefficients close to that of GaAs, whilst for narrower wells, where the carrier wavefunctions are predominantly in the barrier material, the pressure coefficient is reduced towards that of the barrier material.

In InGaAs/GaAs and GaAsSb/GaAs strained-layer quantum wells the reverse trend is observed. A sample with five $In_{.18}Ga_{.82}As$ quantum wells of different width has been studied.[29,30] The pressure coefficients as a function of well width are shown in figure 13. The figure shows that the pressure coefficients depend strongly on well width with decreasing coefficients for wider wells. This is unexpected because the pressure coefficients of the direct gaps of bulk InAs and GaAs are very similar, 10.2[31] and 10.7 meV/kbar respectively. This suggests that the coefficient for $In_xGa_{1-x}As$ should vary little across the alloy range. We would therefore expect the quantum-well pressure coefficients to be rather insensitive to well width. Extrapolation to infinite well width, in figure 13, suggests that the pressure coefficient for bulk strained $In_{.18}Ga_{.82}As$ is in fact 9.4 meV/kbar. The curve shows the best fit to the data taking 9.4 meV/kbar as the pressure coefficient of bulk strained $In_{.18}Ga_{.82}As$ and assuming that the valence-band offset is independent of pressure. The results from GaAsSb/GaAs wells reported earlier are almost identical despite the fact that GaSb has a pressure coefficient of 14.5 meV/kbar[32] and well coefficients would therefore be expected to be greater than GaAs. If a linear interpolation is assumed a value of 11.2 meV/kbar would be expected for wide $GaAs_{.88}Sb_{.12}$ quantum wells.

In the InGaAs system we have also studied the wide well limit as a function of composition or strain.[30] Figure 14 shows the measured pressure coefficients for a sample which contained several wide (100Å) wells of different InGaAs compositions with GaAs barriers. We find the pressure coefficient of the PL for bulk strained $In_xGa_{1-x}As$ varies as (10.7-6.0x) meV/kbar with indium content. Similar low coefficients have been observed in other strained systems including InAs/InP where the coefficient of the well material is estimated to be 8.5 meV/kbar. In each of the cases mentioned above the wells were sufficiently narrow to be below the critical thickness, where plastic strain

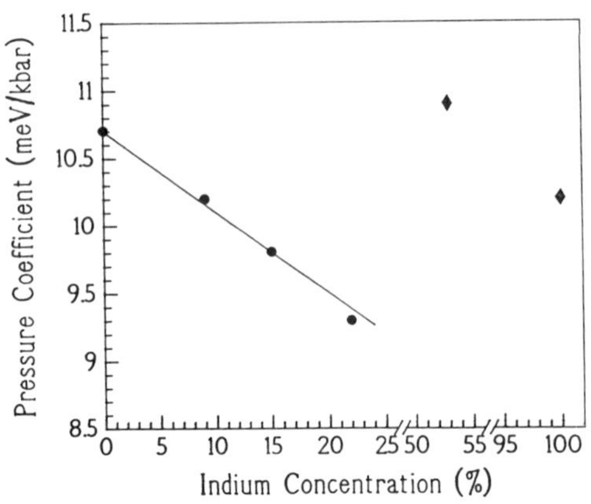

Fig. 14. Pressure coefficients from the structure with 100Å quantum wells of In$_x$Ga$_{1-x}$As of various compositions are plotted against the compositions x (solid circles). Also shown for comparison are the measured pressure coefficients of In$_{.53}$Ga$_{.47}$As and InAs (solid diamonds). The solid line is the linear fit 10.7-6.0x meV/kbar. *(Reproduced from reference 30.)*

relaxation would have commenced. In InGaAs quantum wells the coefficient for the strained In$_{.20}$Ga$_{.80}$As band gap has been confirmed in absorption.[33]   Theoretical investigation of the problem has concentrated on the influence of higher-order elastic constants and strain dependent deformation potentials.[30]   From third-order elasticity theory the predicted variation of the pressure coefficients with composition is given by 10.7-1.7x meV/kbar. While this predicts the trend qualitatively, the decrease is nowhere near sufficient to describe the experimentally observed data. The experimental data could correspond to a composition dependence of the band-gap hydrostatic deformation potential, a, of -7.99(1-0.47x) eV for x<0.25. However, such a strong dependence on composition is not predicted. These results therefore remain anomalous. The magnitude of the pressure coefficients of the indirect, Γ-X gaps of strained InGaAs and GaAsSb are also surprising being -2.6 meV/kbar for In$_x$Ga$_{1-x}$As (x<0.25) and -2.2 meV/kbar for GaAs$_{.88}$Sb$_{.12}$, compared with -1.34 meV/kbar for bulk GaAs.[18]

CONCLUSIONS

The application of high pressure PL to the study of strained-layer quantum well band structures have been described. The importance of Type II transitions in accurate determination of the band offsets has been demonstrated, and a new method for determining the offsets for a solely Type I system has been described and illustrated for the case of InGaAs/GaAs quantum wells. The fractional valence-band offset for InGaAs/GaAs heterojunctions has been measured to be 0.40 ±0.02. The band alignment for $GaAs_{.88}Sb_{.12}$/GaAs quantum wells is found to be Type I with a small conduction-band offset contrary to previous results. Finally a study of the direct band-gap of compressively strained layers under hydrostatic pressure has found pressure coefficients consistently lower than predicted by theory.

ACKNOWLEDGEMENTS

We are grateful to our co-workers for their contributions to the work reported above (see reference list). Also to the Optoelectronics Joint Research Laboratories, Japan, to The Royal Signals and Radar Establishment, UK and to UMIST, UK for the provision of samples. Special thanks are due to L.K. Howard and M.T. Emeny (RSRE) for the supply of many InGaAs samples, to A. D. Prins for much of the high pressure work and to D. J. Dunstan and E. P. O'Reilly for helpful discussions during the course of this work. We are grateful to the Science and Engineering Research Council, UK for financial support.

REFERENCES

1. E. P. O'Reilly, Valence band engineering in strained-layer structures, <u>Semicond. Sci. Technol.</u> 4:121 (1989).
2. A. R. Adams, Hydrostatic pressure investigation of quantum well optoelectronic devices, <u>in</u>: "Frontiers of High Pressure Research" H. D. Hochheimer and R. D. Etters, eds., Plenum (1991).
3. A. R. Adams, Band-structure engineering for low-threshold high-efficiency semiconductor lasers, <u>Elec. Lett.</u> 22:249 (1986).
4. W. Paul and D. M. Warschauer, The role of pressure in semiconductor research, <u>in</u>: "Solids Under Pressure", W. Paul and D. M. Warschauer, eds., McGraw-Hill, New York (1963).
5. R. J. Warburton, R. J. Nicholas, N. J. Mason, P. J. Walker, A. D. Prins and D. J. Dunstan, High-pressure investigation of GaSb and $Ga_{1-x}In_xSb$/GaSb quantum wells, <u>Phys. Rev.</u> B43:4994 (1991).
6. D. J. Wolford, T. F. Keuch, J. A. Bradley, M. A. Gell, D. Ninno and M. Jaros, Pressure dependence of $GaAs/Al_xGa_{1-x}As$ quantum well bound states: The determination of valence band offsets, <u>J. Vac. Sci. Technol.</u> B4:1043 (1986).
7. U. Venkateswaren, M. Chandrasekhar, H. R. Chandrasekhar, B. A. Vojak, F. A. Chambers and J. M. Meese, High pressure studies of $GaAs/Ga_{1-x}Al_xAs$ quantum wells of widths 26Å to 150Å, <u>Phys. Rev.</u> B33:8416 (1986).

8. See, for example, R. Dingle, W. Wiegmann and C. H. Henry, Quantum states of
    confined carriers in very thin $Al_xGa_{1-x}As$ heterostructures, <u>Phys. Rev. Lett.</u>
    33:827 (1974).
9. P. Dawson, K. J. Moore and C. T. Foxon, Photoluminescence studies of type II
    GaAs/AlAs quantum wells grown by MBE, <u>SPIE Quantum Well and
    Superlattice Physics</u> 792:208 (1987).
10. D. J. Dunstan and W. Scherrer, Miniature cryogenic diamond-anvil high-pressure
    cell, <u>Rev. Sci. Instrum.</u> 59:627 (1988).
11. D. J. Dunstan and V. A. Wilkinson, Miniature cryogenic diamond anvil cell, <u>High
    Pressure Research</u> 5:794 (1990).
12. J. D. Barnett, S. Block and G. J. Piermarini, An optical fluorescence system for
    quantitative pressure measurements in the diamond-anvil cell, <u>Rev. Sci.
    Instrum.</u> 44:1 (1973).
13. A. D. Prins and D. J. Dunstan, A determination of the relative bulk moduli of
    GaInAsP and InP, <u>Phil. Mag. Lett.</u> 58:37 (1988).
14. T. Fukunaga, T. Takamoti and H. Nakashima, Photoluminescence from AlGaAs-
    GaAs single quantum wells grown on variously oriented GaAs substrates by
    MBE, <u>J. Crys. Growth</u> 81:85 (1987).
15. G. Bastard, Superlattice band structure in the envelope function approximation,
    <u>Phys. Rev.</u> B24:5693 (1981).
16. P. Lefebvre, B. Gil and H. Mathieu, Effect of hydrostatic pressure on GaAs-
    $Al_xGa_{1-x}As$ microstructures, <u>Phys. Rev.</u> B35:5630 (1987).
17. K. J. Moore, P. Dawson and C. T. Foxon, Observation of luminescence from the 2S
    heavy-hole exciton in GaAs-(AlGa)As quantum-well structures at low
    temperature, <u>Phys. Rev.</u> B34:6022 (1986).
18. D. J. Wolford and J. A. Bradley, Pressure dependence of shallow bound states in
    gallium arsenide, <u>Solid State Commun.</u> 53:1069 (1985).
19. H. Muller, R. Trommer, M. Cardona and P. Vogl, Pressure dependence of the
    direct absorption edge of InP, <u>Phys. Rev.</u> B21:4879 (1980).
20. A. R. Goni, K. Strossner, K. Syassen and M. Cardona, Pressure dependence of
    direct and indirect optical absorption in GaAs, <u>Phys. Rev.</u> B36:1582 (1987).
21. H. Q. Hou, L. J. Wang, R. M. Tang and J. M. Zhou, Pressure dependence of
    photoluminescence in $In_xGa_{1-x}As$/GaAs strained quantum wells, <u>Phys. Rev.</u>
    B42:2926 (1990).
22. V. A. Wilkinson, A. D. Prins, D. J. Dunstan, L. K. Howard, M. T. Emeny,
    Investigation of the band structure of the strained systems InGaAs/GaAs and
    InGaAs/AlGaAs by high-pressure photoluminescence, <u>J. Elec. Mat.</u> 20:509
    (1991).
23. G. Ji, S. Agarwala, D. Huang, J. Chyi and H. Morkoc, Band lineup in
    $GaAs_{1-x}Sb_x$/GaAs strained-layer multiple quantum wells grown by molecular
    beam epitaxy, <u>Phys. Rev.</u> B38:10571 (1988).
24. A. D. Prins, J. D. Lambkin, E. P. O'Reilly, A. R. Adams, R. Pritchard, W. S.
    Truscott and K. E. Singer, Band Offsets in GaAsSb/GaAs strained-layer
    structures from high-pressure photoluminescence, <u>Proc. IVth Conf. on 'High
    Pressure in Semiconductor Physics'</u>, Porto Caras, 933 (1990).
25. C. G. Van de Walle, Band lineups and deformation potentials in the Model-Solid
    theory, <u>Phys. Rev.</u> B39:1871 (1989).
26. R. E. Nahory, M. A. Pollak, J. C. Dewinter and K. M. Williams, Growth and
    properties of liquid-phase epitaxial $GaAs_{1-x}Sb_x$ <u>J. Appl. Phys.</u> 48:1607 (1977).
27. J. D. Lambkin, D. J. Dunstan, E. P. O'Reilly and B. R. Butler, The pressure
    dependence of the band offsets in a GaInAs/InP multiple quantum well
    structure, <u>J. Crys. Growth</u> 93:323 (1988).

28. M. Chandrasekhar, U. Venkateswaren, H. R. Chandrasekhar, B. A. Vojak, F. A. Chambers and J. M. Meese, <u>Proc. XV111th Int. Conf. on the physics of semiconductors</u>, Stockholm (1986).

29. A. D. Prins, J. D. Lambkin, K. P. Homewood, M. T. Emeny and C. R. Whitehouse, Photoluminescence of InGaAs/GaAs strained-layer structures under high pressure, <u>High Pressure Research</u> 3:48 (1990).

30. V. A. Wilkinson, A. D. Prins, J. D. Lambkin, E. P. O'Reilly, L. K. Howard and M.T. Emeny, Hydrostatic pressure coefficients of the photoluminescence of $In_xGa_{1-x}As$/GaAs strained-layer quantum wells, <u>Phys. Rev.</u> B42:3113 (1990).

31. Y. F. Tsay, S. S. Mitra and B. Bendow, Pressure dependence of energy gaps and refractive indices of tetrahedrally bonded semiconductors, <u>Phys. Rev.</u> B10:1476 (1974).

32. R. A. Noak and W. B. Holzapfel, Photoluminescence of GaSb under hydrostatic pressure, <u>Solid State Commun.</u> 28:177 (1978).

33. A. D. Prins, Private communication, (1991).

PRESSURE-TUNED RESONANCE RAMAN SCATTERING STUDIES

ON SUPERLATTICES

Gerasimos A. Kourouklis

Aristotle University of Thessaloniki
Physics Division, School of Technology
Thessaloniki, 54006, Greece

ABSTRACT

Resonant Raman Scattering is a useful tool to probe the electronic band structure of semiconductors, and to study the electron-phonon interaction in them. Pressure-induced resonance Raman scattering in multiple quantum well structures, of even small multiplicity (≈ 10 layers), is a strong enough effect to be useful for probing the electronic structure and characterize the sample. The nature of the resonance mechanisms can be experimentally clarified by the simultaneous measurement of both the photoluminescence and the Raman intensities on the same spot of the sample as a function of pressure. Some recent pressure-induced resonance Raman studies on GaAs and GaAs based superlattice systems are presented and reviewed.

INTRODUCTION

The advancement in epitaxial growth techniques has given the capability of growing a wide range of superlattice systems. The electronic structure of these materials depends on the layer thicknesses as well as on the constituent materials. Because the layer thicknesses and the constituent concentrations can be precisely controlled, it is possible to fabricate superlattices with prescribed electronic band structure. The various superlattice systems show qualitatively different physical behavior because of the different band-edge line-ups, strain conditions, and growth orientations. The flexibility in electronic properties, which is introduced by this design possibility, makes the superlattices useful in several technological applications, including semiconductor diode lasers, electro-optic modulators, nonlinear optical materials and infrared detectors. This wide range of applications has made them one of the most studied class of materials. Many new systems are being fabricated and their properties have not yet been investigated.

The Resonant Raman Scattering (RRS) by optical phonons, plasmons, and single particle excitations has been investigated, theoretically and experimentally, during the last two decades[1-5]. The strong dependence of the Raman efficiency on the incident photon energy is used to provide information on the excitonic spectrum. The resonant Raman scattering profile also contains information on the exciton-phonon scattering mechanisms. Therefore resonant Raman scattering has become a powerful tool to study the optical properties of quantum-well heterostructures[5].

In a resonance Raman experiment, usually the energy of the Raman excitation frequency is varied, while the gap energies of the system remain constant. However, it is also possible to achieve resonance by pressure tuning the electronic gap energies of the material into resonance with the fixed excitation energy ($\hbar\omega_L$)[6-10]. This method is particularly suited for the zinc-blende type of semiconductors, since their direct gaps (E_0) have large pressure coefficients[11] ($dE_0/dP \approx 100$ meV/GPa). Although GaAs is one of the most extensively studied semiconductors, from the point of view of RRS, pressure-induced RRS experiments in bulk GaAs have received much less attention. Semiconductor superlattices and quantum-well structures present a novel situation in having additional features in their electronic structure, which also respond to pressure in a similar way, and therefore, are of interest from the point of view of pressure-induced RRS. Recent pressure-tuned RRS studies on GaAs/AlAs thin layer superlattice[10] and on multiple quantum well (MQW) structures[12,13], have elucidated some aspects related to the nature of the pressure-tuned resonances encountered in these novel semiconducting layered systems. In pressure-induced RRS experiments on a number of compositionally different MQW structures grown on GaAs substrates[12,13] a serious interference problem, both in Raman scattering and in photoluminescence (PL), from the GaAs substrate layer was encountered. With increasing pressure, the layers become transparent to the exciting laser frequency, especially when the latter lies in the red region, and the substrate layer of GaAs dominates the Raman features. This prompted the investigation of the pressure-induced RRS of high quality MBE grown bulk GaAs[14]. This study has revealed a number of interesting new features concerning the interrelationship between electronic structure, luminescence and RRS in bulk GaAs. In this paper the results on resonant Raman scattering induced by pressure on pure MBE grown GaAs as well as on GaAs based superlattice systems will be presented and reviewed.

EXPERIMENTS

The use of the diamond anvil cell has provided us with an effective and reliable way to generate hydrostatic pressure. In working with superlattices based on GaAs, due to their big pressure coefficient of the direct gap, is very important to maintain a high degree of hydrostaticity within the pressure chamber. The pressure transmitting medium plays an important role in fulfilling this requirement. In the experiments reported in this paper hydrostatic pressure was generated using the gasketed Mao-Bell type diamond anvil cell[15], with argon as the pressure medium. High pressure argon filling was carried out using a high pressure gas loading system[16]. The pressure generated was calibrated by the well-known ruby fluorescence technique[17]. The experiments were carried out at room temperature. In our days the possibility of creating and maintaining high pressure at low temperatures (even at liquid helium) is a reality. Despite this fact there are not high pressure RRS studies reported at low temperatures.

The quantum-well samples were grown by molecular bean epitaxy[12,13] usually on a (100) GaAS substrate. For quantum well layer, either In doped GaAs or pure GaAs, was used while for the barrier layer Al doped GaAs was used. Studies were made on samples with various doping levels. Some of the samples used in the experiments were of very low multiplicity, 10 periods of alternating layers, producing quantum well structures of thickness even less than 0.5 μm. This small thickness allows the excitation light to penetrate into the substrate, especially in the transparent regime ($\hbar\omega_L < E_0$) contributing to the scattered intensity as well as giving rise to resonances originating in the substrate. In this paper we are going to present and discuss data taken on a sample with nominal composition $In_{0.15}Ga_{0.85}As/Al_{0.30}Ga_{0.70}As$. To have a better understanding of the resonance mechanisms a study of the resonant scattering in the substrate material (GaAs) under similar conditions was necessary. Therefore RRS studies on high quality epitaxially grown[14] GaAs were also conducted. The GaAs was grown by molecular beam epitaxy on a (100) substrate of GaAs, the epitaxial layer had a 12 μm thickness. The samples have been especially prepared to be used in the high pressure experiments. The substrate side was thinned down to 25 μm, and cleaved rectangular pieces of about 100 μm in dimension were mounted inside the gasket hole of the diamond cell.

All Raman measurements were carried out at room temperature with a Spex double monochromator equipped with a conventional photon counting system. The photon energy (1.916 eV) of the 647.1 nm line of the Kr^+ laser is appropriate for pressure tuned RRS measurements in GaAs-based superlattice systems. It takes only a few GPa to tune the gap to this excitation energy. Excitation with higher photon energies, e.g. the 514.5 and 488 nm of the Ar^+ laser lines, could also be used, especially if one wants to study the electronic structure by means of the RRS technique above the Γ-X cross-over point. Since the samples were (100) slabs and the scattering geometry was close to the backscattering configuration, the TO phonon is forbidden. But a very weak TO is observed, evidently due to deviation from the above geometry. Polarized RRS studies were not attempted, because of the difficulties inherent to such measurements in the diamond cell.

RESULTS

The investigation of high quality bulk GaAs by pressure-induced RRS has revealed a number of interesting new features concerning the interrelationship between electronic structure, luminescence and and RRS.

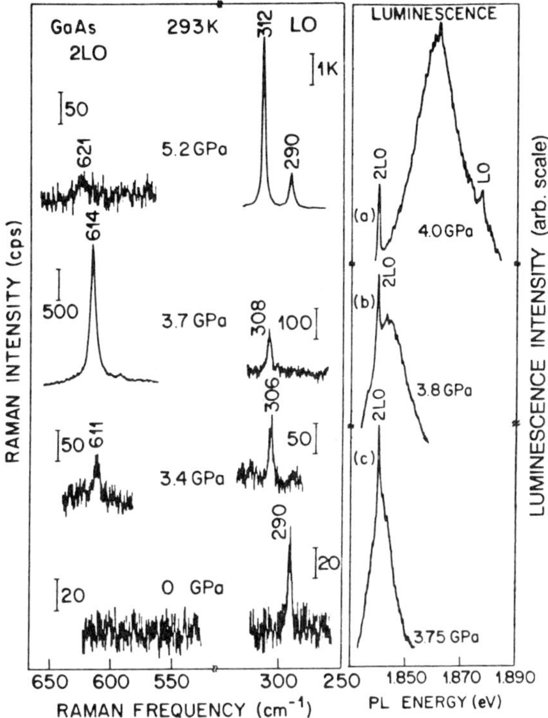

Fig. 1. The Raman intensity of the 2LO-phonon, the LO-phonon and the photoluminescence peak of bulk GaAs at various pressures. The scale for Raman intensities is given in each spectrum. The PL peak is presented at three different pressures close to the 2LO resonance, the spikes on the PL peaks in (b) and (c) are the 2LO phonon, while in (a) the 2LO and LO phonons are symmetrically situated on either side of the PL peak. Excitation is with the 647.1 nm line of the Kr^+ laser at a power level of 20 mW.

In Fig. 1 the 2LO and LO peaks of GaAs recorded with 647.1 nm excitation are shown for ambient pressure and three other pressures (left frame). The strong narrow peak in the 5.2 GPa spectrum is the LO-phonon (312 cm^{-1}) and the weaker one on the right side (290 cm^{-1}) is the TO mode. The TO begins to appear near resonance and is present in all spectra where LO is strong. The photoluminescence peak for three pressures near the 2LO resonance is also shown (right frame). The 2LO peak intensity goes through a pronounced maximum near 3.7 GPa due to pressure-induced RRS. The LO peak intensity also rises sharply between 4 and 5 GPa, but a distinct maximum is not seen. At higher pressures the LO phonon intensity decreases only by a small factor. The photoluminescence peaks are well defined, and correspond to the E_o gap energy. The sharp spikes riding on the PL peaks are due to the 2LO phonon which becomes quite strong; it is, in fact, an order of magnitude stronger than the LO phonon. At 3.75 GPa (see spectrum marked c) the luminescence is strong, and the 2LO phonon sits right on top of the PL peak, and in this situation its intensity has the maximum value. In spectrum (b), a pressure increase of about half a kilobar has created a shift of the PL peak to higher energy relative to the 2LO peak, and the intensity of the latter has dropped by a factor of two. At 4 GPa, spectrum (a) the luminescence peak has shifted further to higher energy relative to the 2LO, and now, both the 2LO and LO phonon peaks can be seen riding on each side of the PL peak. The intensity of the 2LO has dropped further and is about one fourth of its value at resonance. At higher pressure, the luminescence peak shifts to higher energy and ultimately vanishes when $E_o = \hbar\omega_L$ or when Γ crosses X. Concurrently, the 2LO scattering becomes very weak, while the LO gains enormously in intensity.

The 2LO and LO peak intensities are plotted as a function of pressure in Fig. 2. Their pressure-induced RRS behavior is shown clearly for the 2LO phonon which shows a slight asymmetry to the high pressure (high energy) side. Above 5 GPa, the sample appears bright red in color by transmitted light due to the opening up of the E_o gap with pressure.

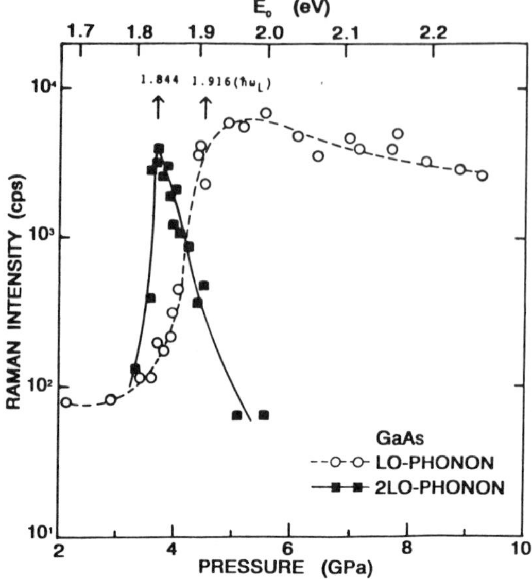

Fig. 2. The intensity of the 2LO and the LO phonon as a function of pressure. Excitation is with the 647.1 nm line, at a constant power level of 20 mW. Notice that the E_0 scale (top) is not linear.

This increase in transparency with pressure is one factor which makes the LO resonance asymmetric to the high energy side.

We proceed to present a set of experimental data which are representative of GaAs based MQW structures. The Raman spectrum from the multiple quantum-well sample $In_{0.15}Ga_{0.85}As/Al_{0.30}Ga_{0.70}As$ is shown in Fig. 3. The 285 and the 294 cm^{-1} peaks are the GaAs-like LO-phonon modes from the AlGaAs layers and the GaAs substrate, respectively. The peak near 380 cm^{-1} is the AlAs-like LO-phonon mode from the AlGaAs layer.

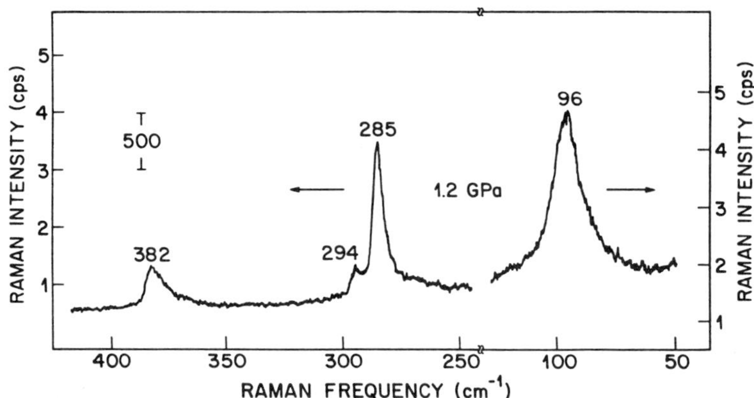

Fig. 3. Raman peaks of the MQW sample $In_{0.15}Ga_{0.85}As/Al_{0.30}Ga_{0.70}As$ recorded in the diamond cell, at 1.2 GPa and ambient temperature. The 285 and 294 cm^{-1} peaks are, respectively, from the AlGaAs and GaAs-InGaAs layers. The InGaAs phonon is too weak and too close to GaAs to be seen separately. The 382 cm^{-1} peak is the AlAs-like mode from the AlGaAs layer. Excitation is with 647.1 nm at 20 mW power.

320

The pressure-dependence of the first-order GaAs LO-like and the AlAs LO-like Raman peaks is shown in Fig. 4 along with some of the second-order peaks. The observed phonon modes and their pressure dependences are listed in Table I. The lowest frequency phonon mode near 100 cm^{-1} (see also Fig. 3) comes out strongly which may be attributed to resonance enhancement. From the observed negative pressure dependence of this mode (see Table I), we conclude that this is a folded TA acoustic phonon.

Table I. Phonon frequencies and their pressure dependence in $In_{0.15}Ga_{0.85}As/Al_{0.30}Ga_{0.70}As$ MQW.

	ω^+ cm^{-1}	$d\omega/dP$ cm^{-1}/GPa
TA	100	-5.32
LO	273	3.02
	281	4.23
	291	4.50
LO^*	377	4.95
2LO	568	9.21
	574	9.73
	585	5.89

$^+$ Extrapolated values to zero pressure.
* AlAs-like LO phonon. The rest are the GaAs-like phonons.

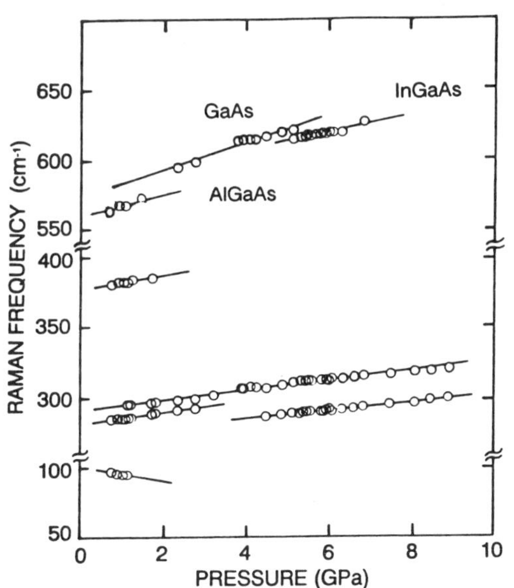

Fig. 4. Pressure dependence of the observed phonon modes in $In_{0.15}Ga_{0.85}As/Al_{0.30}Ga_{0.70}As$ MQW. The solid lines are linear least-square fits to the data.

In GaAs, when the LO phonon comes out strongly, the TO mode also shows up with considerable intensity (see Fig. 1). This could be due to the slight deviation from the geometry in which the TO is forbidden. We also find other features in the second-order spectrum appearing at pressures above 4 GPa, which are undoubtedly from the GaAs substrate.

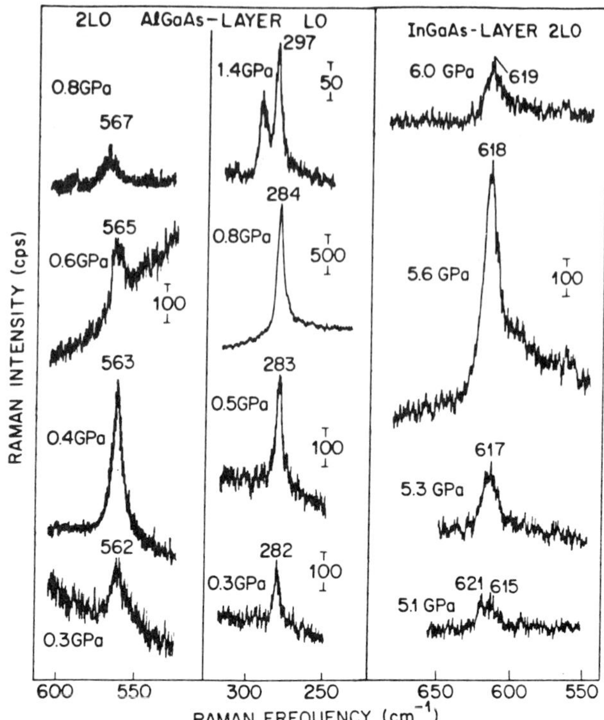

Fig. 5. Raman peaks of the MQW $In_{0.15}Ga_{0.85}As/Al_{0.30}Ga_{0.70}As$. Spectra of the AlGaAs layer showing the 2LO (left) and the LO-peaks (middle) at different pressures near resonance. The 2LO-peak intensity maximizes near 0.4 GPa and the LO intensity near 1 GPa. The 2LO-phonon peak from the InGaAs quantum well (right) maximizes near 5.5 GPa. The frequency of this peak is distinctly lower than that of the GaAs 2LO, confirming its origin from the InGaAs quantum well. Excitation is with 647.1 nm at 20 mW constant power level and ambient temperature.

In Fig. 5, the intensities of both the 2LO and LO-phonon peaks for the AlGaAs and the 2LO of the InGaAs layers are shown at different pressures. The intensity of the phonon peaks increases first and then decreases, and this is due to pressure-induced RRS. The LO-phonon peak from the InGaAs quantum well is totally masked by the GaAs substrate, and hence not reproduced. Since the total thickness of ten InGaAs layers is only 700 A, the 2LO is hardly visible, except near the resonance condition. The pressure dependence provides a way to distinguish this phonon from the GaAs 2LO phonon (see Table I).

In Figures 6 and 7, the 2LO and LO-scattered Raman peak intensities are plotted as a function of pressure, to show their resonant behavior. Three distinct narrow resonance peaks are seen for the 2LO at 0.4, 3.8 and 5.5 GPa, which are, respectively, from the AlGaAs, GaAs and InGaAs layers. At these pressures the 2LO-phonon goes through its maximum intensity and, furthermore the 2LO-peak falls right on top of the PL peak. This is shown in Fig. 8 for the three layers. Identical features are seen for the AlGaAs and InGaAs except that the intensities are weaker and the resonances occur at different pressures.

The LO-resonance profile for the AlGaAs layer is located near 1 GPa and is narrow and sharp. On the other hand, the LO-profile of GaAs has an asymmetric shape and the fall-off on the high energy side is very gradual and small. This behavior is similar to the bulk GaAs (see Fig. 2). This difference in the resonance behavior of the layers can be explained by their difference in thickness. GaAs, being the substrate, is much thicker than the AlGaAs layers in the MQW. Close and above resonance the transparency increases with pressure and so the scattering volume. This affects mostly the thick GaAs layer increasing the LO scatter-ing intensity. This effect competes with the decreasing of the intensity due to going out of resonance. The result of these two processes we believe explains the shape of the high

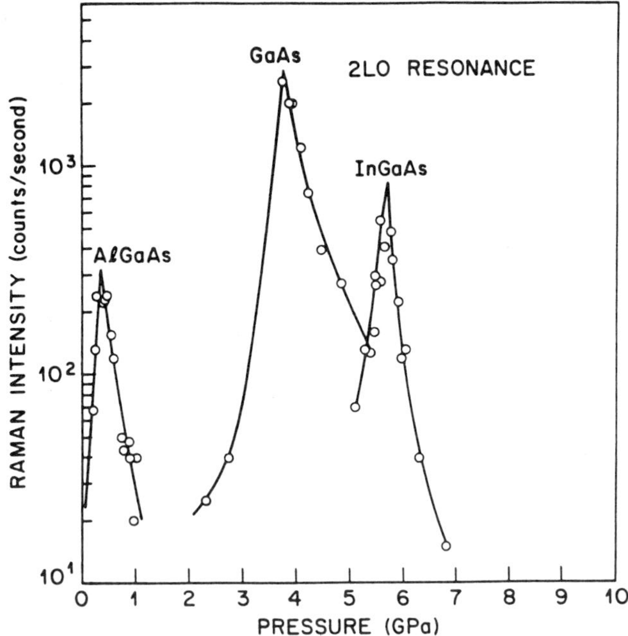

Fig. 6. The 2LO-resonance profile of the MQW $In_{0.15}Ga_{0.85}As/Al_{0.30}Ga_{0.70}As$. Three distinct resonances centered at 0.4, 3.8 and 5.5 GPa are seen, coming from the Al-GaAs, GaAs and InGaAs layers, respectively. The resonances are quite narrow and very sharp. At the above resonance-pressures the LO phonon intensities are an order of magnitude weaker, compared to the 2LO intensity. The solid line is a guide to the eye.

energy side of the LO resonance profile in the GaAs layer. Above 5 GPa, the samples appear bright red in color by transmitted light and this is connected with the opening up of the E_o gap with pressure. It is this increase in transparency with pressure which makes the GaAs PL-peak come out prominently at higher pressures.

DISCUSSION

Resonance Raman scattering experiments in bulk GaAs near the E_o band gap have been reported in earlier publications[6-8,18]. Both the LO and the 2LO phonon resonances have been observed and theoretical studies have been performed[19,20]. The situation in quantum well structures is even more favorable for RRS studies due to the well established fact that they present sharp energy levels[21,22]. Thus, when the excitation energy $\hbar\omega_L$ becomes equal to E_i (an electronic gap energy of the system) or, when $\hbar\omega_L = E_i + \hbar\Omega_{ph}$, resonances in the Raman efficiency become possible. These are, respectively, known as "incoming" and "outgoing" resonances. These resonances have been observed for the LO and 2LO phonons in MQW structures[22-24], and in bulk GaAs at the E_o and $E_o + \Delta_o$ gaps[25,26]. However, we must remark that pressure-tuned RRS studies have been rather limited and the published accounts of them have not adequately pointed out the special features and advantages of this method. For instance, it would be impossible with incident energy tuning to study experimentally the relationship between the PL and the 2LO intensity, such as the one shown in Fig. 1 (right frame), because of the overpowering intensity of the PL. In pressure-tuned RRS studies if one chooses an excitation energy somewhere close to the gap energy at the Γ-X cross-over transition; then becomes possible to monitor concurrently the Raman and the photoluminescence intensities allowing the experimental observation of both the Raman intensity and the energy-gap position simultaneously. With

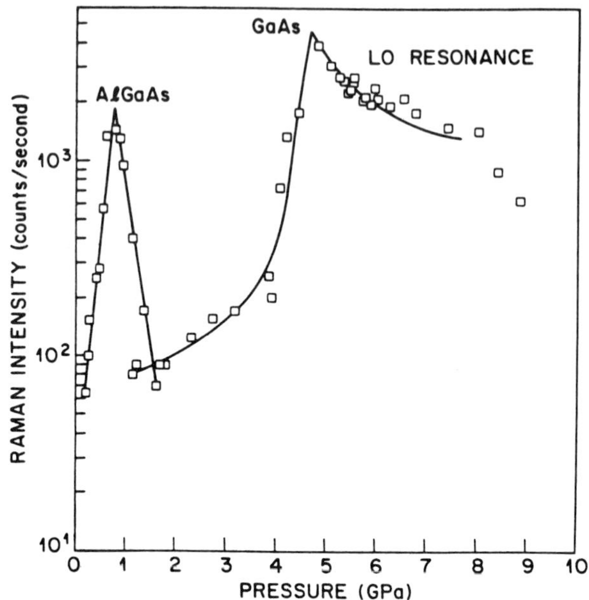

Fig. 7. The LO-resonance profile of the MQW $In_{0.15}Ga_{0.85}As/Al_{0.30}Ga_{0.70}As$. The first narrow resonance is from the AlGaAs layer, which is centered at ≈ 1 GPa. The second highly asymmetric resonance profile is mainly due to the GaAs substrate. The shape of the intensity profile is complicated by the pressure-induced transparency of the sample. The solid lines are drawn to guide the eye.

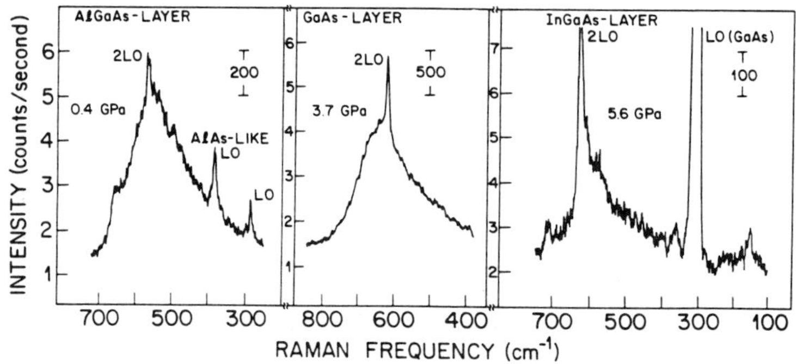

Fig. 8. The 2LO-peaks as observed at pressures very close to the resonance, riding on top of the PL peaks of their respective layers. At the resonance maxima the 2LO-peaks fall right on top of the corresponding PL peaks, revealing that at the 2LO resonance the direct gap energy of the respective layer $E_o = \hbar\omega_S(2LO)$.

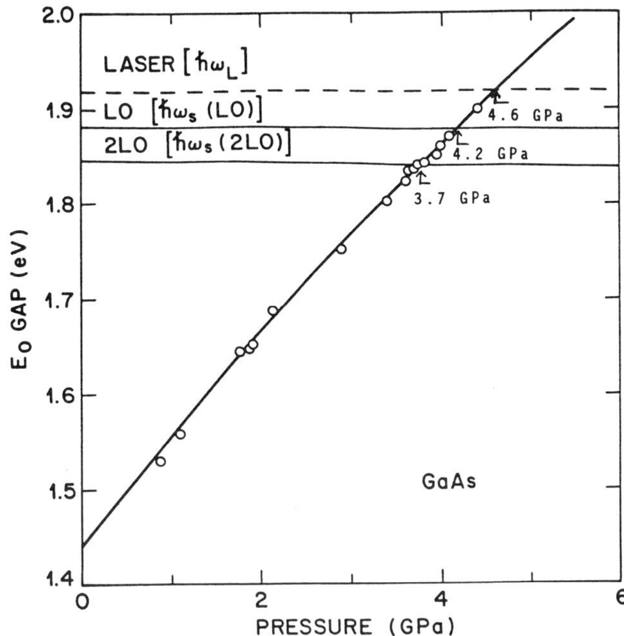

Fig. 9.   The direct gap (E_o) shift as a function of pressure obtained from the PL measurements for GaAs.   The solid line is a second degree polynomial fit to the data.   The three lines are the laser excitation photon energy $\hbar\omega_L$, the LO scattered photon energy $\hbar\omega_S(LO)$ and the 2LO scattered photon energy $\hbar\omega_S(2LO)$ respectively. Their intersections define the incoming ($E_o = \hbar\omega_L$) and the outgoing $E_o = \hbar\omega_S(LO)$ and $E_o = \hbar\omega_S(2LO)$ resonance pressures. The $\hbar\omega_S(LO)$ and $\hbar\omega_S(2\,LO)$ lines are slightly inclined to reflect the pressure shift of the phonon frequency. This shift is two orders of magnitude smaller compared to the E_o shift.

647.1 nm excitation and pressure tuning, the RRS for the 2LO in GaAs occurs at a pressure of 3.75 GPa which is close to the Γ-X crossing. Near that band crossing the PL intensity falls off drastically and as a result the observation of the PL and the 2LO peaks together becomes possible revealing directly that $\hbar\omega_S$ (2LO) = E_0 at the 2LO resonance.

This possibility, of simultaneous monitoring of the PL and the Raman intensity, has given us the opportunity to create diagrams like the ones depicted in figures 9 and 10 which play a key role in understating the pressure tuned RRS processes[12-14]. Fig. 9 is the key diagram to understand pressure-tuned RRS in bulk GaAs. In this figure the E_o gap energy is plotted as a function of pressure. These values were determined from the observed sharp PL peaks (right frame in Fig. 1) and the data are consistent with previous measurements[8]. The three lines drawn on the upper part of this figure represent the incident laser excitation photon energy ($\hbar\omega_L = E_L = 1.916$ eV for the 647.1 nm of the Kr^+ laser used for excitation), the LO phonon scattered photon energy $\hbar\omega_S (LO) = E_{LO} = 1.881$ (eV) - 0.625* 10^{-3}(eV/GPa)*P(GPa), and the 2LO-phonon scattered photon energy $\hbar\omega_S (2LO) = E_{2LO} = 1.846$ (eV) - 1.328*10^{-3}(eV/GPa)*P(GPa), respectively. The intersections of these lines with the E_o gap energy vs. P curve provide us with an insight in the RRS processes.

When pressure is used to tune an electronic level E_i to obtain resonance, the pressure at which resonance occurs is determined by $\hbar\omega_L$ and the pressure dependence of the gap E_i. In most of our experiments, $\hbar\omega_L$ was set at 1.916 eV and the E_i are the direct energy gaps E_o of the AlGas-InGaAs-GaAs system.   In order to characterize the nature of the resonances, we have determined from photoluminescence measurements the variation of the E_o gap energy with pressure for the InGaAs, GaAs, and AlGaAs layers in the MQW structure, these are presented in Fig. 10. Again in the upper part of Fig. 10, the three lines represent the incident photon energy $\hbar\omega_L$, the LO-phonon scattered photon energy $\hbar\omega_S(LO)$ and

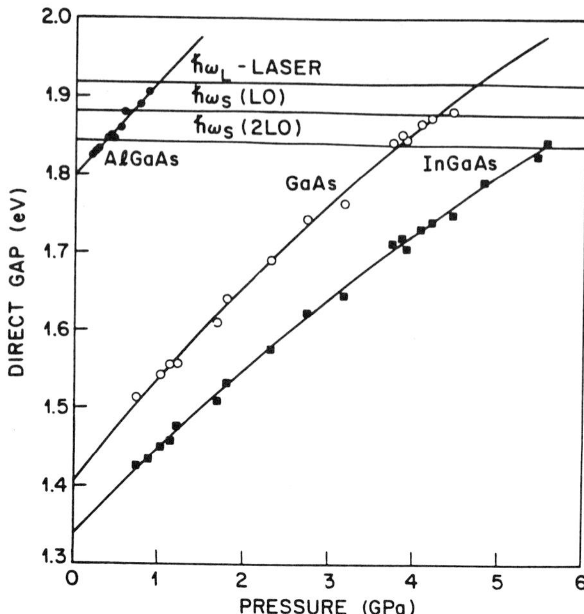

Fig. 10. Shows the pressure dependence of the direct energy gap (E_i) of the different layers in the MQW sample. Solid lines represent fits of the experimental data to a second degree polynomial from which the dE_i/dP for GaAs is obtained as 120 ± 10 meV/GPa. The three lines define the energy of the exciting laser line $\hbar\omega_L$, the LO-phonon scattered photon energy $\hbar\omega_S$(LO) and the 2LO-phonon scattered photon energy $\hbar\omega_S$(2LO) for the AlGaAs layer. The intersection of these lines with the E_i curves define the condition for RRS processes in the respective layer.

the 2LO-phonon scattered photon energy $\hbar\omega_S$(2LO) for the AlGaAs layer. Corresponding lines for the other two layers, if plotted, would have been very close, so they are omitted. Since resonance occurs when these energies become equal to E_i, the intersections of the above three lines with the E_i vs. pressure curves, define the resonance pressure. Accordingly, $E_i = \hbar\omega_S$(2LO) near 0.4 GPa for AlGaAs, 3.8 GPa for GaAs and 5.5 GPa for InGaAs. These are the pressures at which the resonance peaks occur, in excellent agreement with the predictions from PL data. The intersection of the $\hbar\omega_S$(LO) and $\hbar\omega_L$ define the resonance condition for the LO-phonon and, in fact, the observed LO-phonon resonance peak of the AlGaAs layer is very close to the expected pressure (see Fig. 7). We therefore believe that Fig. 10, as constructed, provides the key-diagram to understand pressure-tuned RRS in the MQW structure.

The 2 LO Resonance

The pressure-induced 2LO resonance in GaAs is sharply defined and narrow and with 647.1 nm excitation it peaks near 3.75 GPa (see Fig. 2). The slight asymmetry, we believe is due to the pressure-induced transparency and to the weak incoming channel. We have conducted experiments[27] with a thin layer ($< 1\mu$m) of GaAs on InP substrate to check out the transparency effect. Indeed the 2LO resonance shows both the "incoming" and "outgoing" channels. For the RRS the requirement is, that an electronic gap energy in the system must become equal to the incident radiation $\hbar\omega_L$, or, to the Raman scattered radiation $\hbar\omega_S$. In the case of GaAs, the E_0 gap becomes equal to $\hbar\omega_S$ (2LO) at 3.84 GPa for $\hbar\omega_L = 1.916$ eV, and this is shown by the intersection of the $\hbar\omega_S$ (2LO) line with the E_0 vs. P curve in Fig. 9. This is the condition for the so-called "outgoing" resonance, namely $\hbar\omega_L = E_0 + \hbar\omega_{2LO}$. Furthermore, when $\hbar\omega_S$ (2LO) $= E_0$, the 2LO peak should fall right on top of the PL peak, this is shown in Fig. 1 (right frame). The exciting radiation $\hbar\omega_L = \hbar\omega_S + \hbar\omega_{2LO}$ takes an electron to the conduction band, which is scattered twice in the band (intraband Frolich mechanism) by one LO phonon, before recombining with the hole to give the Stokes radiation $\hbar\omega_S$(2LO). The latter exactly matches with the luminescence and, there, appears as a strong spike on the PL peak in Fig. 1. At resonance, luminescence and Raman scattering become indistinguishable. In a pressure-induced RRS experiment the "outgoing" resonance will occur first, if the gap energy increases with pressure, and this is because $\hbar\omega_S < \hbar\omega_L$. The relevant electron-phonon processes are schematically shown in Fig. 11.

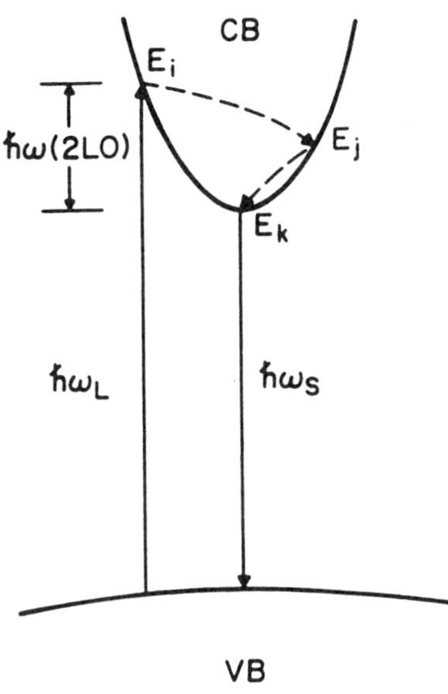

Fig. 11. The 2LO resonant Raman scattering mechanisms in bulk GaAs and MQW are illustrated in this figure. Here $\hbar\omega$, is the incident radiation energy and $\hbar\omega_S$ is the 2LO scattered light energy. E_i, E_j and E_k are the intermediate states which involve electron-phonon scattering through the Frolich interaction.

We have also carried out RRS experiments with other excitation frequencies viz the 568.2 nm line of the krypton laser. The energy of this line is $\hbar\omega_L = 2.182$ eV, and in this case we have observed a sharp pressure-induced 2LO resonance, peaking near 6.75 GPa. From the previous discussion at this pressure the E_0 gap energy should become equal to $\hbar\omega_S$ (2LO) $= \hbar\omega_L - \hbar\omega_{2LO} = 2.10$ eV, and this is indeed the case. Our E_0 vs. P curve shown in Fig. 9, when extrapolated, intersects the $\hbar\omega_S$ (2LO) line for 568.2 nm excitation precisely at this pressure, thereby showing that the scaling with $\hbar\omega_L$ is consistent. From this we conclude that the 2LO resonance in bulk GaAs is a very sensitive and precise way to locate the E_0 gap, far beyond the Γ-X cross-over pressure. Similar results have been reported in pressure-induced RRS measurements[10] on GaAs/AlAs thin layer superlattices.

The very large intensity of the 2LO resonance in bulk GaAs is an indication that this is a double resonance involving the "outgoing" and "intermediate" channels. At a higher pressure a double resonance involving the "incoming" and "intermediate" channels may be expected when $E_0 + \hbar\omega_{LO} < \hbar\omega_L < E_0 + \hbar\omega_{2LO}$. However, we observe only the strong "outgoing" double resonance. In fact, this seems to be always the case, and has been noticed in all experiments where the photon energy has been varied to bring about RRS[22,24].

A striking feature of the pressure-induced 2LO resonances are their sharpness and the large intensity of the 2LO phonon. The latter is much stronger, compared to the LO phonon intensity at the peak resonance pressure. Theory[19,20] predicts that the Frolich-interaction-induced scattering for the 2LO phonons can be stronger than the LO. In fact, this has been observed in GaP[28], CdTe[29], and in second-order scattering by confined optical phonons in a GaAs-AlAs superlattice[30]. The same arguments are valid for the behavior of the 2LO resonance scattering in GaAs-based MQW samples.

The two-phonon scattering process is regarded as an iterated one-phonon interaction, and involves perturbation theory carried to the fourth order. The expression for the Raman scattering amplitude contains three energy denominators[32,2].

$$A_{2LO} \propto \frac{<n|H_{ER}|m><m|H_{EL}|1><1|H_{EL}|k><k|H_{ER}|j>}{[\ \hbar\omega_L - E_{g1}(x)\ ]\ [\ \hbar\omega_i - \hbar\omega_{ph} - E_{g2}(x)\ ]\ [\ \hbar\omega_S - E_{g3}(x)]} \qquad (1)$$

where n and j are the initial and the final states, k, l and m represent the intermediate states, H_{ER} and H_{EL} the electron-photon and electron-lattice Hamiltonians, ω_L and ω_S are the incident and scattered light frequencies, ω_{ph} is the sum of the iterated phonon frequency ($2\omega_{LO}$ in the present case), $E_g(x)$'s represent the pressure dependent energies of the transitions between the valance and conduction band states, in which the wavevector dependence has been neglected. In second order scattering, all phonon wave vectors q are activated and, hence, q vectors which fulfill $E_{g2}(x) = E_{g3} + hq^2/2m^*$ lead to intermediate resonance, simultaneously with either the incoming or outgoing resonances.

The LO Resonance

The LO-resonance profile of GaAs is shown in Fig. 2, this profile is similar to the one observed in the MQW structure attributed to the GaAs substrate. In both cases a shallow peak is observed at $\hbar\omega_L = E_0$. For the LO scattering, the Raman efficiency can be approximated by the expression[33]

$$|R_{LO}|^2 \propto \left|\ A + \frac{B}{[\ \hbar\omega_L - E_0(P) + i\Gamma\ ][\ \hbar\omega_S - E_0(P) + i\Gamma\ ]}\ \right|^2 \qquad (2)$$

where A and B are constants, $\hbar\omega_L$ and $\hbar\omega_S$ are the energies of the incident and scattered photons, $E_0(P)$ is the pressure dependent energy gap and Γ is the damping constant. As the gap is varied by hydrostatic pressure, the incoming resonance should occur when $E_0 = \hbar\omega_L$, and the outgoing resonance when $E_0 = \hbar\omega_S(LO)$. However, only a steep rise of the Raman intensity from the low pressure side is observed. At pressures above the maximum in the LO resonance curve, the intensity decreases gradually by a small factor. A resonant response should exhibit symmetric shape and further the true resonance may be even smaller in magnitude when corrected for the transparency effect. The LO-resonance from the AlGaAs layer is well-defined and has a Lorentzian shape, as would be expected. In this case, however, the total thickness of the AlGaAs layer is less than one micron and, hence, the entire thickness may be expected to contribute to the scattering, thus eliminating any effect attributable to pressure-induced transparency. One observes the true LO-resonance in this case. This is a strong indication that the situation is complicated by the pressure-induced transparency of the sample at pressures above 4 GPa, masking the true resonance profile. Indeed in RRS studies[27] done on a thin (<1μm) MBE grown GaAs on InP substrate, a nearly symmetric LO resonance profile was observed. In the MQW sample, as the gap is varied by hydrostatic pressure, the incoming resonance should occur when $E_0 = \hbar\omega_L$, and the outgoing resonance when $E_0 = \hbar\omega_S(LO)$. The resolution of these two resonances is, however, difficult to achieve. The curve generated by Eq.(2) should be narrow and symmetric. However, the observed resonance profiles are highly asymmetric in the case of thick samples and, this is a reflection of the pressure-induced transparency. In the presence of a thick GaAs substrate layer, it is impossible to observe the true LO-resonance shape of either the GaAs layer, or the InGaAs layer. The LO-resonance from InGaAs is completely masked by the GaAs substrate layer, since it falls at a higher pressure than that of the GaAs. Therefore, it is impossible to observe the InGaAs-LO-resonance, with GaAs as a substrate.

SUMMARY AND CONCLUSIONS

Resonance Raman Scattering can be observed at constant $\hbar\omega_L$ by pressure-tuning the direct electronic gaps, in bulk GaAs and in MQW semiconductor structures the situation is favorable to study the electronic properties of these systems using this technique. The great advantage of pressure-tuned RRS studies is the possibility to choose the constant excitation energy in such a way as to have the resonance happening close to the Γ-X cross-over transition. Then it is possible to monitor concurrently the Raman and the photoluminescence intensities, allowing the experimental observation of both the Raman intensity and the energy-gap position simultaneously.

The pressure-induced 2LO (second order scattering) resonance is very sharp and narrow. It is experimentally verified that the resonance occurs precisely when $E_0 = \hbar\omega_S(2LO)$ and, therefore each layer, depending on its starting E_0 value and its ω_{2LO}, shows its distinct signature in the 2LO resonance. Hence the 2LO resonance can be used to characterize the samples. The key-diagrams to predict and understand pressure-tuned RRS lies in a construction of gap and the phonon scattered photon energies as shown in Figs 9 and 10. At the 2LO resonance, the 2LO-phonon peak falls right on top of the photoluminescence peak (when PL is observable) and, therefore, the Raman scattered light merges with the luminescence. However, high pressure can tell the difference, because the 2LO and PL peaks move in opposite directions when pressure is applied (see Fig. 1).

With pressure-tuning, the LO-resonance occurs when $\hbar\omega_L = E_0$. This resonance is well-defined, only when layer thickness are of the order of one micron. With thick samples, the pressure-induced transparency progressively increases the scattering volume, and this contribution makes it difficult to separate RRS from the transparency effect. We have been able to observe pressure-tuned RRS of the confined LO-phonon from the AlGaAs barrier layer, but the LO resonance from the quantum-well was totally obscured by strong interference from the thick GaAs substrate. In fact, this would be a serious problem with MQW structures or superlattices on thick GaAs or AlGaAs substrates and, therefore, caution is needed to interpret pressure-induced RRS results, whenever such substrates are present.

Pressure-induced RRS in MQW structures, of even small multiplicity (≈ 10 layers), is a strong enough effect to the useful for probing their electronic structure and to characterize samples. Pressure-induced RRS should be generally applicable to all semiconductors and semiconductor heterostructures or superlattices with direct energy gaps.

ACKNOWLEDGMENT

I am very much indebted to prof. A. Jayaraman of AT&T Bell Laboratories for the excellent cooperation and hospitality during most of the course of this work.

REFERENCES

1. M. Cardona, in: "Light Scattering in Solids I," M. Cardona and G. Guntherodt, ed., Springer, Berlin (1975).
2. W. Richter, in: "Springer Tracts in Modern Physics," G. Hoehler, ed., Springer, Berlin (1976).
3. M. Cardona, in: "Light Scattering in Solids II," M. Cardona and G. Guntherodt, ed., Springer, Berlin (1982).
4. G.Abstreiter, M. Cardona and A. Pinczuk, in: "Light Scattering in Solids IV," M. Cardona and G. Guntherodt, ed., Springer, Berlin (1984).
5. B. Jusserand and M. Cardona, Raman Spectroscopy of Vibrations in Superlattices, in: "Light Scattering in Solids V," M. Cardona and G. Guntherodt, ed., Springer, Berlin (1989).
6. R. Trommer, E. Anastassakis and M. Cardona, in: "Light Scattering in Solids," M. Balkanski, R.C. Leite and S.P.S. Porto, eds, Flammarion, Paris (1976).
7. R. Trommer and M. Cardona, Phys.Rev. B17:1865 (1978).
8. P.Y. Yu and B. Welber, Solid State Commun. 25:209 (1978).
9. K. Aoki, A.K. Sood, H. Presting and M. Cardona, Solid State Commun. 50:287 (1984).
10. M. Holtz, U.D. Venkateswaran, K. Syassen and K. Ploog, Phys. Rev. B39:8458 (1989).
11. B.A. Weinstein and R. Zallen, in: "Light Scattering in Solids IV," M. Cardona and G. Guntherodt, ed., Springer, Berlin (1984).
12. G.A. Kourouklis, A. Jayaraman, R. People, S.K. Sputz, R.G. Maines Sr., D.L. Sivco and A.Y. Cho, J. Appl. Phys. 67:6438 (1990).
13. A. Jayaraman, G.A. Kourouklis, R. People, S.K. Sputz, R.G. Maines Sr., D.L. Sivco and A.Y. Cho, High Pressure Research 6:27 (1990).
14. A. Jayaraman, G.A. Kourouklis, R. People, S.K. Sputz and L. Pfeiffer, Pramana-J. Phys. 35:167 (1990).
15. A. Jayaraman, Rev. Mod. Phys. 55:65 (1983).

16. R.L. Mills, D.H. Liebenberg, J.C. Bronson and L.C. Schmidt, Rev. Sci. Instrum. **51**:891 (1980).
17. J.D. Barnet, S. Block and G.J. Piermarini, Rev. Sci. Instrum. **44**:1 (1973).
18. A.K. Sood, W. Kauschke, J. Menendez and M. Cardona, Phys. Rev. **B35**:2886 (1987).
19. R.M. Martin, Phys. Rev. **B10**:2620 (1974).
20. R. Zeyher, Phys. Rev. **B9**:4439 (1974).
21. D.S. Chemla, Helvetica Physica Acta **56**:607 (1983).
22. J.E. Zucker, A. Pinzcuk, D.S. Chemla, A. Gossard and W. Wiegmann, Phys. Rev. Lett. **51**:1293 (1983).
23. J.E. Zucker, A. Pinzcuk, D.S. Chemla, A. Gossard and W. Wiegmann, Phys.Rev. Lett. **53**:1280 (1984).
24. A.K. Sood, J. Menendez, M. Cardona and K. Ploog, Phys. Rev. Lett. **54**:2111 (1985).
25. R. Trommer and M. Cardona, Phys. Rev. **B17**:1865 (1978).
26. D. Olego and M. Cardona, Sol. State Commun. **39**:1071 (1981).
27. A. Jayaraman, G.A.Kourouklis and S. Swaminathan (to be published)
28 B.A. Weinstein and M. Cardona, Phys. Rev. **B8**:2795 (1973).
29. J. Menendez, M. Cardona and L.K. Vadapyanov, Phys. Rev. **B31**:3705 (1985).
30. A.K. Sood, J. Menendez, M. Cardona and K. Ploog, Phys. Rev. **B32**:1412 (1985).
31. A. Alexandrou and M. Cardona, Sol. State Commun. **64**:1029 (1987)
32. A. Alexandrou, M. Cardona and K. Ploog, Phys. Rev. **B38**:2196 (1988).
33. E. Cerdeira, E. Anastassakis, W. Kauschke, and M. Cardona, Phys. Rev. Lett. **57**: 3209 (1986).

OPTICAL INVESTIGATION OF $Cd_xZn_{1-x}Te/ZnTe$ SUPERLATTICES AT HIGH PRESSURE

W. Williamson III, S.A. Lee, Y. Luo and
Y. Rajakarunanayake

Department of Physics & Astronomy
The University of Toledo
Toledo, Ohio 43606   USA

ABSTRACT

We have measured and analyzed the photoluminescence spectra from MBE grown $Cd_xZn_{1-x}Te/ZnTe$ strained layer superlattices under high pressure. In these superlattices, the electrons and heavy holes are confined in the $Cd_xZn_{1-x}Te$ layer (type I) while the light holes are confined in the ZnTe layer (type II). The low temperature photoluminescence is typically dominated by a strong feature due to the decay of the light-hole exciton, and weaker features due to the decay of the heavy-hole exciton and due to luminescence from an excited quantum well state. We report the pressure dependence of the photoluminescence up to 6.8 GPa. Our results are consistent with a small valence band offset between CdTe and ZnTe. We have measured the difference of the hydrostatic valence band deformation potentials.

INTRODUCTION

Current interest in wide band gap II-VI semiconductors is motivated by the possibility of developing optoelectronic devices such as light emitting diodes, semiconductor lasers, etc. which can perform in the green region of the spectrum.[1-3] ZnTe and CdTe are potential candidates for such devices since their low temperature band gaps are 2.39 and 1.61 eV, respectively.[4] Superlattices of these materials (and their

alloys) are of interest since the electronic and optical
properties of such structures can be varied in a controlled
fashion, simply by varying the compositions and thicknesses of
the constituent layers (e.g. tunable band gaps). The
$Cd_xZn_{1-x}Te/ZnTe$ superlattices are also quite interesting
because they have large exciton binding energies compared to
III-V materials such as $GaAs/Al_xGa_{1-x}As$ and there is great
potential for room temperature excitonic effects. Recently,
optically pumped yellow/orange lasers have also been produced
from $Cd_xZn_{1-x}Te/ZnTe$ superlattices.[5]

Due to the lattice mismatch between CdTe and ZnTe, the
superlattices made of these materials and their alloys
manifest effects due to coherent biaxial strain. The most
significant effect of this strain on the band structure is
that the degeneracy of the valence band is lifted due to the
uniaxial component of the strain. In addition, the hydrostatic
component of the strain can shift the average energies of the
valence and conduction bands. In superlattices with built-in
strain (commonly called strained layer superlattices or SLS)
interesting experiments can be performed with high pressure
techniques to precisely manipulate the band structure. High
pressure optical experiments are ideal tools for studying the
band offsets, excitonic effects, etc. in these strained layer
superlattices.

In the particular case of $Cd_xZn_{1-x}Te/ZnTe$ superlattices,
the initial valence band offset is quite small.[6-9] (Although
there is some controversy about the exact magnitude and the
sign of the valence band offset, the common anion rule is
believed to be obeyed quite well for these materials.[10]) In
this case, the electron is confined in the Cd-rich layer by
the substantial conduction band offset. However, the holes are
only confined by strain effects. Since the sign of the strain
is different in the two layers, light holes are confined in
the Zn-rich layer while the heavy holes are confined in the
Cd-rich layer. In this manner, these superlattices can
simultaneously have type I heavy-hole excitons and type II
light-hole excitons. However, whether the ground state of the
valence band is the heavy hole or the light hole is determined

by the exact magnitude of the valence band offsets and the pressure dependences of the band edges.

A high pressure photoluminescence (PL) study of $Cd_{.34}Zn_{.66}Te/ZnTe$ superlattices at about 14 K and up to 6.8 GPa is reported in this paper. A strong PL signal is identified to be due to the decay of the light-hole exciton. A weaker signal is observed at higher energy and is identified to be due to the decay of the heavy-hole exciton. We have also observed an additional PL peak which is believed to involve an excited electronic state of the quantum wells.

EXPERIMENTAL

The samples were grown by molecular beam epitaxy. A more detailed account of the growth of these samples is given elsewhere.[6] The substrate was (001) GaAs with a 600 nm buffer layer of ZnTe to absorb much of the strain due to the lattice mismatch between GaAs and the II-VI compounds. A subsequent buffer layer (380 nm thick) of $Cd_{.10}Zn_{.90}Te$ was grown with the aim of providing an approximately lattice matched buffer layer for the superlattice growth. A 10 nm ZnTe x 7.5 nm $Cd_{.34}Zn_{.66}Te$ superlattice (15 repeats) was then grown, and capped with an additional 10 nm of ZnTe to reduce surface recombination. The samples used in the high pressure experiments were mechanically polished to a thickness of 30 μm and cleaved to about 100 μm x 100 μm square pieces under an optical microscope.

The high pressures were generated with a Merrill-Bassett miniature diamond anvil cell (DAC). The samples were contained in an Inconel X-750 gasket. The pressure-transmitting medium was argon and the pressure was determined using ruby fluorescence. The DAC was mounted in a closed-cycle helium cryostat and the PL measurements were made at about 14 K. Most of the pressure changes were made at low temperature. The 457.9 nm radiation of an Ar-ion laser was used to excite the PL at power ranges of 0.3 to 100 mW. The PL spectra were measured using a double monochromator with standard photon counting electronics.

The large lattice mismatch between CdTe and ZnTe (~ 6.2%) produces a large strain in the superlattice layers. This strain plays the dominant role in determining the valence band offsets. For an accurate description of the band structure, it is important to know the strain conditions present in the individual layers as well as in the overall structure. There are two critical thicknesses of importance for the growth of a SLS on a lattice-mismatched buffer layer.[11] The first is the critical thickness for the entire SLS and the second is the critical thickness of the individual layers. To simplify matters, in these samples the buffer layer was chosen to closely match the free-standing lattice constant of the SLS. (The free standing SLS has the in-plane lattice constant that minimizes the elastic free energy of the constituent layers and the two different layers are oppositely strained.) Also, the thicknesses of individual layers were maintained to be below the critical thickness for strain relaxation, as given by theories of Matthews and Blakeslee[12,13] and Van der Merwe.[14] The effect of applying high pressure to these samples is to add an additional hydrostatic strain.

Fig. 1 shows a schematic band diagram relevant for the samples studied. Throughout this paper, the ZnTe layer will be

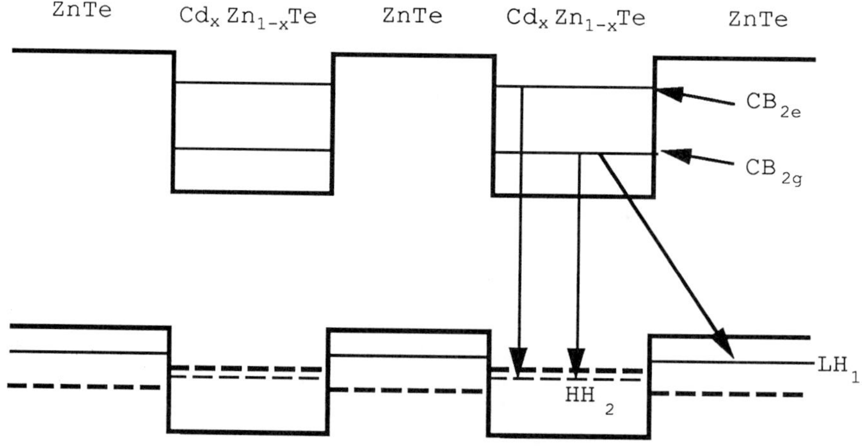

Fig 1. Band diagram of the Cd$_x$Zn$_{1-x}$Te/ZnTe superlattice.

referred to as layer 1 and the $Cd_xZn_{1-x}Te$ layer will be referred to as layer 2. For the purpose of this figure, we have assumed that the initial (unstrained and zero pressure value) valence band offset is zero. The valence band in each material is split due to the uniaxial components of the strain. The confinement energy levels of the electron in the quantum wells are denoted CB_{2g} and CB_{2e} (for the ground and excited levels, respectively). As discussed by Mathieu and co-workers,[15,16] the superlattice band gap in these materials is the transition involving the light hole.

RESULTS AND DISCUSSION

A typical PL spectra at room pressure and high pressure are shown in Fig. 2. The dominant feature of the spectra is a peak which is identified as being due to the decay of the light-hole exciton.[15,16] A less intense PL peak is observed at higher energy and is identified as being due to the decay of the heavy-hole exciton. A higher energy PL peak was also observed from these samples. The energy of this peak is in qualitative agreement with the identification of this peak being due to PL involving the excited state of the quantum well.

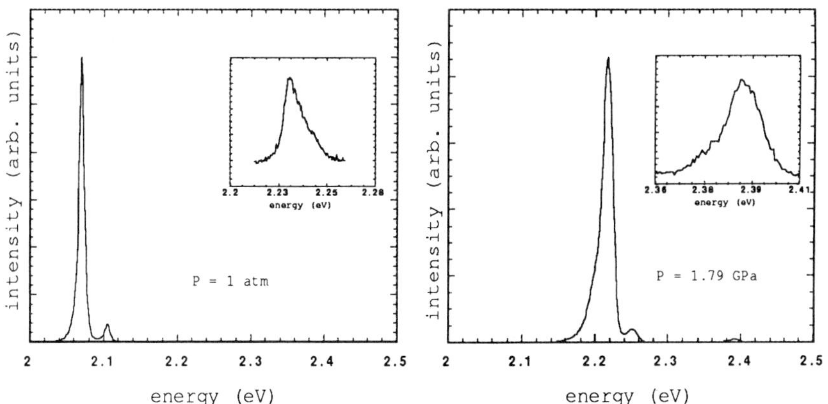

Fig. 2. Typical photoluminescence spectra at 1 atm and 1.79 GPa. The very intense peak is the light-hole exciton and the small peak is the heavy-hole exciton. The inset shows a photo-luminescence peak which might involve an excited state of the quantum well.

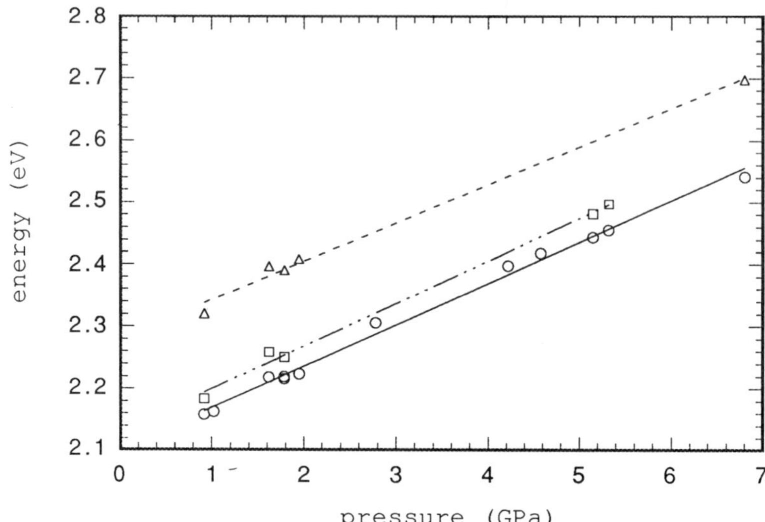

Fig. 3. The pressure dependence of the three photoluminescence peaks. The circles are the data for the light-hole exciton, the squares are the data for the heavy-hole exciton, and the triangles are the data for the transition involving the excited state of the well. Straight line fits are given by E_{LH} = 2.103 + 0.0666 P; E_{HH} = 2.131 + 0.0687 P; and E_{ES} = 2.281 + 0.0617 P. All energies are given in units of eV and pressures in GPa.

As shown in Fig. 1, the PL from the decay of the light and heavy-hole excitons involve valence states in different layers (due to the simultaneous type I and type II nature of this SLS). Consequently, these data can be used to extract information about the differences in the electronic properties of the two layers. In particular, the energy of the heavy hole PL (E_{HH}) is given by

$$E_{HH} = (E_{CB2} + c_2\delta + Q_{e2}) - (E_{V2} + a_2\delta + b_2\varepsilon_2 - Q_{HH2})$$

and the energy of the light hole PL (E_{LH}) is given by

$$E_{LH} = (E_{CB2} + c_2\delta + Q_{e2}) - (E_{V1} + a_1\delta + b_1\varepsilon_1 - Q_{LH1})$$

where E_{CB} and E_V refer to the energies of the conduction and valence bands at zero pressure, respectively. The hydrostatic and uniaxial deformation potentials for the valence band are denoted by a and b while the hydrostatic deformation potential for the conduction band is denoted as c. The cubical dilation is δ and the uniaxial strain is ε. The quantum confinement energies are Q_e, Q_{HH}, and Q_{LH}. Here we neglect the exciton binding energies and the effect of strain on the spin-orbit splitting. Materials 1 and 2 are ZnTe and $Cd_{.34}Zn_{.66}Te$, respectively. Subtracting these two equations leaves only the valence band quantities. Almost all of the pressure dependence will be in the term $(a_1 - a_2)\delta$. Due to the relatively large effective masses of the holes, the quantum confinement energies are relatively small and changes of these energies with pressure will be insignificant. The uniaxial strain will also change as the pressure is increased, but this change is also expected to be small.[16] From the fits to the LH and HH PL data in Fig. 3, we find that the pressure dependence of the difference $(E_{HH} - E_{LH})$ is 2 meV/GPa. In order to convert this to the differences of the valence band deformation potentials, we need to convert the pressure to a dilation. The GaAs substrate is quite thick compared to the superlattice and should dominate the strains in the system. Thus, the bulk modulus of GaAs should be used in the conversion. This yields $a_1 - a_2 \sim 170$ meV.

CONCLUSION

We have performed photoluminescence studies of a $Cd_{.34}Zn_{.66}Te/ZnTe$ strained-layer superlattice at low temperatures up to 6.8 GPa. We have observed the pressure dependence of the photoluminescence of both the heavy-hole and light-hole excitons. No phase transitions have been observed up to the highest pressure of the study. Our results are consistent with type I band alignment for the heavy holes and type II band alignment for the light holes at all pressures.

ACKNOWLEDGEMENT

We would like to thank Dr. T.C. McGill for kindly providing the samples used in this study.

REFERENCES

1.   T. Yao, in *The Technology and Physics of Molecular Beam Epitaxy*, E.H.C. Parker, Ed., (Plenum Press, New York, 1985), p. 313.

2.   H. Fujiyasu, H. Takahashi, H. Shimizu, A. Sasaki, in *Proceedings of the 17th International Conference on the Physics of Semiconductors*, (Springer, New York, 1985), p. 539.

3.   R.H. Miles, G.Y. Wu, M.B. Johnson, T.C. McGill, J.P. Faurie, and S. Sivananthan, Appl. Phys. Lett. **48**, 1383 (1986).

4.   C. Neuman, A. Nothe, and N.O. Lipari, Phys. Rev. B **37**, 922 (1988).

5.   A.M. Glass, K.Tai, R.B. Bylsma, R.D. Feldman, D.H. Olson, and R.F. Austin, Appl. Phys. Lett. **53**, 834 (1988).

6.   Y. Rajakarunanayake, M.C. Phillips, J.O. McCaldin, D.H. Chow, D.A. Collins, and T.C. McGill, SPIE Proc. **1285**, 142 (1990).

7.   T.M. Duc, C. Hsu, and J.P. Faurie, Phys. Rev. Lett. **58**, 1127 (1987).

8.   C.G. Van de Walle and R.M. Martin, Phys. Rev. B **35**, 8154 (1987).

9.   C.G. Van de Walle, K. Shahzad, and D.J. Olego, J. Vac. Sci. Technol. **B6**, 1350 (1988).

10.   J.O. McCaldin, T.C. McGill, and C.A. Mead, Phys. Rev. Lett. **36**, 56 (1976).

11.   R.H. Miles, T.C. McGill, S. Sivananthan, X. Chu, and J.P. Faurie, J. Vac. Sci. Techol. **B 5**, 1263 (1987).

12.   J.W. Matthews and A.E. Blakeslee, J.Cryst. Growth **27**, 118 (1974); **29**, 273 (1975); **32**, 265 (1976).

13.   J.W. Matthews, in *Epitaxial Growth*, E. Kasper and F.J. Grunthaner, Surf. Sci. **174**, 606 (1986).

14.   J.H. Van der Merwe, J. Appl. Phys. **34**, 123 (1963).

15.   H. Mathieu, J. Allegre, A. Chatt, P. Lefebvre, and J.P. Faurie, Phys. Rev. B **38**, 7740 (1988).

16.   B. Gil, D.J. Dunstan, J. Calatayud, H. Mathieu, and J.P. Faurie, Phys. Rev. B **40**, 5522 (1989).

HIGH PRESSURE STUDIES OF IMPURITIES IN SEMICONDUCTORS

R.A. Stradling

Interdisciplinary Research Centre in Semiconductor Materials and
Physics Department,
Blackett Laboratory,
Imperial College, London SW7 2BZ, UK

INTRODUCTION

For many years it has been known that those common semiconducting compounds
with their lowest conduction band minimum located at the centre of the
Brillouin zone (eg GaAs, GaSb, CdTe) have levels resonant with the Γ
conduction band in addition to the normal Γ-associated shallow donors (see
the review by Paul, 1968). These resonant donor levels can be made to
emerge into the forbidden gap by the application of hydrostatic pressure. A
sufficiently large pressure will cause a realignment of the bands where the
Γ-minimum crosses the L-or X-extrema. The resonant levels emerge into the
gap at pressures substantially less than those required to cause band-
crossing. Initially this was thought to be simply due to the larger
effective masses in the higher extrema with each level being rigidly tied
to each critical point in the band structure. This view was supported by
the similarity of the pressure coefficients for the activation energies of
the impurities observed at high pressures with those found for the band
extrema. However, the multiplicity of levels found and the realisation that
the positions of many of the levels with respect to the higher order band
minima was much greater than the respective effective-mass binding energies
showed that the resonant levels are in fact 'deep' and are made up of
admixtures of states taken from all over the Brillouin zone.

Even in the early experiments it was realised that certain of the pressure-
induced localised states became metastable at low temperatures, ie the
population of the states was not in thermal equilibrium with the lattice
but reflected the conditions created at some higher temperature (Litwin-
Staszewska et al,1971; Konczewicz et al, 1977). If these states are emptied
by some non-equilibrium process such as photoexcitation, the electrons can
not be recaptured and the phenomenon of persistent photoconductivity
results.

Alternatively, if the hydrostatic pressure is removed at low temperatures
and the band alignment returns to the normal configuration, then the
population of these localised metastable states cannot change even though
the 'unrelaxed' state concerned is now resonant with the conduction band.
Gas pressure equipment is ideal for this type of experiment as the pressure
is truly hydrostatic and can be changed easily at low temperature.

Very striking metastability is also found with the $Ga_{1-x}Al_xAs$ alloy system with a wide range a donors. This materials system has attracted much attention because of its technological importance, particulartly as the donor-induced metastability has proved to be a major limitation in many applications. The metastable states are traditionally known as DX levels following the first model where it was proposed that the centres consist of a donor and an unknown defect X (Lang, 1979). Following the early observations of metastability, there has been intense speculation concerning its origin and a variety of detailed models have been proposed. The models differ on the questions of whether a large or small lattice relaxation is associated with the metastability and on the charge state of the relaxed state (positive or negative U) (see review by Baraff, 1991). The model currently gaining the most support involves large-relaxation and negative-U. These ideas were initially put forward by Chadi & Chang (1989) and subsequently refined by Zhang & Chadi (1990).

As pointed out by Baraff (1991), the theories differ as to which states, if any, are present in addition to the well-known deep state and the Γ-associated shallow states.

In the case of high-purity InSb, Hall measurements show the presence of two pressure-induced deep levels in high-purity bulk material without deliberate doping. There is strong freeze-out of carriers onto these levels at pressures above about 8kbar and one of these levels exhibits metastability. However, it was unclear from the initial experiments (Litwin-Staszewska et al,1971; Konczewicz et al, 1977) which chemical species of impurity was responsible for the loss of carriers, or indeed whether a single centre was responsible for both levels. Particularly in the case of InSb where the shallow donor energies are extremely low, Hall measurements do not have the resolution in energy to permit the classification of all the levels involved. However magneto-optical measurements have proved to be extremely informative.

HIGH PRESSURE FAR-INFRARED MAGNETO-OPTICAL STUDIES OF DONORS IN HIGH PURITY InSb

Both high pressure and magnetic field result in compression of the shallow donor wavefunctions. In the case of high purity InSb a magnetic field of about 0.3T is needed to bring about a metal-insulator transition and to produce localised shallow states beneath the conduction band. The enhanced interaction of the bound electrons with the atomic cores increases the "chemical shifts" of the electronic transitions for different shallow donor species. Consequently fine structure becomes apparent on the $1s$-$2p_-$ donor transition at fields of the order of 5T and commonly up to four different types of residual donor centres are detected. The application of high pressure results in better spectroscopic resolution and can induce a further and even more dramatic effect on the deepest of these shallow donor states (possibly due to oxygen). As can be seen from fig. 1, a transformation is induced of the shallow Γ-donor ground state for this donor (labelled A in the figure) into a deep state which no longer follows the lowest conduction band minimum. Anticrossing of the ground state with the more localised A_1 state is observed (Wasilewski et al, 1982; Brunel et al, 1986).

When an uncompensated sample is cooled down at pressures greater than 10kbar, the metastable states become saturated for donor A although most of the shallow states remain occupied for the other donors (B, C & D). Because

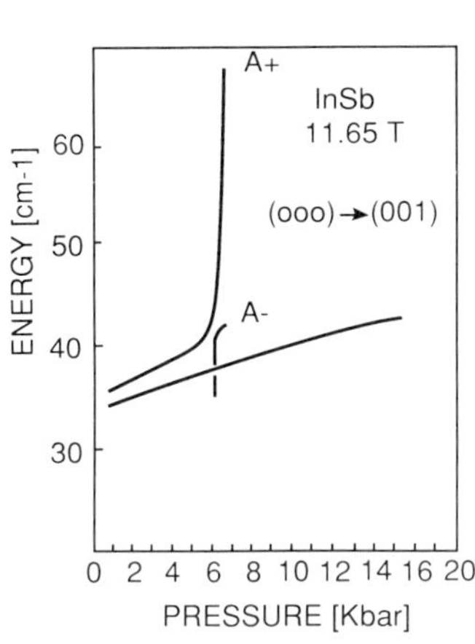
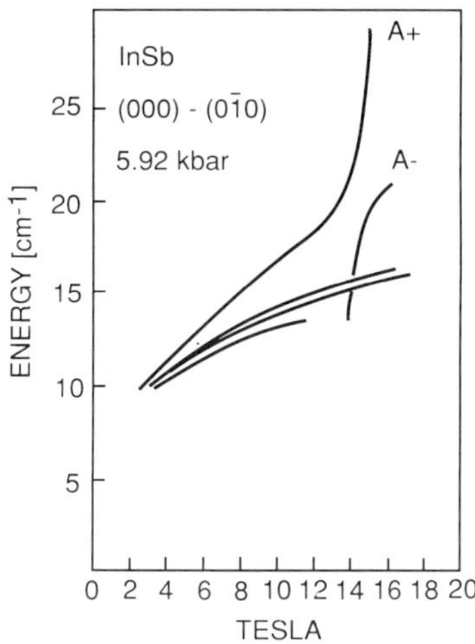

Fig. 1 shows the anticrossing effects observed on the 1s-2p_ (right) and 1s-2p$_O$ lines for donor A in InSb under hydrostatic pressure (Wasilewski et al, 1982; Brunel et al, 1986)

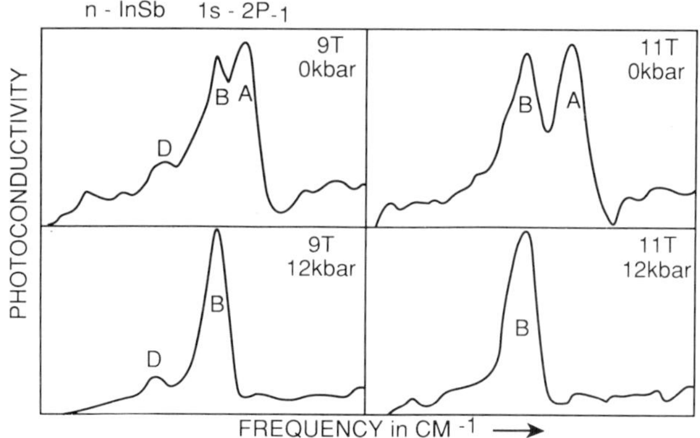

Fig 2 compares the fine structure on the 1s-2p_ line of the shallow donors before (upper two recordings) and after (lower recordings) the pressure treatment described in the text. Occupying the metastable DX-like centres removes donor A from the shallow donor spectrum (Wasilewski et al, 1986a).

the occupancy of the shallow donor levels is not then possible for donor A,
the 1s-2p_ transition for donor A (the strongest of the group of donor
lines under normal conditions) is bleached from the spectrum. If the
pressure is removed, the band gap and effective mass return to their
ambient values, while the remaining lines for donor B to D coincide with
their positions in energy before the sample is subjected to the pressure
treatment (fig.2). The bleaching of the deepest of the Γ-associated
shallow donors (A) from the spectrum after occupancy of the metastable
centres and the anticrossing observed when the non-metastable deep states
are occupied show unambiguously that shallow donor states, non-metastable
A_1 states and metastable states are all possible configurations of the same
donor A but do not occur for the other shallow donors in InSb within the
experimentally accessible range of pressures.

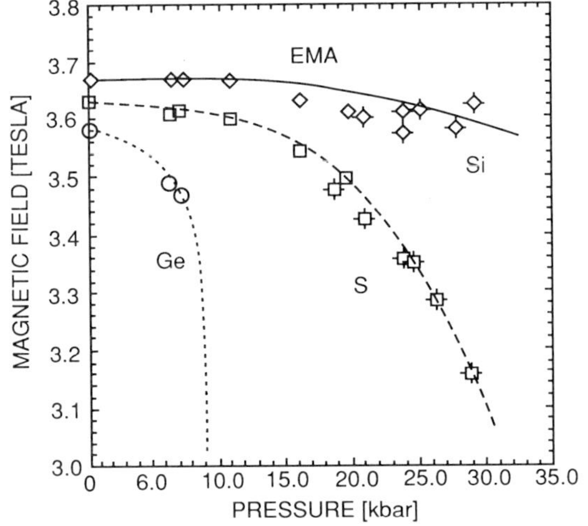

Fig. 3 shows the deepening of the 1s-2p_ line for Ge, S and Si donors in
GaAs as a function of pressure. The pressures above 20kbar involved a
diamond anvil cell (Dmochowski et al, 1990 and 1991a & b).

HIGH PRESSURE FAR-INFRARED MAGNETO-OPTICAL STUDIES OF DONORS IN HIGH PURITY
GaAs

For the first time the diamond anvil cell high pressure technique has been
applied to far infrared magneto-optical experiments (Dmochowski et al, 1990
and 1991a & b). Absorption due to the 1s-2p_+ transition for different
shallow donor species was investigated in transmission experiments using
far infrared laser radiation at pressures up to 40 kbars.

At first the resolution of the central cell components increases relatively slowly with increasing pressure because of the increase in effective mass and decrease of dielectric constant with pressure. Close to a certain pressure which is characteristic of the particular impurity the deepening of ground state becomes dramatic and at a slightly higher pressure the particular donor disappears from the FIR spectrum completely. This critical pressure is approximately 9, 25 and 30 kbar for the Ge, S, and Si donors respectively as can beseen from figure three. Large "chemical shifts", anticrossing effects and fast electron capture at low temperatures demonstrate the coexistence of strong localization, A_1 symmetry and non-metastable (non-DX like) character of the observed deep donor states. The present investigation and earlier results with GaAs (Wasilewski & Stradling, 1986a) and with InSb (as described above) demonstrate that three types of states: shallow and two types of deep states: DX-like and deep A_1 are generally expected to be formed by the donor impurities associated with DX-centres.

PHOTOLUMINESCENCE OF HIGH PURITY GaAs SAMPLES WITH KNOWN RESIDUAL DONOR SPECIES

Shallow donor states are only observed in the FIR measurements when the ground states of the donors are occupied by electrons. This occurs naturally for pressures lower than the critical one at which the deep A_1 donor states enter the gap. At higher pressures and under thermal equilibrium conditions, the deep A_1 states are occupied at low temperatures. The shallow states are not accessible by FIR spectroscopy, as demonstrated by the quenching of the FIR absorption, and the oscillator strength for transitions out of the A_1 states is small except close to the anticrossing condition.

Photoluminescence (PL) techniques can however be used to detect the deep A_1 states above the critical pressure. Since these states effectively capture electrons at low temperatures one can expect that they will contribute to photoluminescence. Pioneering PL experiments with GaAs in diamond anvil cells have been performed by Wolford et al (1985). Deep states were reported but only close to band cross-over. As the nature of the residual contaminants in these samples was unknown, no assignments of the levels to particular impurities could be made.

Diamond cell PL experiments have now been performed with samples where the residual donor states have been positively identified by FIR magneto-optical measurements. The PL spectrum of a high purity Si-doped sample measured at ambient pressure at 4.2K is dominated by a sharp peak at 1.514eV corresponding to unresolved free and donor- and acceptor-bound excitons as well as the neutral donor to valence band transition (Leroux et al, 1986). At lower energy two small peaks at 1.510 eV and 1.505 eV correspond to many overlapping bound excitons involving some unknown linearly polarized defects specific to MBE. Two stronger peaks at 1.493 and 1.489 eV have been assigned to carbon acceptor-conduction band and C acceptor-donor transitions respectively. Finally, small shoulders extending down to 1.465 eV have been attributed to transitions involving several different acceptors. These acceptors are related to unidentified defects which are specific to MBE and show polarization effects also.

Upon applying pressure the spectral features observed at ambient pressure evolve as expected from the changes of band structure. Three main peaks: excitonic, carbon-related (A_1 acceptor) and defect acceptors(A_2)-related,

all shift with pressure following the change of the direct gap energy.
Above 23 kbar new spectroscopic features can be observed which no longer
follow the Γ- band related peaks. These features are due to a nitrogen
bound exciton "N_x" and its phonon replica (Leroux et al, 1986). The
emergence of N_x excitons is accompanied by a drastic reduction in intensity
of the Γ-excitons which practically disappear from the spectrum above 24
kbar, independent of excitation intensity. The Γ-free-to-acceptor
transitions (for both A_1 and A_2 acceptors) can be observed up to 40 kbars.
These transitions are relatively stronger at higher excitations.
Additional spectroscopic features can be observed at pressures higher than
30 kbar. They follow the N_x lines rather than the Γ band and their
strength is higher for lower excitation intensity. These features are
assigned to deep donor acceptor transitions involving both A_1 carbon (DA_1)
and A_2 (DA_2) acceptors. These lines, particularly DA_2, dominate the spectra
at low excitation intensity. Their appearance coincides with the quenching
of the Si-related FIR shallow donor absorption. Thus, we can associate
these features with deep donor states of the Si donor which enter the gap
close to 30 kbar. In the 30-40 kbar range the DA_1 and DA_2 lines correspond
well to donor-acceptor (also called DA_1) and B lines reported (Leroux et
al, 1986) for strongly Si-doped GaAs where it was argued that the
appearance of a deep DA_1 transition coincides with the metal-insulator
transition. The addition of the energy of the acceptor to the observed
luminescence energy enables the position of the deep Si donor state to be
estimated. This energy coincides with some very weak peaks, which can be
distinguished at pressures close to 40 kbar. These peaks can be assigned to
donor-valence band transitions. The estimated position of the deep Si donor
state is shown in fig. four. This level crosses the Γ-conduction band
minimum at 30-31 kbar, in agreement with the observed quenching of the Si
FIR absorption.

The pressure coefficient (-8.6 meV/kbar with respect to the Γ-conduction
band) for the deep Ge state derived from FIR (Wasilewski & Stradling,
1986a) and for the deep Si state observed in luminescence are equal within
experimental error. The energy difference between the Ge and Si deep
levels of approximately 190 meV agrees well with the order of magnitude
estimate for the difference in the 'chemical shifts' for the highly
localised states concerned. The non-DX character of these deep states is
demonstrated clearly. A metastable photopopulation of the shallow states
under and after illumination could not be created in the far-infrared
experiments contrary to the results found with the ionisation of DX-centres
in $Al_xGa_{1-x}As$. A very short capture time is directly demonstrated also by
the observation of the deep state in photoluminescence.

Each deep state entering the gap is specific to a particular impurity since
only one component is removed from the FIR spectrum on passing through the
anticrossing condition. The anticrossing effects provide evidence for their
A_1 symmetry whereas the large 'chemical shifts' observed imply strong
localisation. Since the deep states do not show metastability effects,
they cannot be identified with DX centres.

There is a strong analogy between the effects of hydrostatic pressure on
GaAs and the effects of alloying of GaAs with AlAs. The utility of
pressure measurements is illustrated on comparing the results of the
pressure measurements with GaAs with the variation of donor related
features deduced for the AlGaAs alloy system. In the PL from weakly
silicon-doped or undoped AlGaAs, several features have been reported
(Henning & Ansems; 1987,1988) which follow neither the Γ- nor X-conduction
band minima. The deepest of these features (D_4), which can be seen for
values of x greater than 0.1, has attracted the most attention. Several
authors have suggested that it might arise from the DX state of silicon or
to a deep non-metastable state of silicon (A_1). The height of the A_1 state

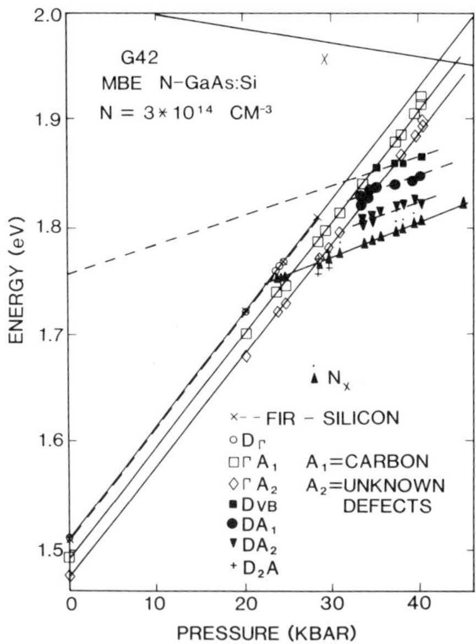

Fig. 4 Pressure dependencies of the different spectroscopic features observed in the photoluminescence from a silicon-doped MBE sample of GaAs. The dotted line shows the energy of the silicon A_1 state estimated by adding the acceptor energy to the DA transition (full circles) - see Dmochowski et al (1991b) for further details

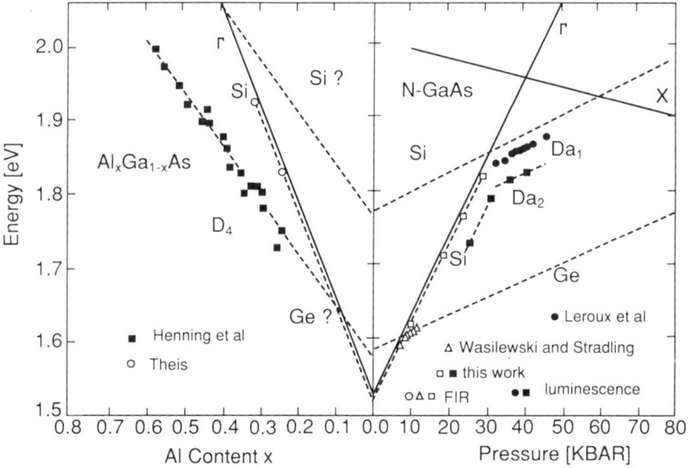

Fig. 5 compares the pressure variation of the Ge and Si A_1 donor states relative to the conduction band edge with the variation in aluminium content of the AlGaAs alloy system. D_4 shows the position of the 'Henning' donor and Si? shows the position of the silicon A_1 state estimated by taking the x=0, P=0 position deduced from the GaAs pressure measurements and assuming the same x dependence as the 'Henning' donor.

of GaAs above the conduction band minimum deduced from the high pressure
measurements (250meV) make it unlikely that this state could emerge into
the AlGaAs band gap at such a low value of x. However, the energy for the
`Henning' donor found by extrapolation of the observed x-variation to x=0
is identical to within experimental error to the position of the Ge donor
found by extrapolating the pressure variation for the Ge energy back to
P=0. We believe therefore that it is extremely likely that the 'Henning'
donor is germanium, which is a common residual contaminant in both MBE and
MOCVD GaAs . As can be seen from figure 5, a similar extrapolation with x
from the P=0, x=0 position of the silicon A_1 energy shows that this state
is unlikely to appear in the gap of AlGaAs until x exceeds 0.4.

ATOMIC PLANE OR DELTA DOPING

A novel development following on the recent advances in epitaxial growth
techniques has been the ability to deposit dopants on single atomic planes.
There has been considerable discussion on whether the dopants remain
localised on single planes but the evidence is that the spreading can be
kept to distances of the order of 1nm provided that the growth temperature
is kept sufficiently low that diffusion is not significant. The diffusion
can be measured with very high accuracy by comparing the relative
occupancies of the different subbands as measured in a Shubnikov-de Haas
measurements with those predicted by self-consistent theory. Frequently the
areal concentrations of donors deposited are sufficiently high that DX
centres can become occupied. In this case hydrostatic pressure can provide
a very useful analytic tool to determine whether carriers are being
captured by DX states or are being lost by some other mechanism such as
donor precipitation. Furthermore, the known pressure-induced changes in
effective mass and dielectric constants change the energies of the
subbands. Consequently, pressure can be used to check on the validity of
the fits between experiment and theory at ambient pressure.

There have been claims that atomic plane or delta doping can enhance the
mobility compared with bulk doping because of the physical separation of
the carriers from the doping plane (the most localised subband is the i=0).
Such experimental evidence as exists in the literature certainly appears to
support a mobility enhancement. However, these claims must be considered
with considerable caution: firstly, because in the samples where a mobility
enhancement has been reported, the dopants almost certainly would have
spread over considerable distances from the doping plane, thereby reducing
the equivalent volume concentration; secondly, because of the differing
spatial extent of the wavefunctions in different subbands, the mobilities
depend very strongly on subband index. Consequently it is essential that a
multicarrier analysis be employed in evaluating any mobility determined in
a Hall experiment. Shubnikov-de Haas measurements provide a precision
method for quantifying these two objections to earlier claims for the
mobility enhancement. As was first demonstrated by the Munich group, the
relative subband occupancies derived in experiments are sensitive to
spreading of the dopant distribution with a spatial resolution in the z-
direction approaching 1nm.

In order to quantify these, ideas a series of samples containing very thin
sheets (2, 5, 10 nm) of silicon donors in MBE GaAs have been investigated
(Skuras et al, 1991). These samples were grown at a very low temperature
(400°C) so that diffusion of the silicon was negligible. Excellent
agreement between theory and experiment wsa achieved whereas the same
concentration donor of silicon ($1x10^{13}cm^{-2}$) deposited on a single plane at 550°C
was found both by Shubnikov de Haas and SIMs to have spread approximately
20nm.

The mobility can also be determined from the Shubnikov-de Haas effect by analysis of the low-field decay of the `oscillations' which are damped by a `Dingle temperature' term. However this analysis cannot be performed directly when more than one subband is occupied because of the interference between the multiple oscillatory series. To overcome this problem we have developed a routine for the analysis of the Fourier transforms of the complex pattern of Shubnikov-de Haas peaks in order to provide quantitative values for the mobilities of the individual subbands. The peak width (in units of magnetic field) is simply given by $\sqrt{3}/2$ times the mobility. The results of this analysis were compared with the values deduced from the magnetic field dependence of the resistivity and Hall effect and good agreement was achieved. The mobility of the $i=0$ subband is typically a factor of three lower than that of the $i=1$ subband and up to an order of magnitude lower than the mobilities of the higher subbands. The systemmatic error which results from a Hall measurement of carrier concentrations without multiple carrier correction is typically 30%. As an example of how misleading mobility measurements can be without correction for multiple carrier conduction, the Hall mobility of the 2nm and 5nm slabs decreases substantially when hydrostatic pressure is applied although the mobility of all the individual subbands increases. This result arises because of depopulation from the higher mobility subbands with $i \geq 1$ into the low mobility $i=o$ subband.

We have also grown and measured a range of delta doped samples in InSb and InAs. The silicon distribution in the InSb samples grown at 350°C or below was less than or the order of 1nm. In contrast the InAs samples grown at 490°C showed spreading of at least 10nm although growth at 440°C produced significant reduction in the width of the dopant profile. The mobility achieved with InSb delta doped at a concentration of 4×10^{11}cm^{-2} was 35,000cm^2/Vs. This corresponds to a volume concentration of 10^{19}cm^{-2} (10,000cm^2/Vs) if the silicon remains on the original doping plane but 10^{18}cm^{-2} (35,000cm^2/Vs) if the silicon spreads a distance of 4nm where the figures in brackets represent the corresponding bulk mobilities. As a spreading of 4nm is inconsistent with the ratio of the subband occupancies, we believe that we have obtained a significant mobility enhancement by delta doping methods.

The mobility changes observed when the population of DX centres in heavily doped samples is varied by changing the pressure can in principle provide information about the charge state of the DX centres. Measurements with bulk samples of GaAs by Maude et al (1987 & 1989) showed up to a factor of two increase in mobility on increasing the pressure and on this basis argued that the DX centre was neutral. However it was shown later that a mobility increase of the required sign and magnitude could result from impurity correlation effects (O'Reilly, 1990; Kossut et al, 1990). Experiments with delta doped samples provide an additional way of investigating this possibility as the extent of the electronic wavefunctions in the z-direction increases rapidly with increasing subband energy. Consequently, if the dipolar fields expected from correlation effects are present, the mobility changes should be greater for the higher order subbands than for the $i=0$ case (which should only 'see' monopolar fields). The electrons are expected to trap out onto donor sites which are favoured energetically and thus have an ionised (positively charged) donor close by. On a negative U model, D^+D^- pairs will form and the dipolar component will be very strong. On a positive U picture, there will be no negative impurities present unless compensating acceptors are introduced during growth. Consequently the dipolar component of the local potential will be very weak. However, a preliminary analysis of the pressure dependent Shubnikov-de Haas data for the thin doping slabs in GaAs did not require the formation of the large number of pairs expected for

correlation. Consequently it can be said that the data for both the bulk
and the planar-doped samples does not favour the negative U model

Pressures up to 20 kbar did not change the free carrier concentrations
significantly with either InSb heavily doped with silicon ($n=5\times10^{18}cm^{-3}$)
or InAs ($n=6\times10^{19}cm^{-3}$), indicating that the DX level for Si was at least
400meV above the bottom of the band in both materials (clearly donor A in
the high purity bulk samples of InSb discussed in section 2 is not
silicon). The fact that the silicon DX levels are much deeper in the
conduction band is almost certainly related to the much greater separation
between the $\Gamma-$ and the higher bands in these materials (eg with InAs $\Gamma-L$ is
1.15eV and $\Gamma-X$ is 1.9eV, Landolt & Bornstein). Nevertheless, El-Sabbahy et
al (1978) and Kadri et al (1985) have reported freeze-out on to
unidentified pressure induced levels in undoped InAs with the levels
entering the gap below 20kbar.

ACKNOWLEDGEMENTS

As always, a wide-ranging review of this nature is based on the
contributions of many individuals. I am particularly indebted to Z.
Wasilewski and S. Porowski for my introduction to Be-Cu clamp and gas cells
and to J.E. Dmochowski who undertook all the diamond anvil cell experiments
described in this paper. J.J. Harris, S.N. Holmes, R. Kumar, E. Skuras,
P.D. Wang, R.L. Williams, W.T. Yuen and many others have contributed to the
experiments.

REFERENCES

Baraff, G. A., _Semicond. Sci. & Tech._ to be published, 1991 in the
Procedings of the Mautendorf Workshop on DX and Other
Metastable Centres

Brunel, L.-C., Huant, S., Baj, M., Trzeciakowski, W., 1986, _Phys.Rev._
B33:6863

Chadi, D. J. & Chang K.J., 1989, _Phys Rev_ B39:10063

Dmochowski, J.E., Wang, P.D. & Stradling, R.A., 1990, Proc. Int Conf
on Physics of Semiconductors (Thesseloniki) 658-661

Dmochowski, J.E., Wang, P.D., & Stradling, R.A., 1991a, _Semicond. Sci
& Tech_ 6:118

Dmochowski, J.E., Stradling, R.A., Wang, P.D., Holmes, S.N.,
McCombe, B.D., & Weinstein, B., 1991b, _Semicond. Sci & Tech_
6:476

El-Sabbahy, A.N., Adams, A.R., 1978, Proc. Int. Conf. on Physics of
Semiconductors (Edinburgh), Inst. of Physics Conf. Series 43
589

Henning, J.C.M., & Ansems, J.P.M., 1987, _Semicond. Sci. & Tech._ 2:1

Henning, J.C.M., & Ansems, J.P.M., 1988, _Semicond. Sci. & Tech._ 3:361

Kadri, A., Aulombard, R.L., Zitouni, K., Baj, M., Konczewicz, L.,
1985, _Phys. Rev._ B31:8013

Konczewicz, L., Litwin-Staszewska, E., & Porowski, S.,1977, Proc. Int
Conf on Physics of Narrow Gap Semiconductors (Warsaw) p211

Landolt & Bornstein New Series III 22a (Ed O. Madelung, pub.
Springer-Verlag) p117

Lang, D.V., 1978, Proc. Int. Conf. on Physics of Semiconductors
(Edinburgh), Inst. of Physics Conf. Series 43:433

Leroux, M., Neu, G., Verie, C., 1986, _Sol.St.Commun._, 58:289

Litwin-Staszewska, E., Porowski, S., Filipchenko, A.S., 1971, _Phys.
Stat. Solidi_ (b) 48:525

Maude, D.K., Portal, J.C., Dmowski, L.,Foster, T.,Eaves, L.,Nathan,
M., Heiblum, M., Harris, J.J., Beall, R.B., 1987,
Phys.Rev.Lett., 59:815

Maude, D.K., Eaves, L., Foster, T.J., & Portal, J.C.; 1989, <u>Phys Rev Lett</u> 62:1922

Paul, W., 1968, Proc Int Conf on Phys Semiconductors (Moscow) p16

Skuras, E., Kumar, R., Williams, R.L., Stradling, R.A., Dmochowski, J.E., Johnson, E.A., Harris, J.J., Beall, R.B., Skierbeszewski, C., Singleton, J., van der Wel, P.J., & Wisniewski, P.: 1991, <u>Semicond. Sci. Tech.</u> 6:535

Wasilewski, Z., Davidson, A.M., Stradling, R.A., Porowski, S., 1982, Lecture Notes in Physics, 177, 233

Wasilewski, Z., Stradling, R.A., 1986a, <u>Semic.Sci.Technol.</u> 1:264

Wasilewski, Z., Stradling, R.A., & Porowski, S., 1986b, <u>Solid State Comm.</u> 57:123

Wolford, D.J., Bradley, J.A., 1985, <u>Sol. St. Commun.</u> 53:1069

Zhang, S. B., & Chadi, D. J., 1990, <u>Phys Rev</u> B42:7174

2D TRANSPORT IN SEMICONDUCTORS UNDER PRESSURE

J.L. Robert

G.E.S. U.S.T.L. URA 357, place E. Bataillon
34095 - Montpellier-Cedex
France

Abstract

Different regimes of conduction in 2D electron gas are examined, emphasizing effects of hydrostatic pressure and external magnetic field on GaAs/GaAlAs heterostructures with a spacer.

As a result of electron transfer in the quantum well controlled by the pressure, various effects can be studied on the same sample by varying the surface electron density.

a) Hydrostatic pressure can be used to reach sufficiently low surface electron densities at which the ultra-quantum limit occurs at available magnetic fields. The binding energy of a magneto donor, composed of atoms and electrons separated each other by a spacer is measured for different surface electron densities controlled by the pressure. A metal non metal transition is observed.

b) In the quantum limit, experiments performed under hydrostatic pressure allow us to analyze the behaviour of the conductivity σ_{xx}. Its dependence on the filling factor as well as the temperature shows evidence for the nearest neighbour hopping process in the zero resistance state.

c) In the weak localization regime, the relevant parameter $k_F l_o$ is varied over two orders of magnitude by applying a pressure (k_F is the Fermi wave vector, l_o is the mean free path). As a result the weak localization is investigated on the same sample between the dirty and pure limits.

I- Introduction

With the development of thin film growth techniques such as MBE and MOCVD during the past decade, high quality heterojunctions[1] have been produced with unusual properties. A quasi two-dimensional electron gas is obtained, associated with sub-band formation from electric quantization. Considerable effort has been devoted to developing the GaAs/GaAlAs system, both because of its basic interest and for its application to new devices.

The problem of forming a conducting channel at the GaAs/GaAlAs interface was solved, using higher mobility electrons and higher drift velocities than in Si-Mosfet.

In 1978, Dingle et al[2] and soon after Stormer et al[3] successfully put Esaki's concept of modulation doping[4] in concrete form. To reduce impurity scattering, Esaki suggested spatially separating the free carrier density and their parent impurities. These heterojunctions were used to produce a new high-speed field transitor called TEGFET, MODFET, HEMT ou SDMT, depending on the manufacturer.

The basic stucture, as shown in figure 1, consists of a buffer layer of undoped GaAs, an undoped GaAlAs layer called the spacer and a highly-doped GaAlAs layer which are sequentially grown on a semi-insulating substrate. Conduction by the 2D electron gas takes place in the undoped GaAs layer and is confined to the GaAs/GaAlAs interface in the quantum well.

Frontiers of High-Pressure Research, Edited by H.D. Hochheimer and
R.D. Etters, Plenum Press, New York, 1991

In this paper, we describe experiments performed on modulation doped GaAs/GaAlAs heterojunctions under hydrostatic pressure in an investigation of their electrical properties. The basic problem is that it is difficult to compare the properties of different heterostructures, since by nature they are unique. The obvious advantage of using hydrostatic pressure is that the electronic properties can be modified without changing the intrinsic characteristics of the material.

In the following (section II-1), we analyze the effect of hydrostatic pressure on impurity states, considering two extremal cases : the hydrogenic model and the large lattice relaxation model. When deep levels are involved in conduction processes, the change in the carrier concentration can be fairly large. As a result, in modulation Si-doped GaAs/GaAlAs heterojunction the surface electron density n_S can be controlled by the effect of pressure on the Si-DX center in GaAlAs (section II-2).
Such a procedure for controlling n_S is used to study some localization effects in a 2D electron gas in presence of a magnetic field (section III).

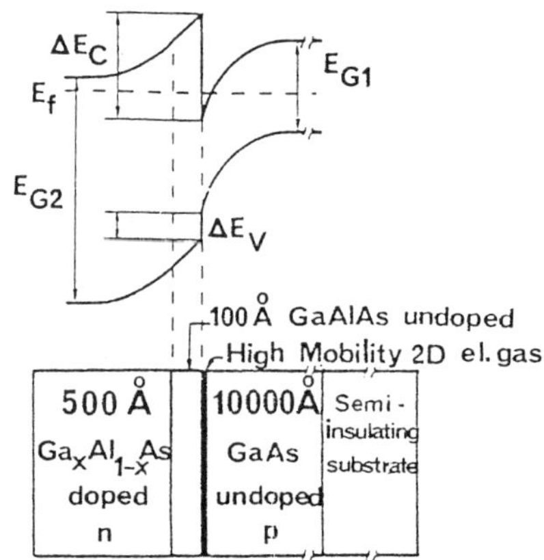

Figure 1. In the energy band diagram, ΔE_c and ΔE_v represent the conduction band and the valence band offsets at the interface between the smaller gap ($E_{\Gamma 1}$) and the larger gap ($E_{\Gamma 2}$). E_F is the Fermi level. E_Γ is equal to $E_{\Gamma 1}$-$E_{\Gamma 2}$. The common well accepted value of $\Delta E_c/\Delta E_\Gamma$ is about 65%.

II - Pressure effects on semiconductors

II-1   Impurity states under hydrostatic pressure

The electronic properties of semiconductors and their electrical behaviour are strongly dependent on the nature of the impurity states which provide the conduction electrons. Two kinds of impurity states may be involved in the conduction processes : shallow impurity states and deep impurity states.

Shallow impurity states are well described by the effective mass theory (EMT). In this model, the wave function of the bound carrier is constructed mainly from the Bloch functions of the appropriate minimum of the conduction band and the wave function of the trapped carrier extends over several lattice constants.

The deep impurity states are due to highly localized potential leading to a strong electron lattice coupling, which depends on the ionicity of the host crystal. The wave function of the trapped electrons is strongly localized. Such levels present a difference between their optical and thermal activation energies which is well described by a configuration diagram (figure 2).

In the case or shallow donors, the ionization energy E_D is the Rydberg energy (Ry):

$$E_D = R_y = \frac{m^* e^4}{32 \pi^2 \varepsilon h^2} = \frac{m^*/m_0}{\varepsilon_r^2} \, 13.6 \, eV \quad (1)$$

In this formula, ε_r is the crystal permittivity, m_0 the electron mass in vacuo, m^* the electron effective mass.

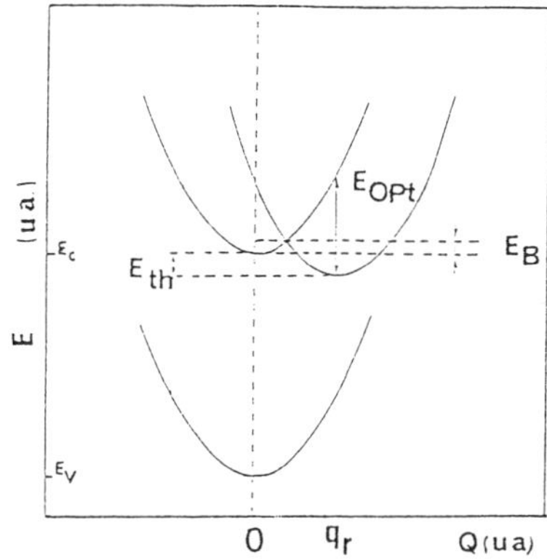

Figure 2. The configuration diagram describes the localized and delocalized states of an impurity coupled to the lattice. The thermal ionization energy E_{th} can be detemined by transport experiments; it differs from the optical energy E_{opt} which can be measured at the photoconductivity edge. The barrier E_B is responsible for the persistent photoconductivity (PPC) observed at low temperature and for the metastable effects.

One can expect an increase of the binding energy due to the increase of the effective mass under pressure. Pressure leads to a change in the lattice parameter which induces an increase of the energy gap and therefore of the effective mass.

The increase of the ionization energy under pressure is all the more large since the energy band gap is low. As an example, the change in the binding energy is rather small and does not exceed 0.1 meV/Kbar in InSb.

In contrast to shallow impurity states, the position of deep impurity states compared to the Γ minimum of the conduction band lowers at a rate of about 10 meV/Kbar. So, under some circumstances, one can observe a strong pressure-induced decrease of the free carrier concentration. This decrease is due to the transfer of conduction electrons to the impurity states when their energetic position is comparable to that of the conduction band (figure 3).

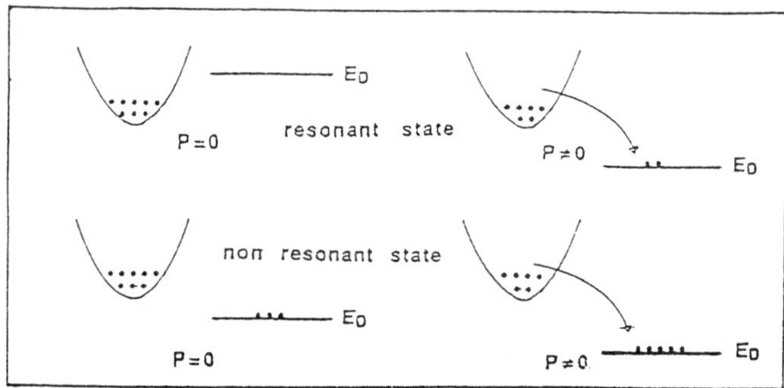

Figure 3. Pressure induced carrier transfer from the direct conduction band minimum into a localized level.

In a binary III-V material, such an occurrence at zero pressure would be fortuitous. For example, such a transfer can indeed be observed at low temperature in InSb, only for pressures higher than 7 kbar, the deep impurity states being located too high above the Γ conduction band minimum (70 meV)[6].

The situation is more favorable in some ternary III-V compounds in which it is possible to tune the position of deep impurity states with respect to the conduction band minimum by changing the alloy composition[7]. In this point of view, the ternary alloy $Ga_{(1-x)}Al_xAs$ which is widely used in opto and micro-electronics is one of the most significant cases.

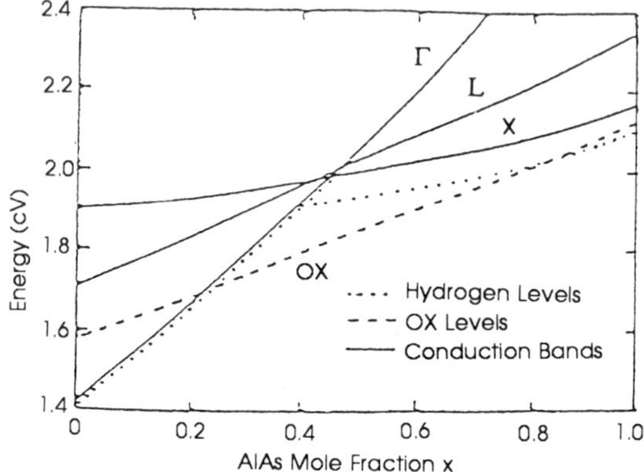

Figure 4. Energy band diagram of $Al_xGa_{1-x}As$. The energetic position of the shallow levels bound to Γ, L and X and of the DX level is also indicated.

The energy band diagram of this alloy (fig. 4) is now well established. This material has a direct energy gap for an aluminium content $x < 0.4$. Furthermore, column IV (Si, Ge, Sn) or column VI (S, Se, Te) doping atoms can be incorporated respectively on Ga and As sites and act as substitutional donors. In addition to the usual shallow levels, these substitutional impurities also give rise to the so-called DX centers which present the characteristics of a deep donor level: their ionization energy relative to the Γ minimum strongly increases when the pressure is applied.

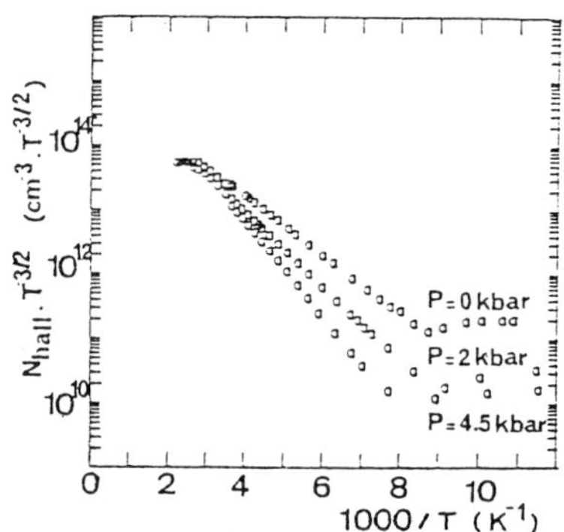

Figure 5. Arrhenius plot of the free carrier concentration in $Al_{0.33}Ga_{0.67}As$ for different pressure values.

The electronic properties of this compound for $x < 0.25$ are controlled by the shallow impurity state limited to the Γ minimum. When the Al content becomes larger than 0.25, the DX level is the lowest impurity level and therefore governs the electronic properties of the material. As a result, one can expect a strong change in the free carrier density in the Γ minimum related to the increase of the thermal activation energy under pressure for Al content $0.25 < x < 0.4$. (fig. 5).

II-2   Properties of modulation doped Ga$_{0.7}$Al$_{0.3}$As/GaAs heterojunction under pressure.

An approximate relationship between the total charge stored in the accumulation layer and the diffusion voltage v_{20} appearing in GaAlAs is given by :

$$n_s = \sqrt{2\varepsilon_2 \, N_d \, \frac{v_{20}}{e^2}} \quad (2)$$

where N_d is the density of ionized donors in semiconductor 2; ε_2 is the dielectric constant; v_{20} is, of course, a function of the quantity $(E_F - E_{c2})$ and e is the electron effective charge (fig.1).

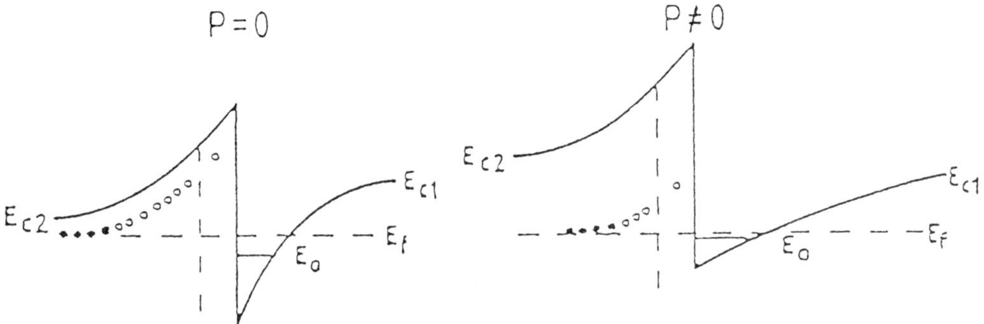

Figure 6. Variation of the energy diagram of GaAs/Ga$_{1-x}$Al$_x$As heterostructure under hydrostatic pressure.

This expression, only valid for non degenerate conditions, has the merit to show that the free carrier density is directly related to v_{20} *i.e.* to the position of the Fermi level in semiconductor 2 compared to the position of the conduction band.

As a result, the surface electron density n_s in GaAs/Ga$_{1-x}$Al$_x$As heterojunctions can be diminished by the effect of the pressure on the Si-donor in Ga$_{1-x}$Al$_x$As which, in the Al content of interest $(0.25 < x < 0.4)$, presents a deep level character.

The situation is shown schematically in figure 6. The electrons from the quantum well in GaAs are transferred under pressure to the Si-donors in GaAlAs.

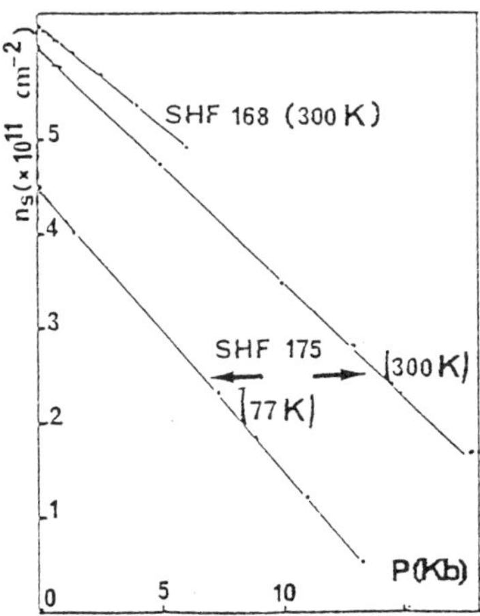

Figure 7. Pressure dependence of n_s.

Typical results are shown in figure 7. The pressure was applied at room temperature, and the carrier density in the well was measured by Hall effect between 4.2K and 300K or by Shubnikov de Haas experiments in the liquid helium temperature range. A linear decrease of n_S is observed under increasing pressure. This change in n_S is also clearly demonstrated by the pressure dependence of the Shubnikov de Haas effect (fig. 8): with increasing pressure, the magnetoresistance maxima shifted towards the lower magnetic fields and the period of oscillations $(\Delta(\frac{1}{B}) = \frac{e}{\pi\hbar}\frac{1}{n_S})$ increased.

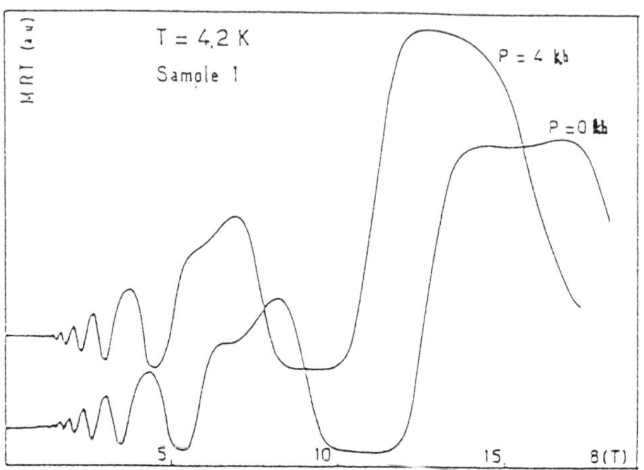

Figure 8. Transverse magnetoresistance R_{xx} (MRT)of GaAlAs/GaAs heterojunction (Table1).

It should be noted that because of the metastable character of the Si impurity states in GaAlAs, the system is frozen when the temperature is lower than 150K. Such a dependence of n_S with pressure is well accounted by assuming that the impurity level has a pressure coefficient of about 11 meV/Kbar[8].

III- Localization effects

The procedure described in the previous section to change the 2D carrier density in a modulation doped heterojunctions is well suited to study the quantum properties of a 2D electron gas and particularly some localization effects in presence of a magnetic field. Several regimes are emphasized, depending on the external magnetic field intensity.

In the ultraquantum limit, the increase of the binding energy of a magnetodonor constituted of donors and electrons separated from each other by a spacer leads to a freeze out of electrons. In this case, pressure is applied to reach sufficiently low surface densities at which the ultra quantum region occurs at available magnetic fields (only the lowest Landau level is occupied).

In the quantum limit, localization effects occur in the zero magnetoresistance state. The observation of hopping processes is interpreted in terms of the existence of an impurity band induced by long range potential fluctuations.

At low magnetic field, the negative magnetoresistance is observed. The weak localization regime is studied, using pressure to vary the relevant parameter of this effect $k_F l_0$ on two orders of magnitude. (k_F is the Fermi wave vector and l_0 the mean free path).

III-1   Localization effects in the ultra quantum limit

Since free carrier densities can easily be lowered to values less than 5.10^{10} cm^{-2} under hydrostatic pressure, studies of magneto-transport in the ultra quantum regime become possible within the limits of magnetic fields available. Several samples have been studied in this regime, in magnetic fields up to 18T.

The carrier concentration n_S in the quantum regime was derived from the measurements of the σ_{xy} and σ_{xx} components of the conductivity tensor, assuming that localized electrons do not participate in the Hall effect.

$$n_S = \frac{1}{e} \; \frac{R_H \, B^2}{\rho^2 + R_H^2 \, B^2} \quad (3)$$

e is the electron effective charge, R_H is the Hall coefficient and ρ the resistivity in a transverse magnetic field B.

The characteristics of the samples grown by MBE and MOVCD techniques are given in table 1.

Table 1. Sample parameters at zero pressure.

	$Ga_{1-x}Al_xAs/GaAs$	Spacer	n_S (10^{11} cm^{-2}) (T = 4.2 K)	μ (cm^2/v.s) (T = 4.2 K)
Sample 1	30 % OMCVD	60 Å	5.2	46 000
Sample 2	25 % OMCVD	150 Å	2.4	129 000
Sample 3	27 % MBE	250 Å	2.08	51 000
Sample 4	30 % MBE	250 Å	3.5	419 000

In figure 9, we report the experimental results obtained on sample 1 at 13.3 Kbars. A magnetic-field induced metal-non metal transition is clearly indicated. In magnetic fields lower than 6 T, the carrier density did not depend on temperature; in magnetic fields higher than 6 T, an activation energy appeared.

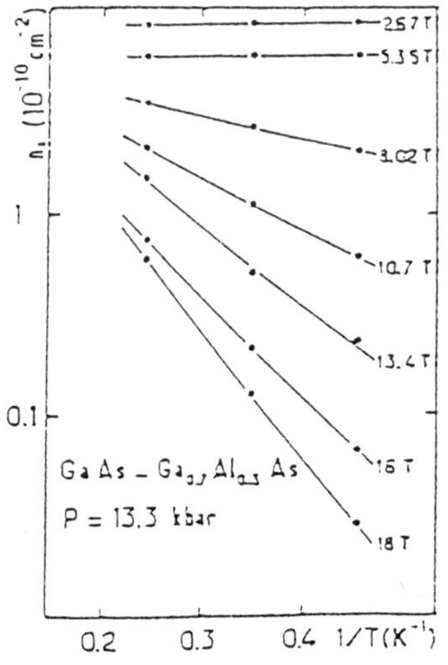

Figure 9. Temperature dependence of the surface electron density n_s for different magnetic fields. Sample 1 under pressure of 13.3 kbar.

The appearance of an activation energy could be explained by considering that the localization is due to Coulomb potential of isolated impurities located in the n-type GaAlAs, which are spatially separated from the GaAs. Bastard[9] has shown that bound states in the quantum well may exist for impurities which are spatially separated from the quantum well. For example, if the spacer is on the order of the Bohr radius in bulk GaAs (100 Å), the activation energy is about 0.1 meV and the extension of the wave function is about five times larger than in the bulk material. It can be seen that this value corresponds to the distance between impurities at the lowest concentration reached in this experiment. The impurities can thus be considered to be isolated.

Since a binding energy exists in zero-magnetic field, this energy must increase in the presence of a magnetic field, as in the classical magnetic-freeze-out observed in 3D systems. This increase is due to the reduction of the orbit radius under increasing magnetic fields.

In this model, the energy must change if the spacer thickness is changed, but it would not change with doping if the density is low enough to verify the criterion of isolated impurity.

To test this model, the same kind of experiment was performed on other samples with different spacer thicknesses. Figure 10 shows the results obtained on several samples with different spacer thicknesses. To compare samples (1) and (2) and to eliminate a possible effect of screening, the pressure was chosen in such a way to obtain approximately the same critical density n_{sc} (6.10^{10} cm^{-2}) and practically the same critical magnetic field B_c (6 T).

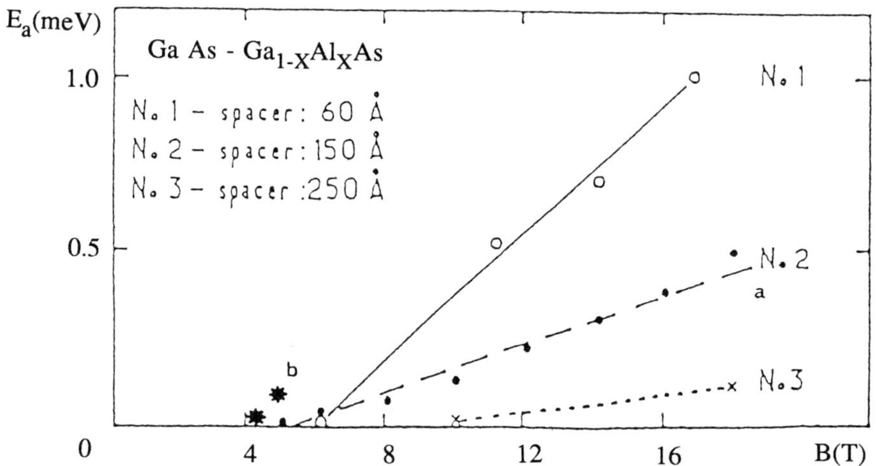

Figure 10. Activation energies E$_a$ of magneto-donors (the free electron density is separated from the donor atom by a spacer) for samples 1, 2 and 3.

We can notice that the effect of screening is observable in figure 10 which shows the activation energy in sample 2 for two pressures. The critical concentration n_{sc} as well as the critical magnetic field B_c decrease when the pressure increases (see Table 2). In the same time, the slope of the activation energy versus the magnetic field B becomes steeper.

Table 2. The samples parameters under pressure.

	n_s 10^{10} cm^{-2}	n_{sc} 10^{10} cm^{-2}	B_c (T)	d_c (Å)	$\sqrt{n_{sc}}\, d_c$	pressure (kbar)
Sample 1	9	6	6	103	0.252	13.3
Sample 2	6.25	5.75	6	103	0.246	9
Sample 3	4.5	4	3	146	0.292	9.58
Sample 4	12.5	11.1	10	80	0.266	5.9

Nevertheless, the main result observed on figure 10 is that the magnetic field dependence of the activation energy decreases with increasing spacer thickness.

A qualitative explanation of this localisation effect[10] is that the ionized donor atoms located in the doped $Ga_{1-x}Al_xAs$ layer at a distance d from the GaAs/$Ga_{1-x}Al_xAs$ interface are responsible for the localization: when d decreases, the coulombic interaction between 2D electrons and Si donors increases and the binding energy E_a increases too (fig.11).

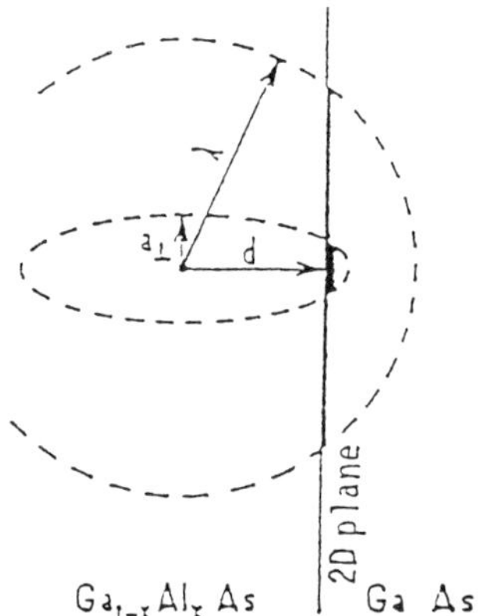

Figure 11. Semiclassical model for the inversion electron in GaAs interacting with the Si donor atom in the doped layer of GaAlAs:d-thickness of the spacer. λ-effective radius of the donor wave function at B=0; a-transverse radius of the orbit of the three dimensional electron bound to a donor at B≠0. This lines indicate intersections of the donor wave functions with the 2D plane for B=0 and B≠0, respectively.

In his theory of the metal insulator transition in ordered systems, Mott argued that as the transition is based on screening, the transition is discontinuous giving a metal when the distance between centers $N^{-1/3}$ drops below $4.a_H$, in which a_H is the Bohr orbit. In the present case, a_H tends towards the orbit radius, (the use of the cyclotron orbit $d_c=(h/eB_c)^{1/2}$ is appropriate because of the low value of the effective Rydberg R_y). In Table 2, the values of the carrier density n_s at B = 0, and the product $\sqrt{n_{sc}}\, d_c$ are also reported. It can be seen that the Mott criterion for the observation of the MNM transition is always verified.

III-2 Localization effects in the quantum limit

In presence of high magnetic field B, perpendicular to the interface, a 2D electron system is completely quantized into discrete levels.

Each level has a degeneracy N = eB/h. If n_S is the 2D electron gas density, the filling factor is given by $\upsilon = n_S/N$. Each Landau level, broadened by disorder, scattering etc..., into a band, contains extended states in the central part, separated from the localized states in the flanks by a mobility edge.

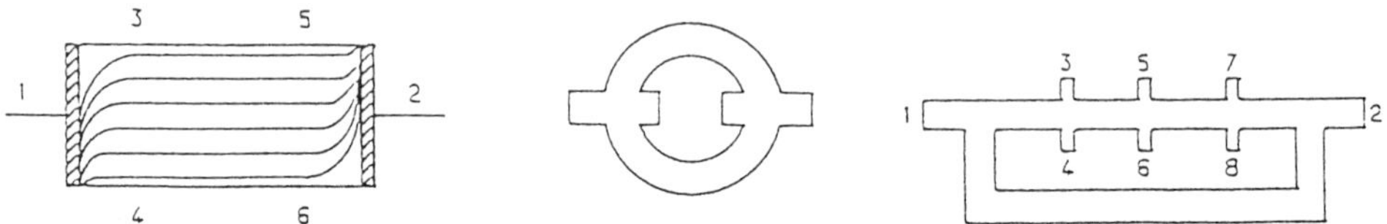

Figure 12. Different geometries for studying the quantum Hall effect and the two-dimensional conductivity. (a) The bridge geometry; (b) the Corbino geometry; (c) the shunted bridge geometry, combining the previous two.

As a consequence of the total quantization of the density of states, the Hall voltage exhibits a series of flat plateaus when the magnetic field is increased. This is the well-known quantum Hall effect[11]. In the quantum Hall regime, the Hall conductivity is equal to ie^2/h (i is an integer equal to the number of filled Landau levels υ).

Simultaneously, the longitudinal resistance vanishes. This situation (the zero magnetoresistance state) can be observed whenever the Fermi level lies in the localized states - or in real gaps in the absence of localized states[12].

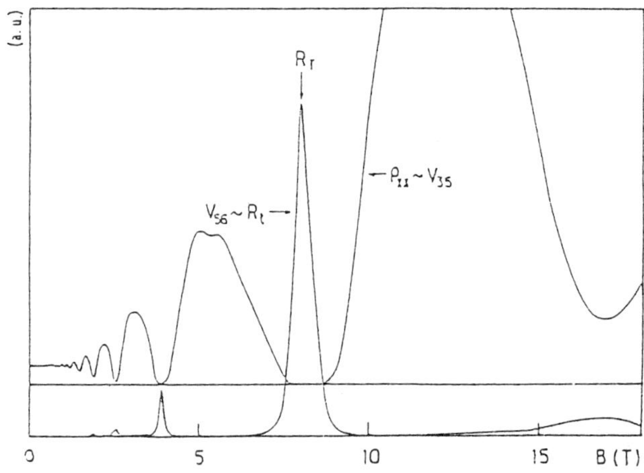

Figure 13. The transverse resistance $R_T \approx 1/\sigma_{xx}$ of GaAs/GaAlAs heterojunction (sample 1 under pressure 400 MPa at T = 4.2 K) near the integer filling factors υ vs. magnetic field. The Shubnikov de Haas magnetoresistivity $\sigma_{xx}/(\sigma_{xx}^2 + \sigma_{xy}^2)$ is also shown on the same magnetic field scale. The resistance R_T has a sharp maximum as the Fermi energy passes the middle of the Landau gap.

The conductivity σ_{xx} of GaAs/Ga$_{1-x}$Al$_x$As heterostructures (the characteristics of which are given in Table 3), have been studied in the Quantum Hall Regime at integer filling factor υ, hydrostatic pressure being employed to vary the 2D electron gas density.

To measure the magnetic field dependence of the σ_{xx} component of the conductivity tensor, samples with a particular shape were used. The shunted bridge configuration was used to carry out the Hall and magnetoresistance experiments and to use the $\rho_{xx} = 0$ nature of the Corbino geometry. (fig.12).

In this geometry, the application of a constant voltage across two opposite probes creates a circulating Hall current practically without dissipation. A direct current being injected between the probes 3-4, the transverse resistance $R_t \approx \dfrac{1}{\sigma_{xx}}$ was directly proportional to the transverse voltage V_{56} in the QHE regime.

As shown in figure 13, the output voltage V_{56} passes through a maximum when the longitudinal resistance ρ_{xx} is at a minimum.

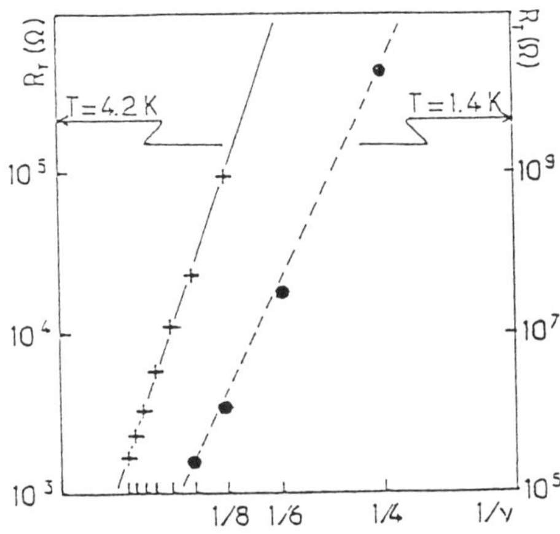

Figure 14. The maximum resistance R_T for sample 2 vs $1/\upsilon$ for integer filling factors υ at two temperatures. The lines are drawn as guides for the eyes.

For each integer values i of the filling factor υ, we determine the maximum R_T of the transverse resistance (fig. 14). A quasi-linear dependence of Ln (R_T) versus $1/i$ is obtained.

As a result of the diminished values of n_S under pressure, the peaks of R_T are shifted towards lower magnetic fields. Figure 15 shows the results obtained on sample 1 at P = 0 and 4 Kbar : the same linear dependence of Log R_T on $1/i$ is observed.

The temperature dependence of R_T for a given value of i is shown in figure 16 : it can be seen that the observed conduction process is thermally activated.

A detailed analysis of the data shows that the resistance R_T obeys the following law:

$$R_T = R_0 \exp\left(\frac{\delta}{i}\right) \exp\left(\frac{\beta B_i - \gamma}{kT}\right) \quad (4)$$

where B_i are the magnetic field intensities for the integer values of υ.

The dependence (4) suggests that we are dealing with nearest-neighbour hopping processes occuring in an impurity band[13,14].

The values of R_0, δ, β and γ are given in Table 3.

Sample	d spacer (Å)	n_s 10^{11} cm^{-2}	μ (m^2/Vs)	R_0 (Ω)	δ	β / k	γ (meV)
1	100	6.1	18.3	337	0.68	3.11	0.23
2	70	6.7	24.9	168	1.23	3.25	0.16

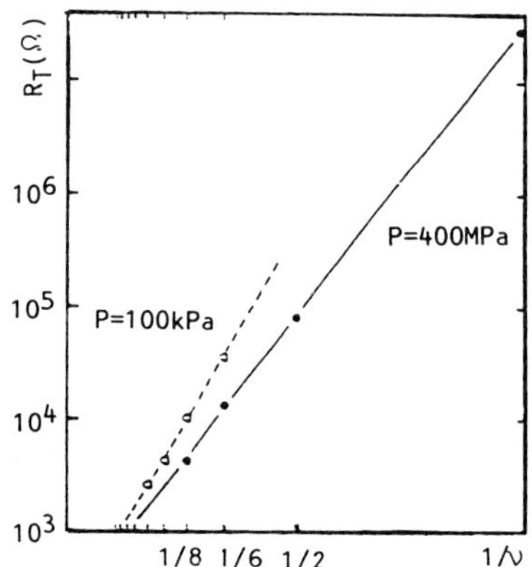

Figure 15. The maximum tranverse resistance R_T for the sample 1 vs. $1/\nu$ for integer ν at two pressures and $T = 4.2$ K. The lines are drawn as guides for the eyes.

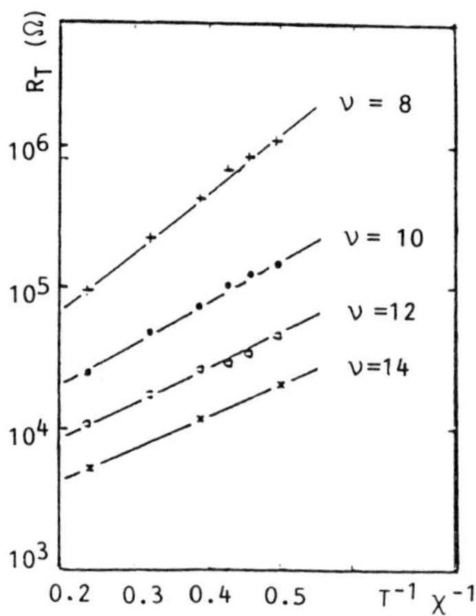

Figure 16. The maximum resistance R_T for sample 2 vs. $1/T$ for different integer filling factors. The lines are drawn as guides for eyes. The thermal activation energy depends linearly on $1/\nu$.

Comparing the parameters for sample 1 and 2, one can see that the twofold increase of R_0 is accompanied by a twofold decrease of δ.

This is in agreement with the expected changes of R_0 and δ which vary in the opposite way with varying number of hopping sites in the nearest neighbour hopping process[15,16].

The difference between the values of R_0 and δ suggests that the density of hopping sites is larger in sample 1 than in sample 2. This is in agreement with the observed lower zero field mobility.

We will now try to explain qualitatively the observed dependence (Eq. 4). We first consider the dependence of resistivity on the magnetic field. For the magnetoresistance governed by jumps in the plane perpendicular to the field Mikoshiba[17] obtains $\rho = \rho_0 \exp(\alpha_0 B)$, which is what we observe.

Now we turn to activation energy for the hopping processes in question. The theoretical description for conductivity in an impurity band, particularly in disordered system, is not still well established but it is believed that the activation energy in the nearest neighbour hopping regime is a measure of the bandwith in the impurity band[18].

The observed linear dependence of the binding energy of the localized states on the magnetic field can be explained assuming that the states possess a Coulomb type character.

In this case, the binding energy is proportional to $\dfrac{e^2}{\varepsilon \eta}$ where η is the average 2D electron radius around the localization center and ε the dielectric constant. For high magnetic fields, the magneto-donor radius becomes equal to the electron radius. In the nth orbital state the Landau radius is $L_n=(2n+1)^{1/2}(\hbar/eB)^{1/2}$. For integer filling factor, the Landau radius is to a good approximation $L_i = i^{1/2} \left(\dfrac{\hbar}{eB_i}\right)^{1/2}$. Since $n_s = \dfrac{i\,e\,B_i}{\hbar}$, so

$$E_b = \frac{e^2}{\varepsilon \eta} = \frac{e^2}{\varepsilon L_\upsilon} = \left(\frac{eB_i}{\hbar}\right)^{1/2} \frac{e^2}{\varepsilon\, i^{1/2}} = \frac{B_i}{\sqrt{2\pi\hbar\varepsilon}\,\sqrt{n_s}}$$

Thus we obtain a linear dependence of the activation energy on the field intensity, which is again what we observe.

III-3   Weak localization regime.

In 2D systems, expected to have a quasi-metallic behavior, a logarithmic increase of the resistivity with temperature can be observed at low temperatures. This feature has been considered to be a striking confirmation of the localization theory of Abrahams, Anderson, Licciardello, Ramakrishman[19].

In a strictly 2D electron gas, the conductivity σ deviates from its Boltzman's value.

$$\sigma_B = \frac{n_s e^2 \tau_{el}}{m^*} = \frac{e^2}{\hbar} \frac{1}{2\pi} k_F l_0 \quad (5)$$

in which n_s is the 2D electron gas density, τ_{el} the elastic scattering time ; m^* and e are respectively the effective mass and the charge of the electron. k_F is the Fermi wave vector and l_0 the mean free path.

According to the scaling theory, the conductivity decreases logarithmically with increasing temperature. At room temperature, the conductivity is given by :

$$\sigma = \sigma_B - \frac{e^2}{\hbar} \frac{s}{2\pi^2} \alpha \, Ln\left(\frac{L}{l_0}\right) \quad (6)$$

where s is the spin degeneracy, L is the linear size of the system, l_0 the mean free path and α is the localization parameter, theoretically expected to be equal to one. At finite temperatures, L must be replaced by the inelastic length $L_{inel} = (D \, \tau_{inel})^{1/2}$ in which D is the diffusivity and τ_{inel} is the inelastic scattering time, proportional to T^p.

In these conditions :

$$\Delta\sigma(T) = -\alpha \frac{e^2}{2\pi^2 \, \hbar} Ln(\frac{\tau_{inel}}{\tau_{el}}) = -\alpha \frac{e^2}{2\pi^2 \, \hbar} p \, Ln(\frac{T}{T_0}) \quad (7)$$

In addition to the weak localization due to disorder, other effects, particularly the electron-electron interactions, have been invoked by Altshuler et al[20] and Fukuyama[21] to explain the logarithmic decrease of the conductance with decreasing temperature.

The contribution of the electron-electron interaction $\Delta\sigma_{int}$ is given by the following expression :

$$\Delta\sigma_{int}(T) = - (1-F) \frac{e^2}{2\pi^2 \, \hbar} Ln(\frac{T}{T_0}) \quad (8)$$

where F is the Hartree parameter and (1-F) is the interaction coefficient.

Recent developments in the understanding of the weak localization show that the application of a transverse magnetic field destroys the localization, resulting in the observation of a negative magnetoresistance.

According to Altshuler et al[22] and neglecting the electron-electron interaction,

$$\Delta\sigma(B,T) = \alpha \frac{e^2}{2\pi^2 \, \hbar} [\psi(a+\frac{1}{2}) - \psi(a'+\frac{1}{2})] \quad (9)$$

where ψ is the digamma function, $a = \dfrac{\hbar}{4e \, L_{el}^2 \, B} = \dfrac{\hbar}{4e \, D \, \tau_{el} \, B}$ and

$$a' = \frac{\hbar}{4e \, L_{inel}^2 \, B} = \frac{\hbar}{4e \, D \, \tau_{inel} \, B}.$$

In the absence of magnetic field, this expression reduces to (7) :

$$\Delta\sigma(0,T) = -\alpha \frac{e^2}{2\pi^2 \, \hbar} Ln(\frac{\tau_{inel}}{\tau_{el}})$$

For a given temperature, the change in the conductivity observed in presence of a magnetic field, is given by :

$$\Delta\sigma = \alpha \frac{e^2}{2\pi^2 \, h} [\psi(a+\frac{1}{2}) - \psi(a'+\frac{1}{2}) + Ln(\frac{\tau_{inel}}{\tau_{el}})] \quad (10)$$

When the following condition is fulfilled :

$$L_{el} < L_B = (\frac{\hbar}{e \, B}) < L_{inel}$$

the asymptotic form of the expression (10) becomes :

$$\Delta\sigma = \alpha \, p \frac{e^2}{2\pi^2 \, \hbar} Ln(\frac{B}{B_e}) \quad (11)$$

We can notice that expressions (7) and (11) can be directly obtained by replacing L in the expression (6) by :

$$L_d(T,B) = [\frac{1}{L_{inel}^2(T)} + \frac{1}{L^2(B)}]^{-1/2}$$

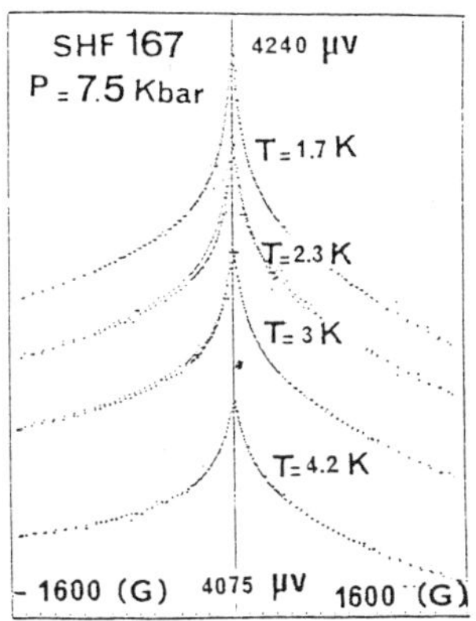

Figure 17. Variation of the magnetoresistance voltage for P = 7.5 kbar at different temperatures.

In the following, experiments are reported[23] in which the weak localization is studied as a function of the values of $k_F\, l_0$. This quantity can be varied on two orders of magnitude by applying a pressure.

As a result, the localization parameter α and the temperature exponent are determined between the dirty limit ($k_F\, l_0 \approx 5.8$, P = 13 Kbars) and the pure limit ($k_F\, l_0 \approx 140$, P = 0). Typical results showing the variation of the magnetoresistance as a function of B are reported for different temperatures on figure 17.

The conductance $\sigma_{xx} = \dfrac{\rho_{xx}}{\rho_{xx}^{2}+\rho_{xy}^{2}}$ as a function of B under different pressures at T = 3 K is given in figure 18.

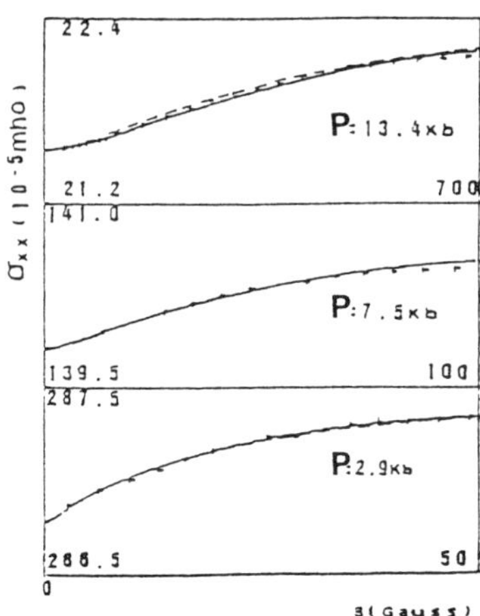

Figure 18. Conductance σ_{xx} as a function of B under different pressures at T=3 K. Solid lines are theoretical fits to eq. (10).

The values of α, p, τ_{inel} deduced from the fit of the experimental results using equation (10) are reported in table 4.

The values of αp and (1-F) are determined by measuring the slope of σ_{xx}=f(LnT) at B = 0.

We can notice that the values of α are always lower than 1, which was the predicted theoretical value.

The same observation has been made by other authors[23] and several hypothesis have been proposed to explain this discrepancy (scattering by the Maki-Thomson process, spin orbit interaction...)[24,25]. But none of them is really convincing.

A new approach of the problem has recently be given by Ravanomanjato[26]. The author has shown that the assumption of an effective dimensionality related to a non

Table 4. Sample parameters (at P = 0, n_s = 4.01 10^{11} cm^{-2} and μ = 7.59 m^2/Vs) Experimental and calculated parameters. $(k_F\, l_0)^{-\varepsilon_0}$ is calculated with $\varepsilon_0 = 0.22$.

P (kbar)	13.4	7.5	5.5	2.9
R_s (10^{11} cm^{-2})	1.03	2.45	2.89	3.52
μ (m^2/Vs)	1.32	3.58	4.34	5.12
$k_F\, l_0$	6	37	52	74
$2\,k_F\,/K$	0.74	1.17	1.24	1.44
τ_{el} (psec)	0.51	1.35	1.65	2.01
α from (9)	.576±.03	.421±.03	.471±.04	.225±.02
$(k_F\, l_0)^{-\varepsilon_0}$	0.698	0.45	0.42	0.38
1-F	1.53	2.95	2.59	5.41
p from (9)	1.09	1.15	1.26	1.39

integer value of the dimension of the system - (d = 2+ε_0 with $\varepsilon_0 \ll 1$) - leads to a new formulation of the conductivity.

Starting from the dimensionless conductance given by Shapiro and Abrahams[27], he obtains
:

$$\sigma_L = \sigma_B \left(1 - \frac{1}{\pi\varepsilon \, (k_F \, l_0)^{1+\varepsilon_0}} (1-\frac{l_0}{L}) \right) \quad (12)$$

Since $\varepsilon_0 \ll 1$ and $l_0 \ll L$, this reduces to :

$$\sigma = \sigma_B - \frac{e^2}{\hbar} \frac{1}{\pi^2} \frac{1}{(k_F l_0)^{\varepsilon_0}} \, Ln \frac{L}{l_0} \quad (13)$$

One can see that α in equation (7) can be identified to $(k_F l_0)^{-\varepsilon_0}$. We must notice that α is equal to 1 only for a strictly 2D system.

In table 4, we show that the experimental results are satisfactorly interpreted in this way. Particularly, the decrease of the disorder parameter which was observed with increasing k_F and remained misunderstood, is well explained by this model.

Acknowledgments

The author would like to thank Dr. A. Raymond, C. Bousquet, J.Y. Mulot, K. Zekentes, S. Ravanomanjato, L. Konczewicz, M. Kubisa, E. Litwin-Staszewska, R. Piotrzkowski, W. Zawadzki for their participation in some experiments or discussions - and Dr. J.P. André, P.M. Frijlink, J.M. Masson, F. Alexandre for providing the samples.

References

1. R.L. Anderson -IBM - J.Res. Develop. 4-283 (1960).
2. R.Dingle, H.L. Störmer, A.C. Gossard, W. Wiegmann, Appl. Phys.Lett. 33, 665 (1978).
3. H.L. Störmer, R. Dingle, A.C. Gossard, W. Wiegmann, M.D. Sturge, Solid State Communication, 29-705 (1979).
4. L. Esaki, R. Tsu, IBM, Res. Note R.C. 2418 (1969).
5. W. Paul, J. Appl. Phys. supplement to volume 32, 2082 (1961).
6. M. Konczykowski, S. Porowski, J. Chroboczek, Proceeding of the 11th Int. Conf. Phys. Semicon., Warsaw 708 (1972).
7. H.E. Lee, L.Y. Juravel, J.C. Wooley and A.J. Springthorpe. Phys. Rev. B21, 659 (1980).
    N. Lifshitz, A. Jarayaman, R.A. Logan and H.C. Card, Phys. Rev. B21, 670 (1980).
    P.M. Mooney, J. Appl. Phys., 67(3) (1990).
8. J.M. Mercy, C. Bousquet, J.L. Robert, A. Raymond, G. Gregoris, J. Beerens, J.C. Portal, P.M. Frijlink, P. Delescluse, P. Chevrier, and N.T. Linh, Surf. Science 142, 298 (1984).
9. G. Bastard, Private Communication.
10. J.L. Robert, A. Raymond, L. Konczewicz, C. Bousquet, W. Zawadzki, F. Alexandre, J.M. Masson,
J. P. André, P.M. Frijlink, Phys. Rev. B 36, 9297 (1987).
11. K. Von Klitzing, G. Dorda, M. Pepper, Phys. Rev. Lett. 45, 449 (1980).
12. R.B. Laughlin, Phys. Rev. B 23, (1981).
13. N.F. Mott, Phil. Mag. 22, 7 (1970).
14. B.I. Shklovskii and A.L. Efros, "Electronic properties of doped semiconductors", Springer Verlag, Berlin (1984).

15.   A.B. Fowler, A. Hartstein, Phil. Mag. B42, 949 (1980).

16.   K.Y. Hayden, P.N. Butcher, Phil. Mag. B38, 603 (1979).

17.   N. Mikoshiba, Phys. Rev. 127, 1961 (1962).

18.   A. Hartstein, A.B. Fowler, M. Albert, Inst.Phys. Conf. Series Nb43, 1001-1004 (1979).

19.   E. Abrahams, P.W. Anderson, D.C. Licciardello, T.V. Ramakrishnan,
       Phys. Rev. Lett. 42, 637 (1979).

20.   B.L. Altshuler, A.I. Khlem'nitzkii, A.I. Larkin and P.A. Lee, Phys. Rev. B22, 5142 (1980).

21.   H.J. Fukuyama, J. Phys. Soc. Jpn. 48, 2169 (1980).

22.   B.L. Altshuler, A.G. Aronov, P.A. Lee, Phys. Rev. Lett. 44, 1288 (1980).

23.   K. Zekentes, S. Ravanomanjato, R. Piotrzkowski, E. Litwin-Stszewska,
       J.L. Robert, Surface Science 228, 449-452 (1990).

24.   B.J.F. Lin, M.A. Paalanen, A.C. Gossard, D.C. Tsui, Phys. Rev. B29 Nb.2,
       927-934 (1984).

25.   P.A. Lee, T.V. Ramakrishnan, Phys. Rev. B26, 4009 (1982).

26.   S. Ravanomanjato, "Thesis " Montpellier (1990), unpublished.

27.   S. Shapiro, E. Abrahams, Phys. Rev. B24, 4025 (1981).

BAND GAPS AND PHASE TRANSITIONS IN CUBIC ZnS, ZnSe AND ZnTe

S. Ves

Max Planck Institut für Festkörperforschung, Heisenbergstraße 1
7000   Stuttgart 80, Germany

Aristotles University of Thessaloniki, Physics Department, Solid
State Physics Section, 540 06 Thessaloniki,   Greece

ABSTRACT

The effect of hydrostatic pressure on the fundamental absorption edge and phase transitions of the zinc chalcogenides is investigated by means of optical absorption and reflection at room temperature. The $E_0(\Gamma_{15}^v \to \Gamma_1^c)$ gap exhibit a sublinear increase under pressure which changes to linear when plotted as a function of the relative change of volume. Various trends, e.g. sublinearity, deformation potentials etc are clearly demonstrated along the sequence ZnS $\to$ ZnTe. A transition to a NaCl - type structure occurs at 15±0.3 and 13.5±0.3 GPa for ZnS and ZnSe, respectively. The NaCl structures are stable at least up to 27.0 and 25.0 GPa for ZnS and ZnSe. For ZnTe an intermediate phase ZnTe-II appears in the pressure range between 9.3 and 12.0 GPa. At 12.0 GPa a second phase transition takes place and the new structure ZnTe-III is stable at least up to 30.0 GPa. The structures of the intermediate and the high pressure range of ZnTe are not known at present, however the NaCl structure is to be excluded.   Absorption and reflectivity data demonstrate conclusively that - contrary to earlier reports - the NaCl - type structure are semiconducting in ZnS, ZnSe as well as the intermediate ZnTe-II phase. The highest ZnTe-III phase is indeed a metallic phase. The experimental results are compared to theoretical calculations based on local empirical pseudopodential and ab initio LMTO calculations.

INTRODUCTION

Recent interest in the pressure behaviour of zinc halcogenides[1] arises in the context of pressure studies of semimagnetic semiconductors[2] and the strain effects in superlattices[3]. Cubic zinc halcogenides crystallize at ambient conditions in the zinc-blende (B$_3$) structure

and are direct gap semiconductors with the $E_0(\Gamma_{15}^v \to \Gamma_1^c)$ gap varying from 2.3 (ZnTe) to 3.7

Frontiers of High-Pressure Research, Edited by H.D. Hochheimer and
R.D. Etters, Plenum Press, New York, 1991

eV (ZnS). By using diamond - anvil - cell techniques (optical absorption and reflectivity as well as x-ray diffraction) the effect of hydrostatic pressure on the fundamental absorption edge and diffraction pattern of ZnS, ZnSe and ZnTe is investigated for pressures over the full stability region of their tetrahedral phases as well as over the high pressure phases of ZnS and ZnTe. In the low pressure phase of these materials the $\Gamma_{15} \rightarrow \Gamma_1$ direct gap is the lowest one lying about 1 eV lower than the $\Gamma_{15} \rightarrow X_1$ or the $\Gamma_{15} \rightarrow L_1$ gaps which stay always above E_0. Hence the optical absorption edges remain sharp with increasing pressure up to pressure just below the phase transitions. Furthermore we investigate the nature of the pressure induced high pressure phases. The high pressure phases of zinc halcogenides are commonly assumed to be metallic[4,5,6] an assumption   which contradicts our experimental results.

In this paper we summarize experimental and theoretical results on the fundamental gaps and pressure induced phase transitions in zinc halcogenides. The aim is (i) to compare the pressure shifts of the direct gap energies in zinc halogenides ; (ii) to find out from optical measurements if their high-pressure phases are indeed metallic and to determine the transition pressures P_{tr} and (iii) to compare the experimental results with theoretical results from semiempirical calculations based on the pseudopotential method (EPM) and ab initio calculations based on density - functional theory in conjunction with the linear - muffin - tin - orbital (LMTO) method.

EXPERIMENT

The experimental studies were performed at room temperature using a gasketed diamond anvil cell (4:1 methanol - ethanol pressure medium) in combination with the ruby luminescence method for pressure determination. The samples for absorption were about 5 μm thick. This was necessary for achieving absorption coefficients of the order of 10^5 cm^{-1}, very close to that of the exciton absorption. Their accurate thickness was determined from the interference patterns observed at energies below the corresponding absorption edges. All the samples were prepared from the same ingots. The details of the used experimental set-ups   have been reported earlier.[7,8,9]

RESULTS AND DISCUSSION

<u>The E_0 band gaps of Zinc Halcogenides under pressure</u>

In Figs. 1 to 3 we display the E_0 gaps of ZnS, ZnSe and ZnTe as a function of pressure. The E_0 values are determined   from the points where the corresponding absorption coefficient curves versus photon energy saturates.   Because an indirect gap is not interfering the energy position of this point remain well defined up to pressure just below the phase transition. The kinks assigned to the gaps do not correspond exactly to the exciton energy but they are shifted to lower energies due to straylight limitations.[10] This effect is expected to be more pronounced in the case of ZnS[9]. As a result , the somewhat incorrect definition of the E_0 should affect slightly the quadratic terms in pressure which should be considered as the lower bounds for the nonlinear coefficients.

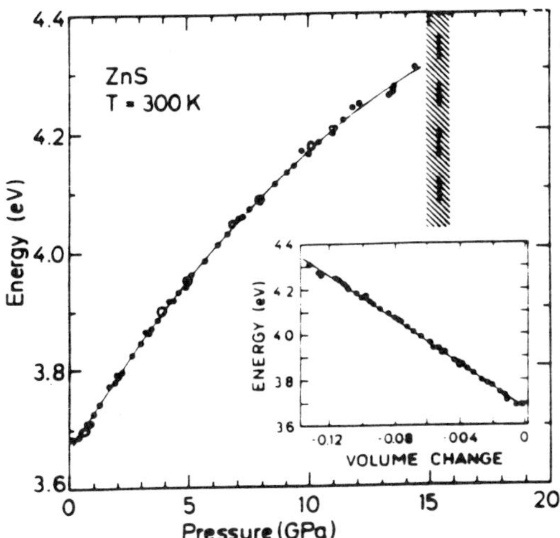

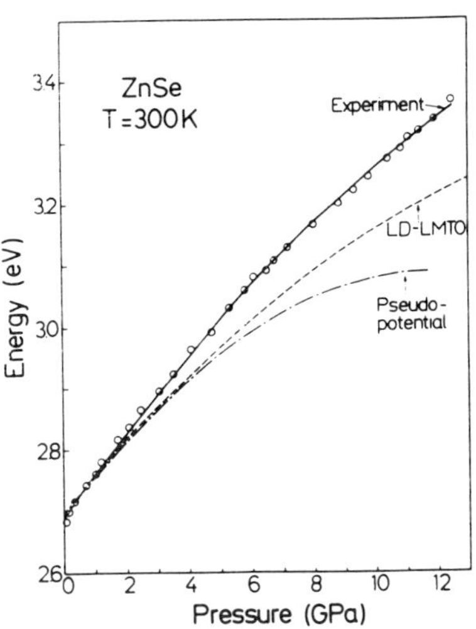

FIG. 1. Lowest direct $E_0(\Gamma_{15}^v \to \Gamma_1^c)$ gap of ZnS as function of pressure at room temperature. The solid lines through the experimental data represent quadratic least square fits. The vertical dashed shaded line refers to the phase transition from the zinc-blende to rocksalt structure while the inset shows the shift of E_0 vs change in relative volume. The open circles represent calculations with the LD-LMTO method. (From Ref. 9)

FIG. 2. The same as in Fig 1. but for ZnSe. The dashed line represents calculations with the LD-LMTO method, while the dashed-dotted line represents calculations based on a local pseudopotential. (From Ref. 7)

TABLE 1. First- (b, β) and second- order (c, γ) pressure and vs relative change of volume coefficients as well as the corresponding deformation potentials (D) and the transitions pressures P_{tr} for the zinc halcogenides ZnS, ZnSe and ZnTe.

	E_0 meV		b meV/GPa		β eV		c meV/GPa^2		γ eV		D eV		P_{tr}	
	Exp.	Ther.	Exp.	Ther.	Exp.	Ther.	Exp.	Ther.	Exp.	Ther.	Exp.	Ther.	Exp.	Ther.
ZnS	3.67[a]	2.06[a]	63.5[a]	62.2[a]	5.4[a]	5.3[a]	-1.31	-1.14[a]	-3.45	-2.0[a]	-5.0[a]	-4.5[a]	15.0[a]	19.5[a]
	3.64[d]	-	67.2[d]	-	-	-	-2.0[d]	-	±0.7[a]	-	-	-3.9[g]	15.0	19.9[h]
ZnSe	2.69[b]	2.69[b]	71.7[b]	65.7[b]	4.8[b]	4.8[b]	-1.5[b]	-2.7[b]	+0.7	-14.0	-4.8[b]	-4.3[b]	13.5[b]	10[h]
	-	1.17[b1]	-	49.0[b1]	-	4.1[b1]	-	-1.0[b1]	±0.4	-3.0[b1]	-	-4.1[b1]	-	28[i]
ZnTe	2.27[c]	2.27[c]	35.0[c]	38.0[c]	5.4[c]	6.4[c]	-2.8[c]	-6.5[c]	0.0	-18.2[c]	-5.4[c]	-6.3[c]	9.3[c]	18.8[e]
	2.25[f]	1.02[e]	38.1[f]	31[e]	-	4.6[e]	-5.0[f]	-2.4[e]	±0.3[c]	+2.0[e]	-	-6.0[c]	-	4.3[h]

[a] Ref. 9 ; Abs. LMTO   [c] Ref. 8; Abs. EMP.   [f] Ref. 11 Abs.   [i] Ref. 23 (Pseudo.)

[b] Ref. 7 ; Abs. EMP   [d] Ref. 12 Abs.   [g] Ref. 13

[b1] Ref. 7 ; LMTO   [e] Ref. 14 LMTO   [h] Ref. 15 (ab initio Pseudop.)

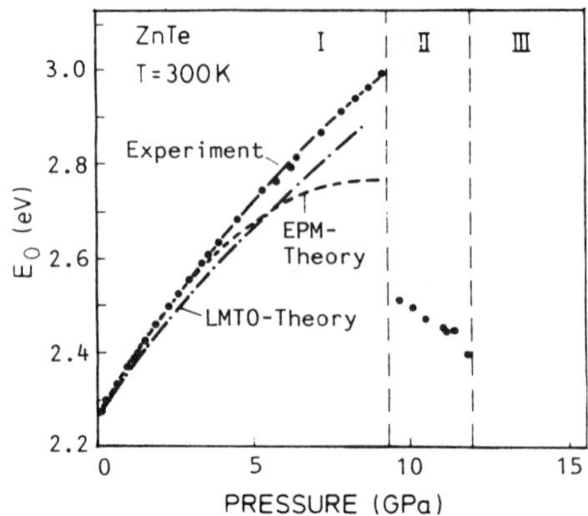

Fig. 3. The same as in Figs. 1. and 2. The vertical lines represent the phase transitions of ZnTe from the zinc-blende phase I to phase II and from that to the monoclinic (?) phase III. (From Ref. 8)

From Figs. 1 to 3 it is clear that in all three compounds a weak sublinearity is present. In these Figs. the LDA - LMTO calculations have been shifted up by the appropriate amount in order to match the experiment at zero pressure. The analytic expressions of the experimental and theoretical pressure dependences are :

$$E_0(p) = E_0\,(p=0) + b\,p + c\,p^2 \;\; = \; E_0 + \beta\,(-\frac{\Delta V}{V}) + \gamma\,(-\frac{\Delta V}{V})^2 \tag{1}$$

where the energy is given in eV and the pressure in GPa. The theoretical curves have been converted from the relative change in volume $(-\Delta V/V)$ to pressure[7-9] by using the Murnaghan's equation of state, except for the LD-LMTO calculations where the theoretical PV-relation has been used. The linear and quadratic coefficients of equation 1 are listed in Table 1. , together with related experimental and theoretical results.

A comparison of the experimental data and the calculated E_0 vs pressure shows a good agreement for the whole stability region of the zinc-blende structure in ZnS while in the cases of ZnSe and ZnTe they show a sufficient agreement for pressures only up to around 3.0 GPa ($\sim$ 0.5 % change in volume). For higher pressures they deviate markedly exhibiting strong sublinearities resulting in both kind of calculations from the fact that the Γ_{15}^v band is increasing faster in energy than the Γ_1^c with decreasing volume.

From Table 1 and inset of Fig. 1 we see that when the experimental E_0 is plotted vs the relative change in the volume, it exhibits an almost linear dependence, suggesting that the main part of this sublinearity originates from the nonlinear PV- relation. In the cases of ZnSe and ZnTe the experimental nonlinear γ coefficients are   almost zero while that of ZnS is slightly negative. The corresponding theoretical coefficients differ significantly depending on the method used for their calculation. For instance, the LD-LMTO calculations gives γ's always smaller(in absolute) than the EMP[16] calculations and closer to the experimental values although a sign reversal is obtained in ZnTe leading to a small supralinearity. However, the LMTO calculations are always affected by the inaccuracies in the local density exchange potential (LDEP) used.[17] While this had been corrected in Ref. 14 at zero pressure through an ad hoc adjustment of the core potentials, there is no guarantee that this correction will work also at higher pressures, especially to the nonlinear terms. The EMP calculations produce always a sublinearity which seems to increase from ZnS to ZnTe.

<u>The high pressure phases of ZnS, ZnSe and ZnTe</u>

At P_{tr} equal to 15 ± 0.4, 13.5 ± 0.3 and 9.3 ± 0.3 GPa the absorption spectra of ZnS, ZnSe and ZnTe respectively, change suddenly. The sharp absorption edge of ZnS drops by about 2.0 eV and becomes broad resembling that of an indirect gap material. In ZnSe the samples above P_{tr} become opaque in the energy region ($h\nu \geq 1.5$ eV). Hence, in ZnSe we have measured the reflectivity spectrum[18]. The corresponding spectra are shown in Figs. 4 and 5. From Figs. 4. and 5. it is clear that the high pressure phases of ZnS and ZnSe are not opaque or metallic as reported or suggested earlier.[5,6,20]

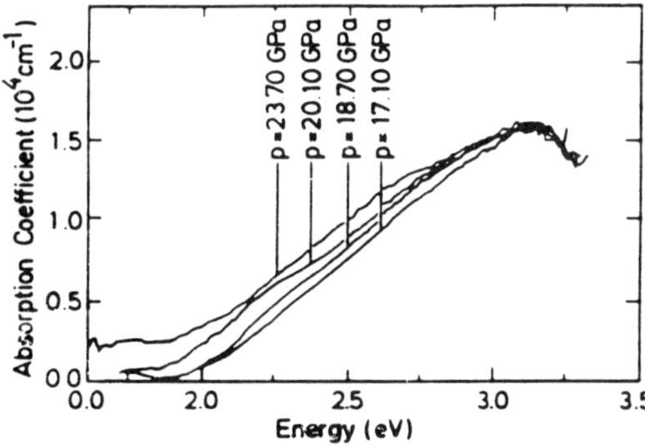

FIG. 4. Absorption spectra of ZnS in the rocksalt phase at various pressures with increasing pressure. (From Ref. 9)

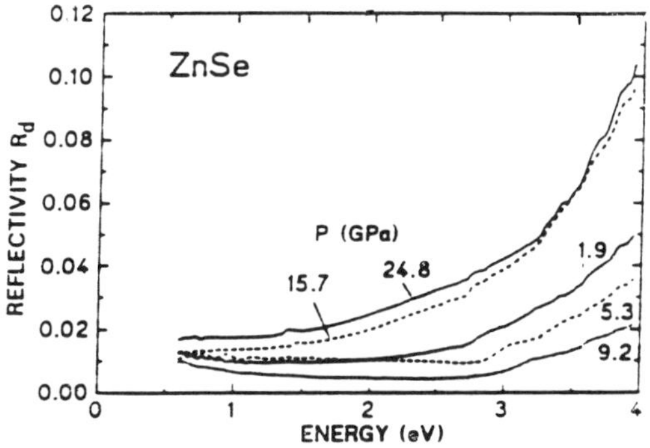

FIG. 5. Optical reflectivity spectra of ZnSe at different pressures. R_d symbolize the absolute reflectivity at the interface between sample and the diamond window of the pressure cell . It is corrected for losses at the outer surface of diamond. (From Ref. 16)

The high pressure phases of ZnS[9,19] and ZnSe[18] are of NaCl-type. The pressure induced phases are stable up to 27 and 25 GPa, for ZnS and ZnSe, respectively whereas it has been suggested that they are metallic[5,6,20] in the sense of an energetic overlap between conduction and valence states. From Figs. 4 and 5. we conclude indeed an overall red shift of the oscillator strength of interband excitations from UV towards the visible range, but the characteristic Drude-like tail of free carrier behaviour is absent in ZnSe while in ZnS a band gap of at least 2.0 eV has been measured. Thus we infer that the NaCl - type structures of ZnS and ZnSe do still remain semiconducting.

In Fig. 6. we display the absorption curves for ZnTe pressures above P_{tr}. Also here the absorption coefficient curves vs. photon energy change shape suddenly (compare Figs. 3. and 6). Their shape and behaviour is very similar to that observed in ZnS. Similarly, the absorption edge becomes also here very broad and resembles that of an indirect gap. In addition it exhibits a jump of about 0.5 eV towards lower energies. With increasing pressure the absorption edges shift uniformly to lower pressures at a rate of about 50 meV/GPa which is considerably higher than that of ZnS (compare Figs. 4 and 6).

Above 12.0 GPa the ZnTe sample becomes suddenly opaque in the energy region ($h\nu \geq 1.5$ eV). From reflectivity measurements under pressure[21] on ZnTe it is clearly shown that the second high pressure phase which is stable up to 30 GPa[22] indeed is metallic.

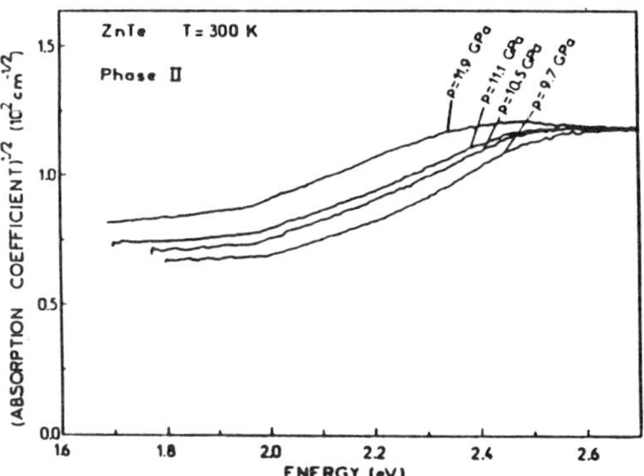

FIG. 6. Optical absorption spectra of ZnTe - II phase (T = 300 K) at different pressures.

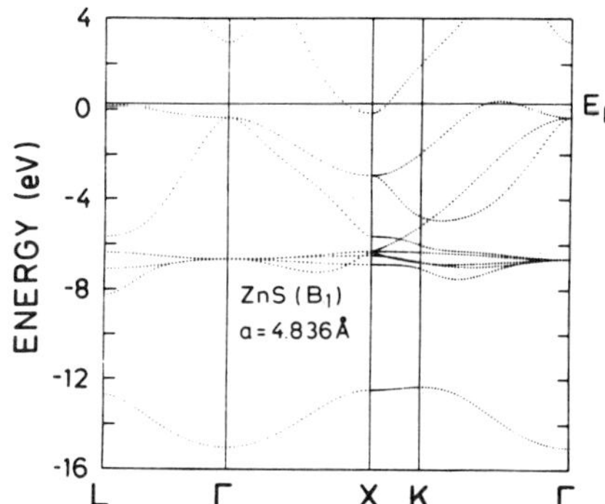

FIG. 7. LMTO - LDA band structure of ZnS in the rocksalt structure. (From Ref. 9)

Concerning the theoretical predictions[14,15,23] about the high pressure structure of zinc halcogenides all of them predict the NaCl structure driven by ion- electron interaction but there appear discrepancies as far as if they become metallic or not. In Fig. 7 the LMTO - LDA band structure of ZnS[9] is displayed which is typical also for the high pressure structure of ther zinc halogenides and of CdTe[14]. In this calculation the valence and conduction bands overlap leading in a metallic character. The band overlapping is caused by a drastic lowering of the X_1^c band with respect to the Γ_1^c and by the trend of LDA method to produce too small gaps. This shift appears to increase along the sequence ZnS → ZnSe → ZnTe. The LDA method predicts for the rocksalt phase of ZnS the gaps $L_3^v \to X_1^c$ and $\Sigma_4^v \to$

X_1^c while for ZnTe the gaps $L_3^v \rightarrow X_1^c$ and $\Gamma_{15}^v \rightarrow X_1^c$. On the other hand, theoretical predictions[23], based on self consistent pseudopodential calculations, suggest that the rocksalt structure is semiconducting although also here still a lowering of the X_1^c state compared to Γ_1^c state is predicted. In particular, for ZnSe a gap of 1.1 eV, caused by transitions $\Gamma_{15}^v \rightarrow X_1^c$, is predicted in Ref. 21 which do not contradict our results. For the sake of completeness, we mention that the same set of gaps has been predicted for the rocksalt phases of copper halides[24] and CdS[25]

X- ray measurements on ZnTe[8,22] showed that indeed phase transitions occur at 9.3 and 12.0 GPa. The phase ZnTe - II has a complicated crystal structure. However, from the diffraction patterns the rocksalt and β- tin structures can be excluded. The second high pressure phase, ZnTe - II, which is stable up to 30 GPa, could be indexed only by assuming a monoclinic structure. The possibility of several intermediate pressure induced phase transitions in ZnTe should not be discarded[26]. For the sake of completeness, we mention that if we replace Zn with Mn ($Zn_{0.7}Mn_{0.3}Te$) the NaCl structure appears while a lowering of the transition pressure occurs[22].

On releasing pressure the ZnS samples become transparent and almost colourless again at 10.3 GPa . A hysteresis of the same order appears also in ZnSe which becomes again transparent at about 10 GPa. The back transformed phase has the zinc-blende structure. Hence, the given transition pressures P_{tr} for increasing pressure correspond to an upper limit. Similar results have been reported also in Ref.[27] for ZnSe, where the thermodynamic equilibrium pressure between the zinc-blende phase and rocksalt phase, has been found to be not more than 11 GPa. Furthermore, from 14 to 10 GPa a coexistence of the two phases on releasing pressures has been found in Ref. 25. Also ZnTe transforms back to the zinc-blende phase as it can be deduced from the absorption spectrum recorded at 4.7 GPa with decreasing pressure. It stays in this structure by increasing the pressure for a second time up to at least 7.3 GPa.   In all three compounds the optical quality of the samples after back transformation is definitely worse than the starting material.

ACKNOWLEDGMENTS

This paper is the result of team work. It is my pleasure to acknowledge the excellent collaboration with K. Strößner, C. K. Kim, W. Hönle, N. E. Christensen, K. Syassen, and M. Cardona.

REFERENCES

1.   S. Adachi and T. Tagushi Phys. Rev. **B41**, 9569, (1991); N. E Christensen, I. C. Gorczyca, O. B. Christensen, U. Schmid, and M. Cardona, J. Ctyst. Growth, **101**, 318, (1990); K. Shahzad, D. J. Olego, and C. G. Van de Wall, Phys. Rev.**B38**,1417, (1987); C. G. Van de Walle, Phys. Rev.**B39**,1871, (1989)

2. W. M. Becker ,in *Semiconductors and Semimetals*, edited by R. K. Willarson and A. C. Beer (Academic Pres New York,1988,), Vol. 25

3. L. Quiroga, F. J. Rodriguez, A. Camacho and C . Tejedor Phys. Rev. **B42**, 11198, (1990)

4. G. A., Samara and H. G. Drikamer, J. Phys. Chem. Solids. **23**, 457, (1962)

5. G. J. Piermarini, H. G. Block Rev. Sci. Instrum. **46**, 973, (1975); B. A. Weinstein Solid State Commun. **24**, 595, (1975)

6   S. R. Tiong, M. Miramatsu, Y. Matsushima, and E. Ito, Jpn. J. Appl. Phys. **28**, 291, (1989)

7. S. Ves, K. Strößner, N. E. Cristensen, C. K. Kim, and M. Cardona Solid State Commun. **56**,479, (1985)

8. K. Strößner, S. Ves, C. K. Kim, and M. Cardona Solid State Commun. **61**,275, (1986)

9. S. Ves, U. Schwarz, N. E. Christensen, K. Syassen and M. Cardona, Phys. Rev. **B42**, 9113, (1990) (and references therein)

10. A. Goni, A. Cantarero, K. Syassen and M. Cardona, Phys. Rev. **B41**, 10111, (1990)

11. B. A Weinstein, R. Zallen, M. L. Slade, and A. de Lozane Phys. Rev. **B24** 4652, (1981)

12. A. Beliveau and C. Carlone Phys. Rev.**B41,** 9860, (1990)

13. A. Blacha, H. Presting, and M. Cardona   Phys. Status Solidi **B126**, 11, (1981)

14. N. E Christensen, and O. B. Christensen Phys. Rev. **B33**, 4739, (1986)

15. J. R. Chelicowsky   Phys. Rev. **B35**, 1174, (1987)

16. S. Ves, K. Strößner, C. K. Kim, and M. Cardona Solid State Commun. **55**, 327, (1985)

17. M. S. Hybertsen and S, G. Lanie, Phys. Rev. Lett. **55**, 1418, (1985)

18. S. Ves, U. Schwartz. N.E. Christensen, K. Syassen and M. Cardona. in *Proceedings of HighPressure in Semicinductor Physics*, Porto Carras, Greece August, (1990), Edited by D. Kyriakos, Suppl. of Annual Scientific Report. of Physics Department of University of Thessaloniki,(1991), Greece, p. 192,

19. P. L. Smith and J. E. Martin , Phys. **19**,541, (1965)

20. B. A. Weinstein Solid State Commun. **24**, 595, (1977)

21. Z. Wang and K. Syassen Private communications

22. K. Strößner,S. Ves, , W. Hönle, W. Gebhardt and M. Cardona, in *Proceedings of the 18th International Conference on the Physics of Semiconductors.*Stockholm, 1986, Edited by E. Engström (Word Scientific, Singapore,1987), p. 1717

23. W. Andreoni and K. Maske   Phys. Rev. **B22**, 4816, (1980)

24. S. Ves, D. Glötzel, H. Overhof and M. Cardona Phys. Rev. **B24**, 3073, (1981); R. K. Singh, and D.C. Kupta, Phys. Rev. **B40**,11278, (1989)

25. W. W. Liu and S. Rabi Phys. Rev. **B13**, 1679, (1976)

26. K. Syassen Private communications; S. Endo, A. Yoneda, M. Ichikawa, S. Tanaka, and S. Kawabe, J. Phys. Soc. Japan, 51, 138, (1982)

27. G. Weil and J. M. Besson Proceedings of XXVIIIth EHPRG Meeting July, 1990, Bourdeaux, France, Publ. in High Pressure Research and Technology. (in press)

HIGHLIGHTS OF THE ROUND TABLE DISCUSSION ON

HIGH PRESSURE AND SEMICONDUCTORS

B. A. Weinstein

State University of New York
Physics Department
Buffalo, New York, USA

Panel: B. A. Weinstein, Chairman, SUNY at Buffalo, NY, USA
A. R. Adams, University of Surrey, Guildford, UK
L. J. Ci, SUNY at Buffalo, NY, USA
V. A. Wilkinson, University of Surrey, Guildford, UK
R. A. Stradling, Imperial College, London, UK

PREFACE

Each of the panel members offered a five minute overview of the
state and future directions of the field, followed by questions and comments from the audience. In the summary presented here, I have tried to
accurately paraphrase (not quote) both the content and spirit of these
remarks based on notes, memory, and a tape recording. I apologize for any
inadvertent omissions or inclusions. (B. A. Weinstein).

A. R. Adams

Recently, there has been a great deal of exciting semiconductor
physics in the area of the low-dimensional structures. Most of this work
has dealt with lattice-matched 2-D (layered) systems, e. g., AlGaAs/GaAs,
InGaAs/InP, etc. Current trends toward strained-layer 1-D quantum wires
or O-D quantum dots present many new opportunities for high pressure research. The combination of lattice-mismatch strain and lower dimensional
confinement should enhance growth efforts and band structure engineering.
High pressure allows one to directly tune and observe the effects of
strain in low-D systems without needing a myriad of different samples.

The research in low-D structures has often concentrated on conduction band engineering. The use of strain for valence band engineering
should be pursued. Device speed could be enhanced by decreasing the hole
effective mass. Also strain can be useful for controlling the light
polarization in optoelectronic applications.

Lattice relaxation effects in mismatched systems may also be studied
by high pressure. Such effects can be very detrimental for the mixed
valent III-V/Si devices desired for integrated optoelectronics. Hydrostatic and uniaxial stress work directed toward understanding and control
of relaxation induced defects could be quite important.

Frontiers of High-Pressure Research, Edited by H.D. Hochheimer and
R.D. Etters, Plenum Press, New York, 1991

<u>L. J. Cui</u>

As a student, my perspective is somewhat limited. However, I feel that current semiconductor research is often device oriented. Sample quality is, clearly, very important, and areas where high pressure research could help to improve device properties should be explored. An example is the formation and role of interface dislocations. This is a limiting factor in the growth of novel mismatched structures. Although the threshold for misfit dislocations is determined by both energetics and kinetics, relatively little work on the latter has occurred. This should be pursued as a function of both high pressure (to tune misfit) <u>and</u> variable temperature. The results can help to control the number and nature of dislocations.

<u>V. A. Wilkinson</u>

Our workshop has presented a fair cross-section of semiconductor pressure studies. My interests are in the electronic properties of heterostructures, and here one of the most important parameters is band alignment. Early research on AlGaAs/GaAs showed how useful pressure could be to determine band alignment, but very few measurements on other systems have followed. There is also a need for more theoretical work to explain the observed band-gap pressure shifts, and how this effects band alignment. These are fruitful areas for future work, especially for strained-layer systems, which after all, can be viewed as structures engineered to exhibit built-in pressure.

<u>R. A. Stradling</u>

There is a need to involve more semiconductor physicists in high pressure experiments by making pressure techniques, especially the diamond-anvil cell, more routine. Pressure is not like other parameters, such as temperature and magnetic field, for which commercial apparatus are commonly available. Yet as a thermodynamic parameters, pressure merits equal attention due to its ability to probe systems. Also, high pressure should be applied to more varied measurement techniques. For example, far infrared studies under pressure have tremendous potential, but offer challenging problems due to diffraction effects; better sources and detectors may be required.

The future role of high pressure for semiconductors will depend on the future of semiconductor physics itself. In this regard, let us keep in mind that only the physics of Si is really mature. The next most studied material, GaAs, has received only 1% of the attention paid Si, and other semiconductors 1% of that. The field is clearly still quite rich. Work on wide gap materials is needed, including diamond, SiC, III-V nitrites, and II-VI's. At the opposite energy range, continued interest in narrow gap compounds is driven by the need for infrared sensors. Real variety in engineered properties is available if we can learn how to control quantum size and impurity effects. This variety can be expanded further by developing hybrid structures, and work should be pursued on III-V/Si and high-T_c/Si combinations. High pressure can contribute significantly in all of these areas.

<u>B. A. Weinstein</u>

It is generally agreed, that semiconductor physics is in the midst of a materials revolution fueled by the remarkable achievements of modern epitaxy. Before long, we can expect quantum wires and dots to be common in the laboratory, if not in the marketplace. Because built-in misfit strain is often unescapable, and sometimes desirable, growth efforts can

benefit importantly from high pressure research. Essentially, applied
pressure enables one to probe epitaxial structures in regions of phase
space normally inaccessible to growth. An example is the GaAs/AlAs
system, which becomes metastable between 12-17 GPa for the same reasons
that allow 1 atm. growth of metastable zincblende MnSe on ZnSe.

Considering the future, I would like to speak to some aspects that
have not yet been raised. In a sense, the study of semiconductors under
pressure is inherently limited, since almost all the interesting materials
become metals by the "modest" pressure of 30 GPa. However, frontier semiconductor physics is today wrestling with many-body effects and when these
are considered, a fascinating range of phenomena opens to the high pressure community. Two examples are the fractional quantum hall effect, and
electron-hole droplets. Indeed, for the latter, some of the physics is
highly able of using pressure to tune the effective masses of both electrons and holes. Let us recall how beneficial pressure studies
(e. g., the work of W. Paul and H. Fritsche) were to single electron band
models; it is likely that similar benefit will accrue for many-body
phenomena. However, the problems are more difficult.

<u>R. A. Stradling</u>

Our group is currently studying the magnitudes of spin splitting and
Landau splitting in factional quantum hall systems up to 8 kbar pressure.
The experiments are difficult because of the need for high pressure trans-
port measurements at millikelvin temperature.

<u>B. A. Weinstein</u>

Future work on many-body phenomena will require considerable technique development. We are already seeing this in the report by S. Klotz
at this workshop on magnetic susceptibility studies with small coils in
the diamond-anvil cell.

<u>H. D. Hochheimer</u>

It is exciting to see how far heterostructure physics has come in
fulfilling the predictions of fundamental and applied phenomena made by
early workers (e. g., G. Dohler, then at MPI Stuttgart). Is it actually
practical to achieve new device properties by applying high pressure to
heterostructures?

<u>A. R. Adams</u>

For the most part, until now, no. Pressure is generally used to
probe the electronic or optical properties of heterostructures under
strain conditions that, because of hydrostatic conditions, cannot exactly
mimic strained-layer growth. It would be advantageous to make combined
uniaxial and hydrostatic pressure measurements -- say 1 GPa uniaxial
stress modulating 1 -2 GPa pressure. One could then achieve quite novel
device properties useful for tuning low-threshold solid state lasers. Of
course, loading intact laser diodes or other electronic chips in the diamond-anvil cell presents nontrivial problems, if one wants to consider
practical devices.

<u>S. Ves</u>

I have used a diamond-anvil cell to produce up to 3 GPa uniaxial
stress in thin films. The films were deposited on one anvil and squeezed
without a gasket. CuCl was studied.

379

B. A. Weinstein

Early work by N. Holonyak (U. of Illinois) in oil bombs showed up to
2000 atm. pressure tuning of GaAs diodes. For diamond-anvil cell work
with semiconductors, a tough problem is how to attach good ohmic contacts
to the samples. The mechanical contacts often used for T_c measurements in
superconductors would severely degrade device properties. By the way, the
combination of hydrostatic and uniaxial pressure is also very desirable
for studying high-T_c materials.

R. J. Wijngaarden

A possible solution of the ohmic contact problem is to deposit the
leads and device structure as a thin film on a diamond-anvil surface.

A. R. Adams

The difficulties associated with adhesion on diamond for such complicated device architectures may surpass those of introducing wires into
the diamond-anvil cell. Also, useful pressure tuning will require reversible tuning, and the diamond-anvil cell has only limited reversibility.

R. A. Stradling

It is worth pointing out that pressure-tuned Pb-salt lasers have
been demonstrated with over a factor of two wavelength tuning in the mid-
infrared.

A. Polian

May I bring up a somewhat different subject? There is a strong need
to study the actual equilibrium thresholds for pressure-induced phase
transitions in semiconductors. These thresholds are generally lower than
the measured onset pressure because of large hysteresis effects. In GaAs
we observe variations of several GPa depending on how the measurements are
done.

B. A. Weinstein

This is a good point. Some of these variations are probably material effects, but a large part depends on the intrinsic transition kinetics,
which is not well understood.

R. A. Stradling

Another interesting set of kinetic problems concerns the details of
epitaxial growth kinetics. There are regimes of temperature and deposition rate where alloys undergo spontaneous ordering into layers (e. g.,
the InAsSb system that I mentioned during the talks), and other regions
where island growth is favored over layer growth. The theoretical difficulties are challenging, since these processes are far from equilibrium.

N. Ashcroft

The adhesion physics of atoms colliding with and "sticking" to a
substrate is, of course, a key factor in epitaxial growth. There is considerable need for ab-initio theoretical work that would consider sticking
mechanisms at the atomic level.

<u>A. R. Adams</u>

Besides MBE (molecular beam epitaxy), other types of epitaxial growth are MOCVP (metal-organic chemical vapor deposition) and LPE (liquid phase epitaxy). Each involves somewhat different chemical and mechanical processes. There is much room for predictive theories.

<u>B. A. Weinstein</u>

In a few cases high pressure has been able to improve growth, as for in the 2 GPa conventional (melt) growth of GaN by the UNIPRESS group.

<u>R. A. Stradling</u>

In a slightly different vein, I would like to comment that many of the opotelectronic applications now under laboratory research appear quite practical. I believe we can expect advances such as optical computing in a reasonable time frame.

<u>B. A. Weinstein</u>

Since the house now grows late, let us end on this optimistic note.

HIGH PRESSURE STUDY OF HIGH TEMPERATURE SUPERCONDUCTORS:

MATERIAL BASE, UNIVERSAL T_c-BEHAVIOR, AND CHARGE TRANSFER

C. W. Chu,* P. H. Hor, J. G. Lin, Q. Xiong, Z. J. Huang, R. L. Meng
and Y. Y. Xue

Department of Physics and
Texas Center for Superconductivity at the
University of Houston
Houston, Texas 77204-5932

Y. C. Jean

Department of Chemistry
University of Missouri
Kansas City, Missouri 64110

ABSTRACT

The superconducting transition temperature (T_c) has been measured in $YBa_2Cu_3O_{6.7}$, $YBa_2Cu_3O_7$, $Y_2Ba_4Cu_7O_{15}$, $YBa_2Cu_4O_8$, $Tl_2Ba_2Ca_{n-1}Cu_nO_{n+4-\delta}$, $La_{2-x}Sr_xCuO_4$, and $La_{2-x}Ba_xCuO_4$ under high pressures. The pressure effect on the positron lifetime (τ) has also been determined in the first four compounds. Based on these and other high pressure data, we suggest that (1) all known cuprate high temperature superconductors (HTS's) may be no more that mere modifications of either 214-T, 214-T', 123, or a combination of 214-T' and 123, (2) a nonmonotonic T_c-behavior may govern the T_c-variation of all hole cuprate HTS's, and (3) pressure can induce charge transfer leading to a T_c-change. The implications of these suggestions will also be discussed.

INTRODUCTION

High pressure has been shown [1] to be very effective in the study of solids by testing theoretical models, providing clues for the search for new compounds, and revealing and/or stabilizing new ground states. Its significant role in the development of high temperature superconductivity (HTS) is equally evident, particularly during the early days. For instance, several excellent attempts have been made [2] to differentiate the various theoretical models proposed by comparing their predictions with the experimental results of the high temperature superconductors (HTS's) of $YBa_2Cu_3O_7$ and $(La_{1-x}A_x)_2CuO_4$ with A = Ba or Sr, under high pressures. An unusually large positive pressure effect on the superconducting transition temperature (T_c) was observed [3] immediately after the report [4] of superconductivity above 30 K in (La,Ba)CuO. It was precisely this observation that signaled the difference between (La,Ba)CuO and other, previously known, superconductors with low T_c, and thus demonstrated [3] the importance

---

* visiting the Geophysical Laboratory of the Carnegie Institution in Washington DC

Frontiers of High-Pressure Research, Edited by H.D. Hochheimer and
R.D. Etters, Plenum Press, New York, 1991

of cuprates in the search for superconductivity above the liquid nitrogen boiling point of 77 K. Later, it was again this positive dT_c/dP that led to the stabilization of superconductivity in YBaCuO [5] and other related compounds [6] with a T_c above 93 K. Recently, $(Sr,Ca)CuO_2$ with a T_c up to 100 K was successfully synthesized [7] under high pressure.

In this paper, we examine the cuprate HTS problems of materials base, universal T_c-behavior, and charge transfer by examining the high pressure data on HTS's obtained here and in other labs. The T_c of $Tl_2Ba_2Ca_{n-1}Cu_nO_{2n+4}$ [Tl22(n-1)n] where n = 1, 2, and 3, $YBa_2Cu_3O_{6.7}$ [Y123(6.7)], $YBa_2Cu_3O_7$ [Y123(7)], $Y_2Ba_4Cu_7O_{15}$ [Y247], and $YBa_2Cu_4O_8$ [Y124], as well as the positron lifetime (τ) in the last four compounds at 300 K have been determined under high pressures. The detailed x-dependence of dT_c/dP of $(La_{1-x}A_x)_2CuO_4$ (214-T) where A = Ba or Sr was also measured to address the question concerning the contradictory results reported [8]. We found that (1) the T_c's of Y123(6.7), Y123(7), Y247, and Y124 increase monotonically under quasi-hydrostatic pressures up to 150 kbar, in contrast to the non-monotonic behavior recently reported [9] in Y123(7); (2) the T_c's of these four compounds appear to merge toward each other asymptotically above ~ 60 kb; (3) dT_c/dP is always positive for all 214-T compounds despite the nonmonotonic variation of T_c with x [10]; (4) dT_c/dP can be either positive or negative for hole-HTS's in contrast to the negative value only [11] for electron-HTS's; and (5) the magnitude of $d\tau/dP$ scales with that of dT_c/dP. Based on these and other high pressure data of HTS's, we conclude that (1) all cuprate HTS's known to date may not be more than mere modifications of one of the 214-T, 214-T', and 123 phases, or a combination of the 214-T' and 123 phases; (2) appropriate shear stress may suppress the instability of HTS and thus lead to higher T_c; (3) the hole-cuprate HTS's may belong to a different group from the electron-HTS's; (4) the T_c of all known hole-cuprate HTS's appear to vary in a universal fashion with a single parameter closely related to the carrier concentration (n); and (5) charge transfer in HTS's can be induced by the application of pressure. It should be pointed out that some of these ideas have been discussed briefly elsewhere [12-13].

EXPERIMENTAL

Polycrystalline samples of Y123(6.7), Y123(7), Y247, Y124, and La214 have been made by the standard solid state reaction technique, followed by repeated pulverization and annealing in an oxygen or a slightly reduced atmosphere. Magnetic measurements using a SQUID magnetometer were made on these samples to assure their high quality prior to high pressure measurements. High pressure studies were carried out only on samples with a single and sharp magnetic transition. The T_c was determined using a standard four-lead resistive technique. Hydrostatic pressures up to 20 kb were generated in a piston-cylinder arrangement while quasi-hydrostatic pressures up to 150 kb were produced by a pair of W-C anvils all housed in a Be-Cu clamp [16]. The pressure media used were flourinert liquid or a mixture of silicon oil and kerosene for the former and steatite for the latter. The τ-measurements at 300 K were performed using a fast-fast coincidence spectrometer on samples under pressures generated by a pair of W-C anvils. The τ-spectra obtained were then analyzed [17] with the PATFIT program.

RESULTS AND DISCUSSION

<u>The Material Base of Cuprate HTS's</u>

One of the most effective approaches towards unravelling the HTS mechanism is to examine a large number of HTS's and to seek out their similarities and dissimilarities. While the similarities tell us the points of focus for the development of theoretical models and give us guidance in the discovery of new, although similar, HTS's, the dissimilarities enable us to generalize the theoretical models to meet constraints set by the various HTS's. By examining the structure and chemistry of the existing cuprate HTS's, many common features have emerged. For instance, they all have [18] layered structures, mixed-valence

Cu-ions, divalent elements, and they can be induced by doping their respective parent insulators. Several classification schemes [19-22] for different purposes have been proposed. However, none has addressed the question if the known HTS's are truly different enough to form a material base sufficiently broad for a comprehensive test of the theoretical model.

In the last four years, more than 75 non-intermetallic compounds with a T_c above 23 K which is the record value for intermetallic compounds. As listed in Table 1, they are

TABLE 1.
Known HTS families with their designations and maximum T_c.

CUPRATES
$(La,A)_2CuO_4$ [214-T (~ 40 K)]; A = Ba, Sr or Ca
$(Ln,R)_2CuO_4$ [214-T' (~ 25 K)]; Ln = Pr, Nd, Sm or Eu and R = Ce or Th
$(La,Sr)(Ln,Ce)CuO_4$ [214-T* (~ 35 K)]; Ln = Nd, Sm or Eu
$(La,Sr)_2CaCu_2O_6$ [212 (~ 60 K)]
$LnBa_2Cu_3O_7$ [123 (~ 98 K)]; Ln = Y, La, Nd, Sm, Eu, Gd, Dy, Ho, Er, Tm, Yb or Lu
$(Ln,Ce)_2(Ln,Ba)_2Cu_3O_{10}$ [223 (~ 40 K)]; Ln = Nd, Sm or Eu
$Pb_2(Ln,Sr)Sr_2Cu_3O_8$ [Pb 2123 (~ 80 K)]; Ln = Nd, Sm, Eu, Gd, Tb, Dy, Ho, Er, Tm
$Y_2Ba_4Cu_7O_{15}$ [247 (~ 70 K)]
$YBa_2Cu_4O_8$ [124 (~ 80 K)]
$Bi_2Sr_2Ca_{n-1}Cu_nO_{2n+4}$ [Bi-22(n-1)n (~ 110 K)]; n = 1, 2 or 3
$Bi_2Sr_2(Ln,Ce)_2Cu_2O_{18}$ [Bi 2222 (~ 40 K)]; Ln = Sm, Eu and Gd
$Tl_mBa_2Ca_{n-1}Cu_nO_{2n+m+2}$ [Tl-m2(n-1)n (~ 125 K)]; m = 1 or 2 and n = 1, 2, 3, 4 or 5
$(Pb_{0.5}Ca_{0.5})Sr_2(Y,Ca)Cu_2O_7$ [Pb1212 (~ 50 K)]
$(Pb_{0.5}Cu_{0.5})(Eu,Ce)_2(Sr,Eu)_2Cu_2O_x$ [Pb1222 (~ 25 K)]
$Pb(Ba,Sr)_2(Y,Ca)Cu_3O_x$ [Pb1213 (~ 50 K)]
$Pb_2(Y,Ca)Sr_2Cu_3O_8$ [Pb2123 (~ 80 K)]

BISMUTHATES
$(Ba,A)BiO_3$ (~ 30 K); A = K or Rb

FULLERITES
A_3C_{60} (~ 30 K); A = K, Rb or Cs

---

cuprates, bismuthates, and fullerites, and the overwhelming majority of these HTS's are cuprates possessing a T_c up to 125 K. We therefore shall confine the following discussions to cuprates only. To group and contrast them, the following bases have been used: (1) the minimum number of elements [19] required to stabilize the basic structure of the HTS's, (2) the number of CuO_2-layers per unit formula [20], (3) four letters [21], k, l, m, and n, where k = number of layers between apexes, l = that next to oxygen at the apex, m = that between bottom faces of the pyramidal CuO_5-layers, and n = that of CuO_2-layers per unit formula, and (4) CuO_2-layers and block layers of different kinds [22]. As more and more new compounds are discovered, the first basis becomes insufficient since many of the elements are present only to serve as a structural stabilizer and play no role in HTS. Regarding the second basis, the number of CuO_2-layers per unit formula appears to correlate well with T_c initially; however, recent results show [23] that even single layer HTS's, such as Tl2201, can have a T_c as high as 90 K. The third basis is rather limited and not inclusive enough, and the fourth is very comprehensive and detailed. Both have been used effectively in the synthesis of new compounds.

In an alternative approach [14], we have chosen to concentrate on the dopability of the various layers. The cuprate HTS's can be written in their respective layer sequences as shown in Table II.

TABLE 2.
Cuprate HTS's in their structural layer sequences.

ONE CuO_2 LAYER

$(La,A)_2CuO_4 \rightarrow [(La,A)O]_2(CuO_2)$

$(Ln,R)_2CuO_4 \rightarrow [(Ln,R)O]_2(CuO_2)$

$2(La,Sr)(Ln,Ce)CuO_4 \rightarrow [(La,Sr)O]_2(CuO_2)[(Ln,Ce)O]_2(CuO_2)$

$Tl_2Ba_2CuO_6 \rightarrow [TlO]_2(BaO)(CuO_2)(BaO)$

TWO CuO_2 LAYERS

$YBa_2Cu_3O_7 \rightarrow (Y)(CuO_2)(BaO)[(CuO)](BaO)(CuO_2)$

$YBa_2Cu_4O_8 \rightarrow (Y)(CuO_2)(BaO)[(CuO)(CuO)](BaO)(CuO_2)$

$Pb_2(Y,Ca)Sr_2Cu_3O_8 \rightarrow (Y,Ca)(CuO_2)(SrO)[(PbO)Cu(PbO)](SrO)(CuO_2)$

$Bi_2Sr_2CaCu_2O_6 \rightarrow (Ca)(CuO_2)(SrO)[(BiO)_2](SrO)(CuO_2)$

$(Pb,Cu)Sr_2(Y,Ca)Cu_2O_7 \rightarrow (Y,Ca)(CuO_2)(SrO)[(Pb,Cu)O](SrO)(CuO_2)$

$Pb(Ba,Sr)_2(Y,Ca)Cu_3O_x \rightarrow$
$(Y,Ca)(CuO_2)[(Ba,Sr)O][(PbO)(CuO)][(Ba,Sr)O](CuO_2)$

$Bi_2Sr_2(Ln,Ce)_2Cu_2O_{18} \rightarrow [(Ln,Ce)O]_2(CuO_2)(SrO)[(BiO)_2](SrO)(CuO_2)$

$(La,Sr)_2CaCu_2O_6 \rightarrow [(La,Sr)O]_2(CuO_2)(Ca)(CuO_2)$

$(Ln,Ce)_2(Ln,Ba)_2Cu_3O_9 \rightarrow [(Ln,Ce)O]_2(CuO_2)[(Ba,Ln)O](CuO)[(Ba,Ln)O](CuO_2)$

THREE OR MORE CuO_2 LAYERS

$Bi_2Sr_2Ca_2Cu_3O_{10} \rightarrow (Ca)(CuO_2)(SrO)[(BiO)_2](SrO)(CuO_2)(Ca)(CuO_2)$

$TlBa_2Ca_2Cu_3O_9 \rightarrow (Ca)(CuO_2)(BaO)[(TlO)](BaO)(CuO_2)(Ca)(CuO_2)$

$TlBa_2Ca_3Cu_4O_{11} \rightarrow (Ca)(CuO_2)(BaO)[(TlO)](BaO)(CuO_2)(Ca)(CuO_2)(Ca)(CuO_2)$

$TlBa_2Ca_4Cu_5O_{13} \rightarrow$
$(Ca)(CuO_2)(BaO)[(TlO)](BaO)(CuO_2)(Ca)(CuO_2)(Ca)(CuO_2)(Ca)(CuO_2)$

Examination of the nature of all layers suggests that there exist only 6 general types, as given below:

$S =$ CuO_2 in a square planar arrangement

$M =$ $[(La,A)O]_2$ with the rock-salt type structure

$N =$ $[(Ln,R)O]_2$ with the fluoride type structure

$A =$ (Y), (Y,Ca), (Ca) or (Ln) with metal atoms only situated between two S-sheets

$B =$ (SrO), (BaO) or [(Sr,Ba)O] with metal and apical oxygen atoms situated next to the S-sheet

$$C = \text{(CuO), (CuO)}_2\text{, (PbO)(Cu)(PbO), (BiO)}_2\text{, (BiO), (TlO)}_2\text{, (TlO) or}$$
[(Pb,Cu)O] with metal and oxygen atoms situated between two B-layers.

Based on these layer-types, all cuprate HTS's can be presented in the following general forms:

SBCB for the one-layered HTS's except the 214-T, -T' and -T*
SBCBSA for all the two-layered HTS's except Bi2222, La212 and 223
SBCBSASA for the three-layered HTS's
SBCBSASASA for the four-layered HTS's
SBCBSASASASA for the five layered HTS's
and
MS for 214-T, NS for 214-T' and MSNS for 214-T*
MSASM for La212
NSBCBS for 223

According to such a grouping scheme, we found that (1) higher S-layer homologues can be achieved solely by adding SA layer pairs, (2) all layers except the B-type are dopable, (3) only NSN is electron-dopable and the rest are hole-dopable, (4) sub-layer-sequences MSM and BSB display CuO_6-octahedron arrays; ASB, ASM, NSB and NSM exhibit CuO_5-pyramid arrays; and ASA and NSN show CuO_4-square arrays, (5) no HTS has more than two S-layers with the CuO_5-pyramid arrangement or more than one S-layer with the CuO_6-octahedron arrangement, and (6) a clear mixture of different HTS families is evident, *e.g.* 214-T + 214-T' $\rightarrow$ 214-T*; partial 214-T + partial 123 $\rightarrow$ La 212 and partial 214-T' + 123 $\rightarrow$ 223. From the above observation, one may easily conjecture that since "MS" and "NS" are HTS's, "AS" can be a HTS, too. Indeed, this was demonstrated recently when $(Sr,Ca)CuO_2$ was synthesized [7] under 60 kb with a T_c up to 100 K. It has been suggested that the S-sheets should be the crucial components of the compounds and the rest serve only as charge reservoirs [18,24] because the anisotropic structural and electrical properties of cuprates HTS's. Since the number of S-layers per unit formula is no longer a good measure of T_c, as was pointed out earlier, the type of S-layers may be important. There are three distinct types of S-layers in cuprate HTS's, namely layers of CuO_2-octahedrons, CuO_5-pyramids, and CuO_4-squares; and each type is clearly associated with one basic family of HTS's, *i.e.* 214-T phase has one CuO_6-octahedron layer per unit formula, 214-T' phase has one CuO_4-square layer per unit formula and 123 phase has two CuO_5-pyramid layers per unit formula. One may therefore ask if all known cuprate HTS's are nothing more than derivatives of 214-T, 214-T', 123 or a combination the 214-T' and 123 phases.

As a first step to examining the above proposition, we have measured the T_c of some of the HTS's with only two CuO_5-pyramid layers per unit cell but with different chemical doping associated with the different intervening layers such as Y123(7), Y247 and Y124, under hydrostatic pressures up to 18 kb and quasi-hydrostatic pressures up to 150 kb. The T_c's of Y123(7), Y247 and Y124 at ambient pressures are 93 K, 88 K, and 79 K, respectively. In spite of these differences, they all increase with pressure and approach ~ 105 K asymptotically as exhibited in Figure 1. The results can be understood in terms of pressure-modified doping due to charge transfers between the CuO_5-layers and their surroundings which consist of the various intervening layers, as will be described in more details later. Under high pressures, *e.g.* > 100 kb, the charge reservoirs of the intervening layers accept the same number of carriers and thus the T_c's of the compounds become almost the same. As a further demonstration, we have determined the T_c of Y123(6.7) under pressure where the chemical doping is the result of O-content instead of the different intervening layers. Regardless, of the low value of T_c of 60 K at ambient pressure, its T_c rises rapidly initially and approaches the T_c of the other three HTS's at high pressures as shown in Figure 1. The observations above therefore imply that the four compounds examined are basically the same, with two CuO_5-layers per unit formula, or that they are derivatives of the 123 phase. Similar studies are being extended to other cuprate HTS's to determine if the proposition is as general as it appears. If it is, the

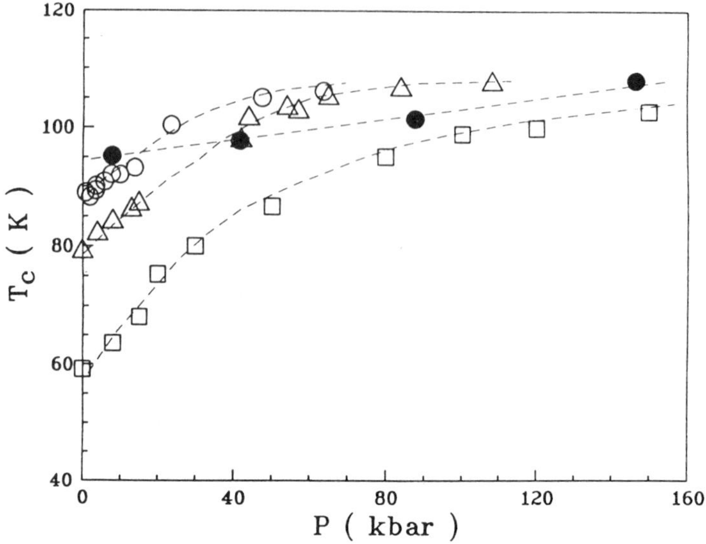

Figure 1. T_c *vs* P for □ Y123 (6.7), ● Y123(7), ○ Y247, and △ Y124.

existing cuprate HTS material base is far narrower than the number of compounds suggests and more truly different new HTS's are needed for the development of a comprehensive theoretical model.

Recently, a T_c-maximum [9] was detected in Y123 hydrostatically at a $P_m \sim 40$ kb and non-hydrostatically [25] at a $P_m \sim 100$ kb, in contrast to the continued T_c-increase under quasi-hydrostatic pressures shown above. The T_c-maximum may be ascribed to the existence of an optimal doping for HTS or an anisotropic dT_c/dP associated with the anisotropic cuprate. Although the large variation in P_m may be attributed to the different samples investigated, the higher P_m for non-hydrostatic condition is puzzling, since it has been shown [26] for the conventional low temperature superconductor (LTS) that the accompanying shear stress usually suppresses T_c. On the other hand, it is not unlikely that shear stress may in fact help stabilize the high-T_c phase in HTS's and thus delay the onset of instabilities which result in the degradation of the T_c of this chemically complicated anisotropic compound system. It will be interesting to see if higher T_c's can be achieved by the continued application of greater pressure with proper shear stress.

A Possible Universal T_c-Behavior of the Hole Cuprate HTS's

The T_c of all LTS's with effective carriers of the d-characters falls under the envelope in the T_c *vs* valence electron/atom (e/a) plot, known as the Matthias empirical rule [27]. It has provided much of the inspiration for the understanding of the LTS's and much of the guidance for the search for superconductors of higher T_c's prior to the discovery of HTS's. Unfortunately, the T_c of HTS's does not vary with e/a in any specific fashion. The discovery of a similar empirical rule for HTS's will be most interesting both scientifically and technologically. According to results of the optical [28] and muon-spin resonance [29] experiments, the T_c of several HTS families studied was found to increase at the beginning almost linearly, continue through a maximum, and then decrease with the plasma frequency square or the muon relaxation rate, both of which are proportional to n/m* with m* being the effective mass of the carrier. Although the maxima of T_c's are different for the different HTS families studied, the linear portions of all T_c *vs* n/m* curves fall into a single straight line, as shown in Figure 2a. By neglecting the nonlinear parts, a

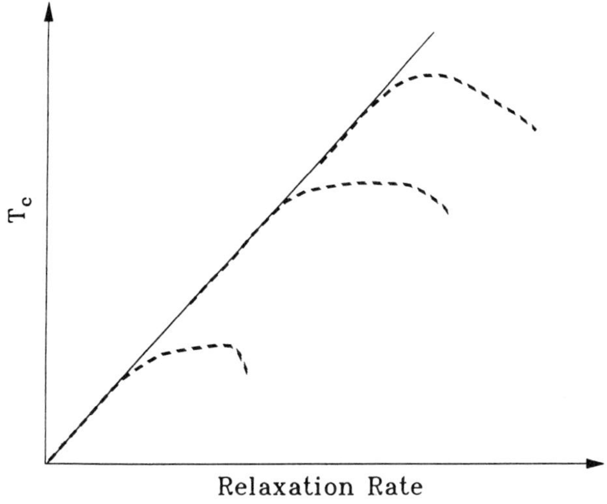

Figure 2a. The schematic T_c-behavior previously proposed: (a) T_c *vs* muon relaxation rate.

universal linear behavior of T_c has therefore been proposed [29] for HTS's. By calculating [30] the Madelung energy difference ΔV_A between the apical oxygen and that on the CuO_2 planes, it was also found that T_c varies with ΔV_A in a universal fashion. T_c first increases rapidly with $-2\ eV \leq \Delta V_A \leq 5\ eV$ and then saturated for $\Delta V_A \geq 10\ eV$ displayed in Figure 2b.

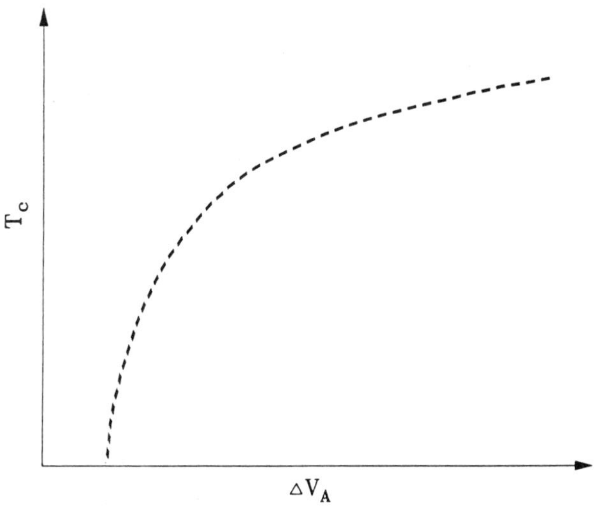

Figure 2b. The schematic T_c-behavior previously proposed: (b) T_c *vs* ΔV_A.

As a first order of approximation where a simple rigid band model is valid, pressure enhances n and a positive dT_c/dP is thus expected for all HTS's according to the universal linear T_c behavior mentioned above. Based on the T_c *vs* ΔV_A-relationship shown in Figure 2b, since pressure brings the apical oxygen closer to the CuO_2-layers, reduces the hole potential on the layer and finally results in an increase in ΔV_A, dT_c/dP is also expected [30] to be positive not just for all hole-cuprates but also for electron cuprates. Indeed, the dT_c/dP's of the 214-T, 123, Bi2212, and Bi2223-phases are generally positive except for $La_{2-x}Sr_xCuO_4$ [8] and $Y_{1-x}Pr_xBa_2Cu_3O_7$ [31] where negative dT_c/dP was also reported for certain x-range. The negative dT_c/dP in $Y_{1-x}Pr_xBa_2Cu_3O_7$ may be understood [31] in terms of the competition among the various pressure effects on n, the valence of Pr, and thus the hybridization between the Pr-4f and Cu-3d states, and the depairing effect of Pr. The situation in $La_{2-x}Sr_xCuO_4$ was not so clear. Because of the possible phase complexity [32] induced by doping near x ~ 0.12, we have examined carefully the dT_c/dP of quality samples of both $La_{2-x}Sr_xCuO_4$ and $La_{2-x}Ba_xCuO_4$ as a function of x. Clearly a T_c-anomaly appears near x = 0.12 in $La_{2-x}Sr_xCuO_4$, as shown in Figure 3, consistent with

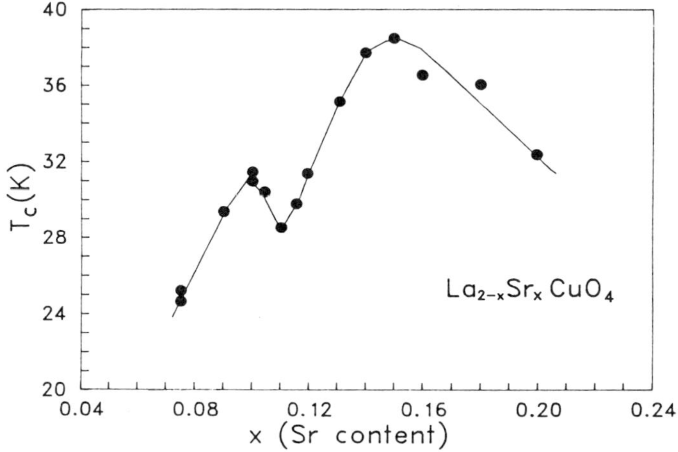

Figure 3a. T_c *vs* x for $La_{2-x}Sr_xCuO_4$.

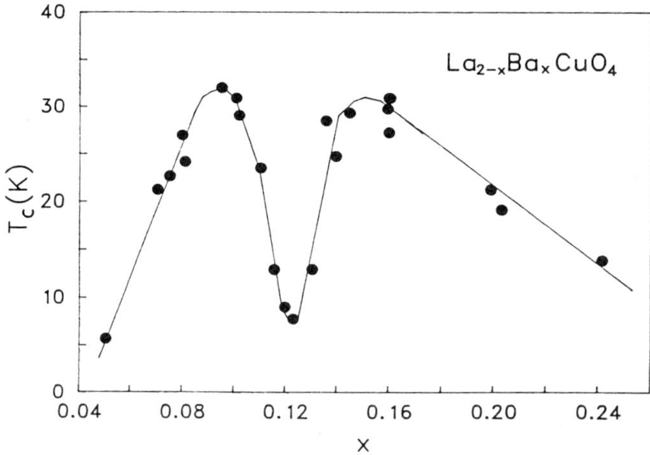

Figure 3b. T_c *vs* x for $La_{2-x}Ba_xCuO_4$.

that previously reported [32] for $La_{2-x}Ba_xCuO_4$. We found that dT_c/dP's can be reproducibly obtained both in sign and magnitude provided the sample is single phase with a sharp magnetic transition as shown in Figure 4. Poor quality samples display a dT_c/dP which can differ from that of a quality sample by a factor of 3 to 4. In spite of the drastic variation associated with the phase complexity near $x \sim 0.12$, dT_c/dP is always positive for the pure 214-T phase, as expected. The detailed difference between the dT_c/dP *vs* x for the two compound systems will be discussed elsewhere [33].

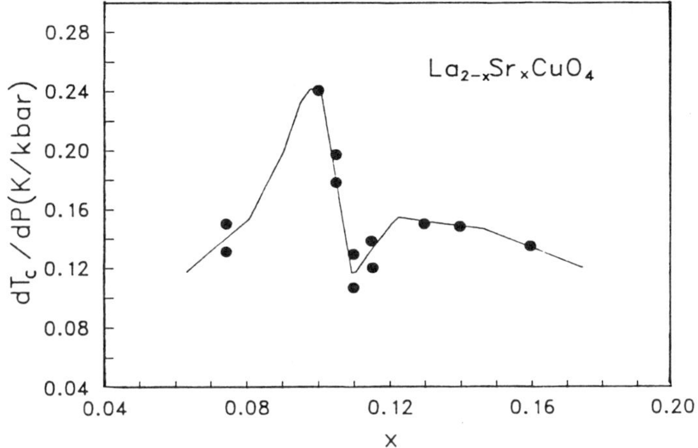

Figure 4a. dT_c/dP *vs* x for $La_{2-x}Sr_xCuO_4$.

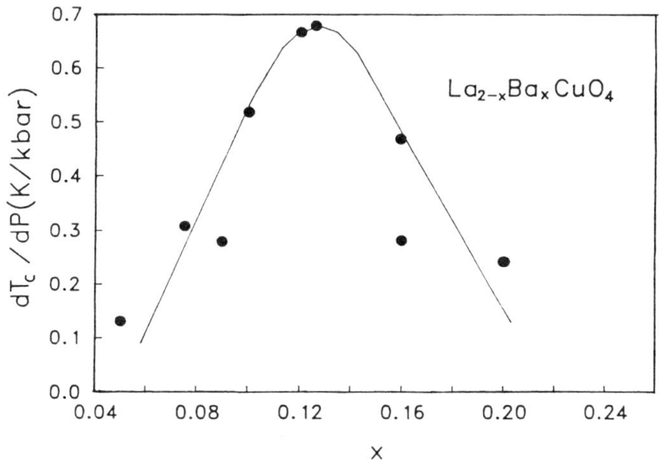

Figure 4b. dT_c/dP *vs* x for $La_{2-x}Ba_xCuO_4$.

Since the $Tl_2Ba_2Ca_{n-1}Cu_nO_{2n+4-\delta}$ HTS-family coves the largest range of T_c, from 0 to 125 K, consists of the widest variation of layer-types from 1 to 5, and for Tl2201, has a large continuous T_c-change from 0 to 90 K with a varying δ, we have measured their T_c's under hydrostatic pressures up to 18 kb. Similar to Bi2212 and Bi2223 [34,35], dT_c/dP's for Tl2212 and Tl2223 are positive. However, dT_c/dP for all Tl2201's with different T_c's are negative, which is in agreement with a recent report [36], although positive Hall constants were detected for all these samples, indicating that they are hole-doped. This is in clear disagreement with predictions by the T_c-behaviors displayed in Figure 2. Since the T_c of Tl2201 increases with reduction or δ-increase, Tl2201 is over-doped as previously suggested [23]. In order to reduce the wide spread of dT_c/dP for cuprate HTS's to a simpler form, we have reviewed all dT_c/dP-data [13] and plotted $dlnT_c/dP$ at low pressure *vs* T_c for all cuprate HTS's investigated. As displayed in Figure 5, all hole-cuprates HTS's fall into a single band whereas the electron-cuprate HTS's another. This suggests that hole cuprate HTS's may be very different from the electron-cuprate HTS's.

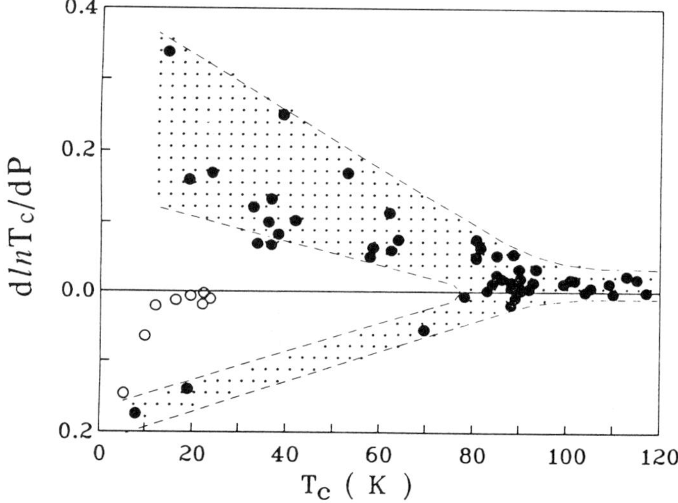

Figure 5. $dlnT_c/dP$ *vs* T_c for cuprate HTS's investigated.

The simple general $dlnT_c/dP$ *vs* T_c relationship implies that there may exist a universal T_c-behavior with a single parameter X which is closely related to n with details yet to be determined as schematically depicted in Figure 6. Such a T_c-behavior is complementary to the universal linear T_c-behavior previously proposed by extending it to larger X [28,29]. In view of the always positive dT_c/dP for the various hole-cuprate HTS systems regardless of their nonmonotonic T_c-variation with n, the deviation from the proposed T_c *vs* X behavior may be attributed to the structural instabilities or other complications of the compounds. Therefore we believe that the single T_c *vs* X relation in Figure 6 should govern the intrinsic superconducting properties for all hole-cuprate HTS's in contrast to the many non-monotonic T_c *vs* n curves, each with a different T_c-maximum, as recently suggested [36]. It would be interesting to see if such deviations can be removed by developing ways to overcome the structural instabilities or other complications of the compounds and to determine if the proposed T_c *vs* X relation is very general by extending the high pressure experiments to other HTS's. Measurements of the Hall constant and the upper critical field will help unravel the nature of X. If the proposed T_c *vs* X shown in Figure 6 for hole cuprate HTS's is proven universal, a T_c higher than ~ 150 K can be achieved only in compounds other than the hole-layered cuprates. At the present time, the possibility of a similar but reduced universal T_c-behavior, *i.e.* (T_c/T_{co}) *vs* (X/X_o) cannot be completely ruled out, with T_{co} the maximum T_c and X_o where T_{co} occurs for each HTS family.

<u>Charge Transfer</u>

It is known that all HTS's families have their respective parent insulators, *e.g.* C_{60} for A_3C_{60} with A = K, Rb, and Cs, $BaBiO_3$ for $Ba_{1-x}K_xBiO_3$, La_2CuO_4 for 214-T, Nd_2CuO_4 for 214-T', $YBa_2Cu_3O_6$ for 123, $La_2CaCu_2O_6$ for 212 and $Bi_2Sr_2YCu_2O_6$ for Bi2212. Through doping, one converts these parent insulators to conductors which at proper doping levels become superconducting upon cooling. This evolution of a metallic and subsequently a superconducting phase from an insulator must be accompanied by a redistribution of charge between various building blocks of HTS's. Preliminary information about charge transfer or charge redistribution has been obtained by examining the chemical bond length [37] between atoms and interatomic layer distance [11], as a function of doping. Indeed, these quantities undergo a sudden change at or near the doping level where superconductivity occurs, as exemplified [18] by the drastic reduction or expansion in interlayer distances shown in Figure 7. These interlayer distance-changes are consistent with an electron transfer from (as for hole-doping) or to (as for electron-doping) the CuO_2-layers in a HTS. In the discussions in Sections III.A. and III.B., we have assumed $dn/dP \neq 0$, implying that pressure can also induce a charge transfer. However, direct evidence is yet to be provided.

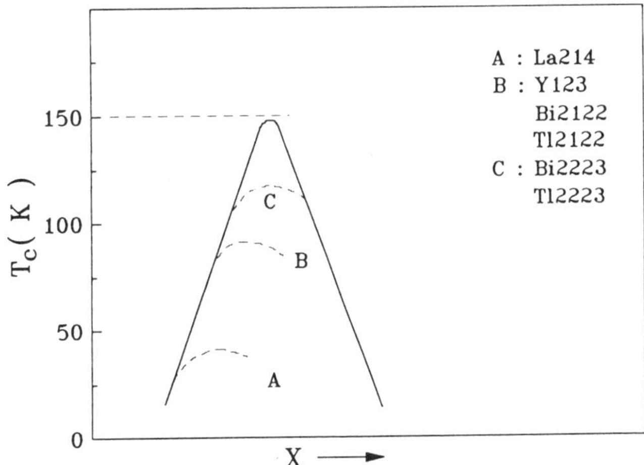

Figure 6. The proposed universal T_c *vs* X.

The τ depends on the charge distribution in the immediate vicinity of the trapping site of the positron. Therefore, τ-measurements can be used as an effective probe to the change of local charge distribution or charge transfer under the influence of pressure provided the trapping sites of positrons are known. Calculations [38] show that the trapping sites of positrons in Y123, Y247 and Y124 are located far away from the CuO_2-layers or near the charge reservoir. We have therefore determined τ of Y123(6.7), Y123(7), Y247, and Y124 at 300 K under quasi-hydrostatic pressures up to 15 kb. The results are displayed in Figure 8. Except for Y123(7) of which the τ is almost pressure-

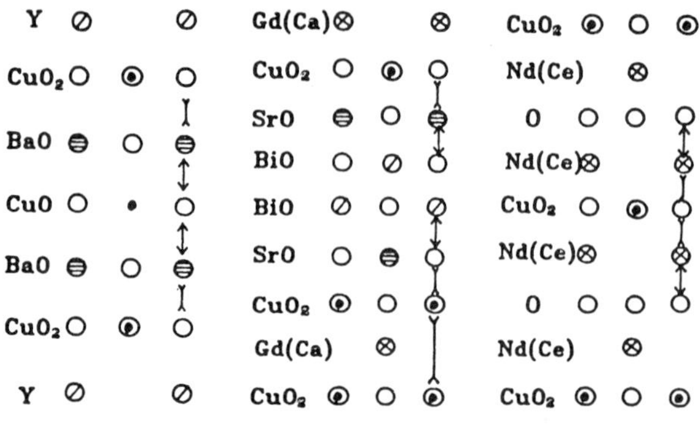

Figure 7. Interlayer distance changes as the insulator parents are made metallic by doping: →← contraction, ↔ expansion.

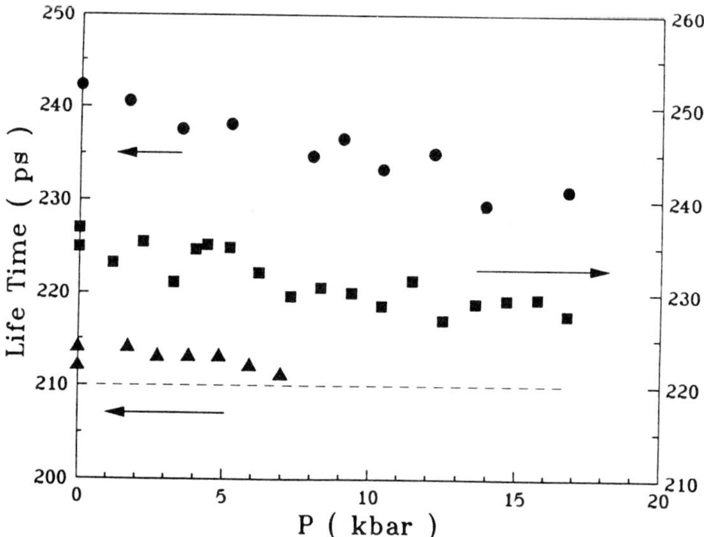

Figure 8. τ *vs* P for ▲ Y123(6.7), ----Y123(7), ■ Y247, and ● Y124.

independent, τ in general decreases with pressure. This shows that the local electron-density at the trapping sites of positron increases under pressure. Since the trapping sites are near the electron reservoir and thus away from the CuO_2-layers, the observation suggests that electrons have been transferred from CuO_2-layers to the reservoir by pressure, resulting in an increase of n and T_c. The magnitude of $d\tau/dP$ seems to be proportional to dT_c/dP of these compounds. We therefore conclude that the application of pressure induces a charge transfer which in turn affects the T_c. This is in qualitative agreement with the conclusion made based on the upper critical field measurements [25] on HTS's under high pressures in this Workshop.

CONCLUSION

We have measured the T_c of Y123(6.7), Y123(7), Y247, Y124, Tl22(n-1)n where n = 1, 2, and 3, La(Sr)214-T, and La(Ba)214-T, under pressures. The τ was also determined for Y123(6.7), Y123(7), Y247 and Y124 under pressures.

We observed that in spite of their different ambient T_c's and different intervening layers in the Y123(6.7), Y123(7), Y247 and Y124, their T_c's tend to merge into one under high pressure. This together with the structural data of existing cuprate HTS's lead us to the suggestion that all known cuprate HTS's may not be more than just modifications of the 214-T, 214-T', 123, or a combination of 214-T' and 123. If this is proven to be true, the present HTS material base may not be as broad as the number of HTS's suggests and more new truly different HTS's are needed for the development of a comprehensive theory.

We also found that dT_c/dP is negative for Tl2201 in contrast to other hole-HTS's and that dT_c/dP for the 214-T phase is positive irrespective of the nonmonotonic variation of T_c with doping. By reviewing all existing dT_c/dP-data, we found that hole-HTS's fall in a simple band in the $d\ln T_c/dP$-T_c plot whereas electron-HTS's in another region. This suggests that hole HTS's may belong to a different group from the electron HTS's. The single band of hole-HTS's implies that a universal nonmonotonic T_c-behavior may exist for all hole cuprates HTS's, with a possible maximum T_c of ~ 150 K for this type of HTS's.

We observed a reduction of τ in Y123(6.7), Y247, and Y124 but no change in Y123(7) under pressures with a magnitude which correlates well with dT_c/dP. It shows that pressure can generate a charge transfer similar to chemical or optical doping, and that the pressure induced charge transfer is responsible for the pressure induced T_c-change.

While the existing high pressure data on HTS's has led to many interesting suggestions mentioned above, definitive confirmations of some of them await further experiments covering more compounds and extending to higher pressures.

ACKNOWLEDGEMENTS

One of us (CWC) would like to thank H. K. Mao and C. T. Prewitt for their hospitality during his visit to the Geophysical Laboratory of the Carnegie Institution in Washington DC where the manuscript was completed. This work is supported in part by the NSF Low Temperature Physics Program Grant No. DMR 86-126539, DARPA Grant No. MDA 972-88-G-002, NASA Grant No. NAGW-977, Texas Center for Superconductivity at the University of Houston, and the T. L. L. Temple Foundation.

REFERENCES

1. See, for example, C. W. Chu and M. K. Wu, *High Pressure in Science and Technology*, Part 1, ed C. Homan, R. K. MacCrone and E. Whalley (Amsterdam: North Holland, 1983), p. 3; and C. W. Chu, P. H. Hor and Z. X. Zhao, Physica B 139-140, 439 (1986).
2. See, for example, R. J. Wijngaarden and R. Griessen, *Studies of High Temperature Superconductors,* vol. 2, ed. by A. Narlikar (Nova Science, 1989) p. 29; High Pressure Research 5, 874 (1990).

3.	C. W. Chu, P. H. Hor, R. L. Meng, L. Gao, Z. J. Huang and Y. Q. Wang, Phys. Rev. Lett. 58, 405 (1987); C. W. Chu, P. H. Hor, R. L. Meng, L. Gao and Z. J. Huang, Science 235, 567 (1987).
4.	J. G. Bednorz and K. A. Müller, Z. Phys. B 64, 189 (1986).
5.	M. K. Wu, J. R. Ashburn, C. J. Torng, P. H. Hor, R. L. Meng, L. Gao, Z. J. Huang, Y. Q. Wang and C. W. Chu, Phys. Rev. Lett. 58, 908 (1987); C. W. Chu, Proc. Natl. Acad. Sci. USA 84, 4681 (1987).
6.	P. H. Hor, R. L. Meng, Y. Q. Wang, L. Gao, Z. J. Huang, J. Bechtold, K. Forster and C. W. Chu, Phys. Rev. Lett. 58, 1891 (1987).
7.	Z. Hiroi, M. Takano, M. Azuma, Y. Takeda and Y. Bando, Proc. M^2S-HTSC III, Kanazawa JAPAN, to be published in Physica C.
8.	J. E. Schirber, E. L. Venturini, J. F. Kwak, D. S. Ginley and B. Marosin, J. Mater. Res. 2, 421 (1987).
9.	S. Klotz, W. Rieth and J. S. Schilling, Physica C 172, 432 (1991).
10.	J. B. Torrance, Y. Tokura, A. I. Nazzal, A. Bezinge, C. T. Huang and S. S. P. Parkin, Phys. Rev. Lett. 61, 1127 (1988).
11.	J. T. Markert, J. Beille, J. J. Neumeier, E. A. Early, C. L. Seaman, T. Moran and M. B. Maple, Phys. Rev. Lett. 64, 80 (1990); Y. Y. Xue, P. H. Hor, R. L. Meng, Y. K. Tao, Y. Y. Sun, Z. J. Huang, L. Gao and C. W. Chu, Physica C 165, 357 (1990).
12.	Z. J. Huang, P. H. Hor, R. L. Meng, Y. Q. Wang, L. Gao, J. Bechtold, Y. Y. Sun and C. W. Chu, APS Bull. 33, 689 (1988).
13.	J. G. Lin, K. Matsuishi, Y. Q. Wang, Y. Y. Xue, P. H. Hor and C. W. Chu, Physica C 175, 627 (1991); J. G. Lin, Y. Q. Wang, P. H. Hor and C. W. Chu, APS Bull. 35, 534 (1990).
14.	C. W. Chu, Proc. AAAS, Washington DC (April 1991).
15.	C. W. Chu, P. H. Hor, Y. Y. Xue, G. Lin, R. L. Meng, K. Matsuishi, Z. J. Huand and C. Diaz, Proc. M^2S-HTSC III, Kanazawa JAPAN, to be published in Physica C.
16.	C. W. Chu, T. F. Smith and W. E. Gardner, Phys. Rev. Lett. 20, 198 (1968).
17.	See, for example, Y. C. Jean, S. J. Wang, H. Nakanishi, W. N. Hardy, M. E. Hayden, R. F. Kiefl, R. L. Meng, P. H. Hor, Z. J. Huang and C. W. Chu, Phys. Rev. B 36, 3994 (1987).
18.	See, for example, C. W. Chu, Vacuum 41, 773 (1990).
19.	C. W. Chu, AIP Conf. Proc. 169: Mod. Phys. in America, ed. by W. Fickinger and K. Kowalski (New York: AIP, 1988) p. 220.
20.	C. W. Chu, Proc. Natl. Acad. Sci. USA 84, 4681 (1987); C. W. Chu, J. Bechtold, L. Gao, P. H. Hor, Z. J. Huang, R. L. Meng, Y. Y. Sun, Y. Q. Wang and Y. Y. Xue, Phys. Rev. Lett. 60, 941 (1988); R. M. Hazen, L. W. Finger, R. J. Angel, C. T. Prewitt, N. L. Ross, C. G. Hadidiacos, P. H. Heaney, D. R. Veblen, Z. Z. Sheng, A. El Ali and A. M. Hermann, Phys. Rev. Lett. 60, 1657 (1988).
21.	S. Adachi, O. Inoue, S. Kawashima, H. Adachi, Y. Ichikawa, K. Setsune and A. Wasa, Physica C 168, 1 (1991).
22.	Y. Takura and T. Arima, J. Jpn. Appl. Phys. 29, 2388 (1990).
23.	C. C. Torardi, M. A. Subramanian, J. C. Calabrese, J. Gopalakrishnan, E. M. McCarron, K. J. Morrissey, T. R. Askew, R. B. Flippen, U. Chowdhry and A. W. Sleight, Phys. Rev. B 38, 225 (1988).
24.	See, for example, P. H. Hor, Y. Y. Xue, Y. Y. Sun, Y. K. Tao, Z. J. Huang, W. Rabalais and C. W. Chu, Physica C 159, 629 (1989) and references therein.
25.	R. J. Wijngaarden, this Workshop.
26.	L. J. Sham and T. F. Smith, Phys. Rev. B 4, 3951 (1971).
27.	B. T. Matthias, Phys. Rev. 92, 874 (1953).
28.	S. Tanaka, *Physics of Low Dimensional Systems*, ed. S. Lundqvist and N. R. Nilsson (Physica Scripta/World Scientific, 1989) p. 7.
29.	Y. I .Uemura, G. M. Luke, B. J. Sterlieb, J. H. Brewer, J. F. Carolan, W. N. Hardy R. Kadono, J. R. Kempton, R. F. Kiefl, S. R. Kreitzman, P. Mulhern, T. M. Riseman, D. Ll. Williams, B. X. Yang, S. Uchida, H. Takagi, J. Gopalakrishnan, A. W. Sleight, M. A. Subramanian, C. L. Chien, M. Z. Cieplak, G. Xiao, V. Y.

Lee, B. W. Statt, C. E. Stronach, W. J. Kossler and X. H. Yu, Phys. Rev. Lett. **62**, 2317 (1989).
30. Y. Ohta, T. Tohyama and S. Maekawa, Physica C **166**, 385 (1990).
31. J. J. Neumeier, M. P. Maple and M. S. Torikachvili, Physica C **156**, 574 (1988).
32. J. D. Axe, H. Mouden, D. Hohlwein, D. E. Cox, K. Mohanty, A. R. Moodenaugh and Y. Xu, Phys. Rev. Lett. **62**, 2751 (1989).
33. Q. Xiong, *et al.*, preprint.
34. C. W. Chu, J. Bechtold, L. Gao, P. H. Hor, Z. J. Huang, R. L. Meng, Y. Y. Sun, Y. Q. Wang and Y. Y. Xue, Phys. Rev. Lett. **60**, 941 (1988).
35. R. J. Wijngaarden H. K. Memmes, E. N. van Eenige, R. Griessen, A. A. Menovsky and M. J. V. Menken, Physica C **152**, 140 (1988); K. Kumagai and M. Kurisu, Jpn. J. Appl. Phys. **27**, L1029 (1988).
36. N. Mori, H. Takahashi, Y. Shimakawa, T. Manako and Y. Kubo, J. Phys. Soc. Jpn. **59**, 3839 (1990).
37. R. J. Cava, B. Batlogg, K. M. Rabe, E. A. Pietman, P. K. Gallagher and L. W. Rupp Jr., Physica C **156**, 52 (1988).
38. See, for example, A. Bharathi, C. S. Sundar, W. Y. Ching, Y. C. Jean, P. H. Hor, Y. Y. Xue and C. W. Chu, Phys. Rev. B **42**, 10199 (1990).

HIGH-T$_c$ SUPERCONDUCTORS UNDER VERY HIGH PRESSURE

Rinke J. Wijngaarden, J.J. Scholtz, E.N. van Eenige and
R. Griessen

Department of Physics and Astronomy, Free University,
De Boelelaan 1081, 1081 HV Amsterdam, The Netherlands

1. INTRODUCTION

High pressure has played a crucial role in the short history of high
T_c superconductors. Soon after the discovery of superconductivity by
Bednorz and Muller[1] in La-Ba-Cu-O, Chu et al.[2] showed that the critical
temperature T_c could be significantly increased by pressure. This
observation led to the discovery of $YBa_2Cu_3O_7$ by Wu et al.[3] with a T_c
above 90 K. Incidentally, this high T_c is probably also due to the fact
that $YBa_2Cu_3O_7$ has two CuO_2 layers per unit cell instead of a single one
in La-Ba-Cu-O.

High pressure is also needed for the synthesis of some of the new
compounds. Although $YBa_2Cu_4O_8$ was first discovered as a contaminant in
$YBa_2Cu_3O_7$ powder[4] its synthesis at first could only be achieved at high
oxygen pressures[5], a technique which is far from trivial. Of course it
later proved possible (using mineralizers) to synthesize $YBa_2Cu_4O_8$ at
ambient pressure[6].

High pressure can be used to change the properties of high-T_c
superconductors by physical means. Later in this chapter we will argue
that pressure changes (among other parameters) the charge carrier
concentration. Of course any theoretical model for high-T_c superconductivity should be able to explain the effect of pressure on the
superconducting properties. This idea was first used by Griessen[7] to
discuss a number of models. In their review on the high pressure
properties of these materials Wijngaarden and Griessen[8] later extended
this work.

Frontiers of High-Pressure Research, Edited by H.D. Hochheimer and
R.D. Etters, Plenum Press, New York, 1991

Pressure experiments were initially done at relatively low pressures: the important measurement by Chu et al.[2] is only up to 1.6 GPa. Later it became clear that the critical temperature changes are not neccessarily monotonous with pressure. A dramatic example is the measurement by Van Eenige et al.[9] and Scholtz et al.[10], who show that the T_c of $YBa_2Cu_4O_8$, which is 78 K at zero pressure, increases at an initial rate of 5 K/GPa to reach 108 K at 10 GPa, after which a decrease to 100 K at 20 GPa is observed. One of the objectives in this chapter is to present and discuss such measurements above 10 GPa.

In section 2 we discuss briefly our experimental technique, in section 3 results up to 50 GPa are presented, and in section 4 we derive the charge carrier concentration under pressure from experiment. In section 5 we find that the effect of pressure on T_c can be separated into several contributions of which one is the pressure induced change in charge carrier concentration. In section 6 the systematics in a plot of $\partial \ln T_c/\partial p$ versus T_c are discussed. We end with a general discussion.

2. EXPERIMENTAL TECHNIQUE

At present we use in our laboratory two different types of diamond anvil cells to obtain pressures up to 50 GPa. One of the cells is of the Silvera-Wijngaarden type and was discussed by Hemmes et al.[12a]. The other cell[12b] can be inserted in a superconducting magnet. Figure 1 shows this cell in its cryostat. The total length is ~ 1 m. The lower thin part of cryostat and cell is inserted in the 54 mm diameter bore of a Thor-Cryogenics 12 T superconducing magnet. The field homogenity is $1:10^5$ in a 1 cm diameter sphere. The sample is centered between the diamonds (12) which are in the middle of this sphere; they are mounted on two concentric tubes, (8) and (10), which can be moved with respect to each other by the force applying mechanism (5), (6), (7). Pressure can be adjusted manually by turning knob (1) which causes the spindle (5) to rotate. The spindle causes a wheel to rotate on which another wheel (6) is excentrically mounted. The movement of this wheel (6) pushes the lever arm (7) to the right, which moves the inner tube (8) downwards.

The bottom part of both tubes (10 and 8') can be separated to mount the sample. Temperature can be adjusted by a heater and the continuous helium flow heat exchanger (14). Temperature is measured by a platinum resistor, calibrated in magnetic fields up to 12 T.

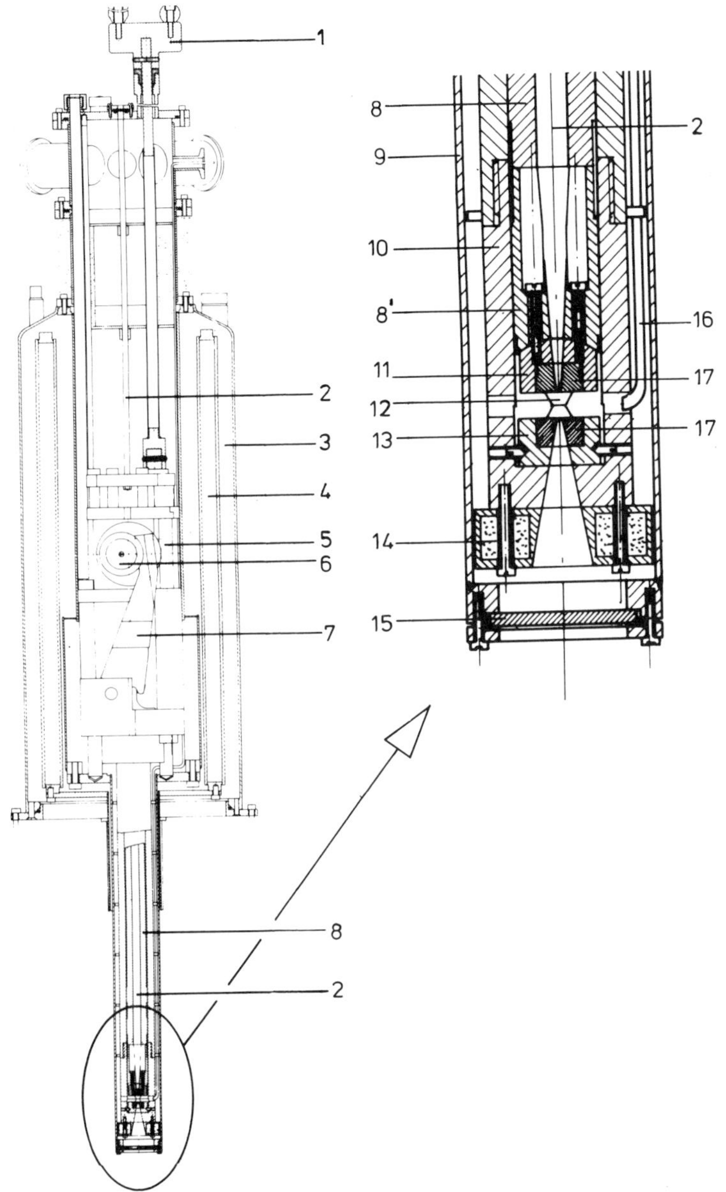

Fig. 1. Diamond cell in its cryostat for use in a 12 T magnet.
(1) Manual knob to change the pressure at any temperature,
(2) optical fiber for tranmittance measurements, (3) vacuum
chamber, (4) liquid nitrogen chamber, (5) spindle,
(6) excentric, (7) lever, (8) piston, (9) inner stainless steel
tail of cryostat, (10) removable lower part of the outer
cylinder, (11) hemisphere, (12) diamonds, (13) translation
table, (14) heatexchanger, (15) window, (16) gas capillary,
(17) tungsten carbide seats.

In figure 2 the sample region is shown. The sample is placed in the hole of a metal gasket without pressure medium. Six electrical wires are pressed on the sample and are insulated from the gasket by a layer of aluminium oxide which is stabilized by epoxy and covered by a thin kapton film. The six wires provide some redundancy for a four point resistivity measurement, which is done using a home-made lock in amplifier, which reverses the current every 0.25 seconds, thus also correcting for thermovoltages. The older design, used for the results to be presented in the next section, has a noise level of 100 nV. Pressure is measured by the ruby fluorescence method[13] using some fine grains of ruby at various positions on the sample.

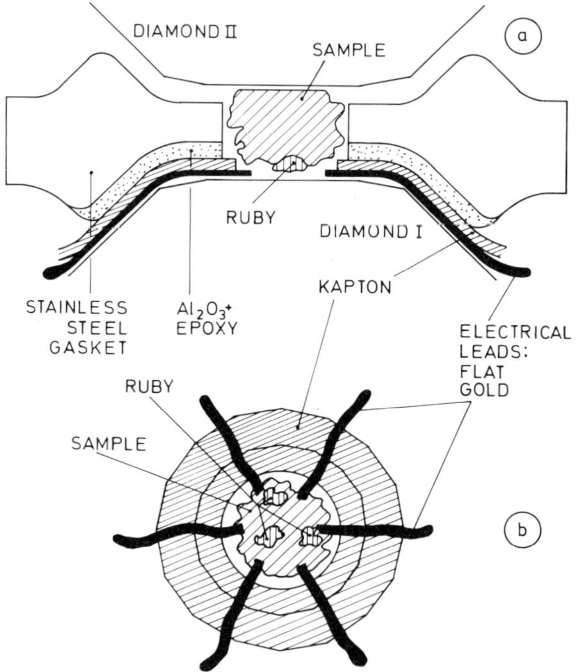

Fig 2.   (a) Cross section through the diamonds and (b) top view of one diamond with the sample, gasket and electrical wires.

Of course the lack of pressure medium (which guarantees good electrical contact with the sample) may cause pressure gradients. To check for these, we measure separately the pressure at various positions on the sample. At 50 GPa the gradient over the whole sample volume was approximately 30 %. Since the electrical wires continue somewhat further towards the centre of the sample, the gradient over the part of the sample that is involved in resistivity measurements is significantly smaller. An independent check on gradients can be made using the width of the resistive transition. In $YBa_2Cu_4O_8$ a large $\partial T_c/\partial p$ exists below 10 GPa. If a large pressure gradient should exist then it would certainly result in a significant broadening of the resistive transition; this is not borne out by experiment. To make sure we do measure the correct pressure, pressure is always measured at the superconducting transition temperature. In any case, our experiments are in very good agreement with hydrostatic measurements[14], at least up to the highest pressures reached in the latter.

3. EXPERIMENTAL RESULTS

High pressure experiments on high-T_c superconductors have been made by quite a large number of authors. A review of most of these results is ref 8, to which some updates have appeared[15]. Although relatively rare, experiments above 10 GPa have been reported by a number of groups. Here, we confine our discussion to some experiments done at our own lab, which illustrate the different behaviour possible.

In figure 3 we show that the T_c of $YBa_2Cu_4O_8$ increases dramatically at an initial rate of 5 K/GPa from 78 K at ambient pressure to 108 K at ~ 10 GPa. Upon further increase of the pressure the T_c decreases again to reach 100 K at 20 GPa. The curve has two branches above 103 K, which correspond to two different samples taken from the same piece of polycrystalline material. Sample preparation and details of the experiment are discussed elsewhere[9,10]. Interestingly, doping with 20 % Ca increases the ambient pressure T_c to ~ 87 K, but it decreases the maximum observed under pressure. Details of this experiment are discussed elsewhere in this volume by Van Eenige et al.. Also shown in figure 1 is a curve for the tetragonal material $CaLaBaCu_3O_7$. In striking contrast with $YBa_2Cu_4O_8$ the T_c is seen to change by only 8 K up to 50 GPa. Sample preparation and details of the experiment are discussed elsewhere[16].

Using our 12 Tesla magnet, the upper critical field H_{c2} of $YBa_2Cu_4O_8$ and $CaLaBaCu_3O_7$ could be determined at temperatures slighly below T_c. Because $H_{c2}(T=0)$ is of order 100 T, it is not possible to determine $H_{c2}(T=0)$ in our experiment directly. Therefore we calculated $H_{c2}(T=0)$ by means of the Werthamer Helfand Hohenberg theory[17] with spin orbit coupling parameter $\lambda_{so}=2$. The result is shown in figure 4. Clearly for $CaLaBaCu_3O_7$ the pressure dependence of H_{c2} is much smaller. As we will show in section 5 this is an important ingredient to understand the low $\partial T_c/\partial p$ of this

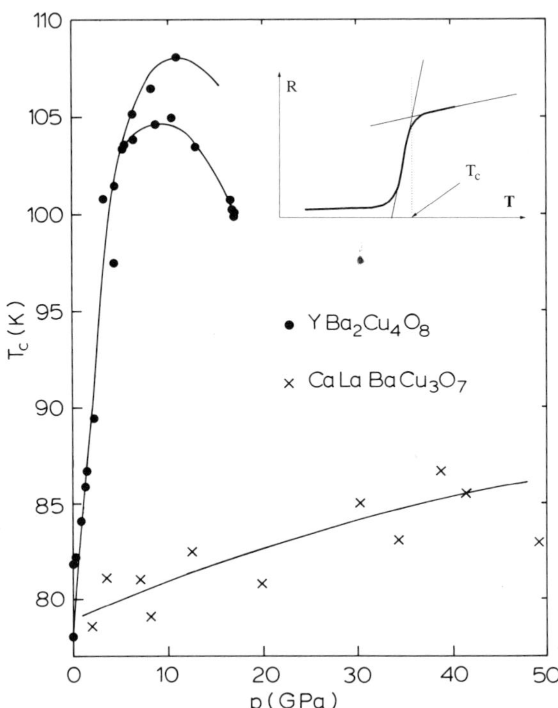

Fig. 3. Onset critical temperature T_c as a function of pressure for the compounds indicated. The lines are a guide to the eye only. The inset shows the definition of T_c

compound. Of course the zero-temperature coherence length $\xi(0)$ can be calculated directly from $H_{c2}(0) = \Phi_0/(2\pi\xi^2(0))$. The result is shown in figure 5.

4. DETERMINATION OF THE CHARGE CARRIER CONCENTRATION UNDER PRESSURE

We will now show that it is possible[18] to find the charge carrier concentration under pressure from our H_{c2} measurements. We define the charge carrier concentration δ as the number of mobile holes in the CuO_2-planes per Cu-atom in these planes. This δ is an important parameter, since T_c is strongly dependent upon δ, as shown e.g. in the review by Shafer and Penney[19]. In figure 6 the curve represents our fit to their curve of $T_c(\delta)$ for materials with two adjacent CuO_2-layers, using the parabolic function $T_c(\delta) = T_0[1-\beta(\delta-\delta_0)^2]$ with $\delta_0 = 0.2$ and $\beta = 60$.

Under high pressure the charge carrier concentration can be determined from measurement of the upper critical field $H_{c2}(0)$, as follows. The upper critical field is related to the coherence length, which is related to the Fermi wavevector, which is determined by the charge carrier concentration. Below this idea will be presented in more detail.

At $T = 0$ the coherence length ξ may be obtained from the upper critical field H_{c2} by

$$H_{c2} = \frac{\Phi_0}{2\pi\mu_0\xi^2} \tag{1}$$

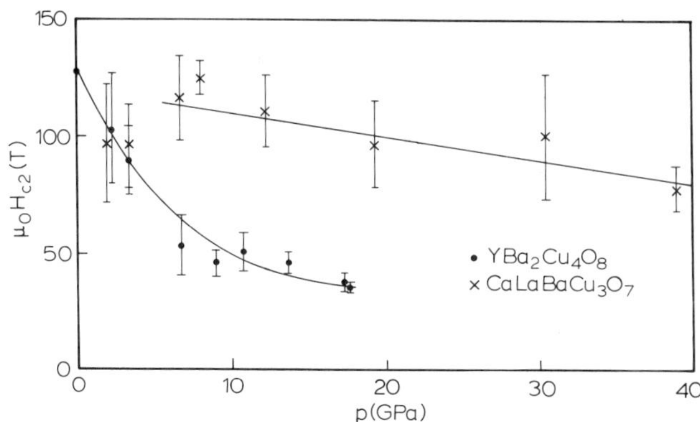

Fig. 4. The upper critical field H_{c2} at $T = 0$ as a function of pressure (found from an extrapolation of resistive determinations of H_{c2} at temperatures close to T_c using the Werthamer Helfand Hohenberg theory (Ref 17) with spin–orbit coupling parameter $\lambda_{so} = 2$.)

where $\Phi_o = 2\times10^{-15}\ Tm^2$ is the flux quantum. Possibly except for the exact value of the numerical constant (which will turn out later to be irrelevant), this relation holds independent of the microscopic model for superconductivity, since macroscopically at H_{c2} the magnetic field outside the sample must equal the magnetic field inside the sample, which is $\sim\Phi_o/2\pi\mu\xi^2$.

From the coherence length we may find the Fermi velocity v_F as follows. Superconductivity is destroyed at a temperature of T_c because the

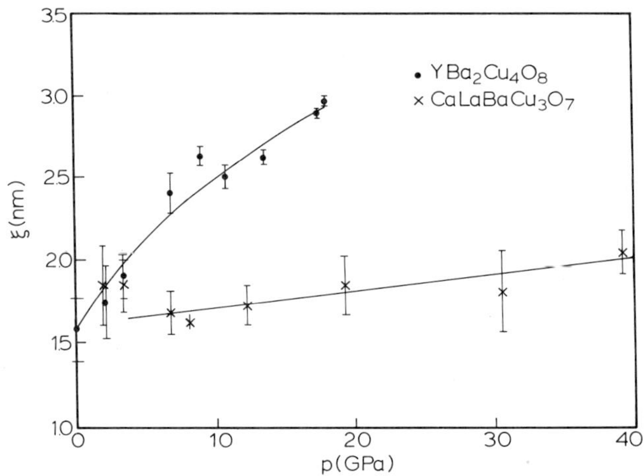

Fig. 5.   Zero temperature coherence length ξ as a function of pressure calculated from the H_{c2} of Fig. 4 using $H_{c2} = \Phi_o/(2\pi\mu_o\xi^2)$.

pairs of electrons can be dissociated by an energy of order $k_B T_c$. This defines the possible spread around the Fermi level in energy ΔE and momentum

$$\Delta p = m\Delta E/p = \frac{ak_B T_c}{v_F}$$

where a is a constant of order unity.
The characteristic length, which is the coherence length is given by

$$\xi = \frac{\hbar}{\Delta p} = \frac{\hbar v_F}{ak_B T_c} \tag{2}$$

The exact proportionality constant may depend on the microscopic theory and the material, the ratio between ξ and v_F/T_c does not. Of course the

Fermi wavevector and Fermi velocity are related through

$$m^* v_F = \hbar k_F \tag{3}$$

where m^* is the effective mass of the electrons. Superconductivity in HTSC is taking place in the CuO_2-layers, with a very short coherence length perpendicular to these layers, of order of the thickness of such a layer. Therefore it seems reasonable to assume a cylindrical Fermi surface corresponding to a real-space radius ξ and height c. The number of charge carriers per unit volume is then given by

$$n = \frac{k_F^2}{2\pi c} \tag{4}$$

The effect of hydrostatic pressure is incorporated in this expression: changes of the a- and b-lattice vectors modify k_F, while the c-lattice vector occurs explicitly. For a normal metal n is the density of the <u>electrons</u> in the lower (valence) band. However, for strongly interacting electrons in the two-dimensional Hubbard model Mattis et al.[20] conclude that in the case of fractional doping ϵ, the Fermi liquid consists of ϵN spin-½ particles and not of the expected $(1\pm\epsilon)N$ particles. In this case n is the number of holes per unit volume, which is adopted for the following analysis, since we feel the Hubbard model is appropriate. If one would interpret n as electron density, the quantitative results of the following analysis will be modified, the qualitative results will remain. The number of holes per planar copper atom, as defined in section 3 can now be found from

$$nV = \delta N_{CuO2} \tag{5}$$

where V is the volume of the sample and N_{CuO2} is the number of planar copper atoms in the sample. Both the right and left hand side of Eqn. (5) give the total number of holes in the sample.

A final ingredient is the dependence of the effective mass m^* on hole concentration. As mentioned by P.W. Anderson[21] for high T_c superconductors the Brinkman and Rice relation

$$m^* \sim 1/\delta \tag{6}$$

holds; an elegant argument can be found in Mott's book[22]. If one should

assume m* independent of δ and pressure, our quantitative results will be
modified, the qualitative results will remain. Combination of Eqns. (1)
through (6) gives:

$$\delta^3 \, \alpha \, \frac{T_c^2 V}{cH_{c2}} \tag{7}$$

The exact proportionality constant in Eqn. (7) can be determined but is
different for (slightly) different models. It is preferable instead to use
logarithmic derivatives:

$$\frac{\partial \ln\delta}{\partial p} = -\frac{1}{3B} - \frac{1}{3}\frac{\partial \ln c}{\partial p} - \frac{1}{3}\frac{\partial \ln H_{c2}}{\partial p} + \frac{2}{3}\frac{\partial \ln T_c}{\partial p} \tag{8}$$

Here $B = (-\partial \ln V/\partial p)^{-1}$ is the bulk modulus, which has been measured,
as well as $\partial \ln c/\partial p$, in X-ray scattering experiments under high pressure.
All terms on the right hand side of eqn. (8) are therefore known
experimentally. For $YBa_2Cu_4O_8$ it was found[23,24], that $B \simeq 140$ GPa and that
$\partial \ln c/\partial p = -3.7 \; 10^{-3}$ GPa^{-1}. For $CaLaBaCu_3O_7$ presently no experimental
numbers are available. We use instead the values for the structurally
equivalent (i.e. tetragonal) $YBa_2Cu_3O_6$ where[25] $B = 125$ GPa and
$\partial \ln c/\partial p = -4 \; 10^{-3}$ GPa^{-1}. Combining these data with least squares fits to
our own results (quoted in section 3) on $T_c(p)$ and $H_{c2}(p)$, eqn. (8) can
now be integrated to calculate $\delta(p)$. The integration constant is
determined by demanding that the point $(\delta(p{=}0),\ T_c(p{=}0))$ lies on the $T_c(\delta)$
phase line as given by the curve in figure 6. The resulting values $\delta(p)$
for $YBa_2Cu_4O_8$ and $CaLaBaCu_3O_7$ are plotted in Fig.7. The curves can be
reasonably approximated by straight lines $\delta = \delta_1 + \alpha p$ where $\alpha = 0.009$
GPa^{-1} for $YBa_2Cu_4O_8$ and $\alpha = 0.0005$ GPa^{-1} for $CaLaBaCu_3O_7$.

A possible explanation[26] for this difference in α is based on the
difference in structure. In the orthorhombic $YBa_2Cu_4O_8$ there are long CuO
chains where the charge carriers have an energy comparable to those in the
CuO_2 layers, enabling charge transfer. In the tetragonal $CaLaBaCu_3O_7$ only
short segments of chain are present, which localize their charge carriers,
which thus have a high momentum, probably well above the Fermi-level of
the planes, making charge transfer more difficult. This idea is
substantiated by electronic structure calculations[27] by Gupta and Gupta,
who find charge transfer for long chains but not for short segments.

Previous experimental and theoretical indications for an increase of

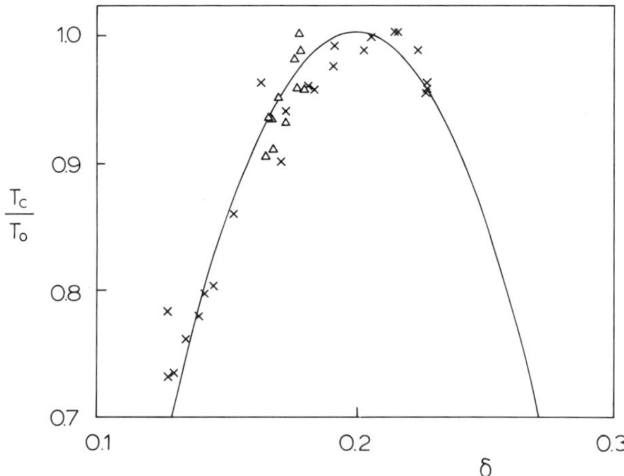

Fig. 6.   The curve is T_c/T_o as a function of the hole concentration δ as given by ref. 19, (see text). Our experimental high pressure data, plotted using eqn. (8) to calculate δ, are indicated by crosses ($YBa_2Cu_4O_8$) and triangles ($CaLaBaCu_3O_7$).

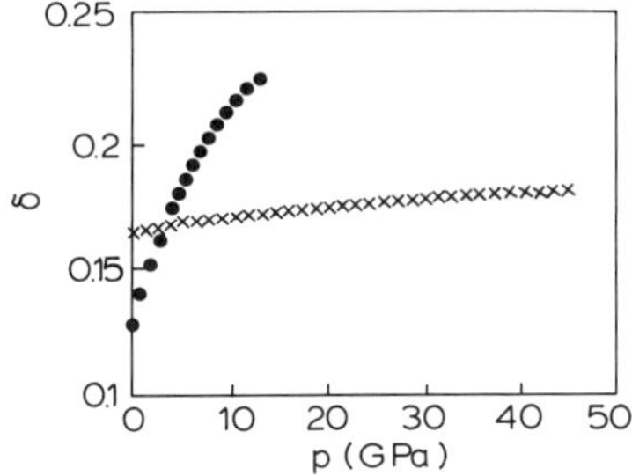

Fig. 7.   Hole concentration δ as a function of pressure, as determined from our experiment using eqn.(8), for $YBa_2Cu_4O_8$ (circles) and $CaLaBaCu_3O_7$ (crosses).

hole concentration with pressure are the following. In neutron spectrocsopic studies of the crystalline electric field by Allenspach et al.[28] an increase of δ with p was observed directly. From calculations of the Madelung energies Yamada et al.[29] found that pressure makes charge transfer towards the CuO_2 layers energetically favourable. A similar result was obtained by bond valence sum calculations by Jorgensen et al.[30].

5. TOWARDS AN EXPLANATION OF THE PRESSURE DEPENDENCE OF T_c

If the only effect of pressure were a change of the hole concentration δ, one would expect a $T_c(p)$ curve very similar to the curve in Fig.6, since δ increases nearly linearly with pressure, as shown in the previous section. Evidently this is not the case: the highest T_c in $YBa_2Cu_4O_8$ under pressure is 108 K, while chemical doping can increase the T_c to 90 K only. (Incidentally, the highest $T_c(p=0)$ in compounds with two CuO_2 layers is[31] 108 K). Nevertheless it is interesting to compare $T_c(p)$ and $T_c(\delta)$. This can be done by scaling all T_c's to the highest T_c observed, which is called T_o, and by transforming our $T_c(p)$ data to $T_c(\delta)$ points by using the relation $\delta(p)$ given in Fig.7. Using this method we plotted our high pressure data in Fig.6. A very nice agreement with the $T_c(\delta)$ curve is obtained. Apart from fixing the p=0 point to the curve, no scaling or fitting on the δ-axis was done. Clearly a very nice agreement is obtained, which shows that for $YBa_2Cu_4O_8$ and $CaLaBaCu_3O_7$ the main effect of pressure on T_c is due to a pressure induced change in the hole concentration, apart from an increase in the maximum T_c. The pressure dependence of T_c can thus be described by the following Eqns.:

$$T_c = T_o[1 - \beta(\delta - \delta_o)^2] \tag{9}$$

$$\delta = \delta_1 + \alpha p \tag{10}$$

Eqn. (9) describes the curve in figure 6 with $\beta = 60$ and $\delta_o = 0.2$. Eqn. (10) describes the experimental finding of Fig.7, from which also the parameters can be read off. Combination gives

$$T_c = T_o\{1 - \beta[(\delta_1 - \delta_o) + \alpha p]^2\} \tag{11}$$

which is a parabolic dependence of T_c on pressure. This formula is used in the paper of Van Eenige et al. in this volume to fit the $T_c(p)$ dependence of Ca-doped $YBa_2Cu_4O_8$.

Up to now a possible pressure dependence of T_o was ignored. We will now discuss the consequences of a pressure dependent T_o, using the measurements by Tanahashi et al.[32] on $La_{2-x}Sr_xCuO_4$ as an example. They find that $T_c(x)$ is a curve similar to the curve in Fig.6, as one would expect. In Fig.8 we have plotted this p=0 curve as a function of hole concentration δ, taking $\delta(p=0) = x$. This assumption may not be exact, in which case the δ-axis should be modified. For our present argument this is not of much importance. If pressure increases the hole concentration one expects that $\partial T_c/\partial p$ is positive if δ is on the left of the maximum T_c, and that $\partial T_c/\partial p$ is negative to the right of the maximum T_c. This is not all observed: $\partial T_c/\partial p$ is always positive, with a minimum value around the maximum T_c. Some authors have even argued that this enigmatic result cannot be interpreted in terms of a modification of the hole concentration by pressure. We will now show that such an interpretation is not only possible, but leads to very useful results.

Tanahashi et al. have also measured the upper critical field H_{c2} as a function of pressure and x. Using the analysis outlined in section 4, we calculated that the pressure induced change in hole concentration is $\partial \delta/\partial p = 0.016$ hole GPa^{-1}, roughly independent of x. The effect of pressure is not only to change δ but also other parameters. All these other parameters are symbolized by the A-axis in Fig.8. Clearly the pressure axis is in the $A\delta$ plane, such that it has projections on both the A and δ axis. The projection on the δ-axis is known, since $\partial \delta/\partial p = 0.016$ GPa^{-1}. This enables plotting $T_c(p)$ for various x in this figure in a quantitative, i.e. non-schematic, way. The lines connecting a circle with a cross are Tanahashi et al.'s $T_c(p)$ measurements for five different values of x. The crosses mark the p = 1 GPa data and hence their locus is $T_c(\delta)$ at 1 GPa. This is the second curve in figure 8. A pressure of 1 GPa is thus found to have two effects on the curve $T_c(\delta)$: (i) the maximum increases at a rate $\partial T_o/\partial p = 3.5$ K GPa^{-1} and (ii) the maximum shifts towards slightly higher δ, i.e. δ_o increases. Also note that the shape of $T_c(\delta)$ is not significantly changed. In a very natural way we have thus explained the enigmatic data by Tanahashi et al. and found the pressure dependence of T_o and δ_o.

6. THE RELATION BETWEEN T_c AND $\partial \ln T_c/\partial p$

As pointed out previously[7,8], in a plot of $\partial \ln T_c/\partial p$ versus T_c certain systematics appear. Compounds with a high T_c have in general a small pressure effect $\partial \ln T_c/\partial p$, whereas for compounds with a low T_c the pressure derivative can be large. In Fig.9ab the symbols represent measurements of $\partial \ell nT_c/\partial p$ at p=0 and T_c by a large number of groups as reviewed in Ref.8 and 15. Compounds with a different number of CuO_2 layers or a different sign of the Hall effect are designated by different symbols. We will now try and explain these systematics.

From Eqn.(11) the pressure derivative of T_c can be found:

$$\frac{\partial T_c}{\partial p} = -2\alpha\beta T_0[(\delta_1 - \delta_0) + \alpha p]. \tag{12}$$

Combining this result with Eqn.(11) yields

$$\frac{\partial \ell nT_c}{\partial p} = 2\alpha[\beta T_0(T_0 - T_c)/T_c^2]^{\frac{1}{2}} \ \text{sgn} \ (\delta_0 - \delta). \tag{13}$$

This equation implies a relation between T_c and $\partial \ell nT_c/\partial p$. Apart from T_0, which we assume, for simplicity, again independent of pressure, only one parameter $\alpha^2\beta$ determines this relation. Taking $\beta=60$, as above, α was fitted to the data points in Fig.9ab. The resulting curves (from eqn.(13)) were plotted in Fig.9ab; the parameters used are indicated in the lower right hand corner. A rather nice fit can be obtained, although in reality T_0 is not pressure independent. Mori has plotted his data[33] in a similar plot, for a number of compounds. In this plot, of which the data points are reproduced in Fig.9c, $\partial \ell nT_c/\partial p$ is plotted versus T_c not only at p=0,

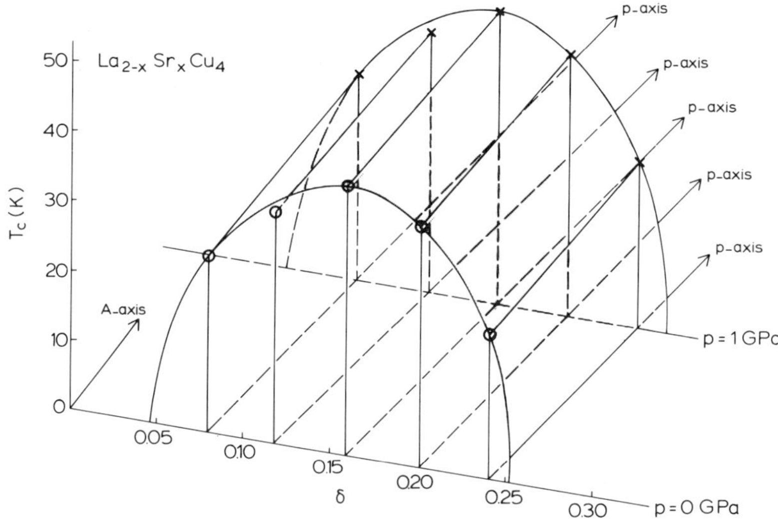

Fig. 8.   Hole concentration δ dependence of T_c at zero pressure (circles) and at 1 GPa (crosses), calculated from the $T_c(p)$ data (lines connection circles and crosses) and $H_{c2}(p)$ data by Tanahashi et al. (Ref. 32).

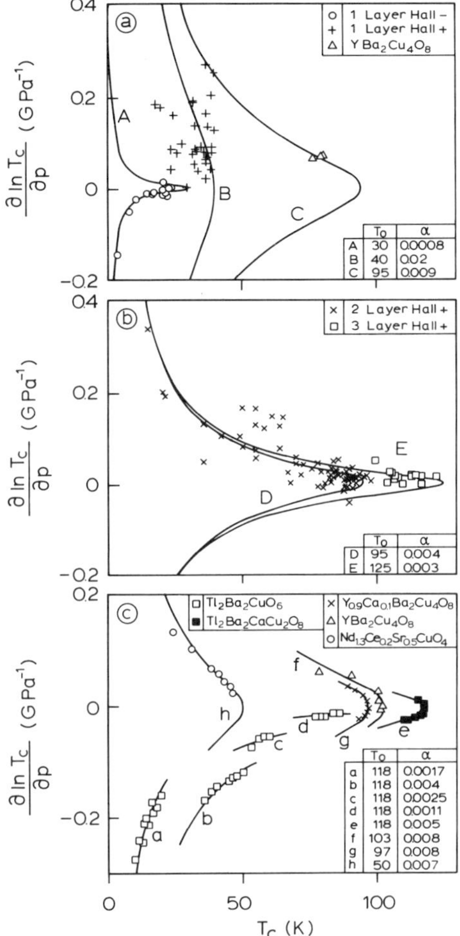

Fig. 9.   (a) and (b) summary of published data (for references see Ref. 8 and 15) on the pressure dependence of T_c at p = 0, plotted as $\partial \ln T_c/\partial p$ versus Tc. For each set of data the number of CuO_2-layers and the sign of the Hall resistivity is indicated. (-: electron-like, +: hole-like).
(c) : Data points of Mori (Ref. 33) on $\partial \ln T_c/\partial p$, not only at p = 0, but also at higher pressures.
<u>curves in abc</u> are fits to the data points using eqn. (13) with β = 60 and α (in GPa⁻¹) and T_0 (in K) as indicated.
Triangles in a and curve C pertain to $YBa_2Cu_4O_8$ (with two CuO_2 layers), for which a perfect fit is obtained using the α obtained from our determination from H_{c2} in section 4.

but at a number of pressures in one experiment. We fitted Eqn.(13) to
these data and the result is shown by the curves in Fig.9c. Clearly also
here a very nice agreement is obtained. The curves are very similar to the
curves in Fig.9ab, as may be expected from the reasoning above: the
relation between T_c and $\partial \ln T_c/\partial p$ holds irrespective whether δ is increased
by pressure or by chemical means.

7. DISCUSSION

We have obtained values for $\partial \delta/\partial p$ in section 5 and 6 using two
different methods (i.e. from H_{c2} vs. p and $\partial \ln T_c/\partial p$ vs. T_c plots). The
results agree very well. For $YBa_2Cu_4O_8$ we obtain exactly the same number
$\partial \delta/\partial p = 0.009$ GPa^{-1} with both methods. Fitting to the plot of Mori yields
0.008 GPa^{-1}, also in good agreement. For $CaLaBaCu_3O_7$ the first method
gives $\partial \delta/\partial p = 0.0005$ GPa^{-1}, which is certainly consistent with the
$\partial \ln T_c/\partial p$ plot, although this plot does not show much variation close to T_o
for such small values of $\partial \delta/\partial p$.

It is interesting to compare with other determinations of $\partial \delta/\partial p$. From
optical reflection measurements on $YBa_2Cu_3O_7$ by Garringa et al.[34] we
estimate for the reflectivity R that $\partial \ln R/\partial p \lesssim 0.01$ GPa^{-1}. From the
Hagen Rubens relation we deduce that $| \partial \ln A/\partial p | \lesssim 0.02$ GPa^{-1}, where
$A = (m*/ne^2\tau)$. From this we find $|\partial \ln \delta/\partial p | \lesssim 0.01$ GPa^{-1}, or
$\partial \delta/\partial p \lesssim 0.002$ GPa^{-1} which is slightly small compared to our values. On the
other hand from Hall Resistivity measurements[35,36] $\partial \ln \delta/\partial p \approx$
$0.1 - 0.2$ GPa^{-1} can be deduced, which yields $\partial \delta/\partial p \approx 0.02 - 0.04$ GPa^{-1},
which is slightly large compared to our values. This may be due to the
fact that in Hall resistivity all mobile charge carriers are measured,
while δ is the hole concentration in the CuO_2 layers only.

Considering everything together our $\partial \delta/\partial p$ values are in a realistic
range. We estimate the error in the $\partial \delta/\partial p$ values determined above to
be ~ 25%. This error may partially improve by a more accurate $T_c(\delta,p=0)$
curve, preferably for the exact compound under investigation.

The knowledge of $\partial \delta/\partial p$ from our results above makes it possible to
adjust the hole concentration to a desired value using pressure. This
sometimes even allows the study of high-T_c superconductors at a doping
unattainable with other means.

Concluding: we have performed electrical transport measurements in

high-T_c superconductors up to 50 GPa and in magnetic fields up to
12 Tesla. From measurements of T_c and the upper critical field we are able
to deduce an experimental value for the hole concentration δ. We found
that δ increases roughly linearly with pressure. We are able to explain the
pressure dependence of T_c, not only the initial slope at low pressure, but
also the full curve at high pressure. We have pointed out and explained
the systematics in the relation between Tc and its pressure derivative. An
independent determination of $\partial\delta/\partial p$ could thus be made. Also, the
measurements of Mori are explained in a consistent manner, using the same
model.

ACKNOWLEDGEMENT

This work is part of the research program of the Stichting voor
Fundamenteel Onderzoek der Materie (FOM) which is financially supported by
NWO.

REFERENCES

1.   J.G. Bednorz and K.A. Müller, Z. Phys. B. $\underline{64}$ (1986) 189.
     J.G. Bednorz and K.A. Müller, Angew. Chem. $\underline{100}$ (1988) 757.

2.   C.W. Chu, P.H. Hor, R.L. Meng, L. Gao, Z.J. Huang, Y.Q. Wang, Phys.
     Rev. Lett. $\underline{58}$ (1987) 405.

3.   M.K. Wu, J.R. Ashburn, C.J. Torng, P.H. Hor, R.L. Meng, L. Gao,
     Z.J. Huang, Y.Q. Wang and C.W. Chu, Phys. Rev. Lett. $\underline{58}$ (1987) 908.

4.   H.W. Zandbergen, R. Grousky, K. Wang and G. Thomas, Nature 331 (1988)
     596.

5.   J. Karpinski, E. Kaldis, E. Jilek, S. Rusiecki and B. Bucher, Nature
     336 (1988) 660.

6.   R.J. Cava, J.J. Krajewski, W.F. Peck, B. Batlogg, L.W. Rupp, R.M.
     Flemming, A.C.W.P. James and P. Marsch, Nature 338 (1989) 328.

7.   R. Griessen, Phys. Rev. $\underline{B36}$ (1987) 5284.

8.   R.J. Wijngaarden and R. Griessen, "High pressure studies" in "Studies
     of high temperature superconductors", A.V. Narlikar ed., Vol.2, Nova
     Science Publ., Commack, 1989, p. 29.

9.   E.N. van Eenige, R. Griessen, R.J. Wijngaarden, J. Karpinski, E.
     Kaldis, S. Rusiecki and E. Jilek, Physica C 168 (1990) 482.

10. J.J. Scholtz, E.N. van Eenige, R.J. Wijngaarden, R. Griessen, J.
     Karpinski, E. Kaldis, S. Rusiecki and E. Jilek, to be published.

11. I.F. Silvera and R.J. Wijngaarden, Rev. Sci. Instrum. 56 (1985) 121.

12a. H. Hemmes, A. Driessen, J. Kos, F.A. Mul, R. Griessen, J. Caro and S.
     Radelaar, Rev. Sci. Instr. 60 (1989) 474.

12b. J.J. Scholtz, A. Driessen, R. v.d. Berg, H. v. Groen, H. Verhoog, J.J. de Kleuver, R.J. Wijngaarden and R. Griessen, High Pressure Research 5 (1990) 874.

13. H.K. Mao, J. Xu and P.M. Bell, J. Geophys. Res. 91 (1986) 4673.

14. D. Braithwaite, G. Chouteau, G. Martinez, J.L. Hodeau, M. Marezio, J. Karpinski, E. Kaldis, S. Rusiecki and E. Jilek, Physica C 177 (1991) 75.

15. R.J. Wijngaarden, E.N. van Eenige, J.J. Scholtz and R. Griessen, High Pressure Research 3 (1990) 105; R.J. Wijngaarden, J.J. Scholtz, E.N. van Eenige and R. Griessen, High Pressure Research (1991), accepted for publication.

16. J.J. Scholtz, E.N. van Eenige, R.J. Wijngaarden, R. Griessen and D.M. de Leeuw, to be published.

17. N.R. Werthamer, E. Helfand and P.C. Hohenberg, Phys. Rev. 147 (1966) 295.

18. R.J. Wijngaarden, J.J. Scholtz, E.N. van Eenige and R. Griessen, submitted for publication.

19. M.W. Shafer and T. Penney, Eur. J. Solid State Inorg. Chem. 27 (1990) 191.

20. D.C. Mattis, Phys. Rev. B 42 (1990) 6787.

21. P.W. Anderson, Science 235 (1987) 1196.

22. N. Mott, Metal Insulator Transitions, Taylor and Francis London, 1974, p. 139. See also M. Cyrot, preprint.

23. H.A. Ludwig, W.H. Fietz, M.R. Dietrich, H. Wühl, J. Karpinski, E. Kaldis and S. Rusiekcki, Physica C 167 (1990) 335.

24. R.J. Nelmes, J.S. Loveday, E. Kaldis and J. Karpinski, Physica C 172 (1990) 311.

25. W.H. Fietz, H.A. Ludwig, T. Wolf, H. Wühl, M. Dietrich, High Pressure Research (1991), accepted for publication.

26. L.F. Feiner, private communication.

27. M. Gupta and R.P. Gupta, Physica C 171 (1990) 465.

28. P. Allenspach, J. Mesot, U. Staub, A. Furrer, H. Blank, H. Mutka, C, Vettier, E. Kaldis, J. Karpinski and S. Rusiecki, Eur. J. Solid State Inorg. Chem. 28 (1991) 627.

29. Y. Yamada, T. Matsumoto, Y. Kaieda and N. Mori, Jap. J. Appl. Phys. 29 (1990) L250.

30. J.D. Jorgensen, S. Pei, P. Lightfoot, D.G. Hinks, B.W. Veal, B. Dabrowski, A.P. Paulikas and R. Kleb, Physica C 171 (1990) 93.

31. R.S. Liu, P.P. Edwards, Y.T. Huang, S.F. Wu and P.T. Wu, J. of Solid State Chem. 86 (1990) 334.

32. N. Tanahashi, Y. Iye, T. Tamegai, C. Murayama, N. Mori, S. Yomo, N. Okazaki and K. Kitazawa, Jap. J. Appl. Phys. 28 (1989) L762.

33. N. Mori, H. Takahashi and C. Murayama, Supercond. Sci. Technol. 4 (1991) S439.

34. M. Garriga, U. Venkateswaran, K. Syassen, J. Humlicek, M. Cardona, Hj. Mattausch and E. Schönherr, Physica C 153-155 (1988) 643.

35. T. Hiraoka, Jap. J. Appl. Phys. 28 (1989) L1135.

36. I.D. Parker and R.H. Friend. J. Phys. C: Solid State Phys. 21 (1988) L345.

PRESSURE DEPENDENCE OF T_c OF $YCa_{0.2}Ba_{1.8}Cu_4O_8$

E.N. van Eenige, R.J. Wijngaarden and R. Griessen

Department of Physics and Astronomy
Free University
1081 HV Amsterdam, The Netherlands

J. Karpinski, E. Kaldis, S. Rusiecki and E. Jilek

Laboratorium fur Festkörperphysik
ETH
CH-8093 Zürich, Switzerland

ABSTRACT

Resistive measurements in a cryogenic diamond anvil cell
show that the critical temperature T_c of $YCa_{0.2}Ba_{1.8}Cu_4O_8$ can be
increased from 90 K to 99 K by applying a pressure of 8.8 GPa.
At higher pressures T_c decreases. An inverted parabola is well
fitted through the data. This behavior can be explained by
assuming that the number of holes in the CuO_2-planes increases
linearly upon applying pressure and that T_c as a function of the
number of holes n_h in the CuO_2-planes follows the
(parabolic-like) $T_c(n_h)$ phase diagram line.

INTRODUCTION

Double chain Y-Ba-Cu-O compounds were shown to have very
high dT_c/dp values (5.5 K/GPa)[1]. Thus we could raise T_c in
$YBa_2Cu_4O_8$ from 80 to 108 K by applying pressure up to 12 GPa[2].
Ca-doping of $YBa_2Cu_4O_8$ increases T_c up to 90 K[3]. These authors
suggested that this is due to the substitution of Y^{3+} by Ca^{2+},
leading to an increase of the hole concentration. Contrary to
that, NQR measurements[4] and Raman investigations[5] show that this
is not the case, so that Ca^{2+} substitutes more Ba^{2+} than Y^{3+}. The
dT_c/dp of $YCa_{0.2}Ba_{1.8}Cu_4O_8$ is 2.5 K/GPa[1,6], and is much lower
than in the undoped material, indicating an internal chemical
pressure effect[1]. In this work we report on the pressure
dependence of T_c of $YCa_{0.2}Ba_{1.8}Cu_4O_8$ measured in a cryogenic
diamond anvil cell.

EXPERIMENTAL

Ca-123 + CuO mixtures corresponding to the formula
$YCa_{0.2}Ba_{1.8}Cu_4O_8$ have been heated up to 1000 °C under high oxygen
pressure. X-ray investigations and very accurate chemical

"

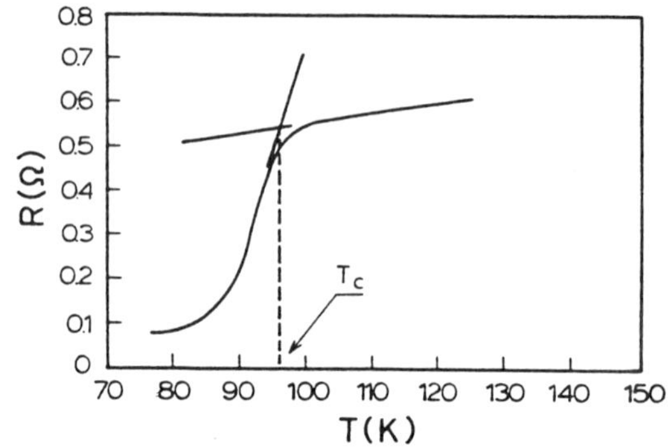

Fig. 1.   Resistance versus temperature curve for
           $YCa_{0.2}Ba_{1.8}Cu_4O_8$ at p = 3.33 GPa (run II). The
           intersection of the tangents defines the onset
           critical temperature T_c.

analysis indicate that, very probably, Ca mainly substitutes Ba
and not Y[7].

The experiment was performed in a cryogenic diamond anvil
cell (DAC) designed for pressures up to 100 GPa. Pressure was
measured using the ruby fluorescence method corrected for
temperature. The sample consisted of one or more grain(s) taken
from a polycrystalline pellet. Four-point resistivity
measurements were made using gold wires. The set-up was similar
to that described by Van Eenige et al.[2]. Out of several runs
with different samples three were successful. In run I no
pressure transmitting medium was used. In runs II and III a
small amount of gypsum was added as a soft filling material.

RESULTS AND DISCUSSION

A typical curve from run II is shown in fig. 1. The lines
in the figure indicate how the onset critical temperature T_c is
determined. The results, T_c versus pressure, are shown in fig.
2. For comparison the curve for $YBa_2Cu_4O_8$ is added. The
$T_c(p)$-curve for run I is lower by 5 K probably because of a
different composition of the grain used in the DAC.

Table 1.   Parameters of the parabolic fits to the formula
           $T_c=T_{cmax}(1-A(p-p_{Tcmax})^2)$.

	T_{cmax} (K)	p_{Tcmax} (GPa)	A (GPa^{-2})
run I	95.5	8.41	0.00141
run II and III	99.1	8.76	0.00108

Through both sets of data (run I: open circles and runs II and
III: closed circles) similar parabola's could be fitted. The
results of the fits are presented in table 1. The line through
the data of $YBa_2Cu_4O_8$ is a guide to the eye but suggests that T_c
will also decrease for pressures over ~ 11 GPa. This decrease
has actually been observed very recently by J.J. Scholtz et
al.[8].

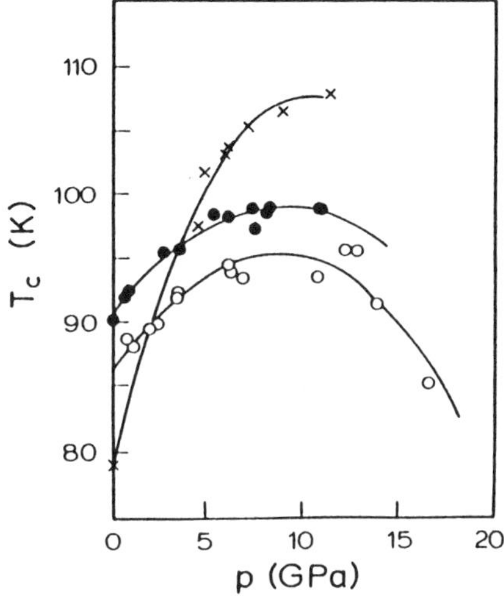

Fig. 2.   Critical temperature T_c versus pressure for $YBa_2Cu_4O_8$
           (crosses; ref. 1) and for $YCa_{0.2}Ba_{1.8}Cu_4O_8$ (open
           circles:run I, closed circles: runs II and III; this
           work). The difference between the curves with the
           open and filled circles is explained in the text.
           The lines through the curves for $YCa_{0.2}Ba_{1.8}Cu_4O_8$ are
           parabolic fits through the data.

Comparing pure and Ca-doped $YBa_2Cu_4O_8$ both differences and
similarities can be noted. Remarkably, T_c can be raised much
more by pressure for $YBa_2Cu_4O_8$ (~ 28 K) than for $YCa_{0.2}Ba_{1.8}Cu_4O_8$
(~ 9 K), although $T_c(p=0)$ for the first material is 10 K lower
than for the second. The pressures at which the maxima are
reached differ by ~ 2 GPa, the maximum for $YBa_2Cu_4O_8$ being found
around 10 GPa. From the literature it is known that T_c as a
function of the number of holes in the CuO_2-planes (the $T_c(n_h)$
line of the phase diagram) has a parabola-like shape[9,10].
Experiments[11,12,13] on $YBa_2Cu_4O_8$ imply further that the number of
holes in the CuO_2-planes increases upon applying pressure. If
this happens in a linear way (as expected in a first order
approximation) the parabolic $T_c(n_h)$ curve implies a parabolic
dependence for $T_c(p)$ as is found experimentally.

The difference in the maxima found for T_c as a function of
pressure indicates that the $T_c(p)$ curve is not only determined
by the variation of n_h with pressure. Changes around the
CuO_2-planes induced by the Ca-substitution have to be taken
into account. We speculate that minor changes around the
CuO_2-planes caused by the Ca-substitution are at the origin of
the difference in the maximum T_c. For a better understanding it
is necessary to know exactly were the Ca-ions can be found in
the lattice.

ACKNOWLEDGEMENTS

We are grateful to H. Huiberts and K. Heeck for their help
with the measurements. This work is part of the research
program of the Stichting voor Fundamenteel Onderzoek der
Materie (FOM) which is financially supported by NWO. The Zurich
group is indebted to the Schulleitung of the ETH and the Swiss
National Fund for Financial Supports.

REFERENCES

1. B. Bucher, J. Karpinski, E. Kaldis and P. Wachter, J. of
    Less-Common Metals 164&165:20 (1990).
2. E.N.van Eenige, R. Griessen, R.J. Wijngaarden, J.
    Karpinski, E. Kaldis, S.Rusiecki and E. Jilek, Physica C
    168:482 (1990).
3. T. Miyatake, S. Gotoh, N. Koshizuka and S. Tanaka, Nature
    341:41 (1989).
4. I. Mangelschots, M. Mali, J. Roos, H. Zimmermann, D.
    Brinkmann, S. Rusiecki, J. Karpinski, E. Kaldis and E.
    Jilek, Physica C 172:57 (1990).
5. T. Heyen, M. Cardona, E. Kaldis, J. Karpinski and S.
    Rusiecki, to be published.
6. N. Mori, H. Takahashi and Ch. Murayama, Supercond. Sci.
    Technology 4:S439 (1991).
7. S. Rusiecki, E. Kaldis, J. Karpinski and E. Jilek, to be
    published.
8. J.J. Scholtz, Ph.D.-thesis, Free University, Amsterdam,
    unpublished and J.J. Scholtz, E.N. van Eenige, R.J.
    Wijngaarden and R. Griessen, to be published.
9. M.W. Shafer and T. Penney, Eur. J. Solid State Inorg. Chem.
    27:191 (1990).
10. M.-H. Whangbo and C.C. Torardi, Science 249:1143 (1990).
11. R.J. Nelmes, J.S. Loveday, E. Kaldis and J. Karpinski,
    Physica C 172:311 (1990).
12. J. Mesot, P. Allenspach, U. Staub, A. Furrer, H. Blank, H.
    Mutka, C. Vettier, E. Kaldis, J. Karpinski and S. Rusiecki,
    J. of Less-Common Metals 164&165:59 (1990).
13. Y. Yamada, J.D. Jorgensen, Shiyou Pei, P. Lightfoot, Y.
    Kodama, T. Matsumoto and F. Izumi, Physica C 173:185
    (1991).

D.A.C. DEVICE FOR THE CHARACTERISATION OF THE PRESSURE-DEPENDENCE OF

SUPERCONDUCTING TRANSITIONS

J. Thomasson, F. Thomas, C. Ayache, I.L. Spain and M. Villedieu

Département de Recherche Fondamentale sur la Matière Condensée,
Service de Physique Statistique Magnétisme et Supraconductivité,
Centre d'Etudes Nucléaires de Grenoble B.P. 85X, 38041 Grenoble
Cédex, France

Abstract

We describe an experimental setup devoted to the study of
superconducting transitions under hydrostatic high pressure. The setup
combines a miniature Diamond Anvil Cell, ruby fluorescence-scale for
measuring the pressure, and helium used as a pressure transmitting medium.
Pressure homogeneity and shear stress free conditions were checked.
Superconducting transitions are detected using an A.C. susceptibility
technique. Two illustrative studies are presented: 1) the comparison of
$T_c(P)$ for lead under hydrostatic and non-hydrostatic conditions and 2)
preliminary results for $T_c(P)$ of the heavy-fermion superconductor
$CeCu_2Si_2$.

1. Introduction

The Diamond Anvil Cell (D.A.C.) is well adapted to generate
hydrostatic pressures and to measure them owing to the ruby
fluorescence-scale. In the following, we describe a miniature D.A.C. for
which these conditions are seen to be met. The D.A.C. is associated with
an A.C. susceptibility bridge using a special geometry, with a very low
filling factor. This setup is devoted to the detection of superconducting
transitions under pressure for which we provide two illustrative
applications.

2. The Diamond Anvil Cell

2.1 Characteristics

We are using a D.A.C. able to achieve pressures between 0.3 and 20
GPa, over a large range of temperature: 0.3 - 300 K. The main
characteristics of our system are clearly the small size (diameter 20mm,

length 40mm, weight 150g) and ability to withstand superfluid ^{4}He loading.

The cell, shown in figure 1, is made of hardened Cu-2%Be alloy. It is a conventional piston-cylinder assembly. One of the diamonds is glued to the center of the piston. A fixed plate screwed to the cylinder supports the second diamond and a soft ring. This ring can be moved during the X-Y diamond alignment procedure (see figure 2). After the second diamond is glued in place, there is no possibility to correct any defect in adjustment. In particular, there is no tilt adjustment mechanism so that miniaturizing the cell has become possible. The piston-to-cylinder clearance is the most critical dimension: it is imperative that it should be less than 20μm. The controlling adjustment of the diamonds after the assembly of the cell is made with a classical optical method, using interference fringes formed between the diamond culets [1]. Belleville spring washers minimise pressure changes due to thermal cycling. The culets are 700μm diameter, bevelled with an angle of 19°. The gasket is made of a 500μm stainless steel sheet. Guide pins on the piston ensure the correct positioning of the gasket relative to the diamonds. A pre-indentation between the two diamonds reduces its thickness to 50μm. Then, the sample cavity is drilled axially with a diameter of 350μm. Optical micro-translators and a microscope allow us to adjust the cavity at the center of the indentation with a precision of better than 20μm.

2.2 Transmitting medium and pressure measurement

The sample and the ruby chips were loaded into the D.A.C., placed in the loading can and cooled down in a superfluid ^{4}He bath [1]. An initial sealing load of 2kN was applied which corresponded to an initial pressure of 0.3 GPa. In our case, this is the smallest load required to avoid any leak from the cavity. Further pressurisations were made at room temperature so that compressions were not applied on the solid phase of ^{4}He , thus avoiding shear-stress effects. The method of measuring pressure was with the ruby luminescence scale [2]. Pressures are systematically determined at 300K and 77K, so that the efficiency of the belleville spring washers can be checked. The maximum pressure change between room temperature and LN_2 temperature was limited to about 1 GPa. We also checked that no significant pressure variation occured between 77K and 4.2K [3]. Two different cryostats were used for pressure measurements at low temperature:

- In the first one, a plastic fibre guides the laser-beam to the cell. Another one picks the luminescence signal up and guides it to the input slit of monochromator. With this system, a mean value of pressure is obtained but, considering the fluorescence line width, it also allowed us to verify the hydrostaticity of the transmitting medium (see figure 3).

- On the other hand, the second cryostat enabled pressure measurements to be made on each ruby chip. This cryostat was built with two quartz windows and the laser beam could be focused through the windows on a single chip. We used it to check the pressure homogeneity in the cavity. Experimentally, we found it to be better than 0.005 GPa at 10 GPa (see figure 4).

3. A.C. Susceptibility Measurement

A.C. susceptibility measurements in the D.A.C. are faced by two shortcomings: 1/ filling factors are very small and 2/ the coils become unbalanced when the temperature is varied due to the thermal gradients in the surroundings and in the coils themselves. In a first step, we

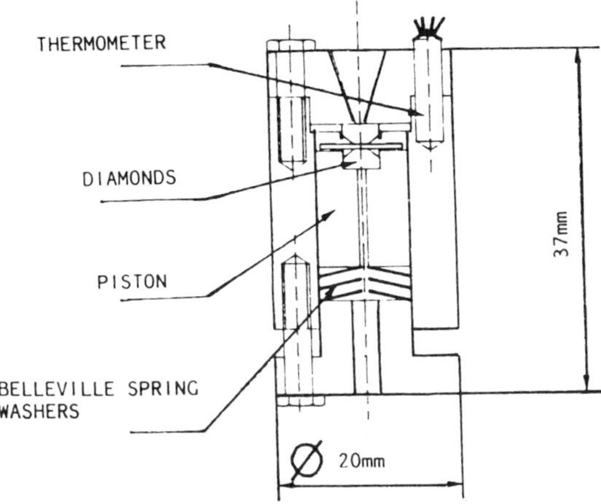

Figure 1. Sectional view of the diamond anvil cell.

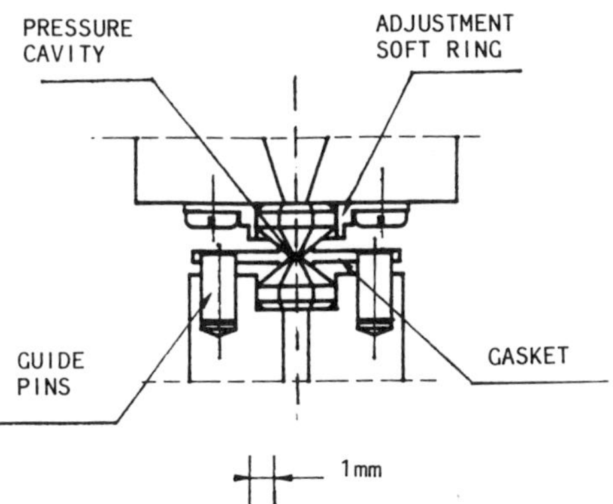

Figure 2. Details of the diamond anvil region.

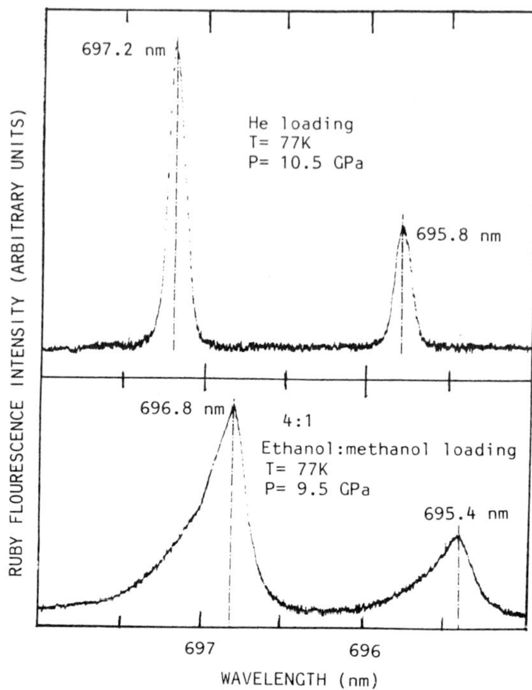

Figure 3. Typical ruby spectra obtained at ~ 10 GPa: a) He, b) 4:1 methanol:ethanol pressure-transmitting media.

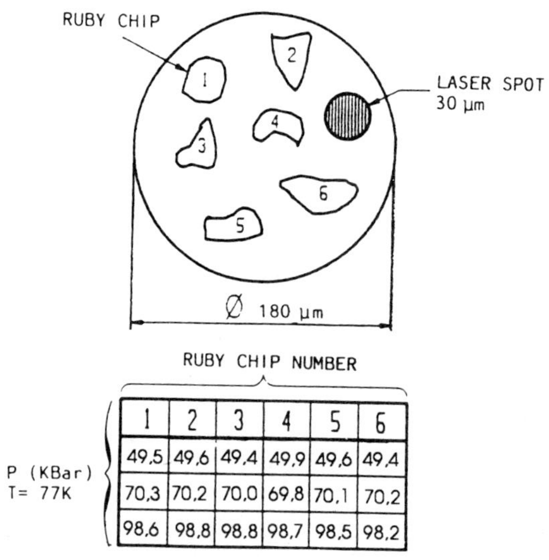

	1	2	3	4	5	6
P (KBar) T= 77K	49,5	49,6	49,4	49,9	49,6	49,4
	70,3	70,2	70,0	69,8	70,1	70,2
	98,6	98,8	98,8	98,7	98,5	98,2

Figure 4. Check of the pressure homogeneity in the cell when it is Helium loaded.

attempted to overcome the first problem. For this we wound a miniature coil with a conical bore so as to fit as close as possible between one diamond and the gasket. However this set-up proved to be rather difficult to balance and exhibited strong drifts with temperature. So we took advantage of the reduced diameter of our cell to put the detection coils outside of it. The dramatic reduction of the filling factor is partly compensated by an increased number of turns for these coils. Figure 5 shows the geometry of the exciting coil and of the two oppositely wound detection coils. Both are immersed in a ^{4}He bath. The cell is fixed in a calorimeter which can be cooled down by pumping either on ^{4}He or ^{3}He. The sample between the two anvils is located at the center of one of the detection coils.

With such a choice, the filling factor, expressed by the ratio sample's volume/coil's volume, is very low ($\approx 10^{-7}$). As a result of the relative smallness of the sample, we can suppose that it acts as a magnetic dipole [4]. The moment of the dipole is given by $M = V_s.\chi.B/\mu_o$, where χ is the magnetic susceptibility and V_s the sample's volume. Assuming that the sample has a negligible interaction with the remote detection coil, the E.M.F. induced by the sample on the other coil is given by:

$$e = 2.\pi.f.n.\alpha.B.V_s.\chi$$

where: f = frequency of the exciting field.

         n = number of turns per meter in detection coils.

$$\alpha = \frac{L}{R}.\sqrt{1+\left(\frac{L}{R}\right)^2}$$ with 2L, R = length and radius of the detection coils respectively (see figure 5).

Assuming that the superconducting transition can be described by $\chi = -1$ for $T < T_c$ and $\chi \approx 0$ for $T > T_c$, the voltage change at the transition becomes: $\Delta e = 2.\pi.f.n.\alpha.B.V_s$

In our experiment n = 10^6 turns/m, L = 3mm, R = 13mm, with typically f = 120Hz and B = 10^{-4}T; this gives Δe = 20nV, in good agreement with experimental measurements.

When the copper coils were not thermally decoupled from the D.A.C. we observed a drift of about 500nV/K at 4K for coils with 1000 turns per mm. We analysed this as mainly resulting from unbalanced resistances of the two detecting coils. Decoupling by immersing the coils in liquid ^{4}He bath confirmed this point, the drifts being strongly reduced. Two further important disturbance factors have also to be considered: 1/ magnetic perturbations may influence the detection coils, especially with the present low filling-factor geometry. Care must be taken to avoid any superconducting or magnetic impurities close to the detection system. In particular, non-magnetic Be-Cu is used for the cell. 2/ non-homogeneous heating of the cell may also produce a thermal gradient in the cell which induces a non homogeneous magnetic response of the cell. At 4K, 50nV/K disequilibrium is observed with a sweep-rate of 5mK/min, when the D.A.C. is heated at one end instead of being heated uniformly.

When care is taken to minimize these different factors, the measured noise level is about 5nV at 120Hz. This corresponds to the Johnson noise of the first amplification stage in the detection chain (see figure 6). The noise voltage due to this tranformer is given by [5]:

$$e = 6.5.10^{-10}.\sqrt{r.\Delta f} \text{ in V.}$$

with: r = source resistance in Ω connected at the input of the transformer, Δf = noise bandwith in Hz of the lock-in amplifier. $\Delta f = 1/4.\tau$ where τ is the time-constant of the lock-in amplifier. With r = 50Ω and τ = 300ms, it comes e = 4.5nV.

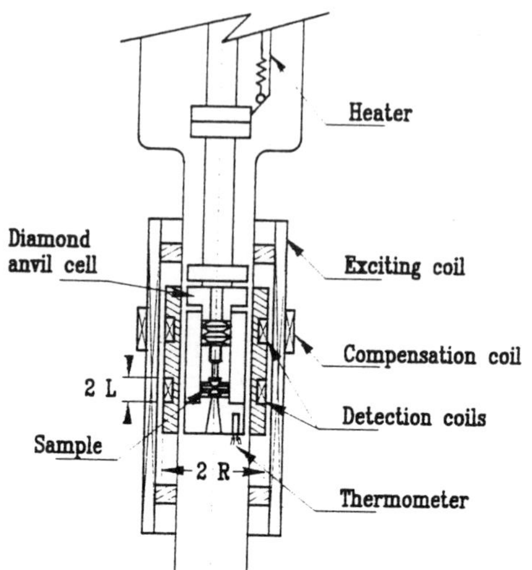

Figure 5. $\chi_{a.c.}$ detection system. The coils are wound outside of the cell. 2L and 2R refer to text.

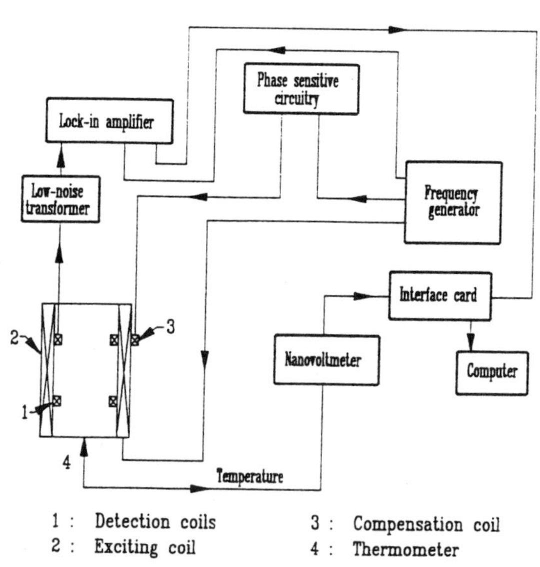

1 : Detection coils   3 : Compensation coil
2 : Exciting coil   4 : Thermometer

Figure 6. $\chi_{a.c.}$ measurement bridge.

On the other hand, the system does not allow us to perform measurements at higher frequencies. The D.A.C. is electrically equivalent to a low-pass filter. Calculations based either on the skin effect [6] or on a L-R circuit model [7] give a cut-off frequency of about 300Hz. This limit is however of little importance since the detection of superconductivity low frequencies are required.

4. Applications

4.1 Is Pb a suitable manometer?

We have reexamined the superconducting transition temperature of lead as a function of pressure. The details of this experiment were published previously [8]. Our main result is a net disagreement below about 14 GPa between measurements using ^{4}He as the pressure transmitting medium and those using 4:1:methanol:ethanol. This disagreement is also obtained in the same pressure range in the recent results of Bireckoven and al. [19]. We ascribed these differences to a sensitivity of T_c to the state of shear stress for the fcc-phase of lead , so that Pb-scale cannot be considered as a suitable substitute for ruby fluorescence-scale, in opaque anvil cells, which are usually not helium loaded. In figure 7 new data obtained in the lower pressure range (full triangles) are added to the previously published data of reference [8].

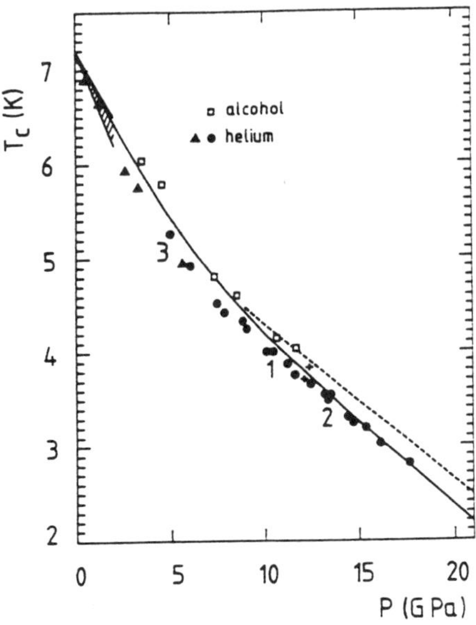

Figure 7. Experimental points for T_c(P) for lead compared with the data of Bireckoven and Wittig (see reference [19]), solid line, and those of Erskine, Yu and Martinez (see reference [20]), dashed line. The hatched region summarizes the dispersion in values obtained by different workers in the low pressure range. Data taken with alcohol as pressure-transmitting medium are represented by squares. Full circles and triangles correspond to those obtained with helium for two different mounting. In this case, the numbers 1-3 indicate the initial non monotonous pressure cycle.

4.2 $CeCu_2Si_2$

Our current work is directed towards measurement of the pressure dependence of the superconducting transition temperature of the ternary compound $CeCu_2Si_2$. At zero pressure, this material shows a heavy fermion behaviour with an extremely large specific heat γ-value of about 1 J/mol.K^2 [9]. The superconducting transition temperature T_c is limited to about 0.7K, depending on the detailed conditions of synthesis [10,11].

This material has already been studied under pressure by several techniques:

- resistivity and thermoelectric power measurements in WC anvil cell and synthetic-diamond anvil cell with steatite as transmitting medium. In this experiment $T_c(P=0) \approx 0.5K$ (Jaccard and al. [12,13]).

- X-ray absorption spectroscopy in D.A.C. without any transmitting medium (Röhler and al. [14]).

- Aliev and al. measured resitivity using a pressure apparatus which was not mentioned in their publications [15,16]. Initially their studied sample did not show the superconducting state but this was restored on applying pressure.

In the low pressure range (up to about 1.5 GPa), specific heat [17] and thermal expansion coefficient [18] were also measured using classical clamps.

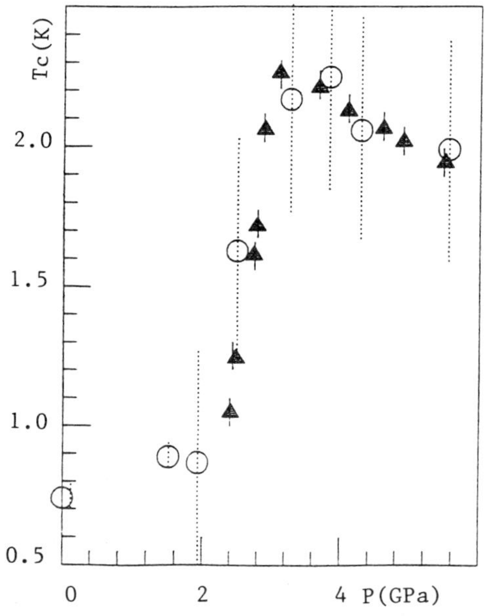

Figure 8. Pressure-dependence of the superconducting transition temperature for a polycristalline sample of $CeCu_2Si_2$. Triangles correspond to our measurements. Data from Jaccard, Mignot, Bellarbi, Benoit, Braun and Sierro (see reference [13]), are represented by circles. The vertical lines on symbols correspond to the transition-widths observed experimentally.

The work by Jaccard and al. [12,13] clearly established that there exists a characteristic pressure P_c (2 GPa$\leq P_c \leq$ 3 GPa). According to the authors, P_c separates two behaviors of $CeCu_2Si_2$ at low temperature: heavy-fermion superconductor behaviour below 2 GPa; normal sd-band superconductor above 3 GPa. On the other hand, Aliev and al. [15,16] noted that the Kondo temperature T_K decreases strongly under pressure between 0 and 1.2 GPa and the Kondo-like behaviour in the resistivity is already suppressed above 4.5 GPa. Röhler and al. [14] found a first order valence transition at 3.8 GPa by measuring the fractional valence of Ce in $CeCu_2Si_2$ under pressure between 0 and 20 GPa. Unfortunately, this last experiment was performed on a non-superconducting sample at zero pressure so that it is difficult to connect this result with those obtained by resitivity measurements.

We decided to study the superconducting transition temperature as a function of pressure by A.C. susceptibility, focusing our measurements in the "critical" region around 2.5 GPa. Measurements were performed on a polycrystalline sample ($T_c \approx$ 0.7K) placed in our D.A.C. with ^{4}He as the transmitting medium. We started at 5.5 GPa and decreased pressure by steps of about 0.5 GPa down to 3.5 GPa; the decreasing steps were reduced to about 0.1 GPa or even less in the lower pressure range. The results we obtained are shown in figure 8. Our results show good overall agreement with those published by Jaccard and al. [12,13]. They even define precisely the sharp change from values of T_c lower than about 1K for $P < P_c$, to values exceeding 2K for $P > P_c$. A.C. susceptibility transition-widths are less than 0.01K in our case, differing from those observed by resitivity measurements [12,13,15,16]. Preliminary checks seem to confirm the reversibility of T_c(P) on increasing and decreasing pressure, in contrast with the irreversible behaviour reported by Röhler and al. [14].

5. Conclusion

Our experimental setup, combining the D.A.C. technique with Helium loading and ruby fluorescence measurement of pressure provides a very accurate method for the precise pressure measurement of solid state physical parameters, even in the low pressure range (starting from 0.3 GPa). Helium is an excellent transmitting medium at low temperature with a very good homogeneity, and no detectable shear stresses, as shown by our investigations, up to about 20 GPa. In summary, the pressure is tunable within about 0.1 GPa, reproducible upon thermal cycling with the same accuracy, and can be measured within 0.05 GPa.

6. Acknowledgments

We thank N. LOCKERBIE for a careful reading of the manuscript and C. GEIBEL and F. STEGLICH for providing the $CeCu_2Si_2$ sample. This work was partly supported by a N.A.T.O. Collaborative Research Grant n° CRG 900262.

References

[1] I.L. Spain and D.J. Dunstan, J. Phys. E: Sci. Instr., 22, 923 (1989).

[2] J.D. Barnett, S. Block and G.J. Piermarini, Rev. Sci. Instr., 44, 1 (1973).

[3] D.M. Adams, R. Appleby and S.K. Sharma, J. Phys. E: Sci. Instr., $\underline{9}$, 1140 (1976).

[4] M. Couach, A.F. Khoder and F. Monnier, Cryogenics, $\underline{25}$, 695 (1985).

[5] C.D. Motchenbacher and F.C. Fitchen, "Low Noise Electronic Design", ed. John Wiley & Sons, 10 (1973).

[6] L.D. Landau, E.M. Lifshitz and L.P. Pitaevskii, "Electrodynamics of continuous media", ed. Oxford Pergamon Press, 208-210 (1984).

[7] Ibid, 121-123.

[8] J. Thomasson, C. Ayache, I.L. Spain and M. Villedieu, J. Appl. Phys., $\underline{68}$(11), 5993 (1990).

[9] W. Assmus, W. Sun, G. Bruls, D. Weber, B. Wolf, B. Lüthi, M. Lang, U. Ahleim, A. Zahn and F. Steglich, Physica B $\underline{165\&166}$, 379 (1990).

[10] M. Ishikawa, H.F. Braun and J.L. Jorda, Phys. Rev. B, $\underline{27}$(5), 3092 (1983).

[11] J.Y. Henry, private communication.

[12] B. Bellarbi, A. Benoit, D. Jaccard, J.M. Mignot and H.F. Braun, Phys. Rev. B, $\underline{30}$(3), 1182 (1984).

[13] D. Jaccard, J.M. Mignot, B. Bellarbi, A. Benoit, H.F. Braun and J. Sierro, J. Magn. Magn. Mat., $\underline{47\&48}$, 23 (1985).

[14] J. Röhler, J. Klug and K. Keulerz, J. Magn. Magn. Mat., $\underline{76\&77}$, 340 (1988).

[15] F.G. Aliev, N.B. Brandt, V.V. Moshchalkov and S.M. Chudinov, Solid State Commun., $\underline{45}$(3), 215 (1983).

[16] F.G. Aliev, N.B. Brandt, V.V. Moshchalkov and S.M. Chudinov, J. Low. Temp. Phys., $\underline{57}$, 61 (1984).

[17] A. Eichler, K.R. Harms and F.W. Shaper, Physica B, $\underline{165\&166}$, 353 (1990).

[18] H. Takakura, K. Iki, H. Okita, Y. Uwatoko, G. Ooni, Y. Onuki and T. Komatsubara, J. Magn. Magn. Mat., $\underline{90\&91}$, 453 (1990).

[19] B. Bireckoven and J. Wittig, J. Phys. E: Sci. Instr., $\underline{21}$, 841 (1988).

[20] D. Erskine, P.Y. Yu and G. Martinez, Rev. Sci. Instr., $\underline{58}$, 406 (1987).

ELASTIC PROPERTIES OF HIGH TEMPERATURE SUPERCONDUCTORS

DERIVED FROM HIGH PRESSURE EXPERIMENTS

W.H. Fietz, H.A. Ludwig, B.P. Wagner, K. Grube, R. Benischke, H. Wühl

Kernforschungszentrum Karlsruhe and Universität Karlsruhe
Institut für Technische Physik
Postfach 3640, D-7500 Karlsuhe , Federal Republic of Germany

INTRODUCTION

Many investigations of the elastic properties of high T_c superconductors (HTSC) have been performed under high pressure. For any type of HTSC the results of these investigations are widely spread. For example the bulk modulus of $YBa_2Cu_3O_7$ varies between 100 GPa and 200 GPa, depending on the experiment - and when calculated from ultra-sound experiments the bulk modulus turns out to be lower than 100 GPa. At the moment these uncertainties cannot be eliminated, but the exact knowledge of the elastic properties of HTSC would be highly interesting, because of the pronounced effect of pressure on the superconducting transition temperature T_c of some HTSC. The pressure effect on T_c is not the main topic of this paper. dT_c/dP has been reviewed in 'High Pressure Studies' by Wijngaarden and Griessen (1989). Our paper will give a review of high pressure experiments on various high T_c superconductors with special regard to elastic properties and a discussion of the uncertainties.

We will start by introducing some basic ideas of structural investigations under high pressure. To demonstrate the calculation of the true bulk modulus ($B_0 = - V\, dP/dV$ at zero pressure) from high pressure experiments, we will include some results on metals and NaF.

Next we will report the results of X-ray and neutron diffraction investigations on La_2CuO_4, $La_{2-x}Sr_xCuO_4$, $La_{2-x}Ba_xCuO_4$, Pr_2CuO_4 , Nd_2CuO_4 doped with Sr and Ce , Bi - HTSC , Tl -HTSC, $YBa_2Cu_3O_{7-\delta}$, $REBa_2Cu_3O_{7-\delta}$ with RE = Eu, Ho, Gd, Yb   and   on $YBa_2Cu_4O_8$.

In the last section we will turn to $YBa_2Cu_3O_{7-\delta}$ and $YBa_2Cu_4O_8$, for there are interesting X-ray and neutron diffraction results on these compounds. We will include some ultrasound investigations to outline some specific problems arising from the comparison of ultrasound , X-ray diffraction and neutron diffraction results.

Frontiers of High-Pressure Research, Edited by H.D. Hochheimer and
R.D. Etters, Plenum Press, New York, 1991

INTRODUCTION IN THE BASIC IDEAS OF STRUCTURAL INVESTIGATIONS UNDER HIGH PRESSURE

The aim of structural investigations under high pressure is to determine the lattice constants and the atomic positions within the unit cell of the material. For these investigations neutron and X-ray diffraction are used.

Pressure Generation

The pressure can be applied by pressing a liquid or a gas into a high pressure chamber, in which the sample must be accessible to the X-ray or neutron radiation. This gives the advantage of true hydrostatic pressure conditions and a large sample volume. If reflection intensities are used for calculating the atomic position within the unit cell of powdered materials, a large sample volume is necessary for obtaining sufficiently good statistics. The disadvantages are the limited pressure range (up to several GPa) and the necessary thick walls of the high pressure chamber, which limit the use of X-rays. In this paper we will cite results obtained by neutron investigations on samples pressurized in high pressure chambers, which were filled with helium or a special liquid (for example Fluorinert) as a pressure transmitting medium.

Another method for pressure generation is to place a small amount of sample material together with a pressure transmitting medium between two or more anvils, which are pressed together by some force generating mechanism (for example a hydraulic press). Several pressure transmitting media can be used, for example special oils, methanol-ethanol-mixture, NaCl or NaF, Fluorinert,...

Very high pressures are accessible by concentrating the force to a very small area. This technique is the basis for pressure generation with diamond anvil cells. The diamonds give optical and X-ray transparency together with the possibility to reach pressures of several hundred GPa. The disadvantage of this kind of investigation is the very small amount of sample material, which give raise to intensity problems. The basic arrangement is shown in fig. 1 . A gasket serves as a pressure seal. A hole ($\varnothing$50µm - 200µm) is drilled into this gasket, which holds the sample with the pressure transmitting medium. The gasket is placed between two opposing diamonds, which are pressed together by some pressure generating mechanism.

Pressure Calibration

Usually a pressure calibrant is placed together with the sample in the high pressure cell. This may be some material with a well known equation of state (for example for NaCl the equation of state was calculated by Decker ,1971). The change of the lattice parameters of the pressure calibrant gives the actual pressure inside the high pressure cell.

Another method of pressure calibration is to use a little piece of ruby within the high pressure cell to measure the linear shift of the ruby fluorescence lines with pressure. This method has been introduced by Forman et al. (1972).

Calculation of the Bulk Modulus and the Linear Compressibilities

The bulk modulus is defined to be $B_0 = -V \cdot dP/dV$ at $P = 0$ Pa. The restriction to infinitesimal small pressures has to be used, as the compressibility of any material changes with increasing pressure. Therefore the determination of B_0 from two data points at $P_0 = 0$ GPa and any higher pressure P_1 will give an incorrect estimate of the true bulk modulus. The error in this estimation will increase with increasing $\Delta P = P_1 - P_0$.

To solve this problem, the volume of the sample is determined at several pressures. These data points can be fitted to several equations of states, which describe the volume decrease with increasing pressure. The different equations of state are obtained with different assumptions. A discussion of several equations of state is for example given by Munro et al. (1984).

In a simple model Murnaghan (1944) assumed the bulk modulus to be pressure dependent in the form $B(P) = B_0 + B' \cdot P$. This results in the following equation

$$P(V) = \frac{B_0}{B'} \left\{ \left(\frac{V_0}{V} \right)^{B'} - 1 \right\} \qquad \textit{Murnaghan equation}$$

Another equation of state was given by Birch (1947), which is often used in calculating the bulk modulus from high pressure experiments. For small pressures (with $V/V_0 > 0.85$), the Birch and the Murnaghan equation give almost identical results. The advantage of the Birch equation is the possibility to calculate analytically the B_0 and B'

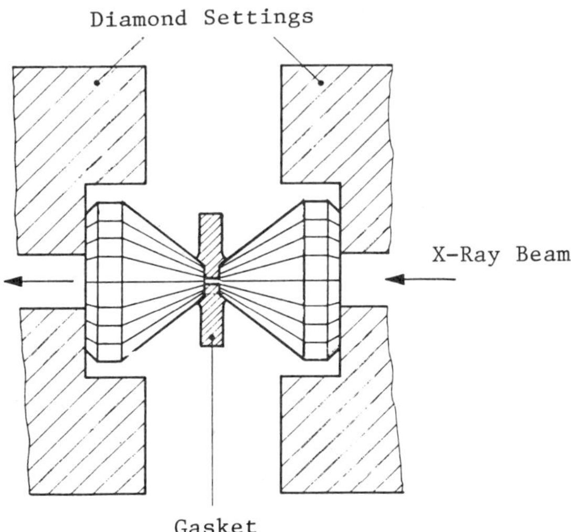

Fig 1 .   Inner section of a diamond anvil high pressure cell

values from experimental data points, without using any fitting procedure. (B′ expresses the change of B_0 with increasing pressure, often B′ is called the 'curvature' of the V(P) - curve).

$$P(V) = \frac{3}{2} B_0 \left\{ \left(\frac{V_0}{V} \right)^{\frac{7}{3}} - \left(\frac{V_0}{V} \right)^{\frac{5}{3}} \right\} \cdot \left\{ 1 - \frac{3}{4} \left(4 - B' \right) \left[\left(\frac{V_0}{V} \right)^{\frac{2}{3}} - 1 \right] \right\} \quad Birch\ equation$$

To demonstrate such a calculation of B_0 from experimental high pressure X-ray data, fig. 2 shows the normalized volume of platinum, tantalum and sodium-fluoride as a function of pressure. These data have been obtained at 300 K with a diamond anvil cell using Mo - Kα radiation.

When the bulk modulus B_0 of NaF is calculated only from the datapoint at ambient pressure and the datapoint at 20.7 GPa , an obviously wrong $B_0 = 96$ GPa is obtained. The correct results are obtained using the Birch equation.

Table 1 gives the bulk moduli of Ta , Pt and NaF, which have been calculated using the Birch equation. For comparison data from literature are included in table 1.

Due to the lack of valid equations of state for highly anisotropic orthorhombic materials we have to use such relatively simple equations to describe high temperature superconductors, although this may include a source of error.

The same care has to be taken in calculating the linear compressibilities. The linear compressibility β_i in direction of the lattice parameter g_i can be calculated in the same way as the bulk modulus :

$$\beta_i = \frac{1}{g_i(0)} \cdot \frac{g_i(0) - g_i(P)}{P}$$

This calculation leads to the same problems as the calculation of the bulk modulus. If the linear compressibility is calculated from the lattice parameter at zero pressure and pressure P, the result will depend on the pressure P.

A possibility to avoid this and estimate the 'true' compressibility at zero pressure is to fit the data to the equation

$$g_i(P) = g_i(0) - \beta_i \cdot g_i(0) \cdot P + \delta_i \cdot g_i(0) \cdot P^2$$

where g_i (P) is the lattice parameter in i-axis direction at the pressure P , β_i is the compressibility in i-axis direction and δ_i is a constant that gives the pressure dependence of β_i.

This was only a brief outline of the basic ideas of structural investigations under high pressure. More information may be found in a review of Jayaraman (1983).

Table 1.   Bulk modulus B_0 and B′ of NaF, Pt and Ta

	B_0 [GPa]	B′	
NaF	46.4	4.9	This work data from fig. 2
	46.2	5.1	Anderson et al., 1966
	46.4	4.9	Sato-Sorensen et al., 1983
Ta	212	2.8	This work data from fig. 2
	196		Bolef et al., 1961
Pt	274	3.5	This work data from fig. 2
	278		Bridgman et al., 1922

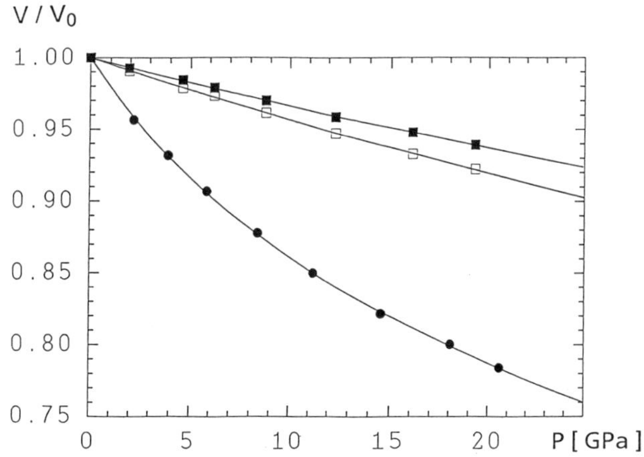

Fig. 2   Normalized Volume of Platinum ■ , Tantalum □ and Sodium-Fluoride ●
at pressures up to 20 GPa.

HIGH PRESSURE EXPERIMENTS ON La$_2$CuO$_4$

La$_2$CuO$_4$ has the K$_2$NiF$_4$ structure and is not superconducting without doping. At temperatures above 420 K it is tetragonal and becomes orthorhombic at lower temperatures (R. Saez Puche et al., 1982) with small distortions of the a- and b-axis lattice parameters and slightly tilted Cu-O octahedra.

After the discovery of high T$_c$ superconductivity in doped La$_2$CuO$_4$ materials (Bednorz and Müller, 1986) there was great interest in the structural behaviour of this parent compound under high pressure.

Akthar et al. (1988) investigated powdered La$_2$CuO$_4$ with an energy dispersive equipment using synchrotron radiation. They used a diamond anvil cell with NaCl as a pressure calibrant . This results are listed in table 2.

Howard et al. (1989) (Nelmes et al., 1990 ; Hatton et al., 1990) performed a neutron diffraction study of La$_2$CuO$_4$ under high pressure. The powdered material was investigated with Fluorinert as a pressure transmitting medium. Unfortunately the pressure could only be estimated with a 5% error from earlier pressure calibration. The investigation was performed at ambient pressure, 0.5 GPa and 0.9 GPa. The results are listed in table 2.

Fietz et al. (1989) reported an angular dispersive X-ray diffraction study on powdered La$_2$CuO$_4$ using a diamond anvil cell, with NaCl as a pressure transmitting medium. The pressure was calibrated using ruby fluorescence. In this experiment data were taken up to pressures of 29 GPa. This allowed to fit the linear compressibility to the expression $g_i (P) = g_i (0) - \beta_i \cdot g_i(0) \cdot P + \delta_i \cdot g_i(0) \cdot P^2$ (with $g_i (0) = g_i$ at ambient pressure) , as we have outlined in the introduction. The results are included in table 2.

Ahsbahs et al. (1989) presented results on X-ray diffraction measurements on a single crystal of La$_2$CuO$_4$, which had a special cut to fit into the gasket of a diamond anvil cell. Therefore they could only measure the a - and the c - axis length, but this gave the possibility to measure atomic positions with X-ray diffraction in a diamond anvil cell. The pressure was measured by ruby fluorescence. The linear compressibility coefficients were calculated from two data points at ambient pressure and at 5.4 GPa. The values are included in table 2.

Table 2.   Bulk modulus and linear compressibilities of La$_2$CuO$_4$

1. Author	Type of invest.	β_a [10^{-3} GPa^{-1}]	β_b [10^{-3} GPa^{-1}]	β_c [10^{-3} GPa^{-1}]	B_0 [GPa]
Akthar	X-ray	1.5	2.8	1.6	150
Howard	neutron	1.8	4.1	1.5	135
Fietz	X-ray	1.7	2.0	1.6	182
Ahsbahs	X-ray	1.53		2.40	

If La$_2$CuO$_4$ is doped with Ba or Sr it is found to be superconducting with transition temperatures up to 40 K (Bednorz and Müller (1986) , Kishio et al. (1987), Cava et al. (1987) , Politis et al. (1987)). This was the starting point of the new high T$_c$ superconductivity. With increasing content x of the dopant M = Sr,Ba   La$_{2-x}$M$_x$CuO$_4$ becomes tetragonal (Fleming et al. ,1987 ;   Tarascon et al. ,1987)   also at low temperatures. For La$_{1.8}$Ba$_{0.2}$CuO$_4$ Chu et al. (1987) reported the strongest pressure effect on T$_c$ of ≈8 K/GPa ever seen.   Therefore, these materials have been highly interesting for structural investigations under high pressure.

Takahashi et al. (1987) investigated powder of tetragonal La$_{1.8}$Sr$_{0.2}$CuO$_4$ by means of X-ray diffraction using kerosen as a pressure transmitting medium. The pressure was determined using the ruby fluorescence. They found a symmetrical lattice compression in a- and c-axis direction. The pressure dependence of T$_c$ was determined to be 1.7 K/GPa. The results are listed in table 3.

Dietrich et al. (1987) performed X-ray investigations of La$_{1.8}$Sr$_{0.2}$CuO$_4$ using a diamond anvil cell with NaCl as a pressure transmitting medium. The pressure was calibrated using the equation of state of NaCl. The Bulk modulus was determined to be 180 GPa at 300 K and 190 GPa at 15 K. The authors pointed out, that the bulk moduli were reduced to 160 GPa and 170 GPa at 300 K and 15 K, respectively, if only low pressure data points are used for calculations. In a later paper (Fietz et al. 1989) the linear compressibilities were also taken from this measurement. The c-axis lattice parameter showed a stronger decrease under pressure than the a- and b-axis lattice parameters. The results are included in table 3.

Terada et al. (1987) presented results on powdered La$_{1.4}$Sr$_{0.6}$CuO$_4$, which have been measured using synchrotron radiation in a cubic anvil system. A methanol-ethanol mixture was used as a pressure transmitting medium. The pressure calibration is not mentioned in their paper. The results show a stronger decrease of the lattice parameter in c-axis direction under pressure. The results are shown in table 3.

Moret et al. (1988) performed X-ray measurements on La$_{1.7}$Ba$_{0.3}$CuO$_4$. They used a diamond anvil cell with a methanol-ethanol pressure transmitting medium. The pressure was determined using AgJ-powder as a pressure calibrant. The results of their investigation are listed in table 3.

Pei et al. (1990) investigated La$_{1.8}$Sr$_{0.2}$CuO$_4$ by neutron powder diffraction. They used a helium high pressure cell which offers perfect hydrostatic pressure conditions and precise pressure determination. They report a bulk modulus of 147 GPa, with almost isotropic linear compressibilities in a- and c-direction. The results are shown in table 3. We have recalculated the linear compressibilities and the bulk modulus in the way we have described in the introduction, using the data of Pei et al. . The result is a bulk modulus of 142 GPa with a B'= 16. The bulk modulus and the linear compressibilities are included in table 3 and indicated by * .

Table 3.   Bulk modulus and linear compressibilities of $La_{2-x}M_xCuO_4$
Values marked by * are obtained by fitting the data of the authors
in the way described in the introduction.

1.Author	Type of invest.	Material	β_a [10^{-3} GPa^{-1}]	β_c [10^{-3} GPa^{-1}]	B_0 [GPa]
Takahashi	X-ray	$La_{1.8}Sr_{0.2}CuO_4$	2.5	2.5	135
Dietrich	X-ray	$La_{1.8}Sr_{0.2}CuO_4$	1.7	2.3	180
Terada	X-ray	$La_{1.4}Sr_{0.6}CuO_4$	1.9	2.26	165
Moret	X-ray	$La_{1.7}Ba_{0.3}CuO_4$	2,0	1.7	179
Pei	neutron	$La_{1.85}Sr_{0.15}CuO_4$	2.29 , 2.4*	2.21 , 2.16*	147 , 142*
Nelmes	neutron	$La_{1.85}Sr_{0.15}CuO_4$	2.08 / 1.65	1.7 / 1.43	170 / 211

Nelmes et al. (1990) (Hatton et al. 1990) carried out a neutron diffraction experiment on powder of $La_{1.85}Sr_{0.15}CuO_4$ at ambient pressure , 0.6 GPa and 1 GPa. Unfortunately the pressure could only be estimated with a 5% error from earlier pressure calibration. The compressibility results depend on the pressure of the data points that are taken for the calculation. The linear compressibility are found to be $\beta_a = 2.08 / 1.65 \cdot 10^{-3}$ GPa^{-1} and $\beta_c = 1.7 / 1.43 \cdot 10^{-3}$ GPa^{-1} for the data points at 0.6 / 1.0 GPa, respectively. Both values for β_i are included in table 3.

HIGH PRESSURE EXPERIMENTS ON Pr_2CuO_4 AND $(Nd_{1-x}Sr_x)(Nd_{1-y}Ce_y)CuO_4$

Pr_2CuO_4 , which has the Nd_2NiF_4 structure, is a material closely related to La_2CuO_4 with the K_2NiF_4 structure. The striking difference between these structures is the lack of Cu-O octahedra in Pr_2CuO_4. This material is not superconducting. When Nd_2CuO_4 , which has the same structure, is doped with Sr and Ce a hybride between the Nd_2CuO_4 and K_2NiF_4 structure is obtained (Izumi,1989). This material is hole doped, superconducting and shows a positiv pressure effect on T_c . In this material only one of the two apical oxygen atoms of the Cu-O-octahedron is missing. Murayama et al. (1989) reported, that the electron doped superconducting $Nd_{1.85}Ce_{0.15}CuO_4$ shows an almost constant T_c versus P.

Fietz et al. (1989) performed an angular dispersive X-ray diffraction study on powdered Pr_2CuO_4 using a diamond anvil cell, with NaCl as a pressure transmitting medium. The pressure was calibrated using ruby fluorescence. The authors pointed out that the lack of the apical oxygens reduce the stability of the structure in the c-axis direction. They found linear compressibilities of $\beta_a = \beta_b = 1.6 \cdot 10^{-3}$ GPa^{-1} and $\beta_c = 3.0 \cdot 10^{-3}$ GPa^{-1}. The results are listed in table 4.

Izumi et al. (1990) carried out neutron diffraction measurements on hole-doped $(Nd_{1-x}Sr_x)(Nd_{1-x}Ce_x)CuO_4$. They used a He-high pressure cell and measured a $dT_c/dP = +4$ K/GPa . They investigated the sample at 6 pressures from ambient pressure up to 0.63 GPa. They calculated the linear compressibilities and the bulk modulus from a

Table 4.   Bulk modulus and linear compressibilities of Pr_2CuO_4 and $(Nd_{1-x}Sr_x)(Nd_{1-y}Ce_y)CuO_4$

1.Author	Material	$\beta_a=\beta_b$ [$10^{-3}GPa^{-1}$]	β_c [10^{-3} GPa^{-1}]	B_0 [GPa]
Fietz	Pr_2CuO_4	1.6	3.0	160
Izumi	$(Nd_{1-x}Sr_x)(Nd_{1-y}Ce_y)CuO_4$	2.25 / 2.47*	2.51 / 2.89*	143 / 124*

least square fit of straight lines, which results in $\beta_a = 2.25 \cdot 10^{-3}$ GPa^{-1}, $\beta_c = 2.51 \cdot 10^{-3}$ GPa^{-1} and $B_0 = 143$ GPa. We have fitted the linear compressibilities with their experimental data as described in the introduction. This results in a bulk moduls of 124 GPa with $B' = 64$. The obtained linear compressibilities are listed in table 4 and indicated by a *.

HIGH PRESSURE EXPERIMENTS ON $(Bi_{1.8} Pb_{0.4})$ $Sr_{1.85}$ $Ca_{2.05}$ Cu_3 O_{10+x} AND $Bi_2Sr_2CaCu_2O_{8+y}$

The Bi-Sr-Ca-Cu-O system, that was discovered by Maeda et al. (1988), shows phases with T_c of 107 K and 85 K. The high T_c phase $(Bi_{1.8}Pb_{0.4})$ $Sr_{1.85}$ $Ca_{2.05}$ Cu_3O_{10+x} has an additional Cu-O-layer in comparison to the low T_c phase $Bi_2Sr_2CaCu_2O_{8+y}$.

Yoneda et al. (1990) performed a X-ray diffraction measurement on the 107K and the 85 K phase of the Bi-superconductor. The dT_c/dP was measured to be $+2$ K/GPa for both compounds. They used a diamond anvil cell with a methanol-ethanol-water mixture as a pressure transmitting medium and calibrated the pressure by the ruby fluorescence method. Due to the very small orthorhobmic splitting, that is only hardly resolvable in X-ray experiments, the linear compressibilities have been calculated assuming a tetragonal unit cell. The obtained data are listed in table 5.

Gavarri et al. (1990) investigated the 85 K compound $Bi_2Sr_2CaCu_2O_{8+y}$ by means of neutron diffraction at ambient pressure and 1 GPa in a cell made of maraging steel. They used an inert fluorcarbon liquid as a pressure transmitting medium, the pressure calibration is not mentioned in their paper. The results are shown in table 5.

Table 5.   Bulk modulus and linear compressibilities of Bi-HTSC

1.Author	Material	β_a [10^{-3} GPa^{-1}]	β_b [10^{-3} GPa^{-1}]	β_c [10^{-3} GPa^{-1}]	B_0 [GPa]
Yoneda	107K Bi-HTSC	3.2	3.2	8.3	73
Yoneda	85 K Bi-HTSC	4.3	4.3	8.3	61
Gavarri	85 K Bi-HTSC	4.9	2.0	9.8	58.8

Table 6.   Bulk modulus and linear compressibilities of $Tl_2Ba_2Ca_2Cu_3O_{10}$

1.Author	Material	$\beta a\,[\,10^{-3}\,GPa^{-1}]$	$\beta c\,[\,10^{-3}\,GPa^{-1}]$	$B_0\,[GPa]$
Fietz	$Tl_2Ba_2Ca_2Cu_3O_{10}$	1.6	3.8	137

HIGH PRESSURE EXPERIMENTS ON $Tl_2Ba_2Ca_2Cu_3O_{10}$

$Tl_2Ba_2Ca_2Cu_3O_{10}$ still is the member of the Tl-HTSC class with the highest superconducting transition temperature known so far. The Tl compounds have a strong pressure effect on T_c (Han et al.,1988; Morosin et al. ,1988) . The dT_c/dP of $Tl_2Ba_2Ca_2Cu_3O_{10}$ was determined to be $+5$ K/GPa by Kubiak et al.(1990) .

Fietz et al. (1990) performed X-ray measurements on $Tl_2Ba_2Ca_2Cu_3O_{10}$ with a diamond anvil cell using NaF as a pressure transmitting medium. The pressure was calibrated using the ruby technique. They found an excessive c-axis decrease with increasing pressure. The results are listed in table 6.

HIGH PRESSURE EXPERIMENTS ON $YBa_2Cu_3O_{7-\delta}$

This compound has been found by following the tremendous pressure effect of $La_{1.8}Ba_{0.2}CuO_4$. The attempt to generate internal stress by replacing the La by the smaller Y atom resulted in a transition temperature beyond the boiling temperature of liquid nitrogen at 77K (Chu et al., 1987) (Zhao et al., 1987). $YBa_2Cu_3O_{7-\delta}$ has been investigated in numerous experiments and is, nevertheless, not well understood. Oxygen doping leads from the non superconducting $YBa_2Cu_3O_6$ to the 93K superconductor $YBa_2Cu_3O_7$ (we neglect the difficulties to reach a perfect oxygenated $YBa_2Cu_3O_7$). The T_c (Cava et al., 1990) and the c-axis lattice parameter (Bucher et al. ,1990) are strongly correlated to the oxygen content. The pressure effect on the transition temperature was object of many investigations (see Wijngaarden et al., 1989) and many controversial results have been reported. These results may be explained by a charge transfer picture (Jorgensen et al. 1990), that is based on a correlation of T_c with the number of holes in the Cu-O_2 planes (Tokura et al., 1988). The problem of determining the true oxygen content makes it difficult to interpret the results of the structural investigations. If not given by the authors, we will try to give a rough estimate of the oxygen content by using the correlation with the c-axis lattice parameter, but this will shurely give only a crude approximation of the true oxygen content.

Takahashi et al. (1987) investigated $YBa_2Cu_3O_{7-\delta}$ by means of X-ray diffraction in a diamond anvil cell with a position sensitive detector. We assume the pressure transmitting medium to be kerosene and the pressure determination to be done by ruby fluorescence, for these information were given in a paper about $La_{1.8}Sr_{0.2}CuO_4$ of the same year. They found a monotonously decrease of all lattice parameters with the linear compressibility $\beta_a=\beta_b=\beta_c=3.3 \cdot 10^{-3}$ GPa^{-1}. The orthorhombic splitting was checked by detecting the (200) / (020) reflection. There is no information about the consideration of the (006) reflection, that is included in the (200) / (020) group. A slight decrease of the a/b ratio was reported. The bulk modulus calculated from this measurement is 103 GPa.

Block et al. (1987) performed X-ray powder diffraction on $YBa_2Cu_3O_{7-\delta}$ in a diamond anvil cell by an energy dispersive method. They used Fluorinert as a pressure transmitting medium. The pressure calibration is not mentioned. They fixed several reflections in the space group Pmmm, neglecting the overlapping of the reflections. Presumable this is the reason why the compressibility in b-axis direction is twice the compressibility in a-axis direction. They obtained a linear compression with increasing pressure and a bulk modulus of 196 GPa.

Fietz et al. (1987) carried out X-ray diffraction measurements on $YBa_2Cu_3O_{7-\delta}$ using a diamond anvil cell with a position sensitive detector. The pressure transmitting medium was NaF, which was also used as a pressure calibrant. The oxygen content may be estimated by the correlation given by Bucher et al. (1990) to be approximately $O_{6.9}$. The authors pointed out, that the small orthorhombic splitting of the a- and b-axis and the fact of c-axis being almost exactly triple b-axis length, causes severe problems in interpreting X-ray powder diffraction data. They observed an anisotropy in the lattice compressibility in b- or c-axis direction, but due to the restrictions in powder X-ray diffraction under high pressure, it is in principle not possible to distinguish between a preferential b-axis or c-axis decrease. Only experiments on single crystals or neutron diffraction under pressure can clarify the uncertainty. This was done in 1988, as we will see later, and it was shown, that the compressibility is enhanced in c-axis direction. The authors reported for a preferential c-axis decrease linear compressibilities of $\beta_a = \beta_b = 1.7 \cdot 10^{-3}$ GPa^{-1}, $\beta_c = 2.3 \cdot 10^{-3}$ GPa^{-1}. The data are included in table 7.

Aleksandrov et al. (1988) investigated single crystals of $YBa_2Cu_3O_{7-\delta}$ by means of X-ray diffraction in a diamond anvil cell. They used Helium as a pressure transmitting medium and calibrated the pressure with the ruby fluorescence. They reported results on 4 crystals with different oxygen content, but the exact value of δ is unfortunately not known. One of the crystals was tetragonal and not superconducting, the others were orthorhombic and superconducting. The oxygen content can be estimated to be $O_{6.0}$ for the non superconducting crystal, $O_{6.78}$, $O_{6.86}$ and $O_{6.6}$ for the superconducting crystals. (We remind to the fact, that this is only a crude estimate, that may be falsified by experimental errors!) This investigation showed, that the c-axis compressibility is clearly enhanced. The bulk moduli were calculated to be in the range 100 GPa to 118 GPa. The results are included in table 7.

Olsen et al. (1989) performed experiments on $YBa_2Cu_3O_{7-\delta}$ powder with a diamond anvil cell. They used methanol-ethanol as a pressure transmitting medium and ruby fluorescence for pressure calibration. The oxygen content of their sample cannot be estimated, as from their lattice parameter at ambient pressure the sample should hardly be superconducting, but they report a T_c of approx. 90 K. The bulk modulus calculated from their investigation is 157 GPa with B'=2.9. The result is included in table 7.

Jaya et al. (1988) report data from an experiment that was done on $YBa_2Cu_3O_{7-\delta}$ powder up to 11 GPa. They used a Bridgman anvil clamp and calibrated the pressure by NaCl. Unfortunately they did not consider the ambiguity of the reflection groups and assigned the faster shifting peak to the (020) reflection. By this wrong interpretation they found a higher linear compressibility in b-axis direction ($\beta_a = 3.0 \cdot 10^{-3}$ GPa^{-1}, $\beta_b = 3.85 \cdot 10^{-3}$ GPa^{-1}, $\beta_c = 3.39 \cdot 10^{-3}$ GPa^{-1}) and ended in a tetragonal phase.

Glazkov et al. (1988) presented results on neutron diffraction investigations on powdered $YBa_2Cu_3O_{6.9}$ and $YBa_2Cu_3O_{6.4}$. They used a high pressure cell with sapphire anvils. The pressure was carefully checked by embedding several pieces of ruby and (for some measurements) an additional piece of lead inside of the high pressure cell. Thus, the pressure distribution inside the high pressure cell could also be observed. In the tetragonal case ($YBa_2Cu_3O_{6.4}$) the linear compressibilities are determined to be $\beta_a = 2.2 \cdot 10^{-3}$ GPa^{-1}, $\beta_c = 4.5 \cdot 10^{-3}$ GPa^{-1}. For the orthorhombic material $YBa_2Cu_3O_{6.9}$ only the c-axis pressure dependence and a mean value for the a- and b- axis pressure dependences are given due to the limited resolution of the experimental setup. The compressibilities are $\beta_{a/b} = 1.7 \cdot 10^{-3}$ GPa^{-1}, $\beta_c = 2.3 \cdot 10^{-3}$ GPa^{-1}. The values are included in table 7.

Akthar et al. (1990) investigated powders of $YBa_2Cu_3O_{7-\delta}$ by means of X-ray diffraction in a diamond anvil cell using synchrotron radiation. They used a methanol-ethanol mixture as a pressure transmitting medium and calibrated the pressure by NaCl. By an apparently arbitrary assignement of the measured reflections to (hkl) values they ended in an excessive b- and c-axis decrease and claimed to see tetragonal $YBa_2Cu_3O_7$ at ≈ 8 GPa. The linear compressibilities they calculated are $\beta_a = 1.33 \cdot 10^{-3}$ GPa^{-1}, $\beta_b = 2.8 \cdot 10^{-3}$ GPa^{-1} and $\beta_c = 3.89 \cdot 10^{-3}$ GPa^{-1}.

Gavarri et al. (1990) / Carel et al. (1990) presented results on a neutron diffraction experiment on powdered $YBa_2Cu_3O_{7-\delta}$. They used Fluorinert as a pressure transmitting medium and the pressure was applied by an hydraulic press. They measured the lattice constants at ambient pressure, 0.5 GPa and 1.05 GPa. The obtained linear compressibilities are $\beta_a = 1.53 \cdot 10^{-3}$ GPa^{-1}, $\beta_b = 1.43 \cdot 10^{-3}$ GPa^{-1}, $\beta_c = 2.31 \cdot 10^{-3}$ GPa^{-1}. The results are shown in table 7.

Jorgensen et al. (1990) performed neutron scattering experiments on $YBa_2Cu_3O_{6.93}$ and $YBa_2Cu_3O_{6.6}$. They used Helium as a pressure transmitting medium, which ensures perfect hydrostatic pressure conditions. The samples were investigated at ambient pressure, 0.102, 0.207, 0.310, 0.416, 0.496 and 0.578 GPa. The values of the linear compressibilities are calculated to be $\beta_a = 2.22 \cdot 10^{-3}$ GPa^{-1}, $\beta_b = 1.65 \cdot 10^{-3}$ GPa^{-1}, $\beta_c = 4.26 \cdot 10^{-3}$ GPa^{-1} for $YBa_2Cu_3O_{6.93}$ and $\beta_a = 2.40 \cdot 10^{-3}$ GPa^{-1}, $\beta_b = 2.14 \cdot 10^{-3}$ GPa^{-1}, $\beta_c = 4.39 \cdot 10^{-3}$ GPa^{-1} for $YBa_2Cu_3O_{6.6}$. These values are given in table 7. We fitted the linear compressibilities in the way we outlined in the introduction with the data points of $YBa_2Cu_3O_{6.93}$. The resulting values are $\beta_a = 2.60 \cdot 10^{-3}$ GPa^{-1}, $\beta_b = 1.9 \cdot 10^{-3}$ GPa^{-1}, $\beta_c = 4.2 \cdot 10^{-3}$ GPa^{-1} and a bulk modulus of $B_0 = 115$ GPa with $B' = 28$. These values are included in table 7 and marked by * .

Oomi et al. (1990) report powder X-ray measurements on tetragonal, non superconducting $YBa_2Cu_3O_{7-\delta}$ and superconducting $YBa_2Cu_3O_{7-\delta}$. The high pressure was generated in an opposed anvil type apparatus using a 65:35 mixture of boron and epoxy-resin as a pressure transmitting medium. The pressure was calibrated using the NaCl equation of state. The data analysis on the superconducting sample was performed in a way, that an excessive b-axis decrease results ($\beta_a = 1.66 \cdot 10^{-3}$ GPa^{-1}, $\beta_b = 3.49 \cdot 10^{-3}$ GPa^{-1}, $\beta_c = 1.91 \cdot 10^{-3}$ GPa^{-1}) . We have already made a comment to the problems of powder X-ray investigations to distinguish between a preferential b- or c-axis decrease. The results of X-ray diffraction on single crystals (Aleksandrov et al., 1988) and neutron diffraction results (Gavarri et al., 1990) , (Jorgensen et al., 1991) show clearly an excessive c-axis decrease. To avoid confusion we do not include the results of the superconducting sample in table 7. The tetragonal sample showed linear compressibilities of $\beta_a = 2.21 \cdot 10^{-3}$ GPa^{-1} and $\beta_c = 2.40 \cdot 10^{-3}$ GPa^{-1}. These data are included in table 7.

Fietz et al. (1990) investigated tetragonal $YBa_2Cu_3O_6$ in a diamond anvil cell with NaF as a pressure transmitting medium. The pressure was calibrated using the ruby technique. The linear compressibilities obtained from these measurements are $\beta_a = 2.0 \cdot 10^{-3}$ GPa^{-1}, $\beta_c = 4.0 \cdot 10^{-3}$ GPa^{-1}. These results are shown in table 7.

Ludwig et al. (1991) remeasured $YBa_2Cu_3O_{7-\delta}$, as the first investigation (Fietz et al., 1987) was done on the first available material in 1987, which might have been of poor quality. This investigation was performed in a diamond anvil cell with NaF as a pressure transmitting medium. The pressure was calibrated using the ruby technique. The linear compressibilities obtained from this measurements are $\beta_a = 2.2 \cdot 10^{-3}$ GPa^{-1}, $\beta_b = 1.8 \cdot 10^{-3}$ GPa^{-1} $\beta_c = 3.4 \cdot 10^{-3}$ GPa^{-1}. These data are shown in table 7.

Within the experimental groups there is a clear tendency , that a higher oxygen content gives a higher bulk modulus. Nevertheless, the results of the different investigations of materials with approx. the same oxygen content are widely spread (for example 188 GPa (Gavarri) , 123 GPa (Jorgensen) for an oxygen content of approx. 6.9)

HIGH PRESSURE EXPERIMENTS ON $REBa_2Cu_3O_{7-\delta}$ with RE = Eu, Ho, Gd, Yb

Ecke et al. (1988) investigated $GdBa_2Cu_3O_{7-\delta}$ in a diamond anvil cell by X-ray diffraction. They used NaF as a pressure transmitting medium and the pressure was calibrated using the ruby technique. The linear compressibilities were calculated from the data points at ambient pressure and 14 GPa to be $\beta_a = 1.6 \cdot 10^{-3}$ GPa^{-1}, $\beta_b = 1.5 \cdot 10^{-3}$ GPa^{-1}, $\beta_c = 2.8 \cdot 10^{-3}$ GPa^{-1}. These data are included in table 8. The bulk moduli at 150K and 15 K are given to be 160 GPa and 165 GPa, respectively.

Olsen et al. (1989) performed X-ray experiments on $EuBa_2Cu_3O_{7-\delta}$ and $Ho_2Cu_3O_{7-\delta}$ powder with a diamond anvil cell. They calibrated the pressure by ruby fluorescence and used methanol-ethanol as a pressure transmitting medium. They don't give linear compressibilities, but they calculated the bulk modulus for the samples. They found $B_0 = 176$ GPa and $B_0 = 167$ GPa for $EuBa_2Cu_3O_{7-\delta}$ and $Ho_2Cu_3O_{7-\delta}$, respectively. The results are included in table 8.

Table 7. Bulk modulus and linear compressibilities of $YBa_2Cu_3O_{7-\delta}$.
Oxygen content marked by § are given by the authors,
oxygen contents marked by $\approx$ are roughly estimated values.
Values marked by * are obtained by fitting the data of the authors
in the way described in the introduction.

1. Author	Type of invest.	Mat. form	Oxygen Content	β_a [10^{-3} GPa^{-1}]	β_b [10^{-3} GPa^{-1}]	β_c [10^{-3} GPa^{-1}]	B_0 [GPa]
Fietz	X-ray	powd.	$\approx$6.9	1.7	1.7	2.3	180
Aleksandrov	X-ray	s.cryst.	$\approx$6.00	2.13	-	5.74	100
Aleksandrov	X-ray	s.cryst.	$\approx$6.60	2.69	2.00	4.25	112
Aleksandrov	X-ray	s.cryst.	$\approx$6.78	2.30	2.10	4.23	115
Aleksandrov	X-ray	s.cryst.	$\approx$6.86	2.30	2.21	4.16	118
Olsen	X-ray	powd.					157
Glazkov	neutron	powd.	§ 6.9	1.7 (a+b)	1.7 (a+b)	2.3	175
Glazkov	neutron	powd.	§ 6.4	2.2	-	4.5	112
Gavarri	Neutron	powd.	$\approx$6.9	1.53	1.43	2.31	189
Jorgensen	Neutron	powd.	§ 6.93	2.22 , 2.6*	1.65 , 1.9*	4.26 , 4.2*	123 , 115*
Jorgensen	Neutron	powd	§ 6.6	2.4 , 2.5*	2.14 , 2.3*	4.39 , 4.9*	112 , 100*
Oomi	X-ray	powd.	$\approx$6.3	2.21	-	2.40	147
Fietz	X-ray	powd.	§ 6.0	2.0		4.0	125
Ludwig	X-ray	powd.	$\approx$6.9	2.2	1.8	3.4	150

Table 8: Bulk modulus and linear compressibilities of $REBa_2Cu_3O_{7-\delta}$

1.Author	Type of invest.	Material	Mat. form	β_a [10^{-3} GPa^{-1}]	β_b [10^{-3} GPa^{-1}]	β_c [10^{-3} GPa^{-1}]	B_0 [GPa]
Ecke	X-ray	$GdBa_2Cu_3O_{7-\delta}$	powd.	1.6	1.5	2.8	155
Olsen	X-ray	$EuBa_2Cu_3O_{7-\delta}$	powd.				176
Olsen	X-ray	$HoBa_2Cu_3O_{7-\delta}$	powd.				167
Fietz	X-ray	$YbBa_2Cu_3O_{7-\delta}$	powd.	1.3	1.5	3.3	155

Fietz et al. (1990) presented results on X-ray investigations on $YbBa_2Cu_3O_{7-\delta}$ in a diamond anvil cell. The pressure was calibrated using the ruby technique and they used NaF as a pressure transmitting medium. The linear compressibilities were calculated to be $\beta_a = 1.3 \cdot 10^{-3}$ GPa^{-1}, $\beta_b = 1.5 \cdot 10^{-3}$ GPa^{-1}, $\beta_c = 3.3 \cdot 10^{-3}$ GPa^{-1}. These data are included in table 8. In addition, this paper gives the bulk moduli of $BaCuO_2$ and $Yb_2Ba_1Cu_1O_5$, which are $B_0 = 114$ GPa and $B_0 = 170$ GPa, respectively.

HIGH PRESSURE EXPERIMENTS ON $YBa_2Cu_4O_8$

The unit cell of $YBa_2Cu_4O_8$ may be derived from the unit cell of $Y_1Ba_2Cu_3O_7$ by doubling the Cu-O chain. Therefore it is closely related to $Y_1Ba_2Cu_3O_7$, but it has some interesting new properties. $Y_1Ba_2Cu_4O_8$ is free of twins (Kaldis et al.,1989). It has one of the largest pressure effects on the transition temperature, 5.5 K/GPa, known so far (Bucher et al., 1989). The oxygen in 124 is strongly bound (Karpinski et al., 1988), and the oxygen content is nearly independent of preparation conditions, if the difficult synthesis has been successful.

The synthesis of $Y_1Ba_2Cu_4O_8$ was first carried out by Kaldis et al. (1989) under high oxygen pressures above 20 MPa. Morris et al. (1989) reported the synthesis at pressures around 3 MPa and even at ambient oxygen pressure with a reaction rate enhancing addition the synthesis is possible (Cava et al., 1989). An increase of T_c up to 90 K by partly substituting Y by Ca is reported by Miyatake et al. (1989) .

Kaldis et al. (1989) performed a high pressure neutron diffraction experiment on $Y_1Ba_2Cu_4O_8$. They used Fluorinert as a pressure transmitting medium and tried to apply 1 GPa pressure to the sample. Unfortunately they had no pressure calibration and they must have estimated the true pressure incorrectly. For example, if the bulk-modulus is calculated from two data points at 165K at ambient pressure and 1 GPa a bulk modulus of approx. 1500 GPa results! The same problem is seen in a later work of the same group (Hewat et al., 1990) on $Y_2Ba_4Cu_7O_{15}$.

Ludwig et al. (1990) showed results on X-ray experiments on powdered $Y_1Ba_2Cu_4O_8$ with a diamond anvil cell. They used NaF as a pressure transmitting medium and the pressure was calibrated using the ruby fluorescence. They found linear compressibilities of $\beta_a = 2.8 \cdot 10^{-3}$ GPa^{-1}, $\beta_b = 1.0 \cdot 10^{-3}$ GPa^{-1}, $\beta_c = 4.5 \cdot 10^{-3}$ GPa^{-1}. The bulk modulus was calculated to be $B_0 = 112$ GPa with $B' = 8$. The results are included in table 9. In a low temperature experiment at 30 K, the bulk modulus was $B_0 = 122$ GPa.

Nelmes et al. (1990) investigated powders of $Y_1Ba_2Cu_4O_8$ in a neutron diffraction experiment. They used a diamond anvil cell with a methanol-ethanol pressure transmitting medium. Ruby fluorescence was used for pressure calibration. They determined the structure of the sample at 5 pressures from 0.2 to 4.65 GPa. Unfortunately they could not determine the structure at ambient pressure. Therefore they had to use literature data to fix the volume at ambient pressure. We calculated the linear compressibilities using their data to be $\beta_a = 3.0 \cdot 10^{-3}$ GPa^{-1}, $\beta_b = 0.97 \cdot 10^{-3}$ GPa^{-1}, $\beta_c = 3.55 \cdot 10^{-3}$ GPa^{-1}. This gives a bulk modulus of $B_0 = 146$ GPa with $B' = 0.77$. The very

Table 9.   Bulk modulus and linear compressibilities of $Y_1Ba_2Cu_4O_8$.
Values marked by * are obtained by fitting the data of the authors
in the way described in the introduction.

1.Author	Type of invest.	Mat. form	β_a [10^{-3} GPa^{-1}]	β_b [10^{-3} GPa^{-1}]	β_c [10^{-3} GPa^{-1}]	B_0 [GPa]
Ludwig	X-ray	powd.	2.8	1.0	4.5	112
Nelmes	neutron	powd.	3.0*	0.97*	3.55*	146*
Yamada	neutron	powd.	3.05 , 3.48*	1.45 , 1.48*	4.00 , 4.34*	118,106*

low B' may be due to the not correctly estimated volume at ambient pressure. The results are included in table 9.

Yamada et al. (1991) presented results on neutron diffraction investigations on $Y_1Ba_2Cu_4O_8$ with helium as a pressure transmitting medium in an aluminium pressure cell. They investigated the sample at 5 pressures from ambient pressure up to 0.632 GPa. They calculated the linear compressibility by performing linear least square fits. They found $\beta_a = 3.05 \cdot 10^{-3}$ GPa^{-1}, $\beta_b = 1.45 \cdot 10^{-3}$ GPa^{-1}, $\beta_c = 4.00 \cdot 10^{-3}$ GPa^{-1} and a bulk modulus of 118. The results are shown in table 9. We have calculated the linear compressibilities and the bulk modulus in the way we have described in the introduction. It results a bulk modulus of 106 GPa with a B'=39. The bulk modulus and the linear compressibilities are included in table 9 and indicated by * .

DISCUSSION OF RESULTS

Because a discussion of all results that have been shown would be a too extensive task, we concentrate mainly on $YBa_2Cu_3O_{7-\delta}$ and $YBa_2Cu_2O_8$ in this section.

The bulk moduli and the linear compressibilities of $YBa_2Cu_3O_7$ derived from the experiments are widely spread. However, if we consider only experiments made in the same group, there seems to be a systematic increase of the bulk modulus with increasing oxygen content. Bulk moduli for the lowest and highest oxygen content from different groups are given as follwos: Aleksandrov et al. (1988) 100 GPa for $O_{6.0}$ and 118 GPa for $O_{6.86}$, Glazkov et al. (1988) 112 GPa for $O_{6.4}$ and 175 GPa for $O_{6.9}$, Jorgensen et al. (1990) 112 GPa for $O_{6.6}$ and 123 GPa for $O_{6.93}$, Fietz et al. (1990) and Ludwig et al. (1991) 125 GPa for $O_{6.0}$ and 150 GPa for $O_{6.9}$. This systematic change may be expected, because of the decreasing c-axis lattice parameter with increasing oxygen content (Tarascon et al., 1987, Roth et al., 1987). The 'precompression' of the unit cell at higher oxygen content will lower the compressibility of $YBa_2Cu_3O_{7-\delta}$. Nevertheless, the differences between results on $YBa_2Cu_3O_{7-\delta}$ with a similar oxygen content are not explainable by this effect. Non-hydrostatic pressure components, errors in the pressure calibration, errors in interpreting the complex spectra of this material, oxygen ordering effects (Claus et al., 1990) or problems with the sample quality may all be explanations for these discrepancies. Experiments at very low pressures can also not solve this problem. Ultrasound investigations, as we will discuss later, are difficult to interpret and give relatively low bulk moduli. Reichardt et al. (1988) measured inelastic neutron scattering on small single crystals of $YBa_2Cu_3O_{6.85}$. From their data they calculated the elastic constants c_{11}, c_{33}, c_{66}, c_{12} and c_{13}. With those constants it is possible to calculate directly the bulk modulus B_0, which gives 128 GPa, a value that is significantly higher than the bulk moduli derived from most ultrasound experiments. So we can easily summarize: the correct bulk modulus of $YBa_2Cu_3O_{7-\delta}$ is not known, but a higher oxygen content seems to give a higher bulk modulus.

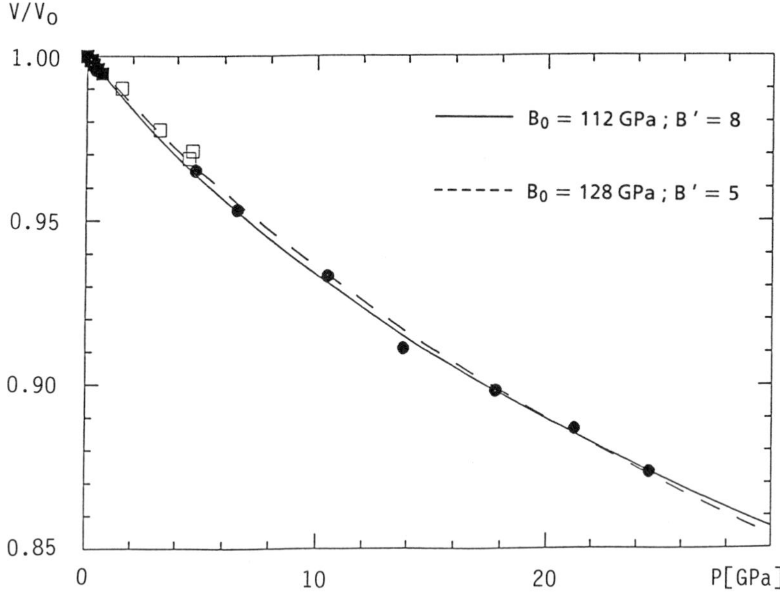

Figure 3: Normalized volume of $YBa_2Cu_4O_8$ as a function of pressure. Data points have been taken from Ludwig et al. (1990) ● , Nelmes et al. (1990) □ and Yamada et al. (1991) ■ . The curves give the normalized volume as a function of pressure according to the Birch equation. The solid curve has been calculated from the data of Ludwig et al. . The broken line has been calculated from all data points included in this figure.

A possibility to avoid the oxygen content problem of $YBa_2Cu_3O_{7-\delta}$ is to switch over to $YBa_2Cu_4O_8$, for the oxygen is strongly bonded in this compound (Karpinsky et al. 1988). There are three interesting investigations on this material, a X-ray investigation (Ludwig et al., 1990) and two neutron scattering experiments (Nelmes et al., 1990 and Yamada et al., 1991). The results of these investigations give 112 GPa, 146 GPa and 118 GPa, respectively. We have plotted the results of these three investigations together in fig.3. The broken line gives the result of the calculation of the bulk modulus according to the Birch equation, using all data points in this figure. This gives a bulk modulus $B_0 = 128$ GPa with B'=5. This high bulk modulus compared with the values given by Ludwig et al. and Yamada et al. results from the data points of the experiment by Nelmes et al. . Considering their problems with the determination of the lattice parameters at ambient pressure, we believe those data to show a too small compressibility. If the lattice parameters at ambient pressure were larger than assumed by Nelmes et al., their data would fit the curve with $B_0=112$ GPa and $B'=8$, that has been derived from the experiment of Ludwig et al. . This is supported by the b-axis lattice parameter reported by Nelmes et al., that has exactly the same value at 0.2 GPa they assumed it to have at ambient pressure. The solid line gives the normalized volume as a function of pressure according to the Birch equation with $B_0=112$ GPa and $B'=8$, which was the result of the X-ray diffraction measurement. In fig. 4 we plotted the same data again up to pressures of 7 GPa. Again, the solid line is given by the results of the X-ray measurement of Ludwig et al. . At first glance the solid line seems to fit the data of Ludwig et al. and Yamada et al. perfectly. But we have to consider, that the experimental error of Yamada et al. is much smaller than the symbols shown in fig. 4. Therefore we used again the Birch equation and fitted the data of Yamada et al. . It resulted a bulk modulus of $B_0=106$ GPa with $B'=39$, represented by the dashed curve in fig. 4. Reasonable values of B' should be in the range from 3 to 8 and it is obvious, that $B'=39$ is a much to high value for the high pressure X-ray data. On the other hand, if we assume the data of Yamada et al. to have very small error bars, $B'=39$ at low pressures must be correct. This would lead to the conclusion, that the compressibility of this high T_c superconductor is extremely pressure dependent at small pressures.

A strong pressure dependence of the elastic properties of high T_c superconductors is also supported by ultrasonic experiments on $YBa_2Cu_3O_{7-\delta}$ (Chang Fanggao et al., 1991; Cankurtaran et al. 1989), but there are some severe problems in interpreting

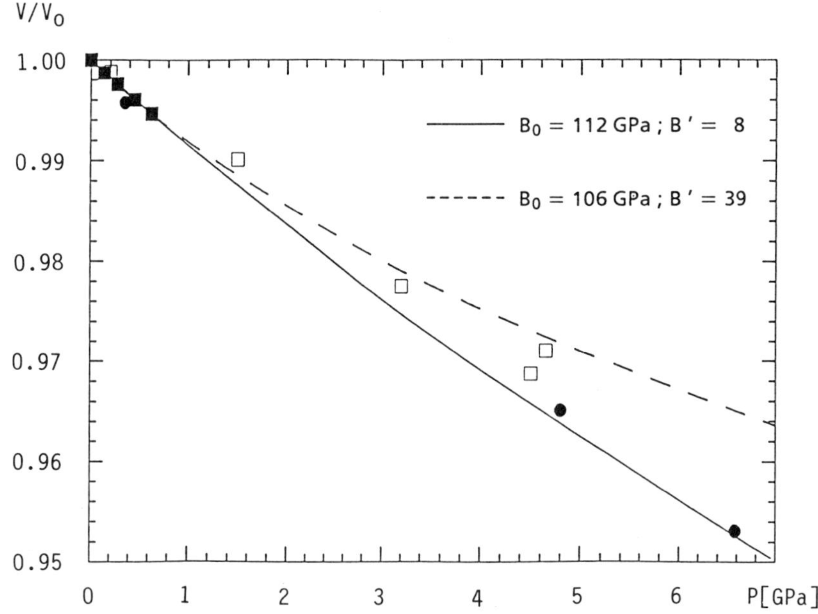

Figure 4.   Normalized volume of $YBa_2Cu_4O_8$ as a function of pressure. Data points have been taken from Ludwig et al. (1990)● , Nelmes et al. (1990)□ and Yamada et al. (1991)□ . The curves give the normalized volume as a function of pressure according to the Birch equation. The solid curve has been calculated from the data of Ludwig et al. . The broken line has been calculated from the data points of Yamada et al..

450

Table 10. Bulk modulus and linear compressibilities of several
HTC materials
Original data of the authors have been fitted
according to the Birch equation

1. Author	Material	B_0 [GPa]	B'
Yamada	$YBa_2Cu_4O_8$	106	39
Pei	$La_{1.85}Sr_{0.15}CuO_4$	142	16
Izumi	$(Nd_{1-x}Sr_x)(Nd_{1-y}Ce_y)CuO_4$	124	64
Jorgensen	$YBa_2Cu_3O_{6.93}$	115	28
Jorgensen	$YBa_2Cu_3O_{6.6}$	100	47

ultrasonic investigations. The results of two investigations at the beginning of high
temperature superconductivity were $B_0 = 38$ GPa (Round and Bridge, 1987) and
$B_0 = 54$ GPa (Lang et al., 1988). These results have been taken from polycrystalline
samples, which have a high porosity. Therefore, the authors of later papers have tried
to correct the results of ultrasonic investigations to a 'non porous matrix' (for example
Ledbetter et al., 1987 ; Bridge and Round, 1989). By these calculations the bulk
modulus was corrected to be, for example, 72 GPa (Ledbetter et al., 1987) or 115 GPa
(Tallon et al.,1988). Ultrasonic velocities measured on single crystals are higher than
sound velocities measured on polycrystalline samples (Baumgart et al., 1990, Saint-Paul
et al., 1989), which supports the necessity of corrections of results on polycrystalline
samples for a 'non porous matrix'. Holcomb and Mayo (1989) have shown that
microcracks lower the value for the bulk modulus, that can be calculated with the
results of the ultrasonic experiments. When for example $YBa_2Cu_3O_7$ is prepared, a
tetragonal to orthorhombic transition occurs on cooling the sample to room
temperature. Therefore, $YBa_2Cu_3O_7$ may include a large amount of microcracks.
Considering these problems, it is not certain, whether the strong pressure dependence
found by ultrasonic investigations results from the microstructure of the sample or
whether it is an intrinsic property of the high T_c superconductors.

If we consider only the data of Yamada et al. on $YBa_2Cu_4O_8$, we cannot exclude
the high $B' = 39$ to be a consequence of an arbitrary error. So we checked other
experiments that give enough high quality data in a low pressure range. We found
experiments on $La_{1.8}Sr_{0.2}CuO_4$ (Pei et al., 1990) , $(Nd_{1-x}Sr_x)(Nd_{1-y}Ce_y)CuO_4$ (Izumi et
al., 1990) and $YBa_2Cu_3O_{7-\delta}$ (Jorgensen et al., 1990) . All experiments were done in the
same group on Argonne's Intense Pulsed Neutron Source. The results of our
calculations using their data are shown in table 10 for simplicity. For all high Tc
superconductors the B' value is very high. Therefore we believe that this is not a
consequence of arbitrary errors. We cannot exclude a systematical error, for instance in
the pressure determination, since all measurements were done within the same
laboratory. But if those data are correct - and we think they should be - we cannot use
the simple Birch equation for describing the volume as a function of pressure over the
whole pressure range up to several ten GPa. A possibility would be to use the second

order Birch equation, that includes B_0 , B' and a constant called B'', that describes the change of B' with pressure. This would require real high quality data. Without these, the constants B_0 , B' , and B'' would be affected very strongly by the experimental errors. On the other hand it is questionable if the volume decrease under pressure of such complex and ansiotropic materials like the high T_c superconductors can be described by a relatively simple equation of state like the second order Birch equation.

REFERENCES

Akhtar, M. J., Akhtar, Z. N., Catlow, C. R. A., J. Phys. Condens. Matter. (9 Apr 1990) V. 2(14) p. 3231-3236

Akhtar, M. J., Catlow, C. R. A., Clark, S. M., Temmerman, W. M., J. Phys., C Solid State Phys. (10 Sep 1988) V. 21(25) p. L917-L920

Aleksandrov, I. V., Goncharov, A. F., Stishov, S. M., JETP Lett. (10 Apr 1988) V. 47(7) p. 428-432

Anderson,O.L., J. Phys.Chem Solids V.27 (1966)p.547

Ashbahs, H., Hellner, E., Soleimani, A. , Superconduct. in power engin. , status, concepts, new aspects, Düsseldorf (Germany, F.R.) VDI-Verl. Ber. 733 (1989) p.347-352

Baumgart, P., Blumenroeder, S., Erle, A., Hillebrands, B., Guentherodt, G., Schmidt, H., Solid State Commun. (Mar 1989) V. 69(12) p. 1135-1137

Bednorz, J.G., Müller, K.A., Z.Phys.B 64, 189 (1986)

Birch, F. , Phys. Rev. 71, 809 (1947)

Block,S., Piermarini,G.J., Munro,R.G., Wong-Ng, W., Adv. Ceram. Mat. V.2 No.3b (1987)

Bolef,D.J., J. Appl. Phys. 32 (1961) 100/5, 102

Bridge, B., Round, R., J. Mater. Sci. Lett. (Jun 1989) V. 8(6) p. 691-694

Bridgman,P.W., Pr. Am. Acad. 58 (1922/23) p.219

Bucher,B., Karpinski,J. , Kaldis,E., Wachter,P. , Physica C 157, 478 (1989)

Bucher, B., Karpinski, J., Kaldis, E., Wachter, P., J. Less-Common Met. (15 Oct 1990) V. 164-165(1/2,pt.A) p. 20-30

Cankurtaran, M., Saunders, G. A., Willis, J. R., Al-Kheffaji, A., Almond, D. P., Phys. Rev., B. (1 Feb 1989) V. 39(4) p. 2872-2875

Carel, C., Gavarri, J.R., Perrin, C., Pena, O., Vettier, C., C.R. Acad. Sci. t 309, II (1989) p.1889-1893

Cava,R.J., van Dover,R.B., Batlogg,B., Rietman,E.A., Phys.Rev.Lett.58 (1987) 408

Cava, R. J., Batlogg, B.,Krajewski, J. J., Peck, W. J. Jr., Rupp, L. W. Jr., Hewat, A. W., Hewat, E. A., Marezio, M.,Rabe, K. M., Physica C. (15 Feb 1990) V. 165 (5/6) p. 419-433

Cava,R.J., Krajewski, J.J., Peck Jr,W.F., Batlogg, B., Rupp Jr,L.W., Fleming,R.M. , James, A.C.W.P.,Marsh,P. , Nature 338, 328-330 (1989)

Chang Fanggao, Cankurtaran, M., Saunders, G. A., Al-Kheffaji, A., Almond, D. P., Ford, P. J., Ladds, D. A., Phys. Rev., B. (1 Mar 1991) V. 43(7,pt.A) p. 5526-5537

Chu , C. W., Hor, P.H., Meng, R.L., Gao, L., Huang, Z.J., Wang,Y.Q. Phys Rev Lett. 58, 4 (1987) 405

Claus, H., Yang, S., Paulikas, A.P., Downey, J.W., Veal, B.W., Physica C 171 (1990) 205-210

Decker, D.L. , J. Appl. Phys. 42, 3239 (1971)

Dietrich, M. R., Fietz, W. H., Ecke, J., Politis, C., Proceedings of the 18. International conference on low temperature physics (LT-18), Kyoto (Japan), 20-26 Aug 1987 Nagaoka, (Japan). Jap. J. Appl. Phys. V. 26(3)(1987) p. 1113

Ecke, J., Fietz, W. H., Dietrich, M. R., Wassilew, C. A., Wuehl, H., Fluekiger, R. , Physica C. (Jun 1988) V. 153-155(pt.1) p. 954-955 (HTSC-M2S), Interlaken, 28 Feb - 4 Mar 1988

Fietz,W.H., Wassilew,C.A., Ludwig,H.A:, Obst,B., Politis,C. Dietrich,M.R., Wühl,H. High pressure research Vol.423 (1990)

Fietz, W. H., Wassilew, C. A., Ludwig, H. A., Obst, B., Dietrich, M. R., Wuehl, H. , Sov. J. Low Temp. Phys. (May 1990) V. 16(5) p. 358-359

Fietz, W. H., Ludwig, H. A., Wolf,T., Wühl,H., Dietrich, M. R., Proceedings of the 3 German Soviet Bilateral Seminar on HTSC, Karlsruhe, Oct. 8-12 (1990) p. 655

Fietz, W. H., Ludwig, H. A., Wolf,T., Wühl,H., Dietrich, M. R., Proceedings of the 28 Congress of EHPRG in Bordeaux, July 8.-13. (1990)

Fietz, W. H., Wassilew, C. A., Ewert, D., Dietrich, M. R., Wuehl, H. , Hochheimer, D., Fisk, Z., Phys. Lett., A. (11 Dec 1989) V. 142(4/5) p. 300-306

Fietz, W. H., Dietrich, M. R., Ecke, J., Z. Phys., B: Condens. Matter. V. 69(1) (1987) p. 17-20

Fleming, R.M., Batlogg, B., Cava,R., Rietman, E.A., Phys.Rev B 35(15) , 7203 (1987)

Forman, R.A., Piermarini, G.J., Barnett, J.D. Block, S. , Science 176, 284 (1972)

Gavarri, J. R., Carel, C., Physica C. (15 Mar 1990) V. 166(3/4) p. 323-328

Gavarri, J. R., Monnereau, O., Vacquier, G., Carel, C., Vettier, C., Physica C. V. 172(3/4) (15 Dec 1990) p. 213-216

Glazkov, V.P., Goncharenko, I.N., Somenkov, V.A., Sov. Phys. Sol. State 30(12) (1988) p. 2127

Han,S., Zhang,Y., Han,H., Liu,Y., Physica C 156, 113(1988)

Hatton, P.D., Howard, C.J., McMahon, M., Nelmes, R.J., Wilding, N. , High Pressure Research, Vol. 4 (1990) p. 450

Hewat, A. W., Fischer, P., Kaldis, E., Karpinski, J., Rusiecki, S., Jilek, E., Physica C. (15 May 1990) V. 167(5/6) p. 579-590

Howard, C. J., Nelmes, R. J., Vettier, C., Solid State Commun. (Jan 1989) V. 69(3) p. 261-264

Izumi, F., Jorgensen, J. D., Lightfoot, P., Pei Shiyou, Yamada, Y., Takayama-Muromachi,E., Matsumoto, T. , Physica C. (1 Dec 1990) V. 172(1/2) p. 166-174

Izumi,F., Takayama-Muromachi,E., Fujimori,A., Kamiyama,T., Asano,H., Akimitsu,J., Sawa, H., Physica C, 158 (1989) 261

Jayaraman, A. , Rev. Mod. Phys. V.55 No 1, p.65-108 (1983)

Jaya, N. V., Natarajan, S., Natarajan, S., Subba Rao, G. V., Solid State Commun. V. 67(1) (Jul 1988) p. 51-54

Jorgensen, J. D., Pei Shiyou, Lightfoot, P., Hinks, D. G., Veal,B. W., Dabrowski, B., Paulikas, A. P., Kleb, R., Brown, I. D. , Physica C. (15 Oct 1990) V. 171(1/2) p. 93-102

Kaldis, E., Fischer, P., Hewat, A. W., Hewat, E. A., Karpinski, J., Rusiecki, S., Physica C. (15 Jul 1989) V. 159(5) p. 668-680

Karpinski,J., Kaldis,E., Jilek,E., Rusiecki,S. , Bucher,B. , Nature 336, 660-662 (1988)

Kishio,K., Kitazawa,K., Kanbe,S., Yasuda,I., Sugii,N., Takagi,H., Uchida, S., Fueki,K., Tanaka,S., Chem.Lett.(1987) 429

Kubiak, R., Westerholt, K., Pelka, G., Bach, H., Khan, Y., Physica C. (1 Apr 1990) V. 166(5/6) p. 523-529

Lang, M., Lechner, T., Riegel, S., Steglich, F., Weber, G., Kim, T. J., Luethi, B.,
Wolf, B., Rietschel, H., Wilhelm, M., Z. Phys., B. (Jan 1988) V. 69(4) p. 459-463

Ledbetter, H. M., Austin, M. W., Kim, S. A., Lei, Ming, J. Mater. Res. (Nov 1987)
V. 2(6) p. 786-789

Ludwig, H. A., Fietz, W. H., Dietrich, M. R., Wuehl, H., Karpinski, J., Kaldis, E.,
Rusiecki, S., Physica C. (1 May 1990) V. 167(3/4) p. 335-338

Ludwig, H. A., Fietz, W. H., Grube, K., to be published

Maeda,H., Tanaka,Y., Fukutomi,M., Asano,T., Jpn.J. Appl. Phys. 27 (1988) L209

Miyatake,T. , Gotoh, S. , Koshizuka,N., Tanaka,S. , Nature 341 41-42 (1989)

Moret, R., Goldman, Alan I., Moodenbaugh, A., Phys. Rev., B. (1 May 1988) V. 37(13)
p. 7867-7868

Morris,D.E., Asmar,N.G., Nickel,J.H., Wei,J.Y.T., Post,J.E., Physica C 159 (3), 287 (1989)

Morosin,B. , Ginley,D.S., Venturini,E.L., Hlava,P.F., Baughman,R.J., Kwaka, J.F.,
Schirber,J.E., Physica C 152, (3), 223 (1988)

Munro, R. G., Block, S., Piermarini, G. J., J. Appl. Phys. (01 Oct 1984) V. 56(7) p. 2174

Murayama,C., Mori,N., Yomo,S., Takagi,H., Uchida,S., Tokura,Y., Nature 339 (1989) 293

Murnaghan, F.D., Proc. Natl. Acad. Sci. U.S. 30, 244 (1944)

Nelmes, R. J., Loveday, J. S., Kaldis, E., Karpinski, J., Physica C. (15 Dec 1990) V. 172(3/4)
p. 311-324

Nelmes, R. J., Wilding, N. B., Hatton,Caignaert, V., Raveau, B., P. D., McMahon, M. I.,
Piltz, R. O., Physica C. (15 Mar 1990) V. 166(3/4) p. 329-333

Olsen, J. S., Steenstrup, S., Johannsen, I., Gerward, L., Z. Phys., B. (Aug 1988) V. 72(2)
p. 165-168

Oomi, G., Suenaga, K., Physica B. (Apr 1990) V. 163(1-3) p. 255-258 (ICPHCES),
Santa Fe, NM (USA), 11-15 Sep 1989

Pei, S., Jorgensen, J.D., Hinks, D.G., Dabrowski, B., Lightfoot, P., Richards, D.R. ,
Physica C 169 (1990) 179-183

Politis,C., Geerk,J., Dietrich,M., Obst,B. , Z.Phys B 66 (1987) 141

Reichardt, W., Pintschovius, L., Hennion, B., Collin, F., Supercond. Sci. Technol. (1988)
V. 1(4) p. 173-176

Round, R., Bridge, B., J. Mater. Sci. Lett. (Dec 1987) V. 6(12) p. 1471-1472

Saez Puche, R., Norton, M. , Glaunsinger, W.S. , Mat. Res. Bull. 17, 1523 (1982)

Saint-Paul, M., Tholence, J. L., Noel, H., Levet, J. C., Potel, M., Gougeon, P.,
Solid State Commun. (Mar 1989) V. 69(12) p. 1161-1163

Saint-Paul, M., Henry, J. Y., Solid State Commun. (Nov 1989) V. 72(7) p. 685-687

Sato-Sorenzen,Y., J. Geophys. Res. V.88 B.4 (1983) p.3543

Takahashi, H., Murayama, C., Mori, N., Ustumi, W., Yagi, T., Yomo, S., Proceedings
of the Jap. J. of Applied Physics 1987 p. 1109-1110

Takahashi, H., Murayama, C., Yomo, S., Mori, N., Kishio, K., Kitazawa, K., Fueki, K.,
Jpn. J. Appl. Phys., Pt. 2. (Apr 1987) V. 26(4) p. 504-505

Tallon, J. L., Schuitema, A. H., Tapp, N. E., Appl. Phys. Lett. (8 Feb 1988) V. 52(6)
p. 507-509

Tarascon, J.M., Greene, L.H., McKinnon, W.R., Hull, G.W., Geballe, T.H., Science,235,
1373 (1987)

Terada, N., Ihara, H., Hirabayashi, M., Senzaki, K., Kimura, Y., Murata, K., Tokumoto,
M., Shimomura, O. , Kikegawa, T., Jpn. J. Appl. Phys., Pt. 2. (Apr 1987) V. 26(4)
p. 510-511

Tokura,Y., Torrance,J.B.,Huang, T.C., Nazzal,A.I., Phys. Rev. B V.38(10) p. 7156 (1988)

Wijngaarden, R. J., Griessen, R., Studies of high temperature superconductors. Advances in research and applications. Narlikar, A. (National Physical Lab., New Delhi (India)) (ed.) Commack, NY (USA): Nova Science Publ. 1989 p. 29-77 of 384 p. Stud. High Temp. Supercond.V. 2

Yamada, Y., Jorgensen, J. D., Pei Shiyou, Lightfoot, P., Kodama, Y. , Matsumoto, T., Izumi, F., Physica C. (1 Feb 1991) V. 173(3/4) p. 185-194

Yoneda, Takuji, Mori, Yoshihisa, Akahama, Yuichi, Kobayashi, Mototada, Kawamura, Haruki, Jpn. J. Appl. Phys., Pt. 2. (Aug 1990) V. 29(8) p. 1396-1398

Zhao, Z.X., Int. J. Mod. Phys. B1 ,187 (1987)

LIGHT SCATTERING IN HIGH T_C SUPERCONDUCTORS

E. Liarokapis

Physics Department, National Technical University
Athens 157 73, Greece

ABSTRACT

Raman and Infrared experimental results are reviewed for the Y-based high T_c superconducting compounds. The effects of elemental substitutions for the Y and Cu sites, the oxygen stoichiometry, and the hydrostatic high pressure, on the lattice vibrations are discussed. In addition, the use of Raman spectroscopy for the assessment of the microcrystalline orientation in bulk systems is outlined.

1. INTRODUCTION

Since the discovery of the high T_c superconductors,[1,2] a large number of light scattering experiments has been carried out,[3,4] in order to elucidate the property of superconductivity. It is now clear that Raman studies can provide important information about the oxygen stoichiometry,[5] the phonon coupling with the carriers,[6-8] the character of the superconducting gap,[9,10] the charge transfer from the chains to the copper plains,[11,12] the transition from the orthorhombic to the tetragonal phase upon elemental substitution,[13] or even about the amount of bulk texture in ceramic superconductors.[14] IR spectroscopy has also been used for the detection of the superconducting gap[15-17] and to determine the preferred orientation of microcrystallites in $YBa_2Cu_3O_{7-x}$ samples.[14]

Elemental substitutions have been investigated extensively[3] in an attempt to clarify or improve the superconducting behaviour, to determine the correct association of the phonons with lattice vibrations,[3,4] for the study of the electron-phonon interaction,[9,10] for the understanding of the role of each element in the new compounds,[3] etc. In connection with high pressure measurements of the transition temperature,[18] they have led to the discovery of the 92K superconductors.[2] On the other hand, light scattering studies of superconductors under high pressure are rather limited,[19-22] partially due to the very weak intensity of the phonon modes. It should be noted here, that internal pressure introduced in the lattice by an elemental substitution or a deficiency in oxygen is quite similar to that developed by a hydrostatic pressure, and therefore, the relative light scattering measurements can be interpreted as induced by an external pressure.[18]

High pressure has also been used to synthesize new compounds which are otherwise unstable at ambient temperatures,[23,24] and for the alignment of microcrystallites in ceramic superconductors.[25] This procedure, when repeated a few times, increases considerably the amount of bulk texture,[26] as was discovered by x-ray[27] and Raman or IR measurements.[14]

In the following, recent data on elemental substitutions for the Cu and Y sites of the YBCO superconductor will be reviewed, in connection with oxygen stoichiometry, high pressure and other measurements. Finally, the polarized Raman technique for the investigation of the bulk texture in the superconductors will be outlined. The method will be described together with some analysis that relates the measured relative intensities to the degree of preferred orientation in the YBCO and Bi-type superconductors.

2. THE YBCO SUPERCONDUCTOR

The YBCO system of superconductors has the general formula $Y_2Ba_4Cu_{6+n}O_{14+n-y}$ with n=0,1,2. As shown in Fig.1, for n=0 (Y123) it consists of a single Cu-O chain, for n=2 (Y124) another chain is added between the BaO planes shifted by b/2 lattice parameter in the b-direction, while the n=1 compound (Y247) corresponds to alternating units of the other two cases.[23,24,28] The single chain system is orthorhombic (usually in a twinned state) at oxygen concentrations y>6.35, and it becomes tetragonal at lower concentrations.[29] It belongs to the crystallographic group Pmmm and, since it has an inversion center, the 33 optical modes at the Brillouin zone center are either Raman ($5A_g+5B_{2g}+5B_{3g}$) or IR active ($7B_{1u}+7B_{2u}+7B_{3u}$).[3,4] Lattice dynamic calculations have shown that all $5A_g$

modes involve motion of the atoms along the c-axis and one of them is a B_{1g}-like mode, with the characteristic $a_{xx}=-a_{yy}$ and $a_{zz}=0$ components of the Raman tensor.[30] This mode, observed at $340cm^{-1}$, involves the out of phase vibrations of the O3 atoms surrounding the Cu2 atom, and was found to soften with temperature due to its strong coupling to the electronic continuum.[7] From the four other modes of A_g symmetry, the $500cm^{-1}$ phonon involves the stretching of the O4 atom along the c-axis, the $440cm^{-1}$ mode is associated with the in-phase vibrations of the O3 atoms around the Cu2 atom, the $150cm^{-1}$ is due to the vibrations of the Cu2 atoms, and the low frequency mode at $116cm^{-1}$ corresponds to the motion of the Ba atoms[4]. Phonons of B_{2g} and B_{3g} symmetry have been observed at 205 and $302cm^{-1}$ (Ref.11), and at 70, 83, 526, $579cm^{-1}$ (Ref.31) but they have not been confirmed. The last mode at $579cm^{-1}$ was also attributed to a defect in-duced IR-active mode.[4] Many IR-active modes have been observed,[3] but only the five at 152 ± 3, 191 ± 8, 277 ± 7, 312 ± 6, and $565\pm14cm^{-1}$ have been unam-biguously identified as lattice modes.[3] The first one was assigned to the Cu1-O1 chain vibration (Fig.1),[32,33] the mode at $191\pm8cm^{-1}$ was shown to involve the Y atoms,[32] the phonons at 277 ± 7 and $312\pm6cm^{-1}$ were found to soften with temperature[34,35] and were attributed to the vibrations of the O2,3 atoms (Fig.1),[3] while the high-frequency mode was associated with the stretching of the O4 oxygen.[32]

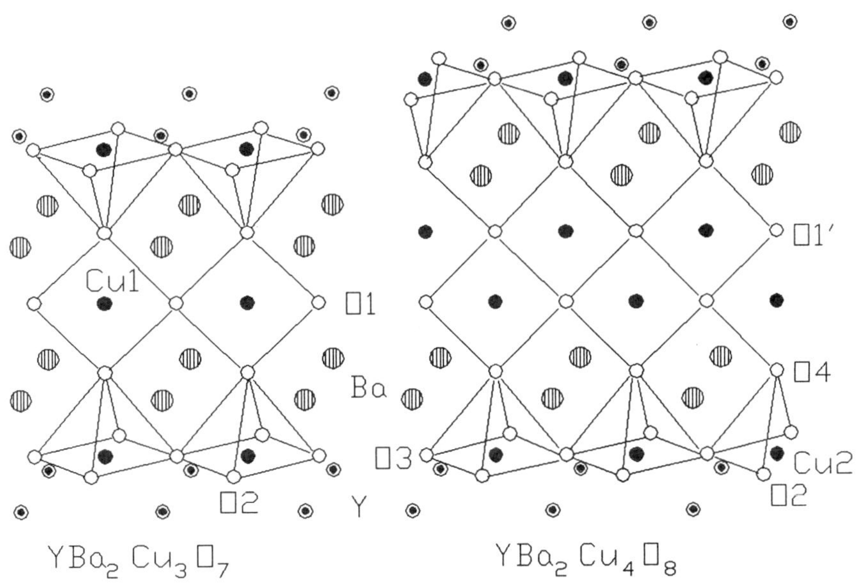

Fig.1. The crystal structure of Y123 (left side) and Y124 (right side) compounds.

The new superconducting phases (n=1,2) have been obtained in bulk form at high oxygen pressures from the study of the P-T-x phase diagram of the YBCO system, for pressures up to 3Kbar and temperatures up to 1700°C (Refs. 23,24). They have transition temperatures at 40K (n=1)[24] and 81K (n=2),[23] and unlike the Y123 system the pure double chain system Y124 is thermally stable up to 850°C.[23] Furthermore, the presence of two chains in the Y124 superconductor stabilizes the oxygen stoichiometry and produces a twin-free structure.[23] The compound belongs to the Ammm group,[36] with a unit cell about twice that of Y123. Besides the lattice modes of the Y123 compound described above, the Y124 stoichiometry has additional modes since the atoms located at the chains are not inversion centers, so are thus involved in Raman active vibrations. It acquires 42 optical modes, from which $7A_g+7B_{2g}+7B_{3g}$ are Raman active and $7B_{1u}+7B_{2u}+7B_{3u}$ IR active.[36] Both n=1,2 compositions have been prepared and analyzed quite recently.[23,24,28,36] The optical measurements of the n=2 untwinned composition has shown some interesting results in connection with elemental substitutions.[37]

Cu substitution by transition metals

The replacement of Cu by transition metals is of particular interest because the Cu-O bonding is an essential feature for the superconductivity, although the presence of two types of copper sites in the planes and the chain (Fig.1) complicates the analysis. It is now understood that elements that substitute on the Cu2 site strongly effect the superconducting transition temperature, while elements that substitute on the Cu1 site have a much less pronounced influence, thus implicating the CuO_2 planes as responsible for superconductivity in this class of materials.[38] It has also become clear that the degree of order on the oxygen sites in the Cu-O chain is of paramount importance, since the disordered structure with overall tetragonal symmetry is not superconducting and only the orthorhombic structures, with well-developed parallel Cu-O chains are superconductors with a T_c depending on the overall oxygen content.[39] However, for certain dopants, in particular Fe, Co, Ga and Al, the structure becomes tetragonal with increasing dopant concentration, although this transformation does not lead to an abrupt loss of superconductivity.[38,40-43] High resolution electron microscopy has shown that in $YBa_2(Cu_{1-x}Fe_x)_3O_y$ the tetragonal structure observed at low dopant concentration by x-ray and neutron scattering actually consists of small orthorhombic microdomains with mutually orthogonal orientations.[44-46]

An independent indication for the existence of an orthorhombic microdomain structure was obtained recently from Raman measurements on Fe substitution samples.[13] A similar study carried out with ir-reflectence measurements,[47] supplied additional information on this subject. In the typical Raman spectra of Fig.2, four out of the five A_g phonons are observed at 116, 150, 440, and $500cm^{-1}$ in the zz-configuration and the fifth phonon at $340cm^{-1}$ in the xx-configuration as expected from their symmetry.[4] From Fig.2a it is observed that the two high frequency phonons at 440 and $500cm^{-1}$ shift towards merging at a different rate and that the

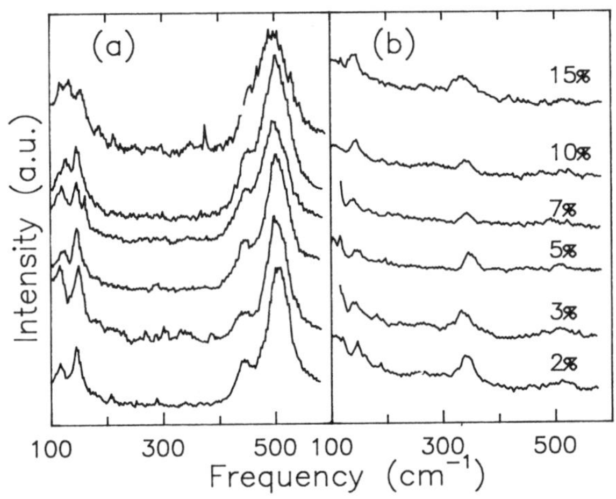

Fig. 2. Polarized microraman spectra of $YBa_2(Cu_{1-x}Fe_x)_3O_y$ for various concentrations and the zz (a) and xx (b) scattering configurations (from Ref.13).

$500cm^{-1}$ mode seems to broaden considerably with iron concentration. The phonon at 340 cm^{-1} seems not to be shifting as in the case of nickel[48] and cobalt[48-50] substitutions. The zz-component of the $150cm^{-1}$ mode does not seem to be shifting also and its intensity remains practically constant (Fig.2a). The xx-component shows a downward shift, a broadening, and possibly a slight increase in its intensity relative to that of the $340cm^{-1}$ mode (Fig.2b). Finally, the $116cm^{-1}$ mode, appears as a double peak with increasing Fe doping (Fig.2a), as it will be discussed below.

Though the shifting of the two high frequency modes is very similar to that of oxygen deficient pure $YBa_2Cu_3O_y$ samples,[51-53] the behaviour of the other modes does not seem to follow the trend associated with oxygen deficiency. Neutron scattering measurements[43,54] have shown that the substitution with iron slightly increases the amount of oxygen due to the higher valency of iron compared to copper.[43] It can therefore be concluded that the strong shift observed for the Cu1-04 stretching mode ($500cm^{-1}$), is not due to oxygen deficiency but to the substitution of the Cu1 atom by Fe (Fig.1). The behaviour of the two phonons at $340cm^{-1}$ and $440cm^{-1}$ upon substitution, cannot be conclusive for the possibility of Fe occupying the Cu2 atomic site, since they involve the vibrations of the 03 atoms (Fig.1). Direct information can only be obtained from the lattice mode associated with the vibration of the Cu2 atom. The basic characteristics of this mode are an unshifted zz-component and a xx-component shifting with x (Fig.2).[13] Compared with the pure $YBa_2Cu_3O_y$ material, these two components behave as the corresponding phonon of the orthorhombic phase as it appears in the zz-configuration and another of the tetragonal phase which shifts to lower frequencies in xx-polarization. This can be described as a two-mode behaviour with different polarization selection rules for the two modes. The conclusion would be that iron substitutes also on the Cu2 sites and that there is a coexistance of two phases, one orthorhombic and another tetragonal.[55] Dufour et al.[47] have arrived at the same conclusion, that Fe substitutes the Cu2 site (Fig.1), on the basis of the changes observed in IR-reflectance spectra with increasing doping for the Cu2-02,3 bending modes at 277 and $312cm^{-1}$. On the other hand, the phonon of Ba vibrations at $116cm^{-1}$ has shown a novel structure in the Raman measurements (Fig.2). Upon substitution with Fe, a new mode appears at $126cm^{-1}$ (for x=5%) in the zz-configuration, increasing in intensity with doping (Fig.2). A similar mode has also been observed in oxygen deficient samples[55] and in Y substitution by rare earth materials.[12]

Substitutions of Y by rare earth materials

Elemental substitution in the YBCO superconductor has been investigated extensively in attempts to clarify or improve the superconducting behaviour of this system.[3] It soon became clear that the substitutions of Y by rare earths (RE) can cause only minor variations in the transition temperature, with the exception of Pr which suppressess superconductivity.[3,56] Nevertheless, RE substitutions have helped to identify the phonons with lattice vibrations,[3,4] and provided the first evidence from Raman measurements of the superconducting gap[9,10] and the charge transfer from the copper plains to the chains.[11,12,57] Upon substitution of the Y atom by another element, distortions might be developed in the lattice. This will mainly happen when Y is replaced by an ion of quite different size that will produce internal strains distributed inhomogeneously at the various atomic sites. The study of the lattice vibrations upon substitution, provides information for these internal strains. Fig.3 shows the dependence of all A_g modes on the ionic radius of the substituting RE

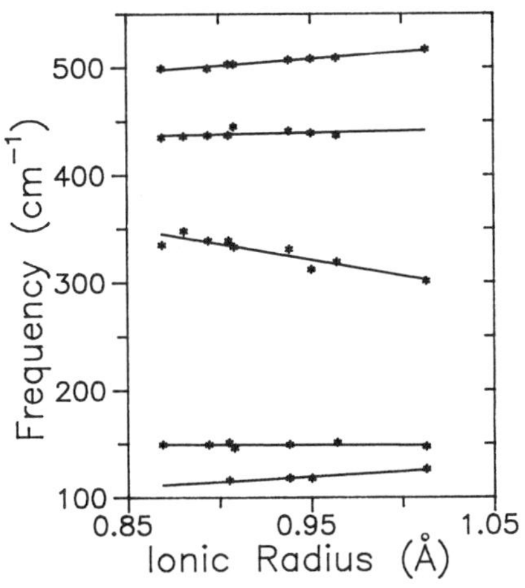

Fig.3. The dependence of the A_g phonon frequencies on the ionic radius of the RE substitution (data from Refs.10,11,57,58).

element.[10,11,57,58] For the low frequency vibration of Ba, the data are limited to only few elements. Based on the variation of the interatomic distances upon substitution,[56] the phonon frequency dependences on the ionic radius of the substituting element and the differences with the high pressure data[19,20] can be anticipated. Hydrostatic high pressure measurements on Y123 have shown a typical hardening of all A_g modes except for the out-of-phase B_{1g}-like phonon which was split into two modes with quite different dependence on pressure.[19,20] Such a splitting was not observed for Pr123 (Ref.20).

The hardening of the phonon that corresponds to the vibration of the O4 atom, agrees quantitatively with the similar behaviour under high pressure, and it is due to the compression of the Cu1-O4 axis.[56] The stability of the 150 cm^{-1} phonon, involving the vibrations of the Cu2 atoms, is rather expected since the internal strains do not change the Cu2-Cu2 distance upon substitution,[56,58] while under a hydrostatic pressure this distance will be decreased.[19,20] On the other hand, the two modes that arise from the motion of the O2,3 oxygens at 340cm^{-1} (out of phase) and 445cm^{-1} (in-phase) do not behave similarly, as it would be expected from their assignment to the vibrations of the same atoms.[4] So, the first mode undergoes a softening while the other is independent of the substitution.[58] By assuming that the Cu-O length increases as the neutron data show,[58] the stability of the in-phase mode cannot be explained. For the Pr substituted samples (but not for Eu123, Ref.57), the same mode splits in two with increasing concentrations, closely reminding a two-mode behaviour.[12] Such a behaviour of the Pr123 compound was explained in Refs.12,57,59 as a signature of hole localization in the Ba atoms, though it is not clear if this is the cause or the effect of another mechanism.

An interesting characteristic is observed in the low frequency mode that involves the Ba motion along the c-axis.[4]   As seen in Fig.2 for $YBa_2(Cu_{1-x}Fe_x)_3O_y$ superconductors, this phonon is split into two modes (at 116 and 126cm^{-1}) showing a typical two-mode behaviour.[13] A quite similar behaviour was observed in oxygen deficient Y123 systems,[60] where in the tetragonal phase of $YBa_2Cu_3O_6$ the mode at 116cm^{-1} disappears and only the mode at 126cm^{-1} remains. While for Eu123 this mode is stable,[57] for Pr123 it seems to be shifting towards 126cm^{-1}, but one cannot be sure if this is a real shift or the appearance of a new mode. In all these three different cases, the appearance of a strong peak at 126cm^{-1} was accompanied by a reduction of T_c or the loss of superconductivity. One could possibly assign this to a defect induced mode, but its appearance even in the tetragonal phase of $YBa_2Cu_3O_6$ case, where all oxygen is lost from the chains and the system is more symmetric, does not agree with this explanation. Another possibility is to assign the 126cm^{-1} mode to the tetragonal phase, based on the data on Fe-substituted[13] and oxygen deficient samples[55], but this assignment would disagree with the light scattering data from the Pr123 compound which remains in the orthorhombic phase. Recently, it was proposed that upon substitution of Y by a larger ion, such as Pr, the Ba atoms move to a slightly different position, due to the strains introduced in the unit cell from the internal pressure.[61]

The same distortion is also expected in oxygen deficient samples[61], and probably in the Fe-substituted compounds. The existence of two positions for the Ba atoms could explain the data, but it should be also related to the pressure dependence of superconductivity in these compounds. The study of the low frequency phonons of Ba substitutions with related elements, of the effects of uniaxial along the c-axis stresses, or of the similar phenomena developed on La, Bi, or Tl-based superconductors, could help test the above hypothesis.

3. BULK TEXTURE DETECTION BY RAMAN SCATTERING

While the transition temperature in the new superconductors has been considerably increased, their technological applications will require a two to three orders of magnitude higher critical current density J_c, than is presently realised in bulk polycrystalline samples (10^2-$10^3 A/cm^2$).[62] The difference from the large value of J_c obtained in single crystals (10^5-$10^6 A/cm^2$)[63] is partially due to the random orientation of microcrystallites in the ceramic samples. Various methods have been examined which would increase the microcrystalline alignment, such as a mechanical pressure that aligns the c-axis of the crystals with the compression axis,[25] a texture growth from the melt[64], an external magnetic field,[65] and the sequential steps of the solid state reaction method.[26] In most of the cases, x-ray measurements have shown some improvement in the texture of the samples, but this was assumed to be confined only to the surface.[66]

Raman spectroscopy has established that, at least the sequential solid state reaction procedure[26] produces bulk microcrystalline orientation.[14] It was found that the orientations of microcrystallites inside the material detected by the Raman[14] and the critical current density[67] measurements track the surface orientations as discovered by the variations in the x-ray intensities.[27] The Raman technique is based on the measurement of the depolarization ratio introduced on the scattered light by the non-random distribution of the microcrystallites.[14] The experimental details are given in Ref.14 with the main characteristics that the spectra were obtained from the freshly cut side faces of the samples, and that the analyzer or the scrambler were used only occasionally for checking the efficiency of the device in the two polarizations, since they were strongly attenuating the weak Raman signal. In the absence of the analyzer and the scrambler, the two polarizations were mixed,[14] due to the difference in the efficiency of the spectrometer for the light polarized in a direction perpendicular or parallel to the slits. Typical Raman spectra for the samples processed with the solid state reaction technique[14] once (B) or twice (C) are shown in Fig.4, where a continuous line was drawn through the data to guide the eye. It is observed that, while the $340 cm^{-1}$ phonon intensity is slightly reduced in the horizontal (defined as the z-axis perpendicular to the slits) polarization of the incoming light, the intensity of the $500 cm^{-1}$ is increased by a factor of two ($I_h/I_v=2.1$) for this polarization. This remarkable increase has been observed on a series of samples processed several times, and it is in agreement with x-ray measurements of the relative Bragg reflection intensities.[67] Calculations show that it cannot be attributed to the difference in the total efficiency of the optical path inside the double spectrometer, which for a truly random orientation of the microcrystals and for the A_g symmetry would amount to <30%, as observed from the front faces of the samples.[14] This is an indication that the microcrystallites are preferrentially oriented both at the surface and in the bulk of the compound, with their c-axis perpendicular to the front compressed face of the sample, as a result of the multiple mechanical treatment during preparation.[27]

Similar variations have been observed in the scattering intensity of the $630 cm^{-1}$ phonon of the Bi-type superconductors,[68] which are due to the same polarization selection rules for the $630 cm^{-1}$ mode, as for the $500 cm^{-1}$ phonon of the YBCO system.[4] This is a clear indication that the microcrystallites in the Bi-type superconductors are also partially oriented but to a degree ($I_h/I_v=1.8$)[68] smaller than in the YBCO superconductors.

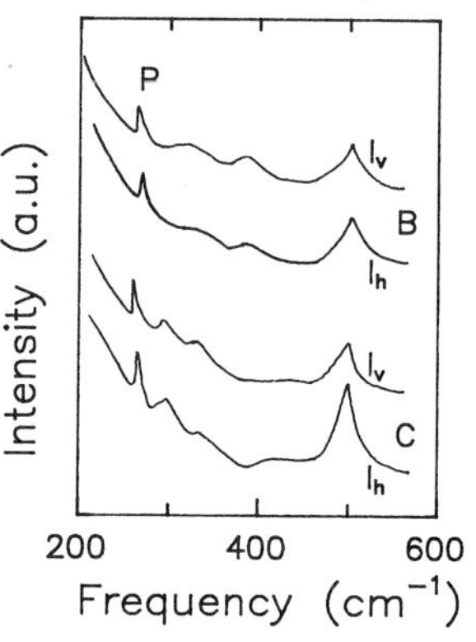

Fig.4   Raman spectra obtained from the side faces of the samples processed once (B) and twice (C), with the solid state reaction technique.[14] I_v, I_h correspond to scattering with incoming light polarized in the vertical (parallel) or horizontal (perpendicular to the slits) axis. P defines plasma lines.

In the case of a polycrystalline scatterer the relative scattering intensity can be evaluated using an averaging procedure over all microcrystals.[69] When the microcrystallites are partially aligned, the scattering intensity will depend on the polarization of the incoming light. For phonons of A_g symmetry the only nonzero components of the Raman tensor in the crystallographic axis will be the a_{xx}, a_{yy}, a_{zz}.[69] By an appropriate transformation to the fixed system of coordinates,[70] the total scattering intensity can be calculated from the incoherent sum of the scattering light from each microcrystallite,[69] taking into consideration the detection efficiency of the detection system for each polarization. The anisotropic distribution of the small platelets in the high T_c superconductors can be described by the angle θ of the c-axis with respect to the coordinate z-axis (the direction of mechanical compression during preparation). As for the distribution function, the commonly used (among x-ray spectroscopists) March model has been assumed,[71]

$$F(\theta) = [G^2\cos^2\theta+\sin^2\theta/G]^{-3/2}$$

though even a simple exponential form did not produce appreciable differences in the final results. In the above equation G is an adjustable parameter related to the amount of preferred orientation[72] (i.e., for G=0.7, the amount of orientation will be 30%). The scattering intensities depend strongly on the relative strength of the Raman tensor components a_{xx}, a_{yy}, a_{zz}. For the 500cm^{-1} phonon of the YBCO systems or the 630cm^{-1} mode of the Bi-based superconductors, the zz-component is much greater than the xx and yy parts, therefore, we can safely set $a_{xx}=a_{yy}=0$, and the scattered intensities in the different polarizations can be calculated,

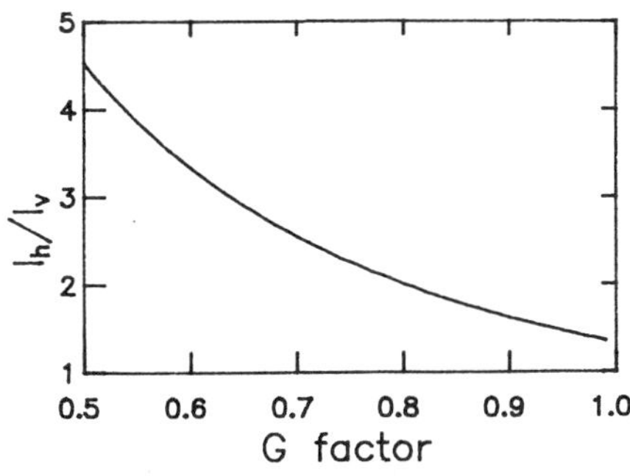

Fig.5 The variation of the relative Raman scattering intensity (I_h/I_v) with the G factor of the March model.[71]

as a function of the parameter G. The light collected by the photomultiplier tube will involve also the detection efficiency which was measured before the run. In the absence of an analyzer in front of the spectrometer, there will be a mixing of the two polarizations (parallel and normal to the slits) in amounts depending on the efficiency of the spectrometer in these directions. Since the relative efficiency of the spectrometer depends on the laser wavelength, the relative intensity in the two directions will also depend on the wavelength. The results of the calculations for the 514.5nm laser line are shown in Fig.5, from which a 25% (G=0.75) partial orientation of the microcrystallites is estimated for the YBCO samples, while for the Bi-type slightly less (20%). In both cases the degree of orientation is referred to the bulk of the material and might differ from the amount detected close to the surface by x-ray measurements. Raman experiments performed from the side face of the samples, but close to the edge (20-30μm), did not reveal any difference in the intensity ratio for the two polarizations, and therefore, in the degree of preferred orientation. If this is correct up to the surface, one should be able to account for the variation of the x-ray intensities[67] based on the Raman measurements.

4. CONCLUSIONS

High pressure measurements in connection with elemental substitutions, can be quite valuable for understanding the mechanism of superconductivity of the new compounds. A more systematic study by light scattering experiments, of the low frequency phonons can help clear up the uncertainty of the Ba position in the unit cell. Furthermore, one has to study the related problems of the oxygen doping and the effect of the external pressure on these compounds. Uniaxial stress studies along the c-axis would be quite informative for the process of charge transfer from the chains to the Cu planes. These studies can be extended to other high-T_c superconductors such as the Bi or the Tl-type to delineate their vibrational modes and check the role of the corresponding atoms in superconductivity.

Finally, the Raman method for detecting the bulk texture in the ceramic compounds should be applied on a series of samples with various degrees of microcrystalline orientation, in order to provide a simple in-situ method for their characterization.

REFERENCES

1. J. G. Bednorz and K. A. Muller, _Zeit. fur Physik_ B64:189 (1986).
2. M. K. Wu, J.R. Ashburn, C.J. Torng, P.H. Hor, R.L. Mong, L. Gao, Z.J. Huang,Y.Q. Wang, and C.W. Chu, _Phys. Rev. Lett._ 58:908 (1987).
3. R. Feile, _Physica_ C159:1 (1989).
4. C. Thomsen and M. Cardona, _in_ "Physical Properties of High-Temperature Superconductors", Ed. D.M. Ginsberg, World Scientific, Singapore (1989).
5. G. A. Kourouklis, A. Jayaraman, B. Batlogg, R. J. Cava, M. Stavola, D.M. Krol, E.A. Rietmann, and L.F. Schneemeyer, _Phys. Rev._ B36:8320 (1987).
6. S. L. Cooper, M.V. Klein, B.G. Pazol, J.P. Rice, and D.M. Ginsberg, _Phys. Rev._ B37:5920 (1988).
7. C. Thomsen, M. Cardona, B. Gegenheimer, R. Liu, and A. Simon, _Phys. Rev._ B37:9860 (1988).
8. M. Cardona, C. Thomsen, R. Liu, H.G. von Schnering, M. Hartweg, Y.F. Yan, and Z.X. Zhao, _Solid State Comm._ 66:1225 (1988).
9. C. Thomsen, M. Cardona, B. Friedl, C.O. Rodriguez, I.I. Mazin, and O.K. Andersen, _Solid State Comm._ 75:219 (1990).
10. B. Friedl, C. Thomsen, and M. Cardona, _Phys. Rev. Lett._ 65:915 (1990).
11. V. G. Hadjiev, M.N. Iliev, C. Raptis, L. Kalev, and B.M. Wanklyn, _Physica_ C166: 225 (1990).
12. In-Sang Yang, G. Burns, F. H. Dacol, and C. C. Tsuei, _Phys. Rev._ B42:4240 (1990).
13. E. Liarokapis, L.T. Wille, Th. Leventouri, L. Martinez, H. Lu, V. Hadjiev, and M. Iliev, _Physica_ C170:419 (1990).
14. E. Liarokapis, E.I. Kamitsos, Th. Leventouri, and F.D. Medina, _Physica_ C157:551 (1989).
15. L. Genzel, A. Wittlin, M. Bauer, M. Cardona, E. Schonherr, and A. Simon, _Phys. Rev._ B40:2170 (1989).
16. R. T. Collins, Z. Schlesinger, F. Holtzberg, and C.F. Feild, _Phys. Rev. Lett._ 63:422 (1989).
17. J. Orenstein, G.A. Thomas, A.J. Millis, S.L. Cooper, D.H. Rapkine, T. Timusk, L.F. Schneemeyer, and J.V. Waszcak, _Phys. Rev._ B42:6312 (1990).
18. R. J. Wijngaarden and R. Griessen, _in_ "Studies of High Temperature Superconductors", Vo.2, ed. A. Narlikar, Nova Science (1989).

19. K. Syassen, M. Hanfland, K. Strossner, M. Holtz, W. Kress, M. Cardona, U. Schroder, J. Prade, A.D. Kulkarni, and F.W. de Wette, <u>Physica</u> C153-155:264 (1988).

20. H. Wilhelm, M. Holtz, and K. Syassen, <u>Verh. Dtsch. Phys. Ges.</u> 25:1269 (1990) and private communication.

21. L. V. Gasparov, O. V. Misochko, M. I. Eremets, E. S. Itskevich, A. V. Lomsadze, V.V. Struzhkin, and A.M. Shirokov, <u>Solid State Comm.</u> 72:465 (1989).

22. O. V. Misochko and V.B. Timofeev, <u>Physica</u> C162-164:1249 (1989).

23. J. Karpinski, E. Kaldis, E.. Jilek, S. Rusiecki, and B. Bucher, <u>Nature</u> 336:660 (1988).

24. J. Karpinski, C. Beeli, E. Kaldis, A. Wisard, and E. Jilek, <u>Physica</u> C153-155:830 (1988).

25. P. Murugaraj, J. Maier, and A. Rabenau, <u>Solid State Comm.</u>66:735 (1988).

26. Th. Leventouri, M. Calamiotou, V. Perdikatsis, and J.S. Faulkner, <u>J. Appl. Phys.</u> 66:3144 (1989).

27. Th. Leventouri, E. Liarokapis, and J.S. Faulkner, <u>Solid State Comm.</u> 74:1103 (1990).

28. R. J. Cava, J.J. Krajewski, J.F. Peck Jr, B. Batlogg, I.W. Rupp Jr, R.M. Fleming, A.C.W.P. James, and P. Marsh, <u>Nature</u> 338:328 (1989).

29. J. C. Phillips, <u>in</u> "Physics of High-T_c Superconductors", Academic Press (1989).

30. R. Liu, C. Thomsen, W. Kress, M. Cardona, B. Gegenheimer, F. W. de Wette, J. Prade, and A.D. Kulkarni, <u>Phys. Rev.</u> B37:7971 (1988).

31. K. F. McCarty, J.Z. Liu, R.N. Shelton, and H.B. Radousky, <u>Phys. Rev.</u> B41:8792 (1990).

32. M. K. Crawford, W.E. Farneth, R. K. Bordia, and E.M. Mc Carron, <u>Phys. Rev.</u> B37:3371 (1987).

33. M. Cardona, L. Genzel, R. Liu, A. Wittlin, Hj. Mattausch, F. Garcia-Alvarado, and E. Garcia-Gonzalez, <u>Solid State Comm.</u> 64:727 (1987).

34. D. A. Bonn, J.E. Greedan, C.V. Stager, T. Timunsk, M. Doss, S.L. Herr, K. Kamaras, and D.B. Tanner, <u>Phys. Rev. Lett.</u> 58:2249 (1987).

35. A. Wittlin, R. Liu, M. Cardona, L. Genzel, W. Konig, W. Bauhofer, Hj. Mattausch, and A. Simon, <u>Solid State Comm.</u> 64:477 (1987).

36. E. T. Heyen, R. Liu, C. Thomsen, R. Kremer, M. Cardona, J. Karpinski, E. Kaldis, and S. Rusiecki, <u>Phys. Rev.</u> B41:11058 (1990).

37. E. T. Heyen, M. Cardona, J. Karpinski, E. Kaldis, and S. Rusiecki, to be published.

38. G. Xiao, M. Z. Cieplak, A. Gavrin, F.H. Streitz, A. Bakhshai, and
    C.L. Chien, <u>Phys. Rev. Lett.</u> 60:1446 (1988).

39. L. T. Wille, <u>Phase Transitions</u> 22:225 (1990).

40. Y. Macno, T. Tomita, M. Kayogoku, S. Awaji, Y. Aoki, K. Hoshino, A.
    Minami, and T. Fujita, <u>Nature</u> 328:512 (1987).

41. E. Takayama-Muromachi, Y. Uchida, and K. Kato, <u>Jpn. J. Appl. Phys.</u>
    26:L2087 (1987).

42. T. J. Kistenmacher, W.A. Bryden, J.S. Morgan, K. Moorjani, Y.W. Du,
    Z.Q. Qiu, H. Tang, and J.C. Walker, <u>Phys. Rev.</u> B36:8877 (1987).

43. J. M. Tarascon, P. Barboux, P.F. Miceli, L.H. Greene, G.W. Hull, M.
    Eibschutz, and S.A. Sunshine, <u>Phys. Rev.</u> B37:7458 (1988).

44. G. Roth, G. Heger, B. Renker, J. Pannetier, V. Caignaert, M. Hervieu,
    and B. Raveau, <u>Z. Phys.</u> B71:43 (1988).

45. P. Bordet, J. L. Hodeau, P. Strobel, M. Marezio, and A. Santoro,
    <u>Solid State Comm.</u> 66:435 (1988).

46. Y. Xu, M. Suenaga, J. Tafto, R. L. Sabatini, A. R. Moodenbaugh, and
    P. Zolliker, <u>Phys. Rev.</u> B39:6667 (1989).

47. P. Dufour, S. Jandl, M. Banville, P. Fournier, and M. Aubin, <u>Physica</u>
    C166:431 (1990).

48. M. Hangyo, S. Nakashima, M. Nishiuchi, K. Nii, and A. Mitsuishi,
    <u>Solid State Commun.</u> 67:1171 (1988).

49. F. U. Hillebrecht, L. Ley, R. L. Johnson, R. Liu, C. Thomsen, M.
    Cardona, Hj. Mattausch, W. Bauhofer, and A. Simon,
    <u>Solid State Commun.</u> 67:379 (1988).

50. M. Kakihana, L. Borjesson, S. Eriksson, P. Svedlindh, and P. Norling,
    <u>Phys. Rev.</u> B40:6787 (1989).

51. G. A. Kourouklis, A. Jayaraman, B. Batlogg, R.J. Cava, M. Stavola,
    D.M. Krol, E.A. Rietmann, and L.F. Schneemeyer, <u>Phys. Rev.</u>
    B36:8320 (1987).

52. D. Kirillov, J.P. Collman, J.T. McDevitt, G.T. Yee, M.J. Holcomb, and
    I. Borovic, <u>Phys. Rev.</u> B37:3660 (1988).

53. C. Thomsen, R. Liu, M. Bauer, A. Wittlin, L. Genzel, M. Cardona, E.
    Schonherr, W. Bauhofer, and W. Konig, <u>Solid State Commun.</u> 65:55
    (1988).

54. P. F. Miceli, J.M. Tarascon, L.H. Greene, P. Barboux, F.J. Rotella,
    and J.D. Jorgensen, <u>Phys. Rev.</u> B37:5932 (1988).

55. G. Baumgartel and K. H. Bennemann, <u>Phys. Rev.</u> B40:6711 (1989).

56. Y. LePage, T. Siegrist, S.A. Sunshine, L.F. Schneemeyer, D.W. Murphy,
    S.M. Zahurak, J.V. Waszczak, W.R. McKinnon, J.M. Tarascon, G.W.
    Hull, and L.H. Greene, <u>Phys. Rev.</u> B36:3617 (1987).

57. In-Sang Yang, G. Burns, F.H. Dacol, J.F. Bringley, and S.S. Trail,
    <u>Physica</u> C171:31 (1990).

58. H. J. Rosen, R.M. McFarlane, E.M. Engler, V.Y. Lee, and R.D.
    Jacowitz, <u>Phys. Rev.</u> B38:2460 (1988).

59. In-Sang Yang, A.G. Schrott, and C.C. Tsuei, <u>Phys. Rev.</u> B41:8921
    (1990).

60. G. Burns, F. H. Dacol, C. Feild, and F. Holtzberg, <u>Solid State Comm.</u>
    77:367 (1991).

61. C. Infante, M. K. El Mously, R. Dayal, M. Husain, S. A. Siddiqi, and
    P. Ganguly, <u>Physica</u> C167:640 (1990).

62. Y. Yamada, N. Fukushima, S. Nakayama, H. Hoshino, and S. Murase,
    <u>Jpn. J. Appl. Phys.</u> 26:L865 (1987).

63. T. R. Dinger, T.K. Worthington, W.J. Gallagher, and R.L. Sandstrom,
    <u>Phys. Rev. Lett.</u> 58:2687 (1987).

64. S. Jin, T.H. Tiefel, R.C. Sherwood, R.B. Van Dover, M.E. Davis, G.W.
    Kammlott, and R.A. Fastnacht, <u>Phys. Rev.</u> B37:7850 (1988).

65. D. E. Farrell, B.S. Chandrasekhar, M.R. DeGuire, M.M. Fang, V.G.
    Kogan, J.R. Clem, and D.K. Finnemore, <u>Phys. Rev.</u> B36:4025 (1987).

66. A. Lusnikov, L.L. Miller, R.W. McCallum, S. Mitra, W.C. Lee, and D.C.
    Johnston, <u>J. Appl. Phys.</u> 65:3136 (1989).

67. Y. S. Hascicek, L.R. Testardi, Th. Leventouri, E. Liarokapis, and L.
    Martinez, <u>J. Appl. Phys.</u> 68:4178 (1990).

68. E. Liarokapis, Th. Leventouri, E. I. Kamitsos, M. Calamiotou, V.
    Perdikatsis, and V. Psiharis, to be published.

69. W. Hayes and R. Loudon, <u>in</u> "Scattering of light by crystals", John
    Willey and Sons, Inc. (1978).

70. E. Anastassakis and E. Liarokapis, <u>Phys. Stat. Sol.</u> b149:K1 (1988).

71. H. R. Wenk, J. Pannetier, G. Bussod, and A. Pechenik, <u>J. Appl. Phys.</u>
    65:4070 (1989).

72. W. A. Dollase and R. J. Reeder, <u>Am. Mineral</u> 71:163 (1986).

APPLICATION OF A DIAMOND ANVIL CELL FOR THE STUDY OF THE MAGNETIC SUSCEPTIBILITY OF CERAMIC SUPERCONDUCTORS UNDER HYDROSTATIC PRESSURE

S. Klotz[1,2], J. S. Schilling[1] and P. Müller[3]

[1]Department of Physics, Campus Box 1105, Washington University
One Brookings Drive, St. Louis, MO 63130–4899, USA
[2]Sektion Physik, University of Munich, Schellingstrasse 4
W–8000 Munich 40, Germany
[3]Walther–Meissner–Institut für Tieftemperaturforschung
W–8046 Garching, Germany

ABSTRACT

We report the development of a diamond anvil cell to study the ac susceptibility of ceramic superconductors to pressures above 10 GPa. Highly densified helium is used as pressure medium. The superconducting transition is detected inductively by a miniature coil system. Results on both $Y_1Ba_2Cu_3O_7$ and $Bi_2Ca_1Sr_2Cu_2O_{8+y}$ with different oxygen content y are presented and discussed within the context of pressure induced charge transfer.

I. INTRODUCTION

It is well known that high pressure experiments have played a major role in high temperature superconductivity since its discovery in 1986. One of the key pieces of information for better understanding these compounds is the dependence of their superconducting transition temperature T_c on pressure, as has been pointed out in several talks at this conference. By far the most data originate from experiments below 2 GPa, a pressure easily achieved by conventional techniques. However, since the compressibility of ceramic superconductors is not very large, most of these experiments are only able to detect a linear dependence between T_c and pressure. To learn more about the change of T_c with the lattice parameters, the pressure range should be increased considerably. This calls for the application of the diamond anvil cell (DAC). A number of groups have studied $T_c(P)$ [1,2] using the DAC technique, but generally under nonhydrostatic pressure conditions. Because of the inherent sealing problems, they use either (1) no pressure medium at all, (2) soft solids like NaCl or (3) fluids which lose their hydrostatic properties under pressure since they solidify. It remains an open question how meaningful such results are in the light of the strongly anisotropic dependence of T_c on lattice parameter anticipated for the layered structure of perovskite superconductors. In a recent paper, Meingast et al. [3] even find that the pressure dependence of T_c in $Y_1Ba_2Cu_3O_7$ has different signs along the a and b axes and is large in magnitude along these directions. They conclude that the small values so far reported from hydrostatic experiments on this system result from cancellation effects. Another system where hydrostatic or nonhydrostatic pressure experiments are

Frontiers of High-Pressure Research, Edited by H.D. Hochheimer and
R.D. Etters, Plenum Press, New York, 1991

known to yield differing signs and magnitudes of dT_c/dP is the La_3(chalcogenides) [4]. It is clear that more care should be taken in future high pressure studies of the oxide superconductors to clearly define the effects of the pressure technique used.

Another important consideration for experiments applying the diamond anvil cell for the study of ceramic superconductors is that T_c normally is detected resistively. This might explain the inconsistent findings of different groups since a percolation path of superconducting material could be mistaken for bulk superconductivity.

Here we present a technique which overcomes the above shortcomings both by (1) using highly densified helium as the pressure medium and (2) detecting bulk superconductivity by an ac–susceptibility measurement. Also we measure on single crystalline material which excludes the granularity effects usually present in sinter samples. Since our experimental technique is rather uncommon, we first want to describe in detail the experiment, discussing the typical problems of such a measurement. Then we will illustrate the method by presenting $T_c(P)$–results on $Y_1Ba_2Cu_3O_7$ and recent measurements on $Bi_2Ca_1Sr_2Cu_2O_8$. On a given crystal we study how increasing the oxygen content effects the $T_c(P)$ dependence. Finally, we discuss these results within the recently proposed model of pressure–induced charge transfer.

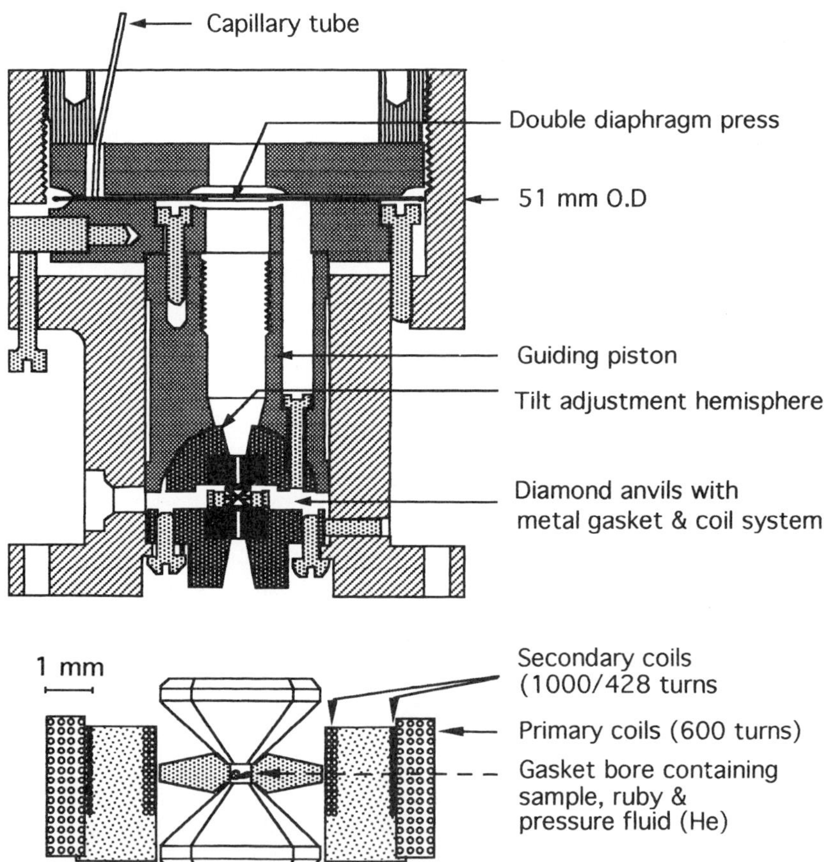

Fig. 1   Diamond–anvil pressure clamp for pressures above 10 GPa. Two 1/6 carat diamond anvils press into a 3 mm O.D. metal gasket with a 0.25 mm dia. bore containing sample, ruby manometer, and liquid He pressure fluid. An enlarged view of the primary(secondary) coil system using 30(16) μm dia. Cu–wire is also shown.

II. EXPERIMENTAL

Figure 1 shows our diamond anvil clamp [5] (DAC). It contains two 1/6 carat type I diamond anvils with 0.5 mm culet diameter mounted on B_4C plates. The rest of the clamp is made completely of copper–beryllium alloy, critical parts are made of magnetically pure binary CuBe. The force pushing the diamond anvils together is provided by a double diaphragm press [6] which can be pressurized to more than 200 bars of helium gas. The coil system used for the susceptibility measurement is placed around a 3 mm O.D. metal gasket as shown in the enlarged view. Primary (field) and secondary (pick–up) coils are made of 30 and 16 µm diameter copper wire, respectively, and are cast in epoxy. The two parts of the pick–up coils with 1000 and 428 windings are compensated to about 1% outside the clamp. Both carbon and platinum thermometers (not shown in Fig. 1) are placed inside the guiding piston, about 10 mm

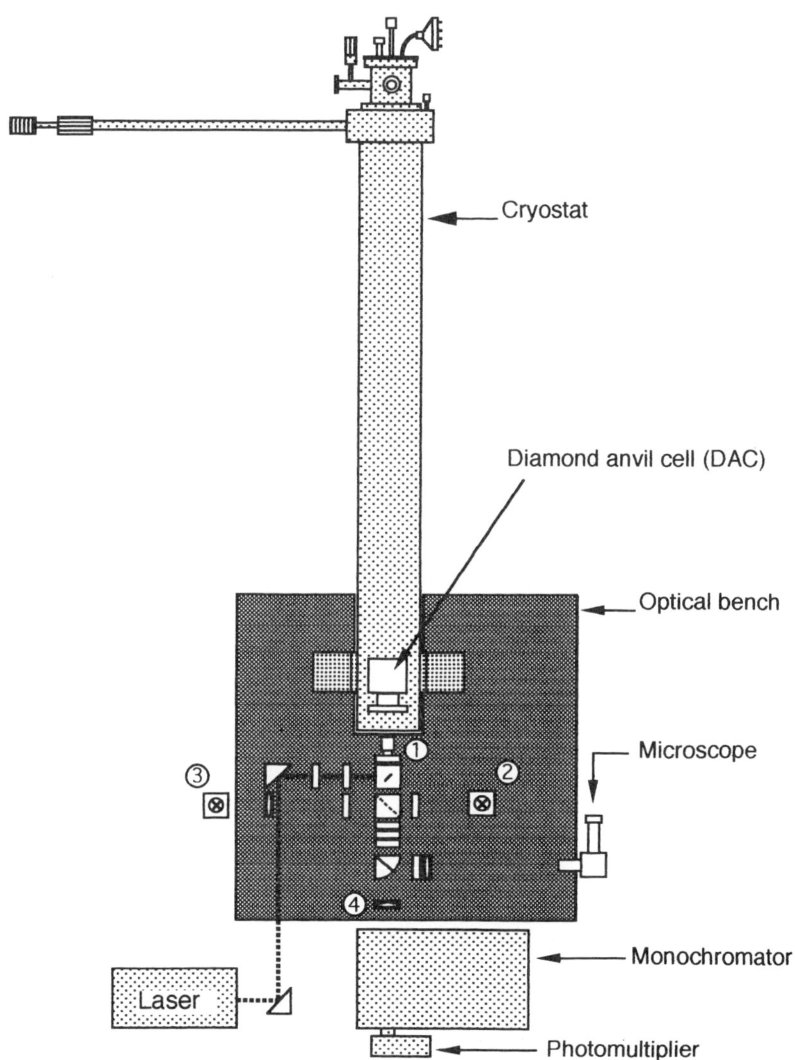

Fig. 2   Cryostat and optical set up for measurements using the diamond anvil cell (DAC). See text for details.

above the sample. A test measurement with four thermometers at different parts in the clamp showed temperature gradients to be less than 0.1 K in a typical measurement cycle between 50 and 150 K. The DAC is mounted on a holder and inserted into a continuous flow cryostat shown in Fig. 2. The top cover contains electrical feedthroughs (including four coax–lines down to the coils), a capillary connector and a feedthrough for a lightguide used to illuminate the cell from above. The cryostat has sapphire/quartz windows at the bottom. An optical set up below provides the possibility of both observing the cell and measuring the pressure via the standard ruby fluorescence technique [7]. The basic parts are: a collimator lens (1) (f = 88 mm) right in front of the window and focussed at the cell, a microscope consisting of an objective (1:10) and a ocular (5x), a white light lamp (2) for illumination of the cell, a neon lamp (3) providing reference lines, a lens (4) (f = 50 mm) focussing the fluorescence light onto the entrance slit (35 μm) of a 1/4 m monochromator and matched to its f/ number as well as some filters and polarizers. The ruby is excited by a 100 mW argon ion laser, but only 5 to 10 mW reaches the interior of the cryostat.

<u>Helium as pressure medium</u>

Helium is the best pressure medium known; it retains its hydrostatic properties in the solid state better than any other compound [8]. Its melting curve up to 300 K is fairly well given by Simon's equation [9]

$$P(GPa) = 0.0016067 \ T(K)^{1.565}$$

which means that it solidifies at room temperature only at about approximately 11.5 GPa. The equation of state (density vs. pressure) of helium is accurately known to 12 GPa from sound velocity measurements [10] as shown in Fig. 3. Unfortunately, it is difficult to load He into the cell at a high initial density. Two methods, high pressure loading at room temperature (typically at 0.1–0.5 GPa) and cryogenic filling near 4.2 K at ambient pressure have been reported so far [7,8]. In the present work we employ the latter technique and fill the cell according to the following procedure: first the cryostat is cooled to 4.2 K and liquid He filled in to cover the DAC. Next, by pumping on the He–bath the temperature of the DAC is lowered to 2 K so that superfluid He can enter the bore of the gasket. Finally, the cell is sealed off by pressing the diamonds together with a force of approximately 1500 N.

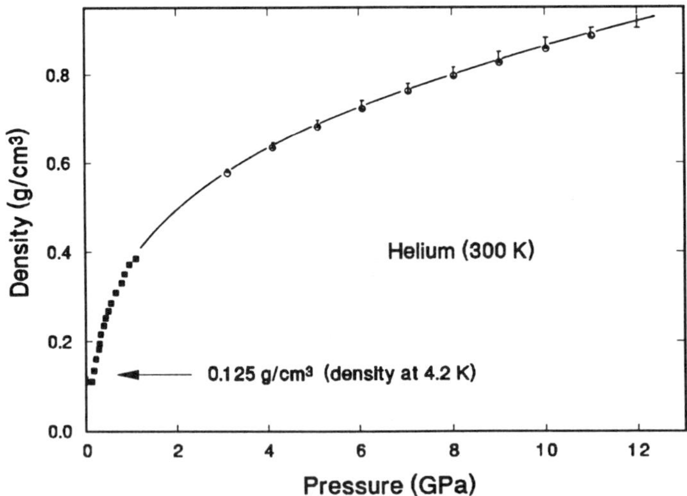

Fig. 3   Density of helium as a function of pressure at room temperature. Data was taken from Ref. 10. The density of the liquid is indicated for comparison.

From Fig. 3 we can deduce important information for the application of He as pressure medium. If liquid He is loaded into the cell at 4.2 K (density 0.125 g/cm), the volume of the cell has to be reduced by a factor of 3 to reach 1 GPa and by a factor of 7 to reach 10 GPa at 300 K. This requires a large bore diameter and a thick gasket at the start of the experiment. In our case we use a gasket with a 250 μm hole diameter and approximately 300 μm thickness preindented to 115–135 μm. Also, from the observed diameter of the cell and the data in Fig. 3 we are able to estimate the thickness of the gasket at a given pressure to help ensure that the diamonds don't squeeze directly onto the sample.

<u>Gasket materials</u>

The following requirements must be fulfilled for the gasket material used in an ac susceptibility measurement: (1) it must be nonmagnetic, (2) it should have a small electrical conductivity even at low temperature to minimize screening effects and, (3) it should possess a high tensile strength to allow measurements with relatively large samples. From the large number of gasket materials reported so far we carried out experiments with the following three: copper beryllium (10 at% beryllium, 0.3 at% Co) hardened for 2 h at 315 °C after preindentation, pure rhenium, and $Ta_{90}W_{10}$ [11]. All three are essentially nonmagnetic; Fig. 4a shows the measured electrical conductivity as a function of temperature. The tensile strengths at 300 K are 1.4 GPa, 2.2 GPa and 1 GPa for CuBe, Re, and $Ta_{90}W_{10}$, respectively [12]. Copper beryllium is the easiest to work with; however, in early experiments with a CuBe gasket and an ethanol–methanol pressure medium we did not reach pressures beyond 7–8 GPa. This might not only be due to the smaller tensile strength compared to Re, but also to the relatively large cell volume used in this experiment. In an attempt to reach higher

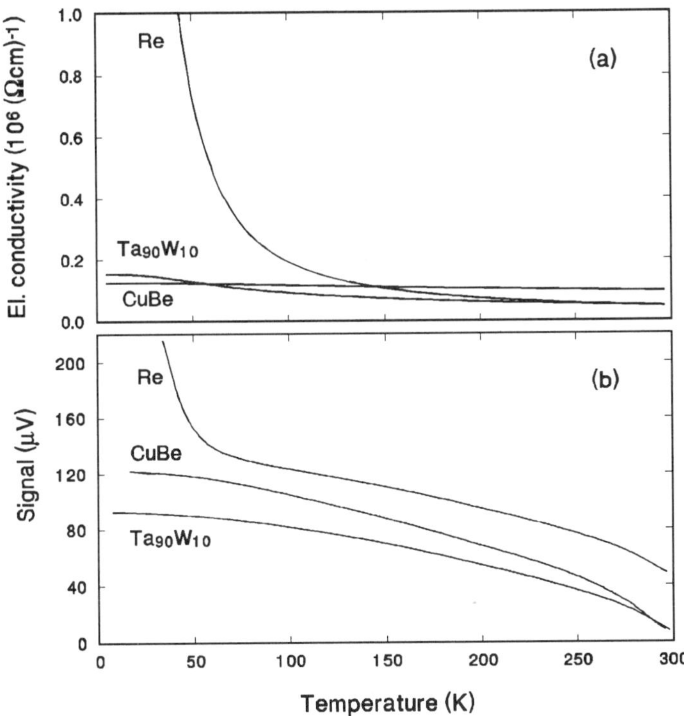

Fig. 4   Temperature dependence of (a) the electrical conductivity and (b) the background signal detected by the pick–up coils in measurements for three different gasket materials (Re, $Ta_{90}W_{10}$, CuBe).

pressures we tried gaskets made of pure rhenium 280 μm thick preindented to 100 μm. It takes considerably more effort to work with Re than with CuBe alloy. In particular, the bore can't be drilled in Re using standard techniques. We were successful in spark–cutting a 200 μm hole in the center of a 3 mm O.D gasket. Another disadvantage of pure rhenium is its relatively high electrical conductvity at low temperatures (Fig. 4a) which prevents measurements below 60 K due to a strong increase in the background signal (Fig. 4b) produced by eddy currents in the gasket. We finally tried a $Ta_{10}W_{90}$ alloy rolled to 290–330 μm thickness which proved to be the most suitable gasket material. This alloy is nonmagnetic, has a relatively small electrical conductivity at low temperature, and, despite its relatively low tensile strength, holds pressure very well due to the fact that it cold–welds to the diamonds. Fig. 5 shows a test measurement to 22 GPa with only helium and a ruby chip in the gasket bore. This measurement was interrupted to check the condition of the DAC; much higher pressures should be possible.

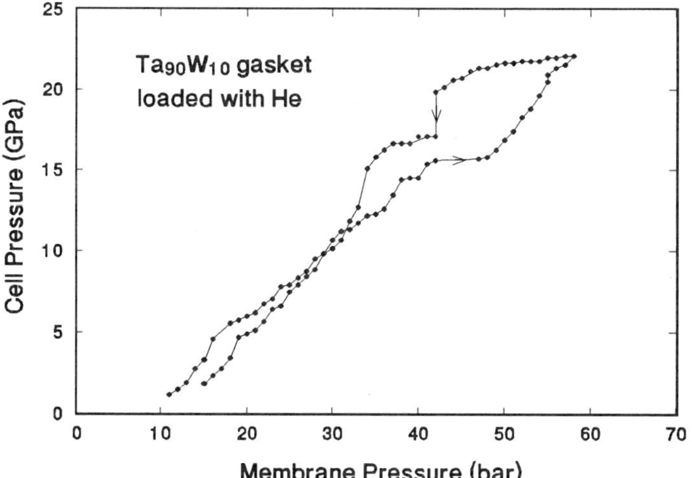

Fig. 5   Pressure in the cell versus pressure in the membrane (see Fig.1) in a test measurement using a $Ta_{90}W_{10}$ gasket loaded with helium. A pressure of 1 bar in the membrane creates a force of approximately 150 N on the diamonds.

<u>T_c by an ac–Susceptibility Measurement</u>

Using a compensated coil system like the one in Fig. 1, the signal from the transition of a perfect superconductor is expected to be given in SI units by [13]

$$U = \pi f B V [1/ (1-D)] [N_i/R_i - N_0/R_0] \quad ,$$

where f is the frequency of the applied ac–field B, V the sample volume, D the demagnetisation factor, and N_i/R_i and N_0/R_0 the number of windings divided by the radius for the inner and outer pick–up coils respectively. For typical values (f = 500 Hz, B = 4 G, V = 1 x 10^{-4} mm³, D = 1/2, N_i = 1000, N_0 = 428, R_i = 1.75 mm, R_0 = 2.4 mm) we find U ≈ 50 nV. To recover such a small signal it is necessary to use a sensitive lock–in amplifier and avoid noise sources by proper shielding and grounding. Fig. 6 shows the circuit diagram of our measurement. The input of the lock–in amplifier (PAR 124) is connected to the pick–up coils via a variable inductance L. Due to the strong coupling of the metallic parts of the clamp and the gasket, the pick–up coils are no longer compensated once the coil system is placed into the clamp. By adjusting the inductance L and the resistor R the resulting offset can be set to zero at any given temperature. The R–C–L unit constitutes a wide angle phase shifter. It has the advantage that both the phase (up to $\pm$ 90° relative to the ac field) and the magnitude of the signal can be varied independently, which is not possible in the classical Hartshorn bridge [14]. The choice of the right preamplifier is critical for noise reduction. Its noise characteristic must match the source impedance and frequency. We use a PAR 116 preamplifier operating in the transformer mode.

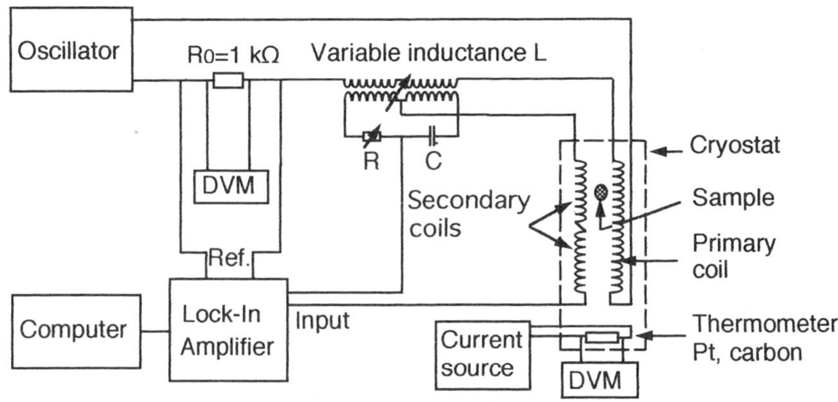

Fig. 6   Circuit diagram for ac susceptibility and temperature measurement systems. See text for details.

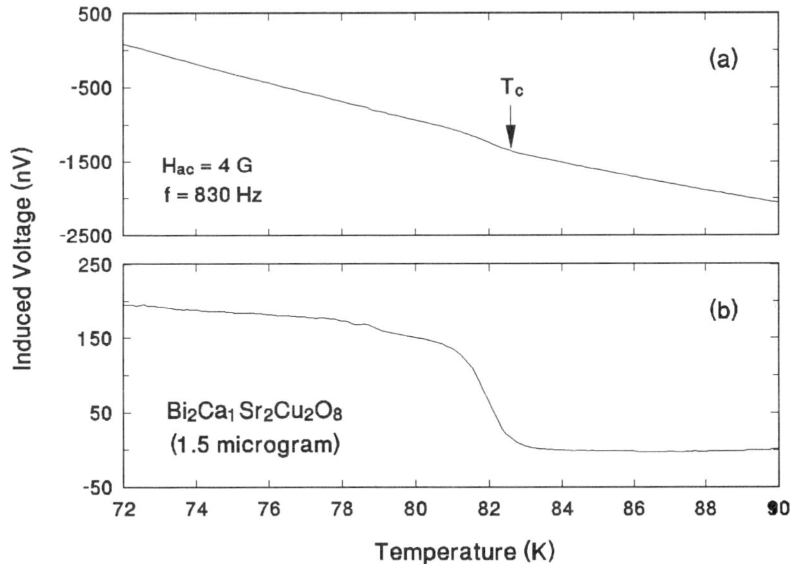

Fig. 7   Magnitude of χ_{ac} versus temperature for $Bi_2Ca_1Sr_2Cu_2O_8$ before (a) and after (b) subtraction of the temperature–dependent background.

In Fig. 7a we show the results of a measurement on a "fairly big" (1.5 µg heavy) single crystal of $Bi_2Ca_1Sr_2Cu_2O_8$ at 4.7 GPa pressure. As mentioned previously, the background signal is a sensitive function of the temperature. Part of this background is due to the proximity of the metallic pressure clamp and an imperfect compensation of the pick–up coils; however, the largest contribution comes from the gasket itself, especially below 60 K. As the temperature is lowered, the conductivity $\sigma(T)$ of the gasket (see Fig. 4a) increases and the penetration depth in SI units $\delta(T) = (1/\pi f \mu_0 \sigma(T))^{1/2}$ [15] of the ac–field at frequency f thus decreases ($\mu_0 = 4\pi$ 10^{-7} Vs/Am). Upon cooling, the gasket expels more and more of the applied ac–field which is detected by the pick–up coils leading to the slope in Fig. 7a. We conclude, therefore, that the temperature dependent background in Fig. 7a is the price which has to be paid in an ac–method whenever the field coils are placed outside the metal gasket as in the present case. After subtraction of the slope in Fig. 7a, the super-conducting transition becomes clearly visible, as seen in Fig. 7b.

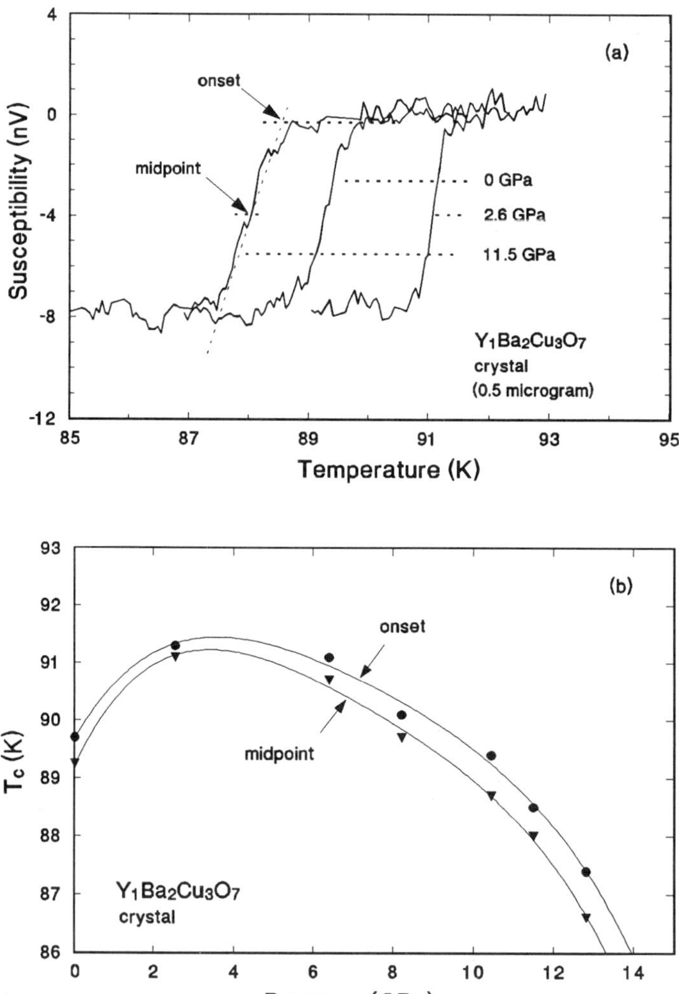

Fig.8 (a) Temperature dependence of ac susceptibility for a tiny crystal of $YBa_2Cu_3O_7$ at three pressures in the diamond–anvil clamp. (b) Superconducting transition temperature of a $YBa_2Cu_3O_7$ crystal versus pressure in the diamond–anvil clamp.

III. RESULTS AND DISCUSSION

One of the first measurements we carried out on high temperature superconductors was on a single crystalline sample of $Y_1Ba_2Cu_3O_7$, as seen in Fig. 8. Details of the experiment are published in Ref. 16. We find an initial increase of T_C with pressure at the rate (0.65 ± 0.15) K/GPa which agrees well with numerous "low pressure" experiments [17]. Among the few experiments reaching more than 10 GPa, there is poor agreement for the pressure dependence of T_C in the upper pressure range [17,18,19]. In our measurement we find that T_C reaches a maximum of 91.5 K at a pressure between 4 and 5 GPa. In an early experiment to 7 GPa using a 4:1 methanol--ethanol mixture as pressure medium we find the same behavior.

What is the explanation for the maximum in T_C as a function of pressure? The essential problem in the interpretation of $T_C(P)$ data is that pressure simultaneously effects several parameters (orbital overlap, phonon frequency, carrier concentration, etc.) which may be crucial for high temperature superconductivity. One of the current beliefs is that the dominant effect responsible for the behavior of T_C under pressure is an increase of the charge carrier (hole) concentration in the CuO_2 planes. This has been derived from neutron scattering data by Jorgensen et al. [20] on $Y_1Ba_2Cu_3O_7$ to 0.6 GPa and by Wijngaarden et al. from recent measurements of H_{c2} and T_C as a function of pressure [21] for the 123 and Y–124 structures, even up to pressures above 10 GPa. A wide variety of experiments indicate that many, if not all, high–T_C superconductors lie on a general phase diagram which exhibits a bell–shaped maximum of $T_C(h)$ as a function of the hole concentration h. Assuming $Y_1Ba_2Cu_3O_7$ exibits such a phase diagram, a sample whose hole concentration is close to its optimal value would then be expected to show the following behavior upon pressure: increasing hole concentration would initially cause T_C to rise until its maximum value (corresponding to the optimal hole concentration) is reached; a further increase of the hole concentration with pressure beyond this point should lead to a decrease of T_C. This behavior is, in fact, observed in Fig. 8. Thus, our measurement would appear to support the idea of a pressure induced charge transfer.

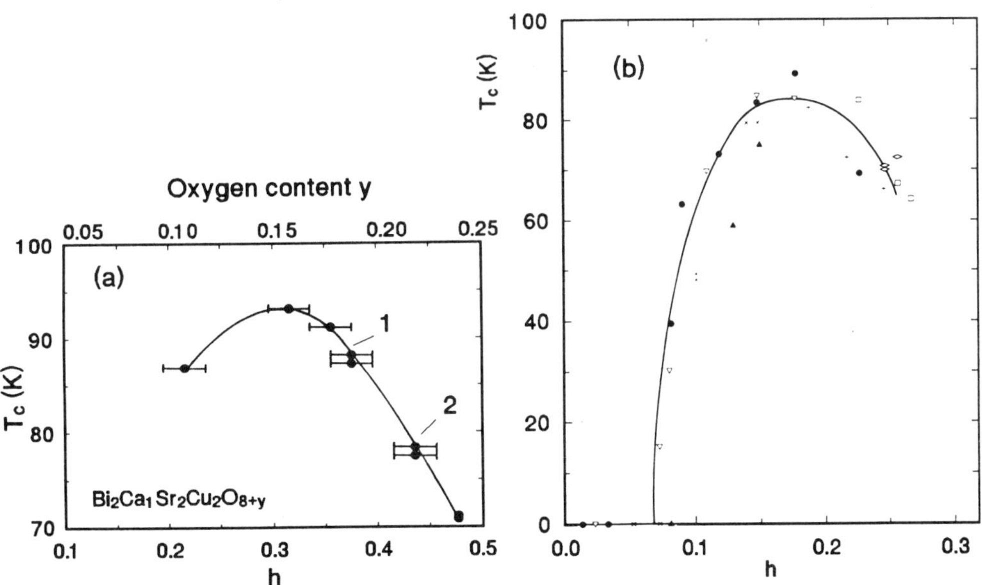

Fig. 9   Superconducting transition temperature T_C of $Bi_2Ca_1Sr_2Cu_2O_{8-y}$ as a function of the hole concentration h according to measurements by (a) Allgeier et al. [22] and (b) Groen et al. [23]. Numbers (1) and (2) refer to sample 1 and sample 2 of the present experiment.

To pursue this idea further, we carried out high pressure experiments on a different system, $Bi_2Ca_1Sr_2Cu_2O_8$. $Y_1Ba_2Cu_3O_7$ is not the ideal system since it shows only the left half of the above $T_c(h)$ phase diagram and the interpretation of data is complicated by the existence of the CuO chains which are a peculiar feature of the 123–structure. The system $Bi_2Ca_1Sr_2Cu_2O_{8+y}$ has two CuO_2 planes, no CuO chains and double BiO layers. As in $Y_1Ba_2Cu_3O_7$, the superconducting transition temperature T_c can be varied over a wide range by changing the oxygen content y. The additional oxygen, interstitially incorporated into the BiO layers, creates holes in the CuO_2 planes and renders the system superconducting. Figure 9 shows a measurement by Allgeier et al. [22] of T_c as a function of the hole concentration h. It is assumed that each oxygen atom creates two holes. In measurements by Groen et al [23] (Fig. 9b) the hole concentration was controlled by substitution of divalent Ca by trivalent Y (hole filling). Note the steep decrease of T_c in the latter experiments. Thus, for $Bi_2Ca_1Sr_2Cu_2O_8$, it seems to be clear T_c exhibits a well pronounced maximum as a function of the hole concentration.

Figure 10 shows the results of our pressure experiments on two samples of $Bi_2Ca_1Sr_2Cu_2O_8$ with different oxygen content. Both samples come from the same single crystal of excellent quality. It was grown from a stoichiometric mixture of the oxides, heated up to 980°C and then cooled down to 800°C at a rate of 1°C/h; this was followed by an anneal in O_2 to increase the oxygen content [24]. As confirmed by Raman scattering [25], the crystals were single phase and close to stoichiometric composition. Their orientation was determined using X–ray diffraction in transmission–Laue geometry. For the as–grown crystal, shielding and Meissner effect measurements with a SQUID magnetometer in a field of 12 mG parallel to the c–axis yielded a transition temperature $T_c \approx 86$ K, a transition width $\Delta T_c(10\%$ to $90\%)$ of 2.5 K, and a Meissner fraction of 42%. Sample 1 was broken off from the as–grown single crystal and measured in the DAC. After this experiment the original large single crystal was annealed for 12 h in flowing oxygen at 600 °C. A measurement in a different SQUID system at 1 G dc field showed a T_c of 80 K with transition width 4 K. Sample 2 was then cut off from this oxygenated single crystal for use in the pressure experiment. This measurement was carried out three times in all using different pieces of the oxygenated single crystal as indicated by different symbols in Figure 10.

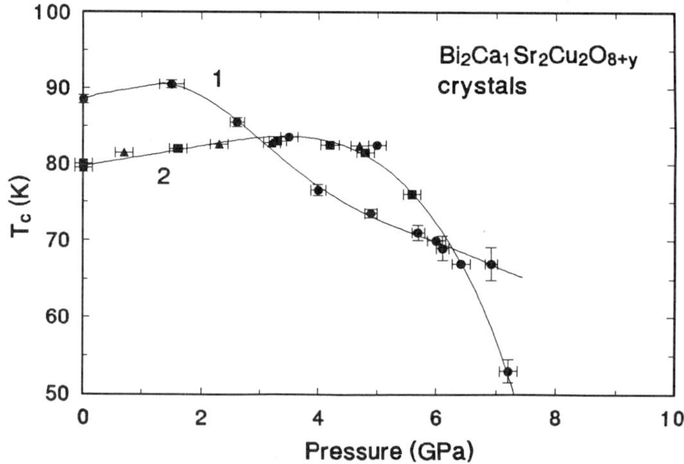

Fig. 10   Pressure dependence of T_c for samples 1 and 2 of $Bi_2Ca_1Sr_2Cu_2O_{8-y}$ for different values of the oxygen content y. The vertical error bars in sample 2 are approximately the same size as the symbols. See Fig. 9 and text for details.

In both samples we observe (see Fig. 10) an initial increase of T_c at a rate $\approx$ +1–2 K/GPa followed by a rather steep decrease after passing through a maximum. These results for the oxygen–rich sample reconcile reports of both a positive pressure dependence to 1 GPa by Goldstein et al. [26] and a negative pressure dependence above 2 GPa by Wijngaarden et al. [27]. The initial increase of T_c is also consistent with earlier results on polycrystalline $Bi_2Ca_1Sr_2Cu_2O_8$ samples with T_c = 90 K and 70 K of Sieburger et al. [28] to 0.6 GPa (dT_c/dP = +1.5 K/GPa) and a measurement by Diederichs on the latter (T_c = 70 K) oxygen–rich sample [29] to 3.2 GPa (+1.2 K/GPa). A positive pressure dependence is also reported by Forro et al. [30] to 2 GPa (+1.5 K/GPa), Kubiak et al. [31] to 1 GPa (+1.6 K/GPa), Beille et al. [32] to 1.5 GPa (+0.2 K/GPa), and Tamura et al. [33] to 1.5 GPa (+1.1 K/GPa) with critical temperatures at ambient pressure of 84 K, 82 K, 80 K and 88.5 K, respectively. No indications of the oxygen content relative to its optimal value were given.

The essential effect of oxygen loading is to shift the maximum from $\approx$ 1 GPa to $\approx$ 3.5 GPa. The overall increase of T_c with pressure is only 2–4 K, with the oxygen-–rich sample being slightly larger. Comparing Figs. 9 and 10, it is tempting to conclude that the effect of pressure is to <u>fill</u> holes so that T_c passes through a maximum by going from right to left in the phase diagram of Fig. 9. This is exactly contrary to $Y_1Ba_2Cu_3O_7$ where pressure is believed to <u>create</u> holes in the CuO_2 planes. However, there are several shortcomings to the picture that pressure fills holes in $Bi_2Ca_1Sr_2Cu_2O_8$. First, we would expect that the increase of T_c with pressure should be much larger in the oxygen–rich sample, i.e. T_c should increase from 80 K to 90 K before decreasing. In fact, we observe an increase of only about 3 K. Second, we would expect a negative pressure dependence for a sample on the left hand side of the phase diagram. In an earlier experiment on a low oxygen polycrystalline sample of $Bi_2Ca_1Sr_2Cu_2O_8$ from a different batch using a helium gas system to 0.6 GPa, Sieburger [28] found that dT_c/dP = +1.4 K/GPa. The simple charge transfer model thus does not appear to be adequate. It can be argued that varying the oxygen content has structural consequences such as oxygen reordering effects. The actual hole concentration in the CuO_2 planes may therefore be quite different to what one expects from simply counting the number of interstitial oxygen in the structure. An experiment on a Ca substituted $Bi_2Ca_{1-x}Y_xSr_2Cu_2O_8$ would help to clarify this question.

We conclude that the present pressure experiments on oxygen–rich $Bi_2Ca_1Sr_2Cu_2O_8$ samples suggest that pressure may fill holes in the CuO_2 planes, in contrast to $Y_1Ba_2Cu_3O_7$ where it creates holes. However, the effect of pressure is clearly more complicated, since the positive pressure dependence for oxygen poor samples is completely unexpected in a charge transfer picture. The different behavior compared to $Y_1Ba_2Cu_3O_7$ might be related to the BiO–layers containing the additional oxygen and acting as charge reservoirs. Band structure calculations [34, 35] show two BiO bands, one contacts the Fermi level E_f from above, the other dips below E_f, leading to the formation of electron pockets in the BiO subsystem. Since the two layers are only loosely bound together, pressure might considerably change these bands and consequently the degree of doping of the CuO_2 planes, as pointed out by Hybertsen and Mattheiss [35].

REFERENCES

1. U. Koch, N. Lotter, J. Wittig, W. Assmus, B. Gegenheimer, and K. Winzer, Solid State Commun. 67, 959(1988).
2. A. Driessen, R. Griessen, N. Koeman, E. Salomons, R. Brouwer, D.G. de Groot, K. Heeck, H. Hemmes and J. Rector, Phus. Rev. B36, 5602(1987)
3. Meingast, T. Wolf, H Wühl, E. Erb, and G. Müller–Vogt, preprint, submitted to Phys. Rev. Letters.
4. R.N. Shelton, A.R. Moodenbaugh, P.D. Dernier, and B.T. Matthias, Mat. Res. Bull. 10, 1111(1975).
5. J.S. Schilling, Mat. Res. Soc. Symp. Proc. 22, 79(1984).
6. W.B. Daniels and W. Ryschkewitsch, Rev. Sci. Instr. 54, 115(1983).
7. A. Jayaraman, Rev. Mod. Phys. 55, 65(1983).

8.   W.F Sherman and A.A. Stadtmuller, in: Experimental Techniques in High Pressure Research (Wiley, New York, 1987).

9.   W.L. Vos, M.G.E. van Hinsberg, and J.A. Schouten, Phys. Rev. B 42, 106(1990).

10. A. Polian and M. Grimsditch, Europhys. Lett. 2, 849(1986).

11. The authors are grateful to D. Schiferl and D. Taylor of Los Alamos National Laboratories for supplying us with material for the Re and $Ta_{90}W_{10}$ gaskets and for giving us advice on their use.

12. Private communication of D. Schiferl.

13. See also, A.W. Webb, D.U. Gubser, and L.C. Towle, Rev. Sci. Inst. 47, 59(1976)

14. E. Maxwell, Rev. Sci. Instr. 36, 553(1965)

15. L.D. Landau and E.M. Lifshitz, in Electrodynamics of Continuous Media (Addison–Wesley, Boston, 1960).

16. S. Klotz, W. Reith, and J.S. Schilling, Physica C 172, 423(1991); this work is part of the PhD thesis work of S. Klotz, University of Munich.

17. R. Griessen, Phys. Rev. B 38, 3690(1988); R.J. Wijngaarden and R. Griessen, in Studies of High Temperature Superconductors, ed. A.V. Narlikar (Nova Science Publisher, N.Y., 1989).

18. I.V. Berman, N.B. Brandt, I.E. Graboi, A.R. Kaul, R.I. Kozlov, I.L. Romashkina, V.I. Sidorov and K. Tsuiun, JETP Lett. 47, 733(1988).

19. M.W. McElfresh, M.B. Maple, K.N. Yang and Z. Fisk, Appl. Phys. A45, 365(1988).

20. J.D. Jorgensen, B.W. Veal, A.P. Paulikas, L.J. Nowicki, G.W. Crabtree, H. Claus, and W.K. Kwok, Phys. Rev. B 41, 1863(1990).

21. R.J. Wijngaarden, J.J. Scholz, E.N. van Eenige and R. Griessen, preprint

22. C. Allgeier and J.S. Schilling, Physica C 168, 499 (1990)

23. W.A. Groen, D.M. Leew and L.F. Feiner, Physica C 165, 55(1990)·

24. F. Steinmeyer, R. Kleiner, P. Müller, Proceedings of the Third International Conference on Materials and Mechanisms of Superconductivity, Kanazawa, 1991, to be published in Physica C; R. Kleiner, F. Steinmeier , G. Kunkel, P. Müller, same proceedings.

25. T. Staufer, R. Nemetschek, R. Hackl, P. Müller, H. Veith, Walther–Meissner- –Institut, to be published.

26. H.F. Goldstein, L.C. Bourne, P.Y. Yu and A. Zettl, Solid State Comm. 70, 321(1989).

27. R.J. Wijngaarden, H.K. Hemmes, E.N. van Eenige and R. Griessen, Physica C 152, 140(1988).

28. R. Sieburger, P. Müller and J.S. Schilling, submitted to Physica C

29. J. Diederichs, Diplom theses, University of Munich, 1990, unpublished.

30. L. Forro, V. Ilakovac and B. Keszei, Phys. Rev. B 41, 9551(1990).

31. R. Kubiak, K. Westerhold, G. Pelka, H. Bach and Y. Khan, Physica C 166, 523(1990).

32. J. Beille, H. Dupendant, O. Laborde, Y. Lefur, M. Perroux, R. Tournier, Y.F. Yen, J.H. Wang, D.N. Zheng and Z. Zhao, Physica C 156, 488(1988).

33. S. Tamura, S. Takekawa and H. Nozaki, J. Phys. Soc. Japan 57, 2215(1988).

34. P.A. Sterne and C.S. Wang, J. Phys. C 21, L949(1988); A.J. Freeman, J. Yu and S. Massidda, Physica C 153, 1225(1988).

35. M.S. Hybertsen and L.F. Mattheiss, Phys. Rev. Lett. 60, 1661(1988); H. Krakauer and W.E. Pickett, Phys. Rev. Lett. 60, 1665(1988).

PRESSURE DEPENDENCE OF THE SUPERCONDUCTING TRANSITION
TEMPERATURE OF Rb_3C_{60} UP TO 20 KBAR

S. L. Bud'ko,[*] R. L. Meng, C. W. Chu and P. H. Hor

Texas Center for Superconductivity at the University of Houston, Houston
TX 77204-5932

ABSTRACT

ac susceptibility measurements of Rb_3C_{60} under hydrostatic pressure up to 20 kbar are reported. The superconducting transition temperature (T_c) decreases linearly under pressure with the pressure derivative $dT_c/dP = -0.78$ K°/kbar.

INTRODUCTION

The recent discovery of superconductivity in the alkali-doped C_{60} compounds, K_3C_{60} and Rb_3C_{60} [1-3], has attracted tremendous experimental and theoretical interest. Pressure studies have provided new information on the superconductivity of these compounds. The unusually large negative linear pressure derivative of the superconducting transition temperature dT_c/dP for K_3C_{60} [4,5] and at initial pressures up to 6 kb for Rb_3C_{60} [6], and the indication of nonlinear behavior of $T_c(P)$ for K_3C_{60} above 15 kb [5] make it important to increase the pressure range in studies of the superconductivity of Rb_3C_{60}.

EXPERIMENT

Rb_3C_{60} samples were prepared by the procedure described in Ref. 3. The samples were pieces of bulk material and were ~ 0.7 to 1.0 mm long. Because the materials are reported to be reactive, special precautions were taken. When the samples were taken from the sealed pyrex tube in a glove box under Ar atmosphere they were immediately covered with vacuum grease to prevent direct contact with air while being transferred to the cell.

The pressure dependence of T_c was determined by ac susceptibility measurements at frequency of 671 Hz and at modulation fields of 3 to 5 G in a piston-cylinder pressure cell similar to that used in Ref. 7, with silicon oil as the pressure medium. The T_c was defined as the onset of inductive signal (Fig. 1). The temperature was measured using a Ge resistor thermometer in contact with the exterior of the cell. There was no thermal lag between sample and thermometer at the typical temperature sweep rate 0.1 K°/min, and the ac susceptibility data were the same as the temperature was increased or decreased.

---

[*] On leave from Institute for High Pressure Physics, USSR Academy of Sciences, Troitsk, Moscow regions, 142092, USSR.

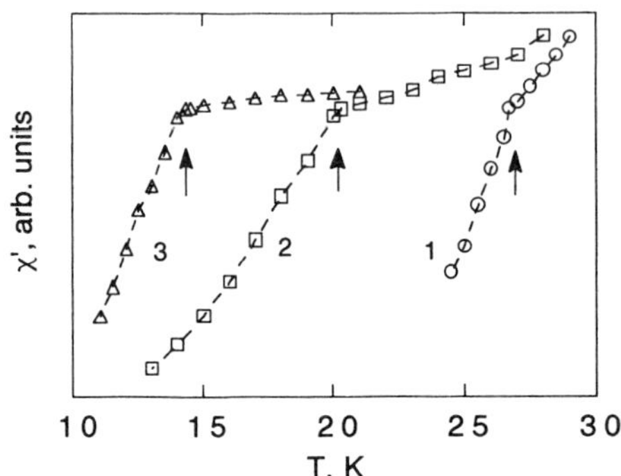

Figure 1. Example of magnetic susceptibility curves for Rb_3C_{60}: (1) P = 1.3 kbar, (2) P = 7.6 kbar and (3) P = 17.5 kbar. Arrows indicate defined T_c (onset).

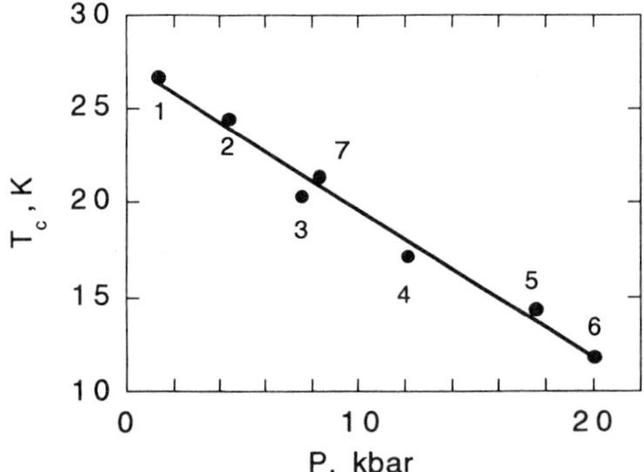

Figure 2. T_c of Rb_3C_{60} *vs* P. The points are numbered in the sequence in which they were taken.

RESULTS AND DISCUSSION

The pressure dependence of T_c is presented in Fig. 2. The extrapolated (P = 0)-value of T_c is 27.4 K and $dT_c/dP = - 0.78 \pm 0.05$ K°/kbar. The pressure dependence of T_c is linear (within the experimental error) throughout the pressure range.

The results of pressure studies cannot be used to determine definitively the mechanism of superconductivity in alkali-doped C_{60}. One of the simplest possibilities [5,8] remains a BCS-like mechanism with a phonon or some other unknown type of pairing mechanism. However, systematic study of the physical pressure and chemical pressure effects should shed some light on these new classes of superconductors.

The value of dT_c/dP in our measurements is close to that for K_3C_{60} [4,5] and is consistent with the results for Rb_3C_{60} for initial pressures [6] and preliminary results reported in Ref. 5. Although pressure measurements up to 21 kbar on K_3C_{60} show traces of nonlinear behavior of $T_c(P)$, it was not observed in our measurements on Rb_3C_{60} up to 20 kbar. If we assume that, in the pressure range under consideration, the changes occur in the intermolecular C_{60} distances (as reported for undoped-C_{60} [9,10]) rather than in intramolecular C-C distances and the compressibility of alkali-doped C_{60} is close to that of undoped-C_{60}, the changes in the Rb-C distance and van der Waals separation between C_{60} molecules (these distances seems to be important for superconductivity of the compounds under consideration) are 10 to 15% at 20 kbar. It is interesting to see that such a large change does not cause nonlinearity in the $T_c(P)$ behavior. It is possible that, in the case of K_3C_{60} at 21 kbar, the C_{60} molecules are close enough to each other to induce nonlinear changes in T_c, since the lattice parameter of K_3C_{60} reaches the value of that of undoped-C_{60} at P ~ 5 kbar.

To our knowledge, no direct measurements of the lattice constant a for Rb_3C_{60} have been published to date. Considering the similar effect of physical and chemical pressure on the T_c of these compounds, we can estimate the value of a (Rb_3C_{60}) using dT_c/dP (from pressure measurement) and compressibility of pure C_{60} [9,10], assuming that it is the same as that of alkali-doped C_{60}. Using the average value reported in Refs. 10 and 11, $d(\ln a)/dP = 2.1 \times 10^{-3}$ kbar^{-1}, we obtain a (Rb_3C_{60}) = 14.6 Å. This result is close to simple estimation of change of lattice parameter due to change of ionic radius of alkali metal (K $\rightarrow$ Rb) in octahedral position [a (Rb_3C_{60}) = 14.56 Å]. With this information, we can estimate both the lattice parameter and T_c for Cs_3C_{60}. Assuming the change of lattice parameter is solely due to the change of K $\rightarrow$ Rb $\rightarrow$ Cs, we estimate that the ionic radii a (Cs_3C_{60}) ~ 14.9 Å and the T_c ~ 38 K.

Recently, superconductivity was reported in $Cs_1Rb_2C_{60}$ [11], $Cs_2Rb_1C_{60}$ [11], and Cs_3C_{60} [12] with T_c = 31, 33, and 30 K, respectively. These results are in qualitative agreement with our estimations that the T_c of Cs-doped C_{60} is, in general, larger than that of Rb-doped C_{60}. However, Cs_3C_{60} is expected to have a higher T_c than all others. Reasons for this might include the following:~

- the size of interstitial sites in C_{60} crystals is small, so three Cs atoms may cause some disturbance in close-packing, which was not taken into account in our estimations; and

- the low superconductive volume fraction and the uncertainty of the exact stoichiometry of the superconductive phase indicates that T_c = 30 K [12] is the lower boundary for Cs-doped C_{60} and the fact that the values of T_c for $Cs_1Rb_2C_{60}$ and $Cs_2Rb_1C_{60}$ in Ref. 11 are higher than that of Cs_3C_{60} in Ref. 12, which is an indirect confirmation of this point.

In summary, we have studied the effects of pressure on Rb_3C_{60} up to 20 kbar. No non-linear pressure dependence was observed and both chemical and physical pressure affect T_c similarly, which is in contrast to the oxide high temperature superconductors.

ACKNOWLEDGEMENTS

This work was supported in part by the NSF Low Temperature Physics Program Grant No. DMR 86-126539, DARPA Grant No. MDA 972-88-G-002, NASA Grant No.

NAGW-977, Texas Center for Superconductivity at the University of Houston and the T. L. L. Temple Foundation.

REFERENCES

[1]   A. F. Hebard, M. J. Rosseinsky, R. C. Haddon, D. W. Murphy, S. H. Glarum, T. T. M. Palstra, A. P. Ramirez and A. R. Kortan, Nature 350 (1991) 600.

[2]   K. Holczer, O. Klein, G. Gruner, S. M. Huang, R. B. Kaner, K. J. Fu, R. L. Wheatten and F. Diederich, Science 252 (1991) 1154.

[3]   M. J. Rosseinsky, A. P. Ramirez, S. H. Glarum, D. W. Murphy, R. C. Haddon, A. F. Hebard, T. T. M. Palastra, A. R. Kortan, S. M. Zahurak and A. V. Makhija, Phys.Rev.Lett. 66 (1991) 2830.

[4]   J. E. Schirber, D. L. Overmyer, H. H. Wang, J. M. Williams, K. D. Carlson, A. M. Kini, M. J. Pellin, U. Welp and W. K. Kwok, Physica C 178 (1991) 137.

[5]   G. Sparn, J. D. Thompson, S. M. Huang, R. B. Kaner, F. Diederich, R. L. Whetten, G. Gruner and K. Holczer, Science 252 (1991) 1829.

[6]   H. H. Wang, A. M. Kini, B. M. Saval, K. D. Carlson, J. M. Williams, M. W. Lathrop, K. R. Lykke, D. H. Parker, P. Wurz, M. J. Pellin, D. M. Gruen, U. Welp, W. K. Kwok, S. Fleshler, G. W. Crabtree, J. M. Schirber and D. L. Overmyer, Inorg.Chem. 30 (1991) 2963.

[7]   S. L. Bud'ko, A. N. Voronovskii, A. G. Gapotchenko and E. S. Itskevich, Sov. Phys.-JETP 54 (1984) 454.

[8]   J. L. Martins, N. Troullier and M. Schabel, Phys.Rev.Lett., submitted.

[9]   J. E. Fischer, P. A. Heiney, A. R. McGhie, W. J. Romanow, A. M. Dedenshtein, J. P. McCauley, Jr. and A. B. Smith III, Science 252 (1991) 1288.

[11] S. J. Duclos, K. Brister, R. C. Haddon, A. R. Kortan and F. A. Thiel, Nature, submitted.

[12] K. Tanigaki, T. W. Ebbsen, S. Saito, J. Mizuki, J. S. Tsai, Y. Kubo and S. Kuroshima, Nature 352 (1991) 222.

[13] S. P. Kelty, C. C. Chen and C. M. Lieber, Nature 352 (1991) 223.

HIGHLIGHTS OF THE ROUND TABLE DISCUSSION ON

TRENDS AND FUTURE DEVELOPMENTS OF THE APPLICATION OF PRESSURE

TO HIGH T_C MATERIALS

C. W. Chu

Texas Center for Superconductivity
University of Houston
Houston, Texas 77204-5932

Panel: C. Ayache, Service Basses Temperatures, Grenoble FRANCE
C. W. Chu (Chairman), University of Houston, Houston TX USA
W. J. Fietz, Kernforschungszentrum, Karlsruhe GERMANY
G. K. Kourouklis, Aristotle University, Thessaloniki GREECE
E. Liarokapis, National Technical University, Athens GREECE
R. Wijngaarden, Vrije Universiteil, Amsterdam THE NETHERLANDS

INTRODUCTION

A discussion of current trends and the future development of the application of pressure to high T_C materials was initiated by a brief presentation by each panel member and followed by an active exchange among all participants in the Workshop. The current status of the study of high temperature superconductivity and the past, current, and future roles of pressure in that study were discussed. They are briefly summarized below:

HIGH TEMPERATURE SUPERCONDUCTIVITY -- PRESENT STATUS

High Temperature Superconductivity (HTS's) not only poses great scientific challenges but also holds unusual technological promises. Consequently, extensive efforts worldwide were devoted to the study of this wonderful class of materials immediately after it was discovered four years ago. During this short period of time, great progress has been made in all areas of high temperature superconductivity research. For instance, the existence of a Fermi surface and superconducting energy gap(s) have been unambiguously established. Many important properties of HTS's in both their normal and superconducting states have been reliably determined as a result of the quality improvement of samples. They show that the HTS system cannot be a simple Fermi liquid. More than 75 non-intermetallic compounds have been found to display a transition temperature (T_C) greater than 21 K. It has been demonstrated that HTS is a rather common phenomenon, occurring not just in cuprates and bismuthates, but also in fullerites. The continued improvement of the HTS material processing techniques has led to an impressive advancement in the critical

Frontiers of High-Pressure Research, Edited by H.D. Hochheimer and
R.D. Etters, Plenum Press, New York, 1991

current density (J_c) of HTS's at 77 K and 0 T, *e.g.* $J_c \sim 6\times10^6$ A/cm^2 for YBa$_2$Cu$_3$O$_7$ thin films and 8×10^4 or 1×10^6 A/cm^2 for bulk YBa$_2$Cu$_3$O$_7$ before or after fast-neutron irradiation. Ribbons of Bi$_2$Sr$_2$CaCu$_2$O$_8$ have been fabricated, exhibiting a $J_c \sim 10^5$ A/cm^2 at 4.2 K with only a slight decrease in a field up to 30 T. The results show that regarding J_c's, HTS's perform well in high magnetic fields, where LTS's fail. The low surface impedance of HTS's at 77 K will ensure an important role for HTS applications in the future of microwave devices. Recently, a multilayer monolithic integrated SQUID magnetometer was successfully fabricated with a performance at 77 K matching that of a commercial SQUID operated at 4.2 K.

HIGH PRESSURE AND HTS's -- PAST AND PRESENT

In spite of the progress made on HTS's as mentioned above, many questions remain, particularly in the occurrence and the mechanics of HTS. High pressure has contributed significantly to the development and understanding of HTS. For instance, it was the unusual high pressure effect on T_c observed in (La,Ba)$_2$CuO$_4$ that first signaled the significance of cuprates in the search for higher T_c and led to the discovery of YBa$_2$Cu$_3$O$_7$ and related compounds. High pressure studies have also been performed to test predictions of various theoretical models that showed that cuprate HTS's cannot be satisfactorily described by a simple BCS theory. New HTS compounds, such as (Sr,Ca)CuO$_2$, have been stabilized under high pressure. The pressure-induced shift in n was semiquantitatively determined by measuring the upper critical field under pressure and thus by-passing the many complications in the direct measurements of n under pressure. This, together with the recent success in extending the HTS experiments to above 150 kb, enables one to gain deeper insight to the close relationship between T_c and n. The attribution of the T_c-change by pressure to a pressure-induced n-shift or charge transfer is consistent with reported positron lifetime measurements under pressure.

HIGH PRESSURE AND HTS's -- FUTURE

While quasihydrostatic and hydrostatic high pressure experiments have generated valuable results on HTS's, it is suggested that uniaxial pressure experiments should be even more effective in unravelling the mechanism and/or dimensionality of HTS by separating the inter-layer from the intra-layer couplings in HTS's. To overcome the technical problems associated with experiments of this kind will be most challenging although rewarding. The thermodynamic analysis of the recent thermal expansion data on HTS single crystals at ambient pressure suggests a possible highly anisotropic pressure effect on T_c along the different crystalline axes. These effects were found to be sensitively to the oxygen-stoichiometry and the impurities present in the compound. On the other hand, caution should be exercised in extracting the pressure-effect on T_c based on a thermodynamic analysis of the thermal expansion results as exemplified by the previous dilemma in the study of A15 compounds. A systematic high-pressure study on T_c with well-characterized samples and well-defined pressure conditions, *e.g.* known stress distribution in the sample, will help obtain quantitative information crucial to the understanding of HTS. The abrupt disappearance of a superconducting transition in some HTS's at pressures above 200 kb is rather intriguing. More work is needed to determine if this is a genuine effect than an artifact. If it is a real effect, it may imply that the evolution of superconductivity from an insulator parent is a discontinuous process through chemical, optical, or pressure doping. Such a discontinuous process should impose significant constraints on any theoretical model of HTS.

The understanding of HTS depends crucially on our understanding of the normal-state properties of the compounds. Until now, high pressure experiments on the normal state properties of HTS's are sparse. In view of recent advances in the band calculation of HTS's, experiments of this type are not only important but timely.

The magnetic phase diagram of HTS's is very complicated and thus provides a fertile ground for the study of phase transition and flux dynamics in a vortex state. Ample experiments have been done at ambient pressure but not under pressure. It is believed that the introduction of pressure to these studies will give us an extra handle in unravelling the detailed nature of the flux pinning, flux dynamics, and phase transition.

The T_c of A_3C_{60} with $A = K$ and Rb was found to be drastically suppressed by pressure. The results can be understood in terms of a band broadening and thus a density of states decrease under pressure, consistent with the continued T_c increase was the increase in lattice parameters through doping C_{60} with K, Rb, Cs, or a combination of these. Information about the transport properties above T_c as a function of temperature at ambient and high pressure will be very interesting. The chemically active nature of the compounds makes the experiments on transport properties difficult at present. Recent calculations showed unusual properties in the molecular liquid of C_{60} if formed. To achieve a liquid state in C_{60} is therefore not only of great interest in its own right but also important in providing a new avenue for the synthesis of new fullerite compounds. One may have to overcome the problems of low temperature sublimation of C_{60} and catalytically assisted disintegration of C_{60} to graphite in order to attain the liquid state of C_{60} under pressure at high temperatures.

High pressure has been showed to hold great promise in the synthesis of metastable HTS's and also in the fabrication of HTS's in usable shapes. The current efforts should be expanded and the technique further exploited to its fullest.

Adams, A. R.
Albert, H. O.
Albinsson, I.
Allan, M.
Ashcroft, N.
Ayache, C.
Barry, S.
Benischke, R.
Bracewell, B. L.
Bradley, R. M.
Bud'ko, S. L.
Chambers, F. A.
Chervin, J. C.
Chu, C. W.
Clark, R. A.
Conradi, M. S.
Cui, L. J.
Däufer, H.
Dollhopf, W.
Donohoe, R. J.
Eckert, B.
Etters, R. D.
Fietz, W. H.
Foggi, P.
Friedman, J. S.
Friend, R. H.
Garcia, M. A. Y.
Gauthier, M.
Griessen, R.
Grube, K.
Halvorson, K. E.
Hochheimer, H. D.
Hor, P. H.

Huang, Z. J.
Jean, Y. C.
Jeanloz, R.
Jilek, E.
Jodl, H.-J.
Jonker, B. T.
Jordan, M.
Kaldis, E.
Karpinski, J.
Klotz, S.
Kourouklis, G. A.
Kuchta, B.
Kulik, A. S.
Lee, S. A.
Liarokapis, E.
Lin, G.
Lorenz, B.
Love, S. P.
Ludwig, H. A.
Luo, Y.
Marsden, I.
Marzke, R. F.
Mellander, B. E.
Meng, R. L.
Michel, K. H.
Müller, P.
Norberg, R. E.
Orgzall, I.
Pasternak, M. P.
Pechhold, W.
Polian, A.
Prins, K. O.
Pruzan, Ph.

Rafaelle, D. P.
Rajakarunanayake, Y.
Robert, J. L.
Rusiecki, S.
Schilling, J. S.
Scholtz, J. J.
Schwarzberger, P.
She, C. Y.
Silvera, I. F.
Spain, I. L.
Stevens, J. R.
Stradling, R. A.
Strauss, M. J.
Swanson, B. I.
Taylor, R. D.
Thomas, F.
Thomasson, J.
Ulug, A. M.
Underhill, A. E.
van Eenige, E. N.
Venkateswaran, U. D.
Ves, S.
Villeudieu, M.
Wagner, B. P.
Weinstein, B. A.
Wijngaarden, R. J.
Wilkinson, V. A.
Williamson, III, W.
Wolf, G. H.
Wühl, H.
Xiong, Q.